SCIFI 97: CONFERENCE ON SCINTILLATING FIBER DETECTORS

SCIFI 97: CONFERENCE ON SCINTILLATING FIBER DETECTORS

Notre Dame, Indiana   November 1997

EDITORS
Alan D. Bross
Fermi National Accelerator Laboratory

Randal C. Ruchti
Mitchell R. Wayne
University of Notre Dame

American Institute of Physics

AIP CONFERENCE PROCEEDINGS 450

Woodbury, New York

Editors:

Alan D. Bross
Fermi National Accelerator Laboratory
MS 121; PO Box 500
Batavia, IL 60510-0500
Email: bross@fnal.gov

Randal C. Ruchti
Department of Physics
408 Nieuwland Science Hall
University of Notre Dame
Notre Dame, IN 46556-5670
Email: ruchti@undhep.hep.nd.edu

Mitchell R. Wayne
Department of Physics
406 Nieuwland Science Hall
University of Notre Dame
Notre Dame, IN 46556-5670
Email: wayne@undhep.hep.nd.edu

TA
1815
.C665
1997

L.C. Catalog Card No. 98-87810
ISBN 1-56396-792-8
ISSN 0094-243X
DOE CONF- 9711182

Printed in the United States of America

CONTRIBUTED PAPERS

SESSION 1: SCINTILLATORS

SESSION 2: PROPERTIES

SESSION 3: STRUCTURES

SESSION 4: PHOTOSENSORS—I

SESSION 5: TRACKING—I

SESSION 6: TRACKING—II

SESSION 7: APPLICATIONS—I

SESSION 8: PHOTOSENSORS—II

SESSION 9: POSTERS

SESSION 10: CALORIMETRY—I

SESSION 11: CALORIMETRY—II

SESSION 12: MEDICAL

SESSION 13: PHOTOSENSORS—III

SESSION 14: ASTROPHYSICS—I

SESSION 15: APPLICATIONS—II

SESSION 16: ASTROPHYSICS—II

PREFACE

The SCIFI97 Conference on Scintillating and Fiber Detectors was held at the University of Notre Dame campus on 2-6 November, 1997. The third conference in a series in which prior meetings were held at Fermilab (SCIFI88) and Notre Dame (SCIFI93), the purpose of SCIFI97 was to provide a forum for scientists and engineers who have an interest in detector development using scintillator and fiber techniques for particle detection, tracking, and calorimetry. Given the substantial progress over the past several years in scintillation detector development, photosensor development, and general interest in fiber detection techniques for a wide variety of applications, the organizing committee felt that 1997 was again an appropriate time to review the state-of-the-art of the field.

At the time of the previous meeting, SCIFI93, experiments were in development or just underway utilizing large scale implementation of scintillating fibers for tracking. UA2 had successfully utilized the technology for tracking and electron identification. Additionally, the technique of tile-fiber calorimetry was being refined. Substantial programs applying scintillating fiber techniques to space physics were presented. Developments of new photosensors were reported - including avalanche photodiodes, vacuum photomultipliers, and VLPCs.

By the time of SCIFI97, fiber tracking was now firmly established as an important experimental technique and utilized and proposed for very large scale experiments at nearly every major experimental laboratory, for non-accelerator based experiments, for space applications in a variety of platforms, and for applications in nuclear medicine. This volume contains timely reports of the latest research and development and experimental experience in all of these areas.

Photosensor development is an integral part of scintillation detector development, and this field has evolved significantly since SCIFI93. Substantially more is known about the operation and performance of VLPCs and new versions with high-speed and still higher quantum efficiency have been produced; APDs with excellent characteristics for scintillation calorimetry are available now; and a wide variety of vacuum photomultiplier structures including hybrid tubes and multichannel devices have been developed. All of these advancements have benefitted from strong cooperation between manufacturers and the research community. The subject of photosensors was given careful attention at SCIFI97.

Surprisingly, developments in scintillator and waveshifter materials and the theoretical framework for predicting new useful structures are somewhat lagging. Although very capable and innovative researchers are working in this area (see the relevant section in this volume), additional resources for research and development are required to accelerate significant advancements over the next several years. There is a considerable need for fast (few nanoseconds), efficient, radiation resistant, and environmentally stable materials.

Over 130 physicists, engineers, technicians and students from the Americas, Europe and Asia participated in SCIFI97. Over eighty papers were presented at the meeting. We are hopeful that the research community will benefit from the information assembled herein.

The Organizing Committee is indebted to support for the conference from several federal agencies, national laboratories and the University of Notre Dame. We especially acknowledge Harriet Baldwin and her staff at the Notre Dame Center for Continuing Education for their many contributions to the success of the meeting, and to Cynthia Sazama, Maja Christensen and Patti Poole of the Fermilab Conference Office for their assistance in assembling these proceedings. Lastly, we gratefully acknowledge Dr. Chris Quigg of Fermilab for his stimulating and inspiring public lecture entitled "Nothing is Too Wonderful to be True". This talk allowed us to draw the general public, and especially high school students and teachers, to the excitement of physics and to share in the enthusiasm of the participants of SCIFI97.

Alan Bross
Randy Ruchti
Mitchell Wayne
For the Organizing Committee

INTERNATIONAL ADVISORY COMMITTEE

K. Arisaka	UCLA
M. Atac	Fermi National Accelerator Laboratory/UCLA
A. Bodek	University of Rochester
R. Clough	Sandia National Laboratory
R. Craig	Pacific Northwest Laboratory
C. D'Ambrosio	CERN
J.-M. Gaillard	Laboratoire de Physique des Particules, Annecy
D. Green	Fermi National Accelerator Laboratory
V. Hagopian	Florida State University
S. Majewski	Thomas Jefferson National Accelerator Facility
M. Mishina	KEK/Fermi National Accelerator Laboratory
R. Mussa	INFN, Ferrara
S. Olsen	University of Hawaii
A. Penzo	INFN, Trieste
S. Reucroft	Northeastern University
A. Skuja	University of Maryland
T. Virdee	CERN
B. Winstein	University of Chicago
R. Wigmans	Texas Tech University

ORGANIZING COMMITTEE

	W. R. Binns	Washington University, St. Louis
Co-chair	A. Bross	Fermi National Accelerator Laboratory
	J. Elias	Fermi National Accelerator Laboratory
	J. Freeman	Fermi National Accelerator Laboratory
	S. Gruenendahl	University of Rochester
	J. Huston	Michigan State University
	J. Kotcher	Brookhaven National Laboratory
	D. Lincoln	University of Michigan
	W. Moses	Lawrence Berkeley National Laboratory
	Y. Onel	University of Iowa
	A. Pla-Dalmau	Fermi National Accelerator Laboratory
	J. Proudfoot	Argonne National Laboratory
Co-chair	R. Ruchti	University of Notre Dame
	R. Rusack	University of Minnesota
Co-chair	M. Wayne	University of Notre Dame

SPECIAL THANKS TO THE FOLLOWING SPONSORS

Fermi National Accelerator Laboratory
University of Notre Dame
U. S. Department of Energy
National Science Foundation
Argonne National Laboratory

Biogeneral, Inc.
Collimated Holes, Inc.
Delft Electron Products B. V.
Electron Tubes, Inc.
Hamamatsu Photonics K. K.
Kuraray Corporation
LeCroy Corporation

CORPORATE PARTICIPANTS

Bicron Corporation
Biogeneral, Inc.
Boeing North America, Inc.
Collimated Holes, Inc.
Delft Electron Products B. V.
E G & G Optoelectronics
Electron Tubes, Inc.
Hamamatsu Photonics K. K.
Kuraray Corporation
LeCroy Corporation
Photek, Inc.
P. M. Manufacturing Services, Inc.
Polymicro Technologies
Radiation Monitoring Devices, Inc.

SESSION 1: SCINTILLATORS

Chair: A. Pla-Dalmau
Scientific Secretary: J. Marchant

Lack of Progress on Waveshifting Fluors for Use in or with Polystyrene Scintillators in High-Energy Particle Detection

Joel M. Kauffman*, Gurdip S. Bajwa and Peter T. Litak

Philadelphia College of Pharmacy & Science, 600 S. 43rd St., Phila., PA 19104-4495

Charles J. Kelley and Ramdas Pai

Massachusetts College of Pharmacy & Allied Health Sciences,
179 Longwood Ave., Boston, MA 02115-5804

Abstract. A number of original proton-transfer fluors were prepared as shifters for green-emitting scintillating fibers of multi-meter lengths. Quantum efficiencies up to 0.52 are reported for the new fluors which showed emission maxima between 478 and 612 nm. None proved clearly superior to 3HF, the standard fluor for scintillating fibers whose light is measured with PM tubes. For the current goal of half-meter length waveshifting fibers with blue-green emission for use with blue-violet-emitting polystyrene tiles, a number of fluors related to TPBD and Bis-MSB were prepared bearing dialkylamino auxofluors. Some of these failed to fluoresce, and those that did were photochemically unstable. A proposal was made on some original conjugated polyenes for waveshifting fibers. In general, accurate predictions could be made on excitation and emission wavelengths, while light output, decay time, photochemical stability and radiation hardness are more difficult targets.

INTRODUCTION

This manuscript includes both original research, and a partial review of the literature. Citations to the SciFi93 Proceedings (1) will be given as: Author, SciFi93, p xx. Two recent reviews show how waveshifting fluors are integrated into overall systems for tracking high-energy particles (2,3).

Five years ago, the general properties of waveshifting fluors were delineated (4). The fluors sought for use in current plastic scintillating devices may be divided into three groups according to their wavelengths of absorption and emission:

Group 1 includes fluors with absorption max 350 ± 10 nm and emission max 540 ± 20 nm for scintillating polystyrene (PS) fibers of multi-meter length containing terphenyl primary fluor and using either green-sensitive photomultiplier (PM) tubes or visible-light photon counters (VLPCs) as detectors (3).

CP450, *SciFi97: Workshop on Scintillating Fiber Detectors*
edited by A. D. Bross, R. C. Ruchti, and M. R. Wayne
© 1998 The American Institute of Physics 1-56396-792-8/98/$15.00

Group 2 includes fluors with absorption max 350 ± 10 nm and emission max 570 ± 30 nm for scintillating polystyrene fibers of multi-meter length containing terphenyl primary fluor and using charge-coupled devices (CCDs) as detectors.

Group 3 includes fluors with absorption max 420 ± 20 nm and emission max 510 ± 20 nm for waveshifting polystyrene fibers of half-meter length for use with blue-violet emitting tiles and using either green-sensitive PM tubes or VLPCs as detectors.

The structure and properties of the most widely used green-emitting fluor of the Group 1 category, 3-hydroxyflavone (3HF), are shown in Figure 1 (5). Although 3HF is thermally and photochemically unstable in air, it is satisfactorily stable to radiation ("rad-hard") once it is incorporated into PS fibers. Also shown in Figure 1 are the properties of 3 other flavones whose emission wavelengths put them on the border of Groups 1 and 2. Flavone 560 is noteworthy for the highest fluorescence

Code	Solvent	Φf	abs.λ max	epsilon	fl.λ max	tau, nsec
3HF	polystyrene	0.31	350 nm	14,000	530 nm	8.8
FLAVONE 560	toluene	0.53	377	52,000	560	5
			395 (sh)	44,000		
FG10	dichloromethane	0.37	380	36,000	560	-
	cyclohexane	-	-	-	-	2.5
FG157	dichloromethane	0.15	350	21,000	566	-
	cyclohexane	0.18	355	-	563	1.7

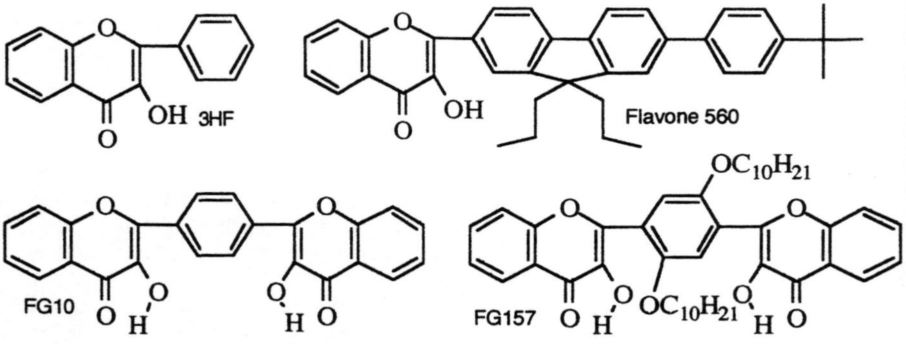

FIGURE 1. Photophysical Properties of 3-Hydroxyflavone Type Fluors

4

quantum yield (Φ_f), 0.55, ever observed among excited-state intramolecular proton transfer (ESIPT) fluors (Kauffman, SciFi93, p353). FG10 possessed the shortest fluorescence decay time (τ), of any flavone with high Φ_f, 2.5 nsec in cyclohexane. A more soluble version of FG10, FG157, had half the light output (Gao et al., SciFi93,361). The poor properties reported for FG157 can be attributed to steric inhibition to the attainment of coplanarity in the excited state. We suggest that placement of solubility-enhancing groups only on the outer benzene rings will retain the fast τ along with other desirable properties.

We are reporting here the preparation and properties of new fluors in each of the 3 Groups described above. We report the preparation of some ESIPT fluors belonging to Groups 1 and 2, and we give some data on photostability for previously reported fluors. In addition we report on the design, synthesis, and some photophysical properties of a series of non-ESIPT fluors of the Group 3 type. Lastly, we propose new fluors which might achieve all the goals of Group 3.

DISCUSSION AND RESULTS

Design and Properties of New ESIPT Fluors

A range of benzazole type ESIPT fluors had been prepared (Figure 2; structures, Figure 3; Kauffman, SciFi93, p 353). The range of emission maxima represented between Imidazole 478 (A) and Oxazole 598F (E) are not fully accessible because these fluors are somewhat photochemically unstable, as judged by the visible light absorption of solutions kept in diffuse daylight for 4 years. Oxazole 3F (B), which possesses an additional absorption peak at 377 nm, was unsuitable for the Group 1 application because of too short an emission wavelength (4); but it is now being re-evaluated as a Group 3 fluor. Oxazole 545M6' (C) was 10% brighter than 3HF, but no faster, and less rad-hard (6). The related Oxazole 550F (D) is expected to be another 10% brighter and faster, and is to be evaluated. The analog of D in which the dibenzofuran oxygen is sulfur (24 in Ref. 7) showed a lower Φ_f (0.36) and exhibited some non-ESIPT absorption at $\approx$ 400 nm. The analogous N-ethylcarbazole (47 in Ref. 7) and Oxazole 598F (E) showed identical emissions at 598 nm, but their light outputs ($\Phi_f = 0.17$ and 0.21) are too low to let them be useful Group 2 fluors.

While 2-(2'-hydroxyphenyl)benzoxazole (HBO) and the corresponding benzothiazole (HBT) have an equally modest Φ_f ($\approx$ 0.12 in PS) (8), substitution of either at the 4' position with a cyano group (Figure 4) enhances Φ_f 3-4-fold. Note that the benzoxazole Cyano-HBO has a higher Φ_f than the -thiazole Cyano-HBT, and Cyano-HBO may now be interest for Group 3 because its emission maximum near 500 nm is now desirable. In an attempt to find an electron-withdrawing equivalent of the cyano group with a larger extinction coefficient, the 4-quinolyl analogs RPI-267 and -8 were prepared. Alas, these had very low Φ_f, but served to confirm that ESIPT benzoxazoles are more efficient fluorescers than the benzothiazoles. The high Φ_f values (0.90-0.94 in ethanol) of certain <u>non</u>-ESIPT coumarin laser dyes substuted in

Code-->	A	B	C	D	E
Name	Imidazole 478	Oxazole 3F	Oxazole 545M6'	Oxazole 550F	Oxazole 598F
Φ_f	0.38	0.41	0.41	0.44	0.21
absorption λmax	342 nm	357 nm	335 nm	389 nm	360, 375 nm
epsilon	24,000	58,000	44,500	54,300	57-, 48,000
emission λmax	478 nm	492 nm	545 nm	545 nm	598 nm
photochemical stability	poor	good	good	good	poor

FIGURE 2. Photophysical Properties of "New" ESIPT Fluors

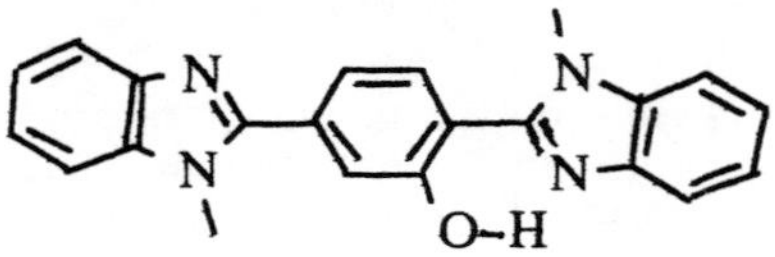

A: Imidazole 478

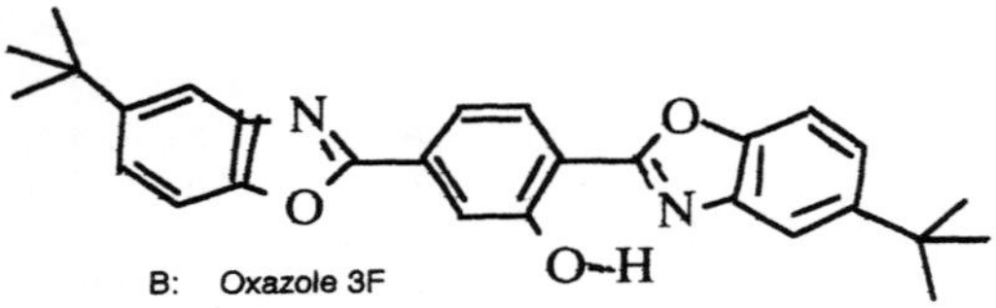

B: Oxazole 3F

C: Oxazole 545M6'

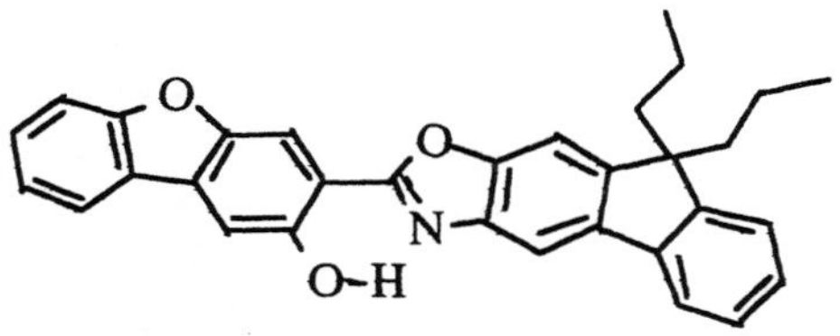

D: Oxazole 550F

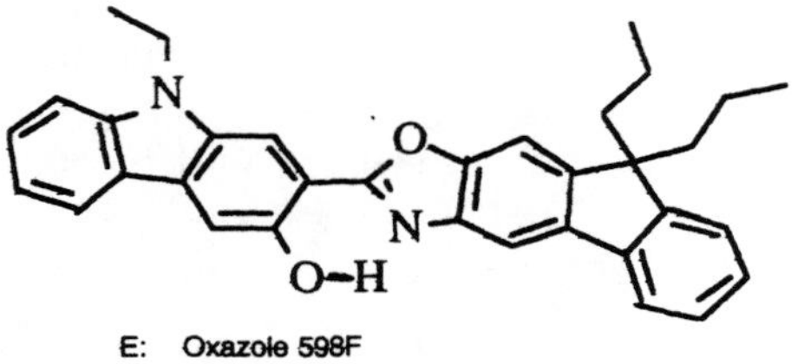

E: Oxazole 598F

FIGURE 3. Structures of "New" ESIPT Fluors

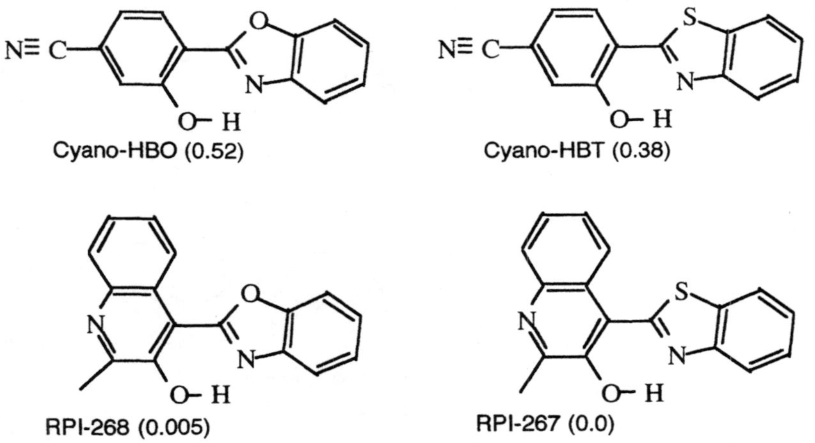

FIGURE 4. Newer ESIPT Fluors Related to HBO and HBT (Φ_f)

the 3-position with benzoxazole or benzothiazole (9) do not indicate that benzothiazoles must always be dimmer than benzoxazoles.

Following our earlier observation that ESIPT "double benzazoles" were fast and robust and had reasonable quantum yields (10), we have prepared dibenzothiazoles and mixed benzoxazole/benzothiazoles as shown in Figure 5. In each compound the proton is transferred to the benzothiazole nitrogen, while the benzoxazoles are present as electron-withdrawing groups which also make a sizeable contribution to the extinction coefficient of the fluors. The Φ_fs in Figure 5 are all less than the $\Phi_f = 0.48$ reported earlier for the parent dibenzoxazole (Kauffman, SciFi93, p353, and Ref. 10). Interestingly, virtually identical Φ_fs were observed for either a phenol or a sulfonamido group serving as the proton donors in the bis(benzothiazoles) CJK-92X-55A and CJK-92X-93B. In the novel bis(benzothiazole) CJK-93-145, the acetamido substituent *para* to the phenol served to shift the fluorescence maximum all the way to 612 nm in poly(vinyltoluene) fiber. The reported Φ_f for CJK-93-145 counted only the light of wavelengths shorter than 750 nm, since the instrument cut off longer wavelengths. We attribute this dramatic bathochromic shift to the electron donation of the acetamido group [$CH_3C(C=O)NH-$], its contribution to the planarity of the fluorophore, and its diminished ability to compete in proton-transfer due to its lower pKa value ($\approx$ 15 vs. 10.5 for the phenol). Weak ESIPT emission ($\Phi_f = 0.04$) was observed even from the benzamido function in CJK-92X-99; but non-ESIPT emission near 410 nm ($\Phi_f = 0.08$) dominated.

7

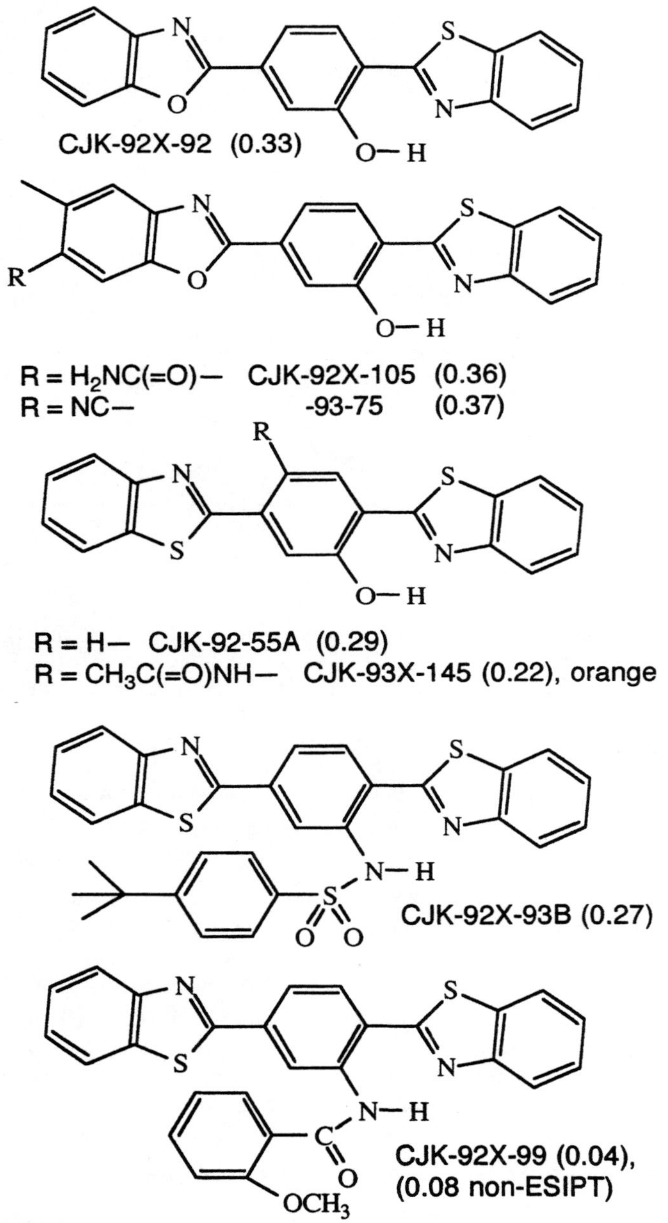

CJK-92X-92 (0.33)

R = H₂NC(=O)— CJK-92X-105 (0.36)
R = NC— -93-75 (0.37)

R = H— CJK-92-55A (0.29)
R = CH₃C(=O)NH— CJK-93X-145 (0.22), orange

CJK-92X-93B (0.27)

CJK-92X-99 (0.04),
(0.08 non-ESIPT)

FIGURE 5. Newer ESIPT Fluors of the Double Benzazole Type (Φ

Design and Properties of New Non-ESIPT Fluors in Group 3 — for Waveshifting Fibers

Stilbenes with Dialkylamino Auxofluors

By reason of their fast emission, high Φ_f and radiation-hardness, oligophenylenes, such as terphenyl, make ideal fluors. The practical long-wavelength limit for the fluorescence of unbridged oligophenylenes is ≈ 395 nm (11). The bridging of oligophenylenes with dialkylmethylene groups between some rings has extended the useful wavelengths of emission to ≈ 430 nm (12). The appendage of auxofluors to the termini of the oligophenylene chromophore can bathochromically shift the wavelength of emission, but frequently diminish the photostability of the fluor. For example, bis(diethylamino)terphenyl and -quaterphenyl, as well as the cyclic aminoterphenyl were found to have 80 nm bathochromic shifts of both absorption and emission and high Φ_f, but diminished photostability (Figure 6).

Since coumarins and rhodamines, which absorb at longer, less energetic wavelengths, show reasonable to excellent photostability while containing alkylamino functions (13), we reasoned that the appendage of dialkylamino auxofluors to aromatic fluorescent molecules absorbing at longer wavelengths than the simple oligophenylenes might lead to fluors with stability acceptable for Group 3 applications.

We chose to modify BDPVB (Figure 7), a fluor which is related to both Bis-MSB and TPBD (14) and exhibits both high Φ_f and good photostability. We hoped that an 80 nm bathochromic shift in both absorption and fluorescence would give a molecule suitable for the Group 3 application. We first prepared the

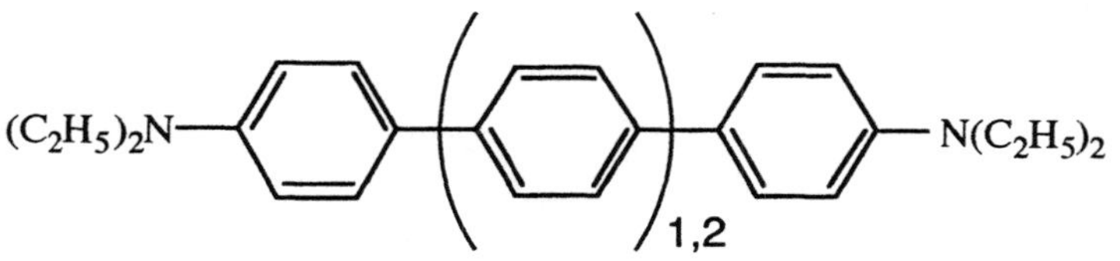

$(C_2H_5)_2N$—⟨benzene⟩—(⟨benzene⟩)$_{1,2}$—⟨benzene⟩—$N(C_2H_5)_2$

T. G. Pavlopoulos, P. R. Hammond, *J. Am. Chem. Soc.* **9 6**, 6568 (1974).

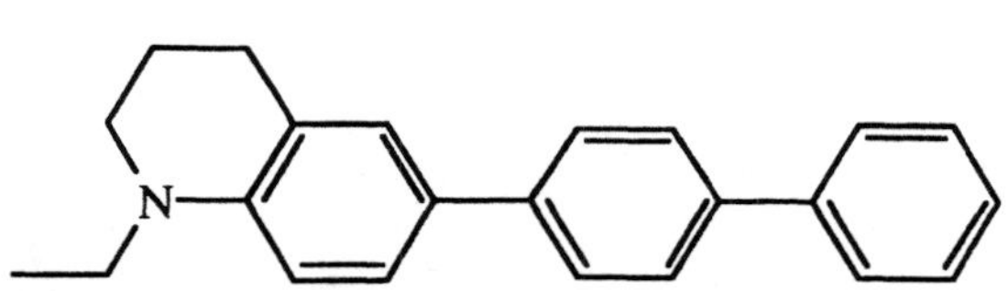

J. M. Kauffman, C. J. Kelley, A. Ghiorghis, E. Neister, L. Armstrong, *Laser Chem.* **8**, 335 (1988).

FIGURE 6. Dialkylamino Groups on Oligophenylenes Shift Wavelengths Bathochromically but Are Unstable

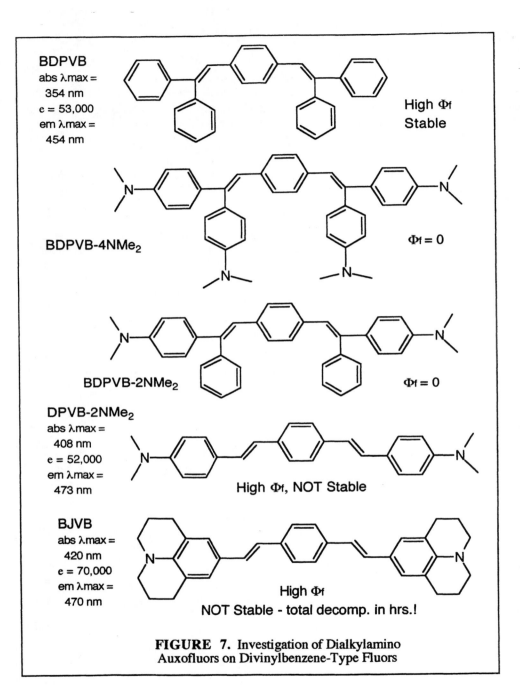

BDPVB
abs λ.max = 354 nm
e = 53,000
em λ.max = 454 nm

High Φf
Stable

BDPVB-4NMe₂

Φf = 0

BDPVB-2NMe₂

Φf = 0

DPVB-2NMe₂
abs λ.max = 408 nm
e = 52,000
em λ.max = 473 nm

High Φf, NOT Stable

BJVB
abs λ.max = 420 nm
e = 70,000
em λ.max = 470 nm

High Φf
NOT Stable - total decomp. in hrs.!

FIGURE 7. Investigation of Dialkylamino Auxofluors on Divinylbenzene-Type Fluors

tetrakis(dimethylamino) derivative BDPVB-4NMe$_2$, which did not fluoresce at all. Subsequent synthesis of a mixture of *(Z)* and *(E)* isomers of BDPVB-2NMe$_2$ gave another non-fluorescent product. Synthesis of the distyrylbenzene analog with 2 auxofluors, DPVB-2NMe$_2$, did give a fluor with high Φ_f, but without the hoped-for photostability.

In a final analogy to the best stability-enhancing substituents in coumarins and rhodamines, the auxofluoric amino groups were incorporated into a double alicyclic ring system at the termini of the distyrylbenzene chromophore. This fluor, BJVB, also displayed high Φ_f, but with no more photostability than the acyclic fluor. Indeed, a dilute solution of BJVB in toluene decomposed completely in a few hours.

Polyenes Proposed as Fluors for Group 3

Among the very few types of promising fluorophores for Group 3 is 1,8-diphenyl-1,3,5,7-octatetraene (Fig. 8). The emission at 518 nm is ideal, and the decay time of 6.2 nsec is good, while the $\Phi_f = 0.09$ is not (14). It is well known that constraining a fluor by bridging will bathochromically shift its absorption maxima while having minmal effect on the emission maxima (12). In addition, two of the sources of instability in alkenes, photochemical cyclization and Diels-Alder reactions would be prevented by constructing a fluor with adequate steric hindrance. We propose to prepare the bridged polyene shown in Fig. 8. We predict that the longwave absorption will lie at 410±20 nm, the emission will remain at 520 nm, the decay time

will remain short, and the Φ_f will be >> 0.09, using as an analogy the fact that a rigidized system such as fluorescein fluoresces well, while the homoelectronic phenolphthalein does not.

CONCLUSIONS

For Group 1, scintillating fibers with green emission, 3HF remains the top choice. The improvements shown by Flavone 560 are not worth 20x the cost. Coumarins, HBO, HBT and HBI derivatives are not sufficiently rad-hard (8, 15, 16). Oxazole 550F is to be investigated.

For Group 2, scintillating fibers with red emission, none of our orange-red emitters are bright or stable enough, and we are not aware of any other satisfactory ones.

For Group 3, waveshifting fibers with blue-green emission, we found that amino group auxofluors on distyrylbenzenes were unstable. Coumarins, HBT and HBI derivatives are not sufficiently rad-hard. A dibenzoxazolylphenol (B in Fig. 3) is being evaluated, and Cyano-HBO should be for this Group. A conjugated polyene is proposed.

Calculations of absorption and emission wavelengths accurate to ± 1000 cm^{-1} (± 25 nm @ 500 nm) can be made (17), while decay time and photochemical stability can be approximated; but light output and radiation hardness predictions are still tenuous.

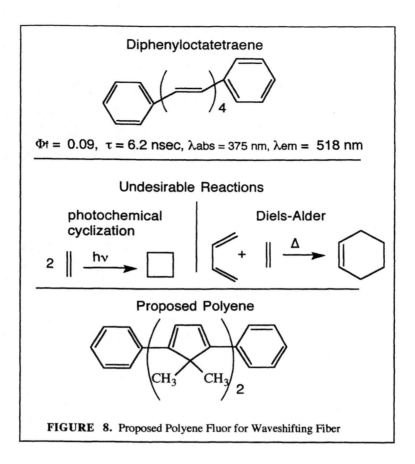

FIGURE 8. Proposed Polyene Fluor for Waveshifting Fiber

ACKNOWLEDGEMENTS

S. Wayne Moser et al., Bicron; Chuck Hurlbut, Ludlum Measurements, Inc.; James K. Walker et al., Nanoptics, Inc., and Univ. Florida Gainesville; Richard N. Steppel, Exciton, Inc.; Randy Ruchti et al., Dept. of Physics , Univ. Notre Dame; Anna Pla-Dalmau and Alan Bross, FERMILAB; U. S. Dept. of Energy.

REFERENCES

1. Proc. *Workshop on Scintillating Fiber Detectors,* Univ. Notre Dame, 24-28 Oct 93, A. D. Bross, R. C. Ruchti and M. R. Wayne, Eds., World Scientific, Singapore, 1995.
2. R. C. Ruchti, "The Use of Scintillating Fibers for Charged-Particle Tracking", *Annu. Rev. Nucl. Part. Sci.* **46**, 281-319 (1996); M. J. Weber, P. Lecoq, R. C. Ruchti, C. Woody, W. M. Yen, R.-Y. Zhu, "Scintillator and Phosphor Materials", Proc. Materials Res. Soc. Symp. 6-8 Apr 94, San Francisco, CA, Materials Res. Soc., Pittsburgh, PA, 1994.

3. R. C. Ruchti, "The Use of Scintillating Fibers for Charged-Particle Tracking", Annu. Rev. Nucl. Part. Sci. **46**, 281-319 (1996); M. J. Weber, P. Lecoq, R. C. Ruchti, C. Woody, W. M. Yen, R.-Y. Zhu, "Scintillator and Phosphor Materials", Proc. Materials. Res. Soc. Symp. 6-8 Apr 94, San Francisco, CA, Materials Res. Soc., Pittsburgh, PA, 1994.

4. J. M. Kauffman, "Review of Progress on Scintillation Fluors for the Detectors of the SSC", *Proc. Intl. Conf. Radiation-Tolerant Plastic Scintillators and Detectors,* 28 Apr - 2 May 92, Florida State Univ., Tallahassee, FL, USA, Radiat. Phys. Chem., **41**, 365-371, 1993.

5. Anna Pla-Dalmau, *Doctoral Diss.,* Northern Illinois University, 1990.

6. A. Pla-Dalmau and A. D. Bross, "Fluorescent Compounds for Plastic Scintillation Applications", in *Proc. Materials Res. Soc. Mtg.,* 4-8 Apr 94, in press.

7. Joel M. Kauffman, Peter T. Litak and Walter J. Boyko, "Syntheses and Photophysical Properties of Fluorescent Dibenzofurans, a Dibenzothiophene, and Carbazoles substituted with Benzoxazole and Hydroxyl Groups to produce Excited State Intramolecular Proton-Transfer", *J. Heterocyclic Chem.* **32**, 1541 (1995).

8. Anna Pla-Dalmau, "2-(2'-Hydroxyphenyl)benzothiazoles, -benzoxazoles, and -benzimidazoles for Plastic Scintillation Applications", *J. Org. Chem.* **60**, 5468 (1995).

9. A. N. Fletcher, D. E. Bliss, J. M. Kauffman, "Lasing and Fluorescent Characteristics of Nine New Flashlamp-Pumpable Coumarin Dyes in Ethanol and Ethanol:Water", *Optics Communs.* **47**, 57 (1983).

10. Joel M. Kauffman and Gurdip S. Bajwa, "Synthesis and Photophysical Properties of Fluorescent 2,5-Dibenzoxazolylphenols and Related Compounds with Excited State Proton-Transfer", *J. Heterocyclic Chem.* **30**, 1613 (1993).

11. H. O. Wirth, F. U. Herrmann, G. Herrmann and W. Kern, "On the Correlations Between Constitution and Scintillation Properties in the p-Oligophenylene Series", *Molecular Crystals* **4**, 321 (1968).

12. Joel M. Kauffman, Peter T. Litak , John A. Novinski, Charles J. Kelley, Alem Ghiorghis and Yuanxi Qin, "Electronic Absorption and Emission Spectral Data and Fluorescence Quantum Yields of Bridged p-Oligophenylenes,Bi- to Deciphenyls, and Related Furans and Carbazoles", *J. Fluorescence* **5**, 295 (1995).

13. G. A. Reynolds and K.-H. Drexhage, "New Coumarin Dyes with Rigidized Structure for Flashlamp-Pumped Laser Dyes", *Optics Communs.* **13**, 222 (1975).

14. I. B. Berlman, *Handbook of Fluorescence Spectra of Aromatic Molecules,* 2nd ed., Academic Press, New York, NY, 1971.

15. A. Pla-Dalmau, G. W. Foster, G. Zhang, "Coumarins as waveshifters in Polystyrene", *Nucl. Inst. Meth. Phys. Res.* A **361**, 192 (1995).

16. D. Denisov, *Instr. Exp. Tech.* **36** (3), 390 (1995); and *These Proceedings.*

17. Walter M. F. Fabian, *These Proceedings.*

Molecular Orbital Computations in Dye Chemistry

Walter M. F. Fabian

*Institut für Organische Chemie, Karl-Franzens Universität Graz, Heinrichstr. 28,
A-8010 Graz, Austria*

Abstract. The possibilities, problems and limitations for the use of various semiempirical molecular orbital methods to predict absorption and emission spectra of various types of organic dyes (laser dyes, e.g. coumarines and carbostyriles, probes for analytical and/or biochemical measurements, e.g. fluoresceins, and dyes for the D2T2 (dye diffusion by thermal transfer photocopying process, donor-substituted aromatics with dimeric malononitrile as acceptor) will be discussed. Although, especially for polycyclic (hetero)aromatic systems, the semiempirical methods give quite encouraging results, several pitfalls requiring a critical assessment of the reliability of any computational method will be presented. Finally, some examples of the effect of a proper treatment of environmental (solvent) effects on the quality of the calculated transition energies will be given.

INTRODUCTION

Traditionally, the term dye has been associated with molecules for which their color or their ability to color other material is the distinguishing property. In the past decades this emphasis has shifted to more unconventional applications (so-called specialty dyes), e.g. laser dyes, dyes for sensors in biological and medical applications, nonlinear optical materials, near infrared absorbing substances, etc.(1) In any case, whether their coloring or other properties, are of importance, it is the interaction of the respective compound with electromagnetic radiation, i.e. transitions between various electronic energy levels, which is the characteristic feature for a dye molecule in the most general definition. Therefore, the aim of applying computational methods to dye chemistry is the calculation of the electronic states (ground and excited) of a molecule. Several methods, ranging from high level ab initio to semiempirical procedures are available for this purpose. The choice depends on the primary goal of the computation, e.g. if it is intended as an aid in spectral assignment or for a decision between two or more possible molecular structures by comparison of the experimental with calculated spectra, the numerical accuracy is the main feature. In such a case the method of choice would be some high level ab initio method, which clearly restricts the applica-

CP450, *SciFi97: Workshop on Scintillating Fiber Detectors*
edited by A. D. Bross, R. C. Ruchti, and M. R. Wayne

tion to some smaller molecules. On the other extreme there are qualitative or semiquantitative concepts like the molecules-in-molecule (MIM) (2) or related approaches (3). These models are intended as tools for an understanding or quantification of concepts used in classical dye chemistry (chromophore, auxochrome, etc.). The main emphasis here is on a reasonable description of spectral trends; numerical values by themselves are of minor importance. Finally, computational methods might be used, like molecular modeling in structure – activity relationships in drug design (4), as an aid in the design of tailor-made dyes for special applications. In such a case, besides high reliability additional requirements are ease of use and little time and computer resource demands. Accurate ab initio calculations, therefore, are ruled out for this purpose. On the other hand, less sophisticated methods lacking the theoretical rigor of ab initio procedures require some assessment of their reliability and accuracy essential for the intended application as a design tool. Therefore, a calibration of such methods by comparison with experimental data is of paramount importance. However, experimental absorption bands refer to transitions between (normally) the vibrational ground state of the electronic ground state and excited vibrational states of excited electronic states. In contrast, calculated transition energies refer to the difference between the minima of the ground state potential energy hypersurface and that for an idealized excited electronic state. There will, therefore, always be some difference between experiment and calculations. The question now is, will this difference be independent of the particular molecule ? In that case a simple additive correction would suffice for a reliable prediction of absorption maxima. Unfortunately, this is not the case; instead, the number of vibrational quanta excited and, thus, the band shape and the position of the maximum depends on the shift between the potential energy hypersurfaces of the two states involved in the transition. Thus, apart from an electronic factor, which in routine application only can be treated, also vibrational factors play a role in determining the exact band position. In judging the reliability of a particular computational method such effects should be kept in mind. In the following successes and limitations of two popular methods, namely the Pariser-Parr-Pople (PPP) (5) and the CNDO/S procedure (and variants, like ZINDO) (6) with special emphasis on their applicability for designing organic molecules with particular absorption/emission characteristics will be discussed.

METHODS

In this section general overviews without going into the methodological details of the PPP and CNDO/S type procedures will be given. The main emphasis is on demonstrating the usefulness as well as possible pitfalls associated with these methods.

The PPP Method

The semiempirical PPP method has become the most popular tool for the dye chemists community. Reviews on this method with special emphasis on practical aspects have been given by Griffiths (1,7). The key features of the PPP method are: (i) it is a π-electron method, i.e. solely π-electrons are treated; (ii) it is a semiempirical method containing adjustable parameters, viz. the valence state ionization potential α, characteristic for a certain atom type, resonance integral β, describing a bond type, and one- and two-center electron – electron repulsion integrals γ_{ii} and γ_{ij}. The parameters α, β, and γ_{ii} normally are treated as adjustable parameters (hence the criticism of lacking theoretical rigor) whereas γ_{ij} is calculated by some empirical formula from γ_{ii} and γ_{jj} and the distance r_{ij} between atoms i and j. The major drawback, therefore, is that from a purist's point of view it should only be applied to planar conjugated systems. Many dyes, however, contain formal single bonds around which some torsion is possible (e.g. biphenyl). Strictly speaking, the PPP method should not be used for such compounds. To overcome this restriction, an approximate treatment of nonplanar systems is done by weighting the resonance integral β of the twisted bond with the torsional angle φ (equ. (1))

$$\beta_\varphi = \beta_0 * \cos\varphi \tag{1}$$

Absorption Spectra

To illustrate what one can expect from such a procedure in Figure 1 a plot of experimental excitation energies (8) for a rather diverse series of simple donor/acceptor substituted aromatic molecules, heterocyclic laser dyes, and merocyanine type compounds as dyes for thermographic printing, D2T2 (dye diffusion by thermal transfer) dyes, etc. vs. those calculated by PPP is given. A standard error (~100 data points) of ~1300 cm^{-1} is obtained. The results shown in Figure 1 were obtained by using a standard set of parameters (7) applicable to a diverse set of molecules.

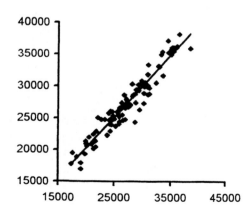

FIGURE 1. Experimental Excitation Energies (cm^{-1}) vs. PPP Results.

γ_{ij} is the only parameter explicitly, although only moderately, dependent on the molecular structure. In contrast to CNDO/S based methods (see below), therefore, the choice of the geometrical parameters is of minor importance with the PPP procedure. Alternating bond lengths, e.g. as found in conjugated polyenes, are usually simulated by choosing different ß-values for formal single and double bonds (7). Such a distance dependence of the resonance integral ß can be exploited for the calculation of fluorescence spectra.

Fluorescence spectra

As transition from the vibrational ground state of the (normally) first excited electronic state to vibrationally excited states of S_0, the simulation of fluorescence spectra requires knowledge of the S_1 state geometry. Within the framework of the PPP method geometrical changes accompanying electronic excitation may be approximated via the distance dependence of the resonance integral ß (9,10). Excited state bond lengths are estimated from calculated bond orders P_{ij} by utilizing a bond order – bond length relationship (equ. (2)):

$$R_{ij} = a + b * P_{ij} \tag{2}$$

These bond lengths then are used to evaluate new ß – values (equ. (3)):

$$\beta_{ij} = -k_1 * exp(-k_2 * R_{ij}) \tag{3}$$

In an iterative procedure, until convergence of the first transition energy, fluorescence spectra are obtained. Using the same molecules as above fluorescence transitions are obtained with a similar standard error (1400 cm^{-1}, 70 data points) as for UV/VIS spectra (8). A plot of experimental vs. PPP calculated fluorescence maxima is depicted in Figure 2.

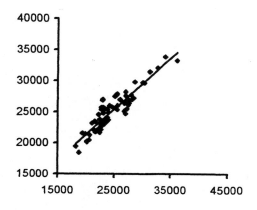

FIGURE 2. Experimental Fluorescence Transitions (cm^{-1}) vs. PPP Results.

Some specific examples (8) (heteroaromatic molecules (**1** – **8**, see structures below) with possible application as laser dyes) illustrating the usefulness and/or limitations of this approach are given in Table 1

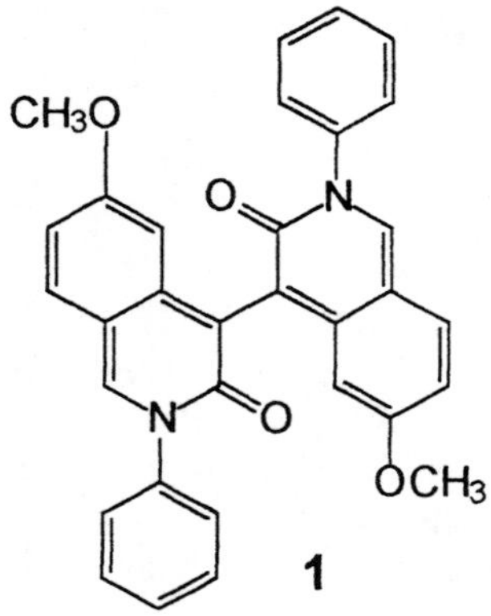

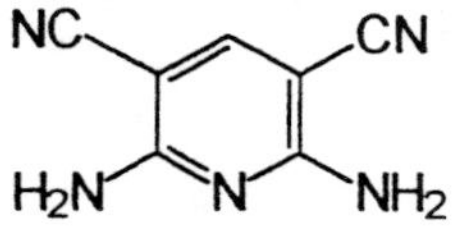

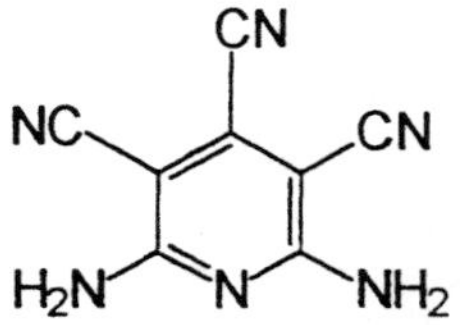

TABLE 1. Experimental and Calculated (PPP) Absorption and Fluorescence Transitions (cm^{-1}) for Compounds **1** – **8** (8).

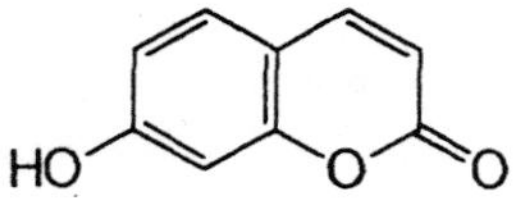

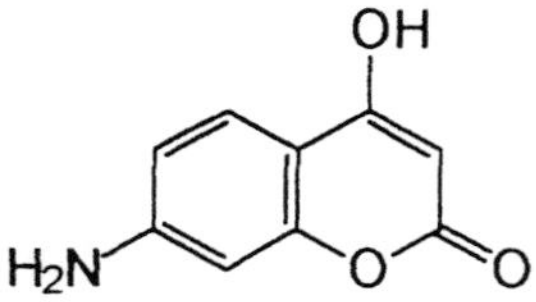

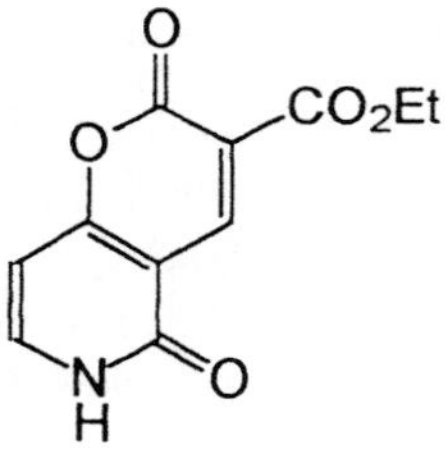

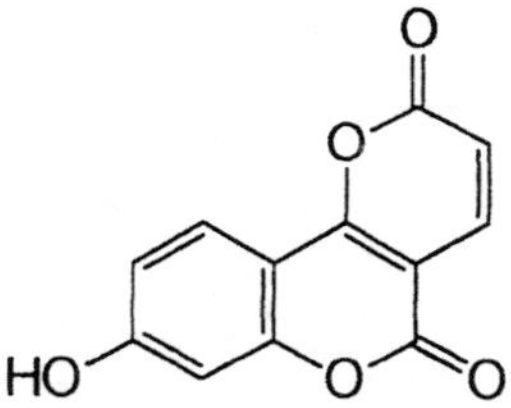

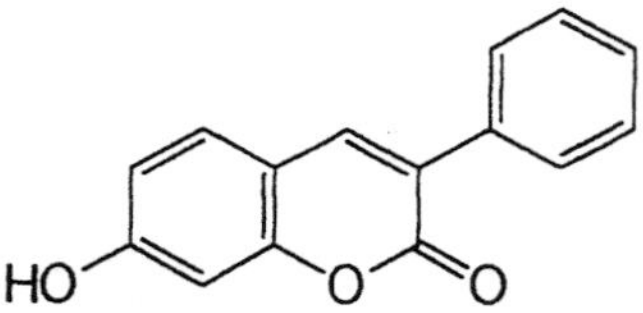

compound	ν(abs)$_{exp}$	ν(abs)$_{PPP}$	ν(flu)$_{exp}$	ν(flu)$_{PPP}$
1	19800	20000	18800	18500
2	30300	29800	26700	26400
3	25800	25200	23300	23000
4	30900	31800	26000	27000
			(20600)^a	(19400)^a
5	28600	28700	25600	26000
6	26200	25400	23000	23200
7	29400	26900	23800	23800
8	29800	30200	23700	23400

^{a} fluorescence of the phototautomer.

CNDO/S Type Methods (6)

A more recent review on semiempirical quantum chemical methods in general has been given by Zerner (11). The main difference between CNDO/S based methods and the PPP procedure is that in the former the σ-skeleton is treated explicitly. Although applications of INDO/S to rather large systems have been published, e.g. calculations on the bacterial photosynthetic reaction center (12), for applications in the design of new materials this places some restrictions on the size of treatable molecules. In contrast to the PPP method transition energies obtained by CNDO/S procedures, e.g. ZINDO (13), can be considerably more sensitive to the choice of molecular geometry. For instance, the pyrazolone derivative **9** – a thermosensitive dye – may exist in two main conformations ((Z) and (E), respectively, with respect to the exocyclic double bond) (14):

(Z)-9   **(E)-9**

Each one of these two conformations can adopt two rotameric structures depending on the torsional angle ϕ ($\sim 0^0$ and $\sim 180^0$). INDO/S calculated excitation energies for these four structures are collected in Table 2. Differences of up to 1300 cm^{-1} are obtained.

TABLE 2. Effect of Molecular Geometry on Calculated Transition Energies v (cm^{-1}) and Oscillator Strengths f.

(Z)-9 $\phi = 28^0$		$\phi = 155^0$		(E)-9 $\phi = 15^0$		$\phi = 180^0$	
v	f	v	f	v	f	v	f
26200	0.183	26500	0.177	24900	0.285	25900	0.519
27400	0.492	27400	0.518	27300	0.269	27800	0.036

An interesting example is provided by the furan derivative **10** which can exist in three conformations (*trans-cis*, *cis-cis*, and *trans-trans*, respectively, Table 3, Figure 3):

TABLE 3. Ground and Excited State Conformational Energies (kcal mol^{-1}) and Transition Energies (cm^{-1}) of the Three Conformations of Compound **10**.

	$E(S_0)$ (kcalmol$^{-1)}$	v(abs) / cm^{-1}	$E(S_1)$ (kcalmol^{-1})	v(flu) /cm^{-1}
trans - cis	143.8	26889	224.8	20712
cis–cis	142.0	27091	228.8	18672
trans-trans	146.6	27966	223.9	19667
exp		25773		19305

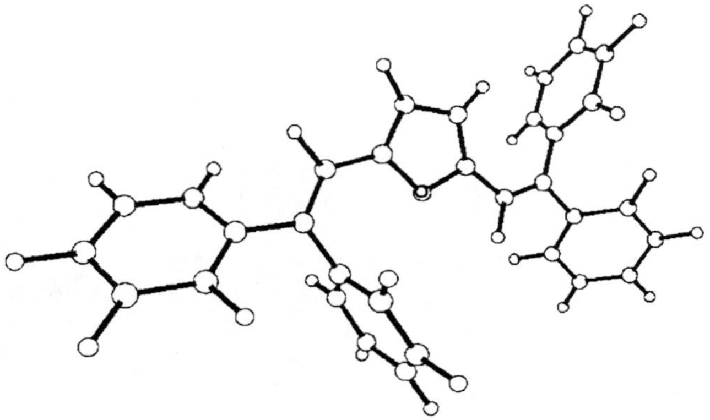

trans-cis - 10

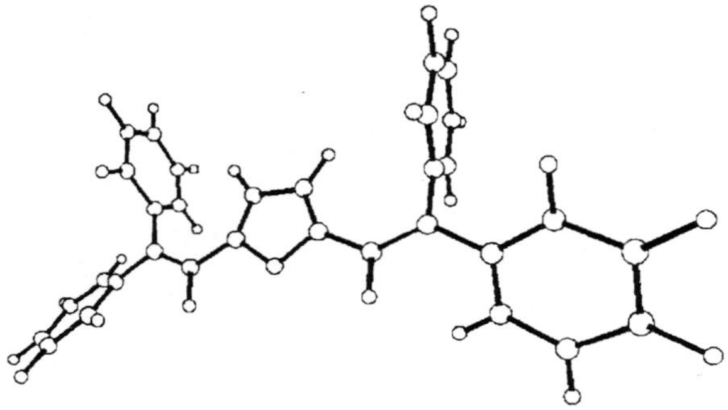

cis-cis - 10

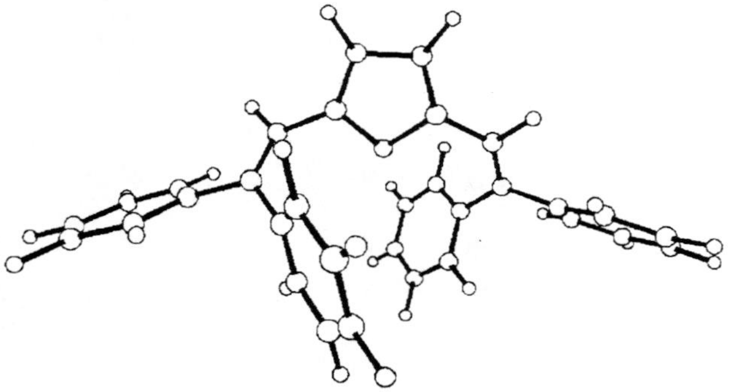

trans-trans – 10

FIGURE 3. Calculated Structures of the Three Conformations of Compound **10**.

In the ground state the *cis-cis* conformation is the most stable one, in S_1, however, it is the *trans-trans* structure. Transition energies between the different conformations can vary up to 2000 cm^{-1}. In other compounds, e.g. **11**, which also may exist in three different rotamers with respect to the two fluoren – furan single bonds, there is hardly any influence discernible (< 100 cm^{-1}).

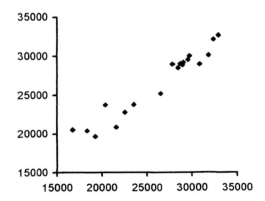

11

The preferred procedure is to optimize the geometry by some computational method, e.g. the semiempirical AM1 method (15), and to use this geometry as input for the ZINDO calculation. Solvent effects are also more important in CNDO/S type calculations than for PPP. The most frequently used procedure for an approximate treatment of solvent effects is the self-consistent reaction field model (16). For instance, in a series of push – pull (methoxy – dicyanovinyl) diarylpolyenes (8i) inclusion of the solvent via the SCRF model improved the agreement between experimental and calculated UV/VIS spectra by a factor of 2. In contrast, for benzoannulated fluorescein derivatives (see below) (17) gas phase and SCRF results were essentially identical.

Fluorescence spectra

To compute $S_1 \rightarrow S_0$ transition energies by CNDO/S procedures, an explicit optimization of the geometry of the first excited electronic state is necessary. For a series of oligophenylenes and related compounds (18) (Figure 4) this procedure leads to a stan-

FIGURE 4. Experimental and Calculated (ZINDO) Fluorescence Spectra (18).

dard error of ~ 1500 cm^{-1} (25 data points). Amino- and, even more so, dialkylamino groups belong to the substituents with the largest bathochromic effect. It appears to be a general shortcoming of ZINDO to grossly underestimate this effect. The largest deviations from the trend line in Figure 4 are exactly for molecules containing such a substituent.

X = O, NEt

Similarly, in the excited state proton transfer dyes **12** and **13** (19) both absorption as well as fluorescence are predicted at considerably too short wavelengths for the carbazol derivative (X = NEt) as compared to the dibenzofuran. The effect of excited state intramolecular proton transfer (phototautomerism) on the fluorescence of these compounds is, however, reasonably well described (see Table 4).

As a final application the design of a fluorescence-based optical pH sensor (17,20) suitable for measurement of pH in biological material, e.g. whole cells or, specifically, inside the body, is presented. Requirements were both excitation as well as emission maxima well outside the background absorption and fluorescence of the biological material (e.g. blood) and a pH-dependent fluorescence with a pK–value in the range 7 to 8. Possible candidates are fluorescein derivatives. By suitable benzoannulation it was expected to achieve the necessary red shift of the excitation and emission characteristics. The predictions from computations with respect to the most promising mode of annulation are summarized in Figure 5.

CONCLUSION

From the results presented above one can expect an accuracy for the prediction of both absorption and fluorescence maxima of about 1300 cm^{-1}. There seems to be little difference between the PPP and ZINDO methods, at least in the case of molecules for which application of the PPP method is justified. The choice of molecular geometry appears to be rather important for CNDO/S based methods, especially if a compound can exist in different conformations. The most efficient procedure then is to optimize

TABLE 4. Experimental and Calculated Absorption and Fluorescence of Compounds **12** and **13** and Their Phototautomers.

	Absorption		Fluorescence	
	exp	calc	exp	calc
X = O (**12**)	26596	27971	18349	20442
X = NEt (**13**)	23810	27828	16722	20552

22

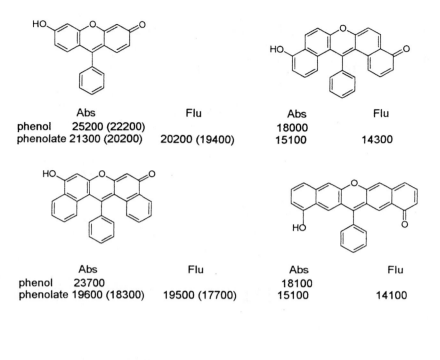

	Abs	Flu		Abs	Flu
phenol	25200 (22200)			18000	
phenolate	21300 (20200)	20200 (19400)		15100	14300

	Abs	Flu		Abs	Flu
phenol	23700			18100	
phenolate	19600 (18300)	19500 (17700)		15100	14100

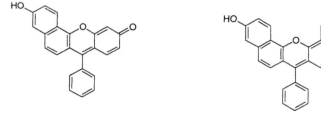

	Abs	Flu		Abs	Flu
phenol	21200 (18400)			20400 (19500)	
phenolate	18900 (18600)	17900 (16000)		17100 (16700)	16400 (15200)

FIGURE 5. Effect of Benzoannulation on Absorption and Fluorescence (cm^{-1}) of Fluorescein Derivatives (Experimental Data in Parentheses).

the structure by the semiempirical AM1 method (structures calculated by e.g. the TRI-POS force field (21) or with INDO/1 (as provided by the ZINDO package (13))) generally yield transition energies which are considerably too high except for pure aromatic hydrocarbons (18). Proper treatment of solvent effects, especially for merocyanine type structures is of particular importance within the framework of ZINDO. Although there remain several discrepancies, for example the gross underestimation of the bathochromic shifts induced by amino groups semiempirical molecular orbital calculations appear to be a useful addition to the chemists armory used for the design of new dyes with special properties.

REFERENCES

1. Griffiths, J., *Chem. Britain* **22**, 997-1000 (1986).
2. Longuett-Higgins, H. C. and Murrell, J. N., *Proc. Phys. Soc. A* **68**, 601-611 (1955).
3. (a) Polansky, O. E. and Derflinger, G., *Int. J. Quant. Chem.* **1**, 379-401 (1967). (b) Baba, H., Suzuki, S., and Takemura, T., *J. Chem. Phys.* **50**, 2078-2086 (1969). (c) Fabian, J., *J. Signal AM* **4**, 307-321 (1978), *ibid.* **7**, 67-89 (1979).
4. Bohm, G., *Biophys. Chem.* **59**, 1-32 (1996).
5. (a) Pariser, R. and Parr, R.G., *J. Chem. Phys.* **21**, 466-471, 767-776 (1953). (b) J. A. Pople, *Trans. Faraday Soc.* **49**, 1375-1385 (1953).
6. (a) Del Bene, J. and Jaffe, H.H., *J. Chem. Phys.* **48**, 1807-1813, 4050-4055 (1968). (b) Del Bene, J. and Jaffe, H.H., *J. Chem. Phys.* **49**, 1221-1229 (1968). (c) Ridley, J.E. and Zerner, M.C., *Theor. Chim. Acta* **32**, 111-134 (1973).
7. Griffiths, J., *Dyes and Pigments* **3**, 211-233 (1982).
8. (a) Fabian, W., *J. Mol. Struct.* **69**, 227-231 (1980). (b) Koitz, G., Fabian, W., Schmidt, H.-W., and Junek, H., *Monatsh. Chem.* **112**, 973-985 (1981). (c) Fabian, W., *J. Mol. Struct. (Theochem)* **90**, 249-264(1982). (c) Fabian, W., *Z. Naturforsch.* **40a**, 719-725 (1985). (d) Fabian, W., *Z. Naturforsch.* **41a**, 1425-1428 (1986). (e) Fabian, W., *Dyes and Pigments* **9**, 275-282 (1988). (f) Nakatsuji, S., Matsuda, K., Uesugi, Y., Nakashima, K., Akiyama, S., Katzer, G., and Fabian, W., *J. Chem. Soc., Perkin Trans. 2*, 861-867 (1991). (g) Nakatsuji, S., Matsuda, K., Uesugi, Y., Nakashima, K., Akiyama, S., and Fabian, W., *J. Chem. Soc., Perkin Trans. 1* 755-758 (1992). (h) Reidlinger, C., Dworczak, R., Fabian, W. M. F., and Junek, H., *Dyes and Pigments* **24**, 185-204 (1994). (i) Dworczak, R., Fabian, W. M. F., Pawar, B. N., and Junek, H., *Dyes and Pigments* **29**, 65-76 (1995).
9. (a) Fratev, F., Hiebaum, G., and Gochev, A., *J. Mol. Struct.* **23**, 437-447 (1974). (b) Fratev, F., *Z. Naturforsch.* **30a**, 1691-1695 (1975). (c) Fabian, W., *Dyes and Pigments* **6**, 341-348 (1985).
10. Mehlhorn, A., Schwenzer, B., and Schwetlick, K., *Tetrahedron* **33**, 1483-1488 (1977).
11. Zerner, M. C., in Lipkowitz, K. B. and Boyd, D. B. (Eds.), *Reviews of Computational Chemistry*, New York: VCH Publishing, Vol. 2, pp. 313 – 366.
12. Scherer, P. O. J. and Fischer, S. F., *Chem. Phys.* **131**, 115-127 (1989).
13. ZINDO, a comprehensive semiempirical quantum chemistry package: Zerner, M. C., Quantum Theory Project, University of Florida, Gainesville, FL, USA.
14. Dworczak, R., Fabian, W. M. F., Sterk, H., Kratky, C., and Junek, H., *Liebigs. Ann. Chem.* 7-14 (1992).
15. Dewar, M. J. S., Zoebisch, E. G., Healy, E. F., and Stewart, J. J. P., *J. Am. Chem. Soc.* **107**, 3902-3909 (1985).
16. Karelson M. M. and Zerner, M. C., *J. Phys. Chem.* **96**, 6949-6957 (1992).
17. Fabian, W. M. F., Schuppler, S., and Wolfbeis, O. S., *J. Chem. Soc., Perkin Trans. 2* 853-856 (1996).
18. Kauffman, J. M., Litak, P. T., Novinski, J. A., Kelley, C. J., Ghiorghis, A., and Qin, Y., *J. Fluorescence* **5**, 295-305 (1995).
19. Kauffman, J. M., Litak, P. T., and Boyko, W. J., *J. Heterocyclic. Chem.* **32**, 1541-1555 (1995).
20. (a) Kotyk, A. and Slavik, J., *Intracellular pH and its measurement*, Boca Raton: CRC Press, 1989. (b) Haugland, R. P., *Handbook of Fluorescent Probes*, Eugene, OR: Mol. Probes, 1995, 6th ed., ch. 4. (c) Lee, L. G., Berry, G. M., and Chen, C.-H., *Cytometry* **10**, 151-164 (1989). (d) Whitaker, J. E., Haugland, R. P., and Prendergast, F. G., *Anal. Biochemistry.* **194**, 330-344 (1991).
21. SYBYL 6.0, Tripos Associates, St. Louis, MO, USA.

SCINTILLATING CRYSTALS FOR PRECISION CRYSTAL CALORIMETRY IN HIGH ENERGY PHYSICS

Ren-yuan Zhu[1]

California Institute of Technology, Pasadena, CA 91125, USA

Abstract. Scintillating crystals in future high energy physics experiments face a new challenge to maintain its performance in a hostile radiation environment. This paper discusses the effects of radiation damage in scintillating crystals. The importance of maintaining crystal's light response uniformity and the feasibility to build a precision crystal calorimeter under radiation are elaborated. The mechanism of radiation damage in scintillating crystals is also discussed. While the damage in alkali halides is found to be caused by the oxygen/hydroxyl contamination, it is the structure defects, such as oxygen vacancies, cause damage in oxides. Material analysis used to reach these conclusions are presented in details.

INTRODUCTION

Total absorption shower counters made of inorganic scintillating crystals have been known for decades for their superb energy resolution and detection efficiency. In high energy and nuclear physics, large arrays of scintillating crystals have been assembled for precision measurements of photons and electrons. Recently, several crystal calorimeters have been designed for the next generation of high energy physics experiment. These include a CsI calorimeter for the KTeV experiment at Fermi Lab [1], two CsI(Tl) calorimeters for B Factory experiments: the *BaBar* experiment at SLAC [2] and the BELLE experiment at KEK [3], and a lead tungstate ($PbWO_4$) calorimeter for the Compact Muon Solenoid (CMS) experiment at the Large Hadronic Collider (LHC) [4]. Each of these calorimeters requires several cubic meters of high quality crystals.

The unique physics capability of the crystal calorimetry is the result of its superb energy resolution, hermetic coverage and fine granularity [5]. In future high energy experiments, however, scintillating crystals face a new challenge: the radiation damage caused by the increased center of mass energy and luminosity. While the dose rate is expected to be a few rad per day for CsI(Tl) crystals at two B Factories, it would reach 15 to 500 rad per hour for $PbWO_4$ crystals at LHC. This paper discusses effects of radiation damage in scintillating crystals, several importance issues related to the precision of a crystal calorimeter in radiation, and the cause and cure of the radiation damage in crystals.

---

[1] Work supported in part by U.S. Department of Energy Grant No. DE-FG03-92-ER40701.

CP450, *SciFi97: Workshop on Scintillating Fiber Detectors*
edited by A. D. Bross, R. C. Ruchti, and M. R. Wayne

COMMONLY USED CRYSTAL SCINTILLATORS

Table 1 lists the basic properties of some heavy crystal scintillators: NaI(Tl), CsI(Tl), undoped CsI, BaF_2, BGO and $PbWO_4$. All these crystals have been used in high energy and nuclear physics experiments, and are available in large quantity. The CsI(Tl) crystal is known to have high light yield, and is widely used in experiment. Its photon yield per MeV energy deposition exceeds that of NaI(Tl). The relative light yield of 45% listed in the table is due to the quantum efficiency of the Bi-alkali cathode at its emission peak (7%), which is about 1/3, as compared to 20% of NaI(Tl). The $PbWO_4$ crystal is distinguished with its high density and small radiation length and Moliere radius, so was chosen by the CMS experiment to construct a compact crystal calorimeter. The low light yield of $PbWO_4$ can be overcome by gains of the photo-detector, such as the PMT and avalanche photodiode (APD).

TABLE 1. Properties of Some Heavy Crystal Scintillators

Crystal	NaI(Tl)	CsI(Tl)	CsI	BaF_2	BGO	$PbWO_4$
Density (g/cm^3)	3.67	4.51	4.51	4.89	7.13	8.3
Melting Point (°C)	651	621	621	1280	1050	1123
Radiation Length (cm)	2.59	1.85	1.85	2.06	1.12	0.9
Moliere Radius (cm)	4.8	3.5	3.5	3.4	2.3	2.0
Interaction Length (cm)	41.4	37.0	37.0	29.9	21.8	18
Refractive Index[a]	1.85	1.79	1.95	1.50	2.15	2.2
Hygroscopicity	Yes	slight	slight	No	No	No
Luminescence[b] (nm) (at Peak)	410	560	420 310	300 220	480	510 510
Decay Time[b] (ns)	230	1300	35 6	630 0.9	300	50 10
Light Yield[bc]	100	45	5.6 2.3	21 2.7	9	0.3 0.4
d(LY)/dT[bd] (%/°C)	~0	0.3	−0.6	−2 ~0	−1.6	−1.9

[a] At the wavelength of the emission maximum.
[b] Top line: slow component, bottom line: fast component.
[c] Relative and measured with a PMT with a Bi-alkali cathode.
[d] At room temperature.

RADIATION DAMAGE IN SCINTILLATING CRYSTALS

All known crystal scintillators suffer from radiation damage. The most common damage phenomenon is the appearance of radiation-induced absorption bands caused by color center formation. The absorption bands reduce crystal's light attenuation length (LAL), and hence the light output. The color center formation, however, may or may not cause a degradation of the **light response uniformity**. The radiation also causes phosphorescence (afterglow), which leads to an increase of readout noise. Additional effect may include a reduced intrinsic scintillation light yield (damage of the scintillation mechanism), which would lead to a reduced light output and a deformation of the light response uniformity. The damage may recover under room temperature, which leads to the **dose rate dependence** [6]. Finally, thermal annealing and optical bleaching may be effective in eliminating color centers in the crystal.

In previous studies, we conclude that the scintillation mechanism of BGO [7], BaF_2 [8], CsI(Tl) [9] and $PbWO_4$ [10] are not damaged, and the consequence of the radiation-induced phosphorescence is negligible for BaF_2 [8] and $PbWO_4$ [11]. It is also known that the damage in BGO [7], BaF_2 [8] and $PbWO_4$ [11] are thermally annealable, as well as optically bleach-able, while it is not in CsI [9].

Radiation-Induced Absorption

The most common damage in scintillating crystals is the radiation-induced color center formation, which reduces the light attenuation length and thus the light output. Depending on the type of the defects in the crystal, the color centers may be electrons located in anion vacancies (F center), or holes located in cation vacancies (V center), or interstitial anion atoms (H center) or ions (I center). Radiation-induced color centers may be observed by measurement of crystal's transmittance. The left side of Figure 1 shows the longitudinal transmittance of three full size (~30 cm) CsI(Tl) samples [9]. The radiation-induced absorption bands are clearly identified in samples BGRI-2 and SIC-4, but is not seen in sample SIC-5, which was grown with a special scavenger in the melt to remove the oxygen contamination. The distinguished absorption bands centered at 850, 560, 520 and 440 nm are commonly attributed to F, F', F'' and F''' centers, respectively. The right side of Figure 1 shows the longitudinal transmittance of four large size (~20 cm) $PbWO_4$ samples [10]. The radiation-induced absorption bands in $PbWO_4$ are broad, and can be seen in all four samples BTCP-767, 1015, 1018 and SIC-66. A more detailed discussion of color centers can be found in references [12].

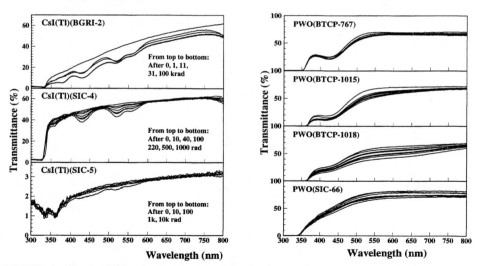

FIGURE 1. Longitudinal transmittance of CsI(Tl) (left) and $PbWO_4$ (right) samples, showing radiation-induced absorption bands.

Light Response Uniformity

An adequate light response uniformity profile is the key to the precision of a crystal calorimeter. The effect of the light response uniformity to the energy resolution has been studied by full GEANT simulations for crystal calorimetry. The left side of Figure 2 [8] shows a GEANT prediction for the energy fraction (top figure) and the intrinsic energy resolution (bottom figure) calculated by summing the energies deposited in a 3 × 3 sub-array, consisting of tapered crystals of 25 radiation length, as a function of the light response uniformity. In this simulation, the light response (y) of the crystal was parametrized as a normalized linear function:

$$y = y_{mid}[1 + \delta(x/x_{mid} - 1)], \tag{1}$$

where y_{mid} represents the light response at the middle of the crystal, δ represents the deviation of the light response uniformity, and x is the distance from the small (front) end of a tapered crystal. The right side of Figure 2 is a schematic illustrating the light response uniformity and its measurement by shooting a collimated γ-ray source at difference distance along the crystal axis.

While the degradation of the amplitude of the light output can be inter-calibrated, the loss of energy resolution, caused by the degradation of light response uniformity is not recoverable. To preserve crystal's intrinsic energy resolution, the light response uniformity must be kept within tolerance. According to the simulation, the δ value is required to be less than 5% so that the constant term of the energy resolution is below 0.5%. The dose profile caused by energy deposition of particles is not uniform. A damage to the scintillation mechanism thus would cause a distortion of the light response uniformity and a unrecoverable degradation of the energy resolution. The radiation-induced absorption, however, may or may not cause a severe distortion to the light response uniformity.

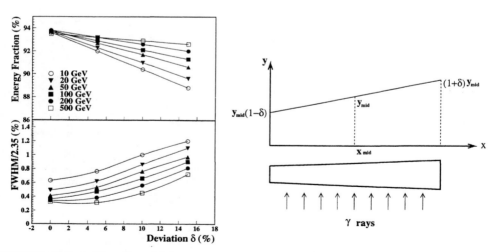

FIGURE 2. Left: The effect of light response uniformity predicted by a GEANT simulation. Right: A Schematic showing light response uniformity definition.

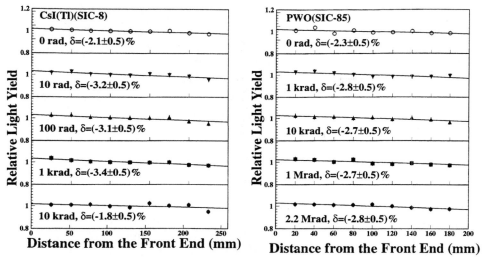

FIGURE 3. The light response uniformities are shown as a function of integrated dose for full size CsI(Tl) (left) and PbWO$_4$ (right) samples.

Figure 3 shows the light response uniformity as a function of accumulated dose for full size CsI(Tl) (left) and PbWO$_4$ (right) crystals. The pulse heights measured in nine points evenly distributed along the longitudinal axis of the crystal is fit to Equation 1, showing clearly that the slope (δ) does not change up to 10 krad for the CsI(Tl) sample and to 2.2 Mrad for the PbWO$_4$ sample. The result of CsI(Tl) sample confirms that the CsI(Tl) scintillation mechanism is not damaged, since only the front few cm of the sample was subject to the radiation dose in this experiment. Although the PbWO$_4$ sample was subject to a uniform irradiation, an experiment of irradiating only the middle 10 cm of a PbWO$_4$ crystal led to the same result, confirming that the PbWO$_4$ scintillation mechanism is also not damaged. The result of this laboratory measurement was later confirmed by a beam test for a PbWO$_4$ crystal array at CERN. Intensive electron beam was used to irradiate crystals. No degradation of energy resolution was found for a sample SIC 69 before and after beam irradiation with an integrated dose of 650 rad [13].

These results may be understood, as the intensity of all light rays attenuates equally after passing the same radiation-induced absorption zone in the crystal. A ray-tracing simulation was carried out. The exact geometry of a full size PbWO$_4$ crystal, i.e. $2^2 \times 23 \times 2.3^2$ (cm), was used in this simulation. The refractive index of PbWO$_4$ [14] was assumed to be a tensor with the optical axis at $15°$ to the crystal axis, and the polarization of light is taken into account. The simulation was carried out for two photo-detectors: a PMT and an APD. While the PMT covers the full back face of the crystal, only 3.7% of the back area of the crystal was covered by the APD of ϕ5 mm diameter. The optical coupling between photo-detectors and the crystal was assumed to have a refractive index of of 1.5. The crystal was assumed to be wrapped with Tyvek paper for all surfaces except the part covered by the detector. The interface between Tyvek paper and crystal surface was assumed to have a thin air gap and a reflectivity of 0.9, and a defused reflection with

TABLE 2. Result of a Ray-Tracing Simulation for a Full Size PbWO$_4$ Crystal

LAL (cm)	20	40	60	80	200	400
	Full Face Photo-detector, Covering 100% Back Face					
y_{mid} (%)	9.5±0.2	15.7±0.4	19.2±0.5	21.6±0.6	26.9±0.7	29.1±0.8
δ (%)	23±1	-5±1	-11±1	-15±1	-15±1	-16±1
	$\phi5$ mm Photo-detector, Covering 3.7% Back Face					
y_{mid} (%)	.38±0.04	.74±0.08	1.1±0.1	1.4±0.2	3.0±0.3	5.1±0.3
δ (%)	23±4	-4±4	-12±4	-16±4	-17±3	-16±2
$\frac{y_{mid}(\phi5mm)}{y_{mid}(Full)}$ (%)	4.0	4.7	5.7	6.5	11	17.5

the reflection angle smeared with a Gaussian width of 0.5 radian. Photons were generated isotropically in the crystal and were traced all the way. The cumulated path length of the photon was used together with the assumed light attenuation length to calculate the weight of the photon. Photons were followed until either absorbed by the photo-detector or escaped from the crystal, or its weight is less than e^{-5}. Finally, the light collection efficiency (y) was obtained by averaging all weights of photons absorbed by the photo-detector, and was fitted to the Equation 1.

Table 2 lists the numerical result together with statistical errors of the simulation. As can be seen from this table, the slope of the light response uniformity (δ) is entirely determined by the crystal geometry, if the light attenuation length is long enough. The table also shows that the light output (y_{mid}) of a detector covering a smaller fraction of the back face (APD) is more sensitive to the degradation of the light attenuation length, or radiation induced color centers.

Since the degradation of the amplitude of the light output can be inter-calibrated with physics events, or by a light monitoring system if it is caused by optical absorption, to preserve crystal precision *in situ*, thus, two necessary conditions must be satisfied: (1) crystal's scintillation mechanism must not be damaged and (2) crystal's initial light attenuation length must be long enough and the density of radiation-induced color center must be low enough, so that the light response uniformity does not degrade *in situ*.

Damage Recovery

The radiation damage in crystal may recover at room temperature with the recovery speed depending on the depth of the traps. A slow recovery would ease the calibration requirement, so is preferred. It is known that damage in BaF$_2$ [8] and CsI [9] recovers at an extremely slow speed, while BGO [7] and PbWO$_4$ [11] do recover at a speed of orders of few hours to few weeks.

The left side of Figure 4 shows the recovery speed of less than 0.5% per day for three full size CsI(Tl) samples after a 100 krad irradiation, which indicates that the radiation-induced color centers in CsI(Tl) are caused by deep traps. The thermal annealing did not accelerate the recovery when below 300°C, and made sample BGRI-2 turned milky and produced no more scintillation light when beyond 300°C. This was explained by the drift of the thallium luminescence centers out of the CsI(Tl) lattice at 300°C. Because of this

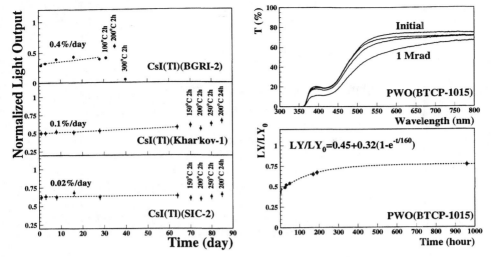

FIGURE 4. The recovery of radiation damage is shown for three CsI(Tl) crystals after 100 krad irradiation (left) and a PbWO$_4$ crystal after 1 Mrad irradiation (right).

slow recovery, the accuracy of inter-calibration may be provided by physics *in situ*, e.g. Bhabha events in a few hours.

The right side of Figure 4 shows the damage recovery for a PbWO$_4$ sample BTCP-1015 after a dose of 1 Mrad. The top plot shows the recovery in the transmittance as a function of wavelength. The bottom plot shows normalized light output as a function of time after irradiation. The solid line in the light output plot is a fit to an exponential recovery with a time constant of 160 hour, as shown in the figure. Because of this damage and recovery, special attention should be paid to the stability of a PbWO$_4$ calorimeter *in situ*.

Dose Rate Dependence

If both annihilation and creation coexist, the color center density at the equilibrium depends on the dose rate applied. Assuming the annihilation speed of color center i is proportional to a constant a_i and its creation speed is proportional to a constant b_i and the dose rate (R), the differential change of color center density when both processes coexist can be written as [15]:

$$dD = \sum_{i=1}^{n} \{-a_i D_i + (D_i^{all} - D_i) \, b_i R\} dt, \tag{2}$$

where D_i is the density of the color center i in the crystal and the summation goes through all centers. The solution of Equation 2 is

$$D = \sum_{i=1}^{n} \{\frac{b_i R D_i^{all}}{a_i + b_i R} \left[1 - e^{-(a_i + b_i R)t}\right] + D_i^0 e^{-(a_i + b_i R)t}\}, \tag{3}$$

31

where D_i^{all} is the total density of the trap related to the center i and D_i^0 is its initial density. The color center density in equilibrium (D_{eq}) thus depends on the dose rate (R).

$$D_{eq} = \sum_{i=1}^{n} \frac{b_i R D_i^{all}}{a_i + b_i R}, \qquad (4)$$

The left side of Figure 5 shows the entire history of the light output measurement for a PbWO$_4$ sample SIC 115-1, which lasted for 20 days. The measured light output was normalized to that before irradiation, and was plotted as a function of the time of the measurement. The dose rate and integrated dose are also shown in the figure. There were periods when dose rate was shown as zero, indicating a recovery test. As shown in the figure, the level of PbWO$_4$ damage is indeed dose rate dependent, and the fast recovery (in order of a few hours) was observed only when dose rate is 15.7 krad/h.

If the recovery speed (a) is slow, however, the dose rate dependence would be less important, the color center density in the equilibrium would less depends on the dose rate. The extrema is no dose rate dependence, as shown in Equation 5

$$D'_{eq} = \sum_{i=1}^{n} D_i^{all}. \qquad (5)$$

The right side of Figure 5 shows the transmittance as a function of wavelength for a BaF$_2$ sample before and after 100, 1k, 10k, 100k and 1M rad irradiations (from top to bottom) at a fast (a) and a slow (b) dose rate, as shown in the figure [8]. While the fast rate is up to a factor of three higher than the slow rate, the damage levels for the same integrated dose are identical, showing no dose rate dependence. This is expected, since no recovery at room temperature was observed for BaF$_2$.

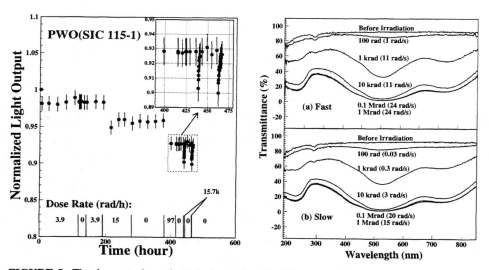

FIGURE 5. The dose rate dependence is shown for a PbWO$_4$ sample SIC 115-1 (left), and no dose rate dependence was observed for a BaF$_2$ sample (right).

DAMAGE MECHANISM IN ALKALI HALIDES

Material analysis would help to identify the mechanism of radiation damage in crystal scintillators, and thus improve crystal's quality. Glow Discharge Mass Spectroscopy (GDMS) analysis was used in Charles Evans & Associates [16], looking for correlations between the trace impurities in the CsI(Tl) crystals and their radiation hardness. Samples were taken 3 to 5 mm below the surface of the crystal to avoid surface contamination. A survey of 76 elements, including all of the lanthanides, indicates that there are no obvious correlations between the detected trace impurities and crystal's susceptibility to the radiation damage.

The oxygen and hydroxyl contamination is known to cause radiation damage in alkali halides. In BaF$_2$ [8], for example, hydroxyl (OH$^-$) may be introduced into crystal through a hydrolysis process, and latter decomposed to interstitial and substitutional centers through a radiolysis process, as shown in Equation 6

$$OH^- \rightarrow H_i^0 + O_s^- \ or \ H_s^- + O_i^0, \tag{6}$$

where subscript i and s refer to interstitial and substitutional centers respectively. Both the O_s^- center and the U center (H_s^-) were identified [8].

Following the BaF$_2$ experience, effort was made to reduce oxygen contamination in CsI(Tl) crystals at SIC. An approach was taken: using a scavenger, to remove oxygen contamination, similar to the Pb for BaF$_2$. A significant improvement of the radiation hardness was achieved [9]. The left side of Figure 6 shows the light output as a function of accumulated dose for full size CsI(Tl) samples, compared to the *BaBar* radiation hardness specification (solid line). While samples SIC-5, 6, 7 and 8 (with scavenger)

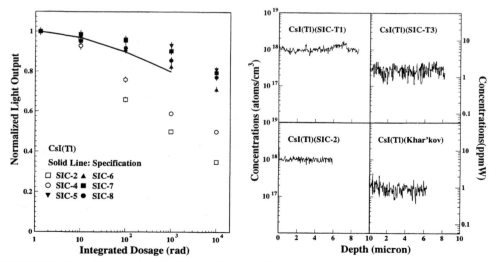

FIGURE 6. Left: The progress of CsI(Tl) radiation hardness is shown for full size CsI(Tl) samples from SIC together with the rad-hard specification of *BaBar* experiment. Right: The depth profile of oxygen contamination is shown for two rad-soft CsI(Tl) samples (left) and two rad-hard samples (right).

satisfied the *BaBar* specification, samples SIC-2 and 4 did not. It is understood that the function of the scavenger is to form oxide with density less than CsI, so that the oxide will migrate to the top of the ingot during the growing process, similar to the zone-refining process. By doing so, both oxygen and scavenger are removed from the crystal.

Further analysis was carried out to quantitatively identify the oxygen contamination in CsI(Tl) samples. We first tried the Gas Fusion (LECO) at Shiva Technologies West, Inc. [17], and found that the oxygen contamination in all CsI(Tl) samples is below the LECO detection limit: 50 ppm. We then tried Secondary Ionization Mass Spectroscopy (SIMS) at Charles Evans & Associates. A Cs ion beam of 6 keV and 50 nA was used to bombard the CsI(Tl) sample. All samples were freshly cleaved prior before being loaded to the UHV chamber. An area of 0.15×0.15 mm^2 on the cleaved surface was analyzed. To further avoid surface contamination, the starting point of the analysis is at about 10 μm deep inside the crystal. The right side of Figure 6 shows depth profile of oxygen contamination for two rad-soft (left) and two rad-hard (right) CsI(Tl) samples. Crystals with poor radiation resistance have oxygen contamination of 10^{18} atoms/cm^3 or 5.7 ppmW, which is 5 times higher than the background count (2×10^{17} atoms/cm^3, or 1.4 ppm). The radiation damage in CsI(Tl) is indeed caused by oxygen contamination.

DAMAGE MECHANISM IN OXIDES

Similarly, the GDMS analysis at Charles Evans & Associated found no specific correlations between the detected trace impurities in PbWO$_4$ crystals and crystal's susceptibility to the radiation damage. The result of GDMS analysis, however, revealed that

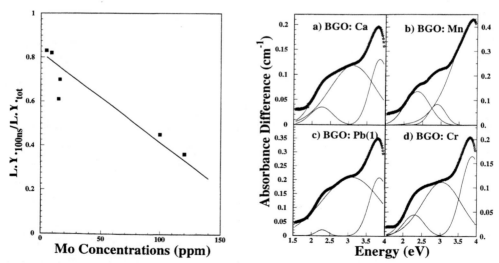

FIGURE 7. Left: The correlation between Mo contamination and the fraction of fast scintillation component. Right: Radiation-induced absorption spectra of four doped BGO samples are decomposed to three common absorption bands.

the origin of the slow scintillation component in PbWO$_4$ is the Mo contamination, as shown in the left side of Figure 7, confirming an early observation by M. Kobayashi [18]. This correlation is easy to be understood, since lead molybdate (PbMoO$_4$) is a known scintillators with slow ($\sim$10 μs) decay time.

Crystal defects, such as oxygen vacancies, is known to cause radiation damage in oxide scintillators. In BGO, for example, three common radiation induced absorption bands at 2.3, 3.0 and 3.8 eV were found in a series of 24 doped samples [7], indicating defect-related color centers, such as oxygen vacancies. The right side of Figure 7 shows the radiation induced absorption spectra for four BGO samples doped with Ca, Mn, Pb and Cr, respectively. Although the shape of these radiation-induced absorption bands are rather different, a decomposition showed three common absorption bands at the same energy and with the same width.

Following the BGO experience, effort was made to reduce oxygen vacancies in PbWO$_4$ crystals at SIC. An approach was taken: oxygen compensation through post-growth thermal annealing in an oxygen-rich atmosphere. Significant improvement of radiation hardness was achieved for PbWO$_4$ crystals produced taking this approach. The left side of Figure 8 shows the normalized light output at the equilibrium as a function of the dose rate for 6 pairs of 5 cm long PbWO$_4$ samples, which were treated with different oxygen compensation conditions. The result shows that samples annealed under different conditions have much different behavior. Samples, which were not annealed, have the worst radiation hardness. Samples annealed in oxygen is more radiation hard than that in air. Samples, which were annealed under the optimized oxygen conditions, are the best. Samples treated under the optimized oxygen annealing are really radiation hard. The right side of Figure 8 shows the normalized light output as a function of integrated

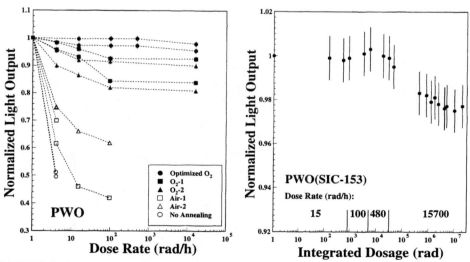

FIGURE 8. Left: Normalized light output at the equilibrium is shown as a function of the dose rate for six pairs of 5 cm PbWO$_4$ samples treated with different post-growing thermal annealing conditions. Right: The normalized light output of 5 cm PbWO$_4$ sample SIC 153 is shown as a function of integrated dose.

35

dose up to 20 Mrad for sample 153. This sample showed no degradation in light output under a dose rate below 480 rad/h, and had only 2% degradation under an extremely high dose rate of 15.7 krad/h.

Further analysis was carried out to quantitatively identify the stoichiometry deviation and oxygen vacancies in $PbWO_4$ samples. We first tried Particle Induced X-ray Emission (PIXE) and quantitative wavelength dispersive Electron Micro-Probe Analysis (EMPA) for $PbWO_4$ crystals in Charles Evans & Associates. Indeed, crystals with poor radiation hardness were found to have a non-stoichiometric W/Pb ratio [11]. However, both PIXE and EMPA did not provide oxygen analysis.

The direct observation of oxygen vacancies requires a measurement of the complete stoichiometric ratio of Pb:W:O in $PbWO_4$ samples. X-ray Photoelectron Spectroscopy (XPS) was tried at Charles Evens & Associates. It, however, was found to be very difficult due to the systematic uncertainty in oxygen analysis [19]. An effort to directly identifying oxygen vacancies by Electron Paramagnetic Resonance (ESR) and Electron-Nuclear Double Resonance (ENDOR) through observing unpaired electrons was also found to be difficult to reach a quantitative conclusion.

By using Transmission Electron Microscopy (TEM) coupled to Energy Dispersion Spectrometry (EDS), a localized stoichiometry analysis was possible to identify oxygen vacancies. A TOPCON-002B Scope was first used at 200 kV and 10 μA. Samples were made to powders of an average grain size of a few μm, and then placed on a sustaining membrane. With a spatial resolution of 2 Å, the lattice structure of $PbWO_4$ crystals was clearly visible. The left side of Figure 9 shows a TEM picture taken for a sample with poor radiation hardness. Black spots of a diameter of 5 – 10 nm were clearly seen in the picture. On the other hand, samples with good radiation hardness show stable TEM

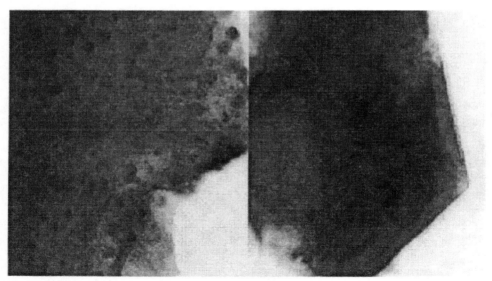

FIGURE 9. TEM pictures of a $PbWO_4$ crystal of poor (left) radiation hardness, showing clearly the black spots of ϕ5–10 nm related to oxygen vacancies, as compared to that of a good one (right).

TABLE 3. Result of a Localized (ϕ0.5 nm) Stoichiometry Analysis by TEM/EDS [20]

Element	Black Spot	Peripheral	Matrix$_1$	Matrix$_2$
		As Grown Sample		
O	1.5	15.8	60.8	63.2
W	50.8	44.3	19.6	18.4
Pb	47.7	39.9	19.6	18.4
Element	Point$_1$	Point$_2$	Point$_3$	Point$_4$
	The Same Sample after Oxygen Compensation			
O	59.0	66.4	57.4	66.7
W	21.0	16.5	21.3	16.8
Pb	20.0	17.1	21.3	16.5

picture with no black spots, as shown in the right side of Figure 9.

By employing a TEM with EDS system, a localized stoichiometry analysis was carried out at SIC [20]. The system is a JEOL JEM-2010 scope and a Link ISIS EDS. The spatial resolution of this system allows a localized stoichiometry analysis in a region of a diameter of 0.5 nm [20]. An as grown sample was first analyzed, and black spots were observed. Points inside and surrounding the black spots were analyzed as well as points far away from the black spots. The uncertainty of the analysis is typically 15%. The resultant atomic fractions (%) at these areas are listed in Table 3. A clear deviation from the atomic stoichiometry of O:W:Pb = 66:17:17 was observed in the center of these black spots, pointing to a severe deficit of oxygen component. In the peripheral area, the oxygen deficit was less, but still significant. There was no oxygen deficit observed in the area far away from the black spots. As a comparison, the same sample after oxygen compensation was re-analyzed. No black spot was found. The result of the analysis is also listed in Table 3. In all randomly selected points no stoichiometry deviation was observed. This analysis thus clearly identified oxygen vacancies in $PbWO_4$ samples of poor radiation hardness.

SUMMARY

Possible effects of radiation damage in scintillating crystals include (1) radiation induced absorption; (2) radiation induced phosphorescence, and maybe (3) damage in scintillation mechanism. The predominant radiation damage effect in crystal scintillators is the radiation induced absorption, or color center formation, not the loss of scintillation light yield.

It is understood that crystal scintillators for precision calorimetry must preserve their light response uniformity under irradiation, which requires a long enough initial light attenuation length and a low enough color center density. With such crystals, a precision light monitoring may function as inter-calibrations.

The radiation damage in alkali halides is understood to be caused by the oxygen and/or hydroxyl contamination, as evidenced by the SIMS analysis and the effectiveness of the usage of a scavenger in removing oxygen contamination in CsI(Tl) crystals. It is also understood that the stringent specifications for CsI(Tl) crystals to be used to construct calorimeters for two B Factories can be meet by mass-produced crystals.

The radiation damage in oxides is understood to be caused by stoichiometry-related defects, e.g. oxygen vacancies, as evidenced by localized stoichiometry analysis by TEM/EDS, and the effectiveness of the oxygen compensation for $PbWO_4$ crystals. It is expected that rad-hard $PbWO_4$ crystals will be developed through the systematic research and development program carried out by CMS physicists in collaboration with crystal manufactures.

ACKNOWLEDGEMENTS

Measurements at Caltech were carried out by Mr. Q. Deng, H Wu, D.A. Ma, Z.Y. Wei and T.Q. Zhou. Part of the $PbWO_4$ related work was carried out by Dr. C. Woody and his group at Brookhaven National Laboratory. Prof. Z.W. Yin, Drs. G. Cheng and P. Lecoq provided samples described in this report. Many inspiring and interesting discussions with Drs. M. Kobayashi, P.J. Li, J.Y. Liao, D.Z. Shen, C. Woody, and Z.W. Yin are also acknowledged.

REFERENCES

1. Arisaka K.*et al.*, *KTeV Design Report*, **FN-580**, January (1992).
2. *BaBar* Collaboration, *Technical Design Report*, **SLAC-R-95-457** (1995).
3. BELLE Collaboration, *Technical Design Report*, **KEK Report 95-1** (1995).
4. CMS Collaboration, *The Electromagnetic Calorimeter Technical Design Report*, CERN/LHCC 97-33 (1997).
5. Zhu R.Y. *et al.*, *Nucl. Phys.* **B44** 547 (1995).
6. Zhu R.Y., *IEEE Trans. Nucl. Sci.* **NS-44** 468 (1997).
7. Zhu R.Y. *et al.*, *Nucl. Instr. and Meth.* **A302** 69 (1991),
8. Zhu R.Y., *Nucl. Instr. and Meth.* **A340** 442 (1994).
9. Zhu R.Y. *et al.*, in *Proc. of the 6th Int. Conf. on Calorimetry in High Energy Physics*, ed. A. Antonelli *et al.*, Frascati Physics Series (1996) 589.
10. Zhu R.Y. *et al.*, in *Proc. of the 6th Int. Conf. on Calorimetry in High Energy Physics*, ed. A. Antonelli *et al.*, Frascati Physics Series (1996) 577.
11. Zhu R.Y. *et al.*, *Nucl. Instr. and Meth.* **A376** 319 (1996); Woody C. *et al.*, in *Proceedings of SCINT95 Int'l Conf.* Delft, August 1995 and *IEEE-NUCL-S* **V43** (1996) 1585.
12. There are many literatures discussing color centers in solids, for example, Crawford J.H. and Slifkin L.M., *Point Defects in Solids*, Plenum Press, New York (1972).
13. Auffray E. *et al.*, *Nucl. Instr. and Meth.* **A385** (1997) 425.
14. Bakhshiva G. & Morozov A., Sov.J.Opt.Technol. **44**(9), Sept, 1977.
15. Ma D.A. and Zhu R.Y., *Nucl. Instr. and Meth.* **A332** 113 (1993) and Ma D.A. *et al.*, *Nucl. Instr. and Meth.* **A356** 309 (1995).
16. Charles Evans & Associates, 301 Chesapeake Drive, Redwood City, CA 94063.
17. Shiva Technologies West, Inc., 16305 Caputo Drive Suite C, Morgan Hill, CA 95037.
18. Kobayashi M. *et al.*, *Nucl. Instr. and Meth.* **A373** (1996) 333.
19. Lazik C. of Charles Evans and Associates, private communications.
20. Yin Z.W. *et al.*, in *Proceedings of SCINT97 Int'l Conf.*, Shanghai, September 1997.

Glass and Phosphor Scintillators
for X-Ray Imaging

John Ellis

Collimated Holes, Inc.
460 Division Street, Campbell, CA 95008

Abstract. In x-ray imaging, many combinations of scintillators and detectors are used for a variety of reasons. This paper reviews several scintillators currently used in x-ray imaging systems in medical and non-destructive testing (NDT) applications.

X-RAY IMAGING SIMPLIFIED

A typical x-ray imaging system consists of a source of x-rays, an object under study, and a detector which records an image by reading the variation in intensity of the x-rays across an image plane. The variation results from differences in the transmission of x-rays through the object. The detector and readout portion, the camera, is the most interesting part of the system. Not surprisingly, there are many of different types of cameras used today. Though based on similar principles, there are important differences in the implementation of the various types.

X-ray imaging system designers would consider all aspects of the system important: from the energy spectrum, the collimation (or lack thereof) of the x-rays, the absorption characteristics of the object under test, and the efficiency and resolution of the detector and readout system.

In the simplest system, the object under test is placed directly on a piece of film, relying on this proximity (and a reasonably parallel beam, depending on the thickness) to record an image. This is how x-rays were first discovered, and the imaging technique still works today. But this method requires a high dose of x-rays and is generally limited to NDT on smaller parts. In this scheme, the film is the detector, readout, and storage mechanism. The principal drawback is that it is relatively inefficient since the film is so thin that most of the x-rays do not deposit their energy in the film. Also, if many images are required, it can be difficult to process, store, and

CP450, *SciFi97: Workshop on Scintillating Fiber Detectors*
edited by A. D. Bross, R. C. Ruchti, and M. R. Wayne
© 1998 The American Institute of Physics 1-56396-792-8/98/$15.00

retrieve images efficiently. The benefits are it is easy to do, and it is well understood.

Newer technologies include direct-reading image-capture devices include CCD imagers and amorphous silicon detectors. These can be fabricated to read out x-rays directly, like film. Here, the principal advantage lies in digital processing, where the images are captured only momentarily on the silicon device, then transferred to computer memory for viewing and storage. This frees the camera to capture another picture in a relatively short time (vs. moving the film). The major disadvantages are low detection efficiency and, in the case of the CCD, radiation damage effects.

Scintillator Advantages

In order to increase the detection efficiency of these cameras, scintillators are employed as a transducer to convert x-rays to visible light, which can then be detected more efficiently than x-rays directly. In some cases, the scintillator functions alone as the x-ray converter. In other systems, it works to supplement the film or silicon device. At lower energies, scintillators can be dense enough to stop all of the x-rays, thus protecting a silicon (or other) device from damage. With this increased absorption, a scintillator can reduce the radiation dose required to get a good image.

A good example is medical and dental imaging (30—120 keV). It is common to use a thin phosphor screen in close proximity to film to increase sensitivity to the point where the x-ray dose can be reduced to a safe level. Dental x-rays are commonly made this way. Film/phosphor sheets are made large enough to do chest x-rays and even larger areas. These materials are produced in large quantities by at least two major industrial companies and, due to good availability, can be used economically for other general purpose applications. Computerized Tomography (CT) scanning is made possible by the use of a large array of scintillator crystals mounted on silicon photodetectors. Virtually all medical x-ray imaging uses a scintillator in one form or another.

In NDT applications (120keV—up), scintillating glasses in fiber optic or plate form are used in addition to phosphors and other crystals. In this application, radiation dose is generally not a limitation, and generating the best resolution at a particular energy is a paramount concern. Higher energy systems require thicker scintillators to convert enough photons to receive an image. Sometimes speed can be an issue. Film has been traditionally used, but the NDT industry has been quick to recognize the advantages of digital imaging. A scintillating fiber optic coupled with a CCD imager offers the highest resolution, but covers a limited image area.

PHOSPHOR

There are many types of phosphors available with a wide range of characteristics. "Equivalent" types from different manufacturers vary enough so they may not be interchangeable. Matching a phosphor to an application can be difficult and time consuming, but is generally required only in special circumstances. For x-ray imaging, $Gd_2O_2S{:}Tb$, (green emitter), or $Gd_2O_2S{:}Eu$, (red emitter), are typically considered bright and fast enough for general purpose use. Adding small amounts of Ce, $\leq 1\%$, reduces the decay time of the light pulse, for slightly faster response. Phosphors have good stopping power, but are used in thin coatings. Generally, matching the light output of the phosphor with the spectral response of the detector yields the best results. Phosphors are relatively inexpensive and can be applied to a wide variety of substrates. Phosphors are applied in thin coatings, 30—100 mg/cm^2. Thinner phosphor layers produce better resolution than thicker ones, at the expense of x-ray absorption. Coatings can be applied to large, ≈ 1 m^2, areas. Resolution is typically in the 5—20 line pairs per millimeter range. There are commercially available products where the phosphor is adhered to a plastic backing material so it can be cut to any desired size. Another property of phosphors is that they may "charge up", resulting in longer decay times in higher radiation doses. Notably, there are some very fast phosphors ($CaWO_4$, $1/e < 25$ μs), a blue emitter, and some very dense ones ($GdTaO_4$, 8.75 g/cm^3), which have a very high stopping power for low energy (≈ 10 kV) x-rays.

GLASS SCINTILLATORS

Glass can be host to many scintillators: various oxides of Ce, Tb, Pr, Eu, Gd, & other elements. Ce and Tb doped glasses were tried as imaging detector targets at CERN and Fermilab in the 1980's. Poor transmittance and high cost limited the usefulness and scalability of Ce glass targets. Ce glass with ^6Li works well as a neutron detector. Over years of experimentation, glasses were dropped by particle physicists as a primary focus, but Tb-doped glass emerged as a good performer in x-ray imaging. Tb-doped glasses are somewhat slow but fine for x-ray imaging, with decay times from 0.1—0.5 sec typical. Some Tb-doped glasses are suitable for fiber optic manufacturing, withstanding the repeated heating during fabrication without devitrification and concomitant transmission loss. Glass scintillators are useful over a wide energy range, from 50 keV up to 15 MeV. The primary advantage of the glass over phosphor is that it can be made quite thick, (several cm) for increased output and stopping power, without compromising resolution

Solid Glass Scintillators

Monolithic scintillating glass plates are typically used to inspect large objects at higher energies in the MeV range, with resolution the 2-4 line pairs per mm. Systems generally employ a 45° mirror to reflect the image in the plate to a camera facing perpendicular to the beam located behind suitable shielding. Solid plates are less expensive than fiber optic structures, produce more light than a phosphor screen, and make much better use of a reflector, usually applied to one side.

Fiber Optic Scintillators

Fiber optic scintillators are used in the same manner and offer the advantages of a glass scintillator, but with high resolution, up to 25 line pairs per mm. Fibers have a 10 μm core, 11 μm pitch, with statistical (replacement of fibers 1:9) EMA for increased contrast. Best resolution is at lower energies, with excellent results from 70 keV to 450 keV. Above this range x-ray scattering begins to limit resolution, until at some point a glass plate may be preferred. Fiber optic plates have good stopping power at lower energies, with about 90% of the energy from a 70 kV dental x-ray source stopped in a 2 mm thickness. Fiber optics can be mounted directly on a CCD imager, or can be made quite large— up to 12 inches square. There are two glass types available. Historically, one is based on a patent issued to Corning. The more recent formula (LKH-6) is based on a Lockheed patent now owned by Collimated Holes. LKH-6 is preferred with its higher output at higher energies, higher NA (0.58), and lower afterglow (faster decay time). The old formula is about 20% less expensive and may be slightly brighter at low energies.

HYBRID FIBER/PHOSPHOR SCINTILLATORS

A phosphor screen on a fiber optic scintillator, described in another Lockheed originated patent now owned by CHI, offers the best of both types, with good resolution over a wide energy range. The resolution is typically that of a phosphor alone, 10-15 line pairs/mm. The phosphor extends the usefulness of the fiber optic scintillator down to the 30 keV range. Development of these plates is ongoing, with some prototypes being provided to potential users for testing at the present time.

ACKNOWLEDGEMENTS

The information reflects the practical experience of many years in the x-ray imaging field from the perspective of Collimated Holes, Inc., and includes contributions from Mr. Richard Mead, president. Mr. Harold Berger, of Industrial Quality, Inc. also provided valuable insight from his many years of experience. In addition, thanks to all of Collimated Holes, Inc.'s customers and suppliers who shared their technical expertise and experimental results over the past two decades.

SESSION 2: PROPERTIES

Chair: A. Bross
Scientific Secretary: E. Popkov

B-FIELD EFFECTS ON SCINTILLATOR AND CALORIMETRY

Shuichi Kunori

Department of Physics
University of Maryland
College Park, MD 20742, USA

(CMS Collaboration)

Abstract. Effects of strong magnetic fields on scintillators and sampling calorimeter have been studied. Light yield in scintillators increased with increasing magnetic field until saturation at about 2 tesla. The maximum increase was between 6% and 8%, depending on the plastic composition. The sampling calorimeter saw an additional effect due to changes in shower particle path in the magnetic field. An additional 20% (8%) increase was observed in the energy measurement for electrons (pions) in 3 tesla. This additional effect strongly depends on geometry of the sampling calorimeter and orientation of the field.

INTRODUCTION

The CMS hadron calorimeter for the LHC at CERN is a sampling calorimeter made of copper absorbers and scintillator tiles with optical fiber read out system [1,2]. It will be operated in a 4 tesla solenoidal field.

Prior to CMS, several experiments have measured effects of B fields on scintillators [3] and sampling calorimeters [4]. They reported that the light yield of scintillators continuously increased as the B field increased (scintillator brightening effect). Those measurements were limited to maximum fields of 2 tesla and showed no saturation in the scintillator brightening. In the measurements of the B field effect on the sampling calorimeter, no significant effect was observed over the scintillator brightening effect.

In this paper, we review results on B field effect on the scintillator light yield up to 14 tesla and report test beam results with a CMS hadron calorimeter prototype in 3 tesla field. We also discuss a Monte Carlo simulation study.

CP450, *SciFi97: Workshop on Scintillating Fiber Detectors*
edited by A. D. Bross, R. C. Ruchti, and M. R. Wayne
© 1998 The American Institute of Physics 1-56396-792-8/98/$15.00

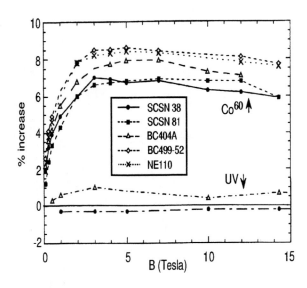

FIGURE 1. Relative light yield increase in scintillators and WLS fibers as a function of magnetic field.

EFFECTS ON SCINTILLATORS

Various plastic scintillators were placed in B fields up to 14 tesla, and dependence of the light yield on the B fields was measured. Details of the experiments have been described elsewhere [5].

Figure 1 shows responses of five scintillators, SCSN38 and SCSN81 by Kuraray, BC404A and BC499-52 by Bicron, and NE110 by Nuclear Enterprises as a function the B field. A wavelength shifting (WLS) fiber was placed in a groove on each scintillator for read out. With a Co^{60} source, all five scintillators showed very similar behavior. The light yields increased with the B field up to ~2 tesla and saturated at 6%~8% above that value.

The light yield change could be due to the primary excitation of the polymer in the scintillators, or excitation of the fluors in the scintillators or the WLS fibers. Short wave length (300nm) UV light was used to excite the fluors and the light yield change is consistent with zero within the measurement error of ≤1%. This result was consistent with result from an independent measurement of light yield changes of the WLS fiber(Y11) in the B field, which also showed no effect. These indicate that the effect was due to the primary excitation of the scintillators.

Several thicknesses of scintillators and different magnetic field orientations were also tested with the Co^{60} source. No dependence on thickness or the field orientation was observed.

EFFECTS ON SAMPLING CALORIMETER

Strong magnetic field changes trajectory of shower particles in sampling calorimeter and may influence its energy measurement in addition to the scintillator brightening effect discussed in the previous section.

The CMS hadron calorimeter prototype was placed in a super conducting magnet in the H2 test beam line at CERN. The maximum B field of the magnet was 3 tesla. The test module consisted of copper absorber plates and scintillator layers (SCSN38) with WLS fiber (Y11) for read out. The transverse size of the module was 64×64 cm^2 and total depth was ~ 11 interaction lengths. Each scintillator layer was read out by a separate photomultiplier through clear fiber connected to the WLS fiber. Gain for each layer was calibrated by muon beam. Movable radio active source Cs^{137} was instrumented on the scintillator layers and used to monitor the scintillator brightening effect in the B field. Measurements were made with two configurations corresponding to early design of the CMS endcap module and barrel module. [1]

Endcap configuration- $\vec{B} \perp SamplingLayers$

The endcap configuration consisted of a compartment (HAC1) with 9 layers of 5 cm sampling followed by a second compartment (HAC2) with 10 cm sampling. The B field was perpendicular to the sampling layers.

Figure 2 shows the relative responses to electrons and pions and the Cs^{137} source. All responses were consistent with the brightening effect discussed earlier, and no additional effect was seen.

Barrel configuration- $\vec{B} \parallel SamplingLayers$

The barrel configuration consisted of HAC1 with one 2 cm sampling plus 6 layers of 3 cm sampling and HAC2 with 6 cm sampling. The B field was parallel to the sampling layers.

Figure 3 shows relative response to electrons and hadrons as a function of the B field. The data points were normalized by muon signal so that the brightening effect had been removed from those responses. The responses increased with the B field and depend on the type of incident particles. At 3 tesla, the increases were 20% for electrons and 8% for pions. As shown in the figure, a GEANT simulation reproduced the effect very well.

Discussion

In order to understand the source of the additional B field effect in the sampling calorimeter, we simulated the barrel configuration and varied air space in front of

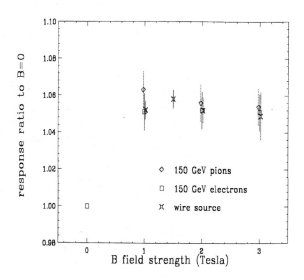

FIGURE 2. Relative response of CMS prototype calorimeter as a function of B field perpendicular to scintillator layers

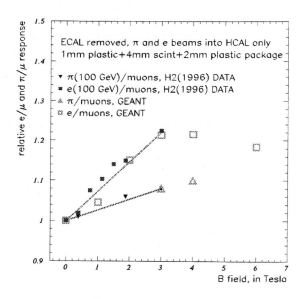

FIGURE 3. Relative response of CMS prototype hadron calorimeter as a function of B field parallel to sampling layers. All responses for data and GEANT simulation are normalized by muon response to remove the scintillator brightening effect.

the scintillator packages, which was made of 4 mm scintillator with plastic cover plates (1 mm in front 2 mm on back), while keeping air space behind the package constant.

Figure 4 shows the calorimeter response to 10 GeV electrons as a function of the air space with and without the B field. In the 3 tesla field, the response decreased as the gap increased in width because low momentum (a few MeV/c) shower particles coming out of the upstream absorbers curled in the air gap and did not reach the scintillator. Note that the radius of 1MeV/c particle is ~1 mm in 3 tesla. This also worsened the energy resolution because of loosing statistics in sampling. There was no such effect in cases of no field or the field perpendicular to the sampling layer.

For pion beam, mainly electromagnetic component in the hadronic shower (i.e. electrons and positron) see the effect, resulting in smaller net effect comparing to the electron beam. Since the effect is produced by mainly low energy electrons and positrons around sampling layers, the effect is strongly dependent on the geometry and the materials around the layer and on the field orientation. More simulation results can be found in a paper. [6]

This additional effect introduces a requirement for placement of scintillator packages in gap between absorbers in the CMS barrel hadron calorimeter(HB). Since calibration data will be taken in the calibration beam line without magnetic field, it will be very desirable to have minimal extrapolation from the calibration beam data (in 0 tesla) to the CMS data (in 4 tesla). In addition, gravity may push down the scintillator packages toward the front in absorber gaps at the top of the barrel calorimeter and increase the HB response, while at the bottom of HB, it would be push them toward the back and thus decrease the response. First we choose the scintillator package orientation with the thicker plastic cover plate (2 mm thick) in front giving larger distance between scintillator and front absorber. Then by forcing the package toward the back, we can limit the effect in HB to less than 2%. [2]

CONCLUSION

Two magnetic field effects were measured for scintillator sampling calorimeter. The light yield of the plastic scintillators increased with the B field and reached a plateau between 6% and 8% above ~2 tesla. In addition to this effect, the CMS calorimeter prototype data showed ~20% and ~8% increase in the responses to electrons and pions, respectively. The additional effect was due to change in the trajectory of low momentum (a few MeV/c) particles coming out of absorbers and into the scintillators.

REFERENCES

1. Technical Proposal, CMS Collaboration, CERN/LHCC/94-38,LHCC/P1,December 1994.

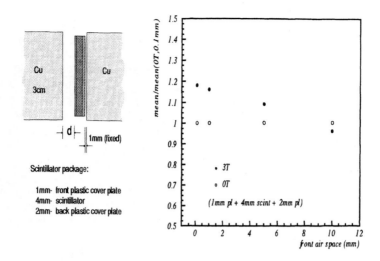

FIGURE 4. GEANT simulation of calorimeter response to 10GeV electrons as a function of air gap (**d**) between scintillator package and front absorber.

2. The CMS Hadron Calorimeter Technical Design Report, CERN/LHCC/97-31, June,1997.
3. S.Bertoucci et al. *Nucl. Instrum. Methods* A **254**, 561 (1987), J.P.Cumalat et al., *Nucl. Instrum. Methods* A **293**, 606 (1990), D.Blomker et al., *Nucl. Instrum. Methods* A **311**, 505 (1992).
4. J. Mainusch et al., *Nucl. Instrum. Methods* A **312**, 451 (1992). P. Aspell et al. CERN-PPE/95-152(1995)
5. D.Green et al., Fermilab-TM-1937, June 1995.
6. V. Abramov, *Nucl. Instrum. Methods* A **374**, 34 (1996).

Radiation Damage of Fibers

Vasken Hagopian and Ian Daly

Florida State University
Tallahassee, Florida, 32306-4350 USA

Abstract. Optical fibers are used extensively in high energy physics, as well as in communication. Fibers in medium and high radiation environments show considerable damage. This damage can be quantified as a reduction in the transmission length, which is a function of the transmitted wavelength. The primary observed damage in a fiber is due to the cladding material. Damage is observed even at low levels of irradiation. Quartz fibers damage less than plastic fibers.

Introduction

Three types of optical fibers are used in high energy physics experiments[1]: scintillating fibers for tracking, wavelength shifting (WLS) fibers which are embedded inside scintillating tiles and transfer light signals from the tiles to the outside of the enclosure, and a third type made of either clear plastic or quartz. The clear plastic fibers have longer transmission lengths and are used to carry out optical signals. The quartz fibers are used in high radiation areas where plastic fibers damage by turning brown. Another use of quartz fibers in high energy physics, is particle detection, where Cerenkov radiation in the quartz fiber is used to measure the passage of charged particles. Quartz fibers come in two flavors: those with a less expensive plastic cladding, and those with a quartz cladding, which is more radiation hard but costs five to ten times more. Quartz fibers with quartz cladding are normally used in very high radiation areas. In general, the longer the wavelength of transmitted light, the longer its attenuation length and the less the radiation damage affects its transmission.

To study the damage due to radiation, we irradiate the fibers and scintillators at our laboratory with a 3MeV electron accelerator. The fibers are attached on a 50cm radius rotating wheel. The first 20cm and the middle 10cm are shielded from radiation. The first 20cm are then used for normalization purposes, while the protected middle 10cm remains unirradiated and is used to measure the light yield loss due to the damage of the fluor. The scintillating tiles are similarly irradiated on a rotating wheel, where the electron beam is swept in the horizontal direction as the wheel moves the tiles in the vertical direction. In the CMS collider detector, which is now being designed for the LHC accelerator at CERN, the expected radiation levels are 100krad or less in the hadron calorimeter. This is where the scintillating tiles and fibers will be deployed. In

CP450, *SciFi97: Workshop on Scintillating Fiber Detectors*
edited by A. D. Bross, R. C. Ruchti, and M. R. Wayne
© 1998 The American Institute of Physics 1-56396-792-8/98/$15.00

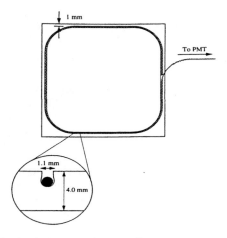

FIGURE 1. A square tile with a keyhole groove for 1mm fiber.

25cm x 25cm Tile - Fiber System

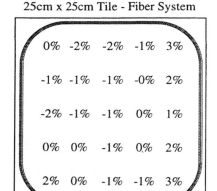

FIGURE 2. Light yield deviation from the average of a tile-fiber system.

the forward calorimeter, made of quartz fibers embedded in a metal, the radiation levels are about 500Mrad over a 10 year period of operation.

Scintillating Fibers

Scintillating fibers usually emit light in the blue part of the visible spectrum. The signal is relatively fast, typically a few nanoseconds. Tracking hodoscopes use scintillating fibers for position determination of charged particles. Green scintillating fibers are also used in tracking of charged particles. These green fibers are typically doped with the 3HF fluor, have longer transmission lengths and show less damage to irradiation. On the negative side, these green fibers produce fewer photons than the blue fibers. The green fibers are better suited to the inner tracking chambers of major collider detectors, where radiation levels are higher.

Wavelength Shifting (WLS) Fibers

WLS fibers are typically used with blue scintillators in high energy physics experiments. A common application is calorimetry in collider detectors, where the individual tiles are small and hermiticity does not allow for ordinary light guides. Typically, a rectangular tile is grooved with a keyhole shape near the edge of the tile, and the WLS fiber is inserted into the groove. The WLS fiber is fused to a clear fiber at one end and mirrored at the other end by aluminum sputtering. The advantage is that only a single 1mm fiber needs to come out, which allows a hermetic design. Some designs call for clear fibers as much as 10 meters long. When this R&D effort was started about nine years ago, the number of photons obtained from a 10cm x 10cm x 0.4cm tile fiber system was only one photoelectron for a minimum ionizing particle. The various improvements have increased the photoelectron number by a factor of five to ten.

Substantial improvement has been made in WLS fibers over the past decade or so. Double cladding is now standard and attenuation lengths have been improved to over 250cm when used with green enhanced bi-alkali phototubes, such as the Hamamatsu R580-17. Fibers are now more flexible, and can accommodate a smaller radius of curvature around the corners of tiles. Even the fluors have improved substantially. The newer WLS fiber BCF92, for example, has a shorter decay time of 4nsec as compared to other WLS fibers, such as BCF91A or Y11, which use the K27 fluor with a decay time of the order of 10nsec.

Figure 1 shows a typical scintillating tile with a single groove cut in the tile, near the edge. The light yield from such a tile fiber combination is very uniform. Figure 2 shows the uniformity of light yield, over the whole surface area for a 25cm x 25cm tile-fiber system.

In a radiation environment, both the tile and fiber suffer damage. In both fibers and scintillating tiles alike, the damage primarily affects the transmission of the light through the plastic. Under severe radiation, the fluors themselves also damage. Since the damage affects more strongly the transmission of shorter wavelengths, the effects

TABLE 1. Radiation Damage Constant D_0 of Various Scintillator tiles.

Type	BC412	BC448	BC408	BC404	SCSN38	SCSN81
D_0(Mrad)	4.9	6.2	4.6	5.1	4.6	6.6

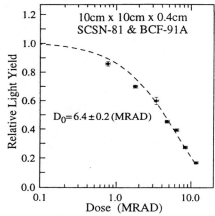

FIGURE 3. Relative light yield of a tile-fiber system as a function of radiation dose.

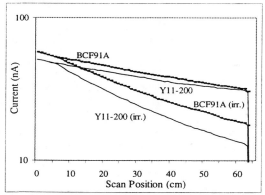

FIGURE 4. Light yield of two WLS fibers, before and after 8Mrad radiation.

are more severe for the blue light than for green light. Figure 3 shows the damage to a tile fiber system as a function of radiation dose. The loss of light yield follows an exponential form exp(-Dose/D_0), where D_0 is a measure of the radiation hardness of the system. The higher the D_0, the more radiation hard the system. Table 1 lists various scintillator tiles made either by Bicron (prefix BC)[2] or Kuraray (prefix SCSN)[3].

56

Orange and red WLS fibers can also be used with green scintillating tiles. These types of systems are more radiation hard, but they yield fewer photons and have longer decay times. Most scintillating, WLS, and clear plastic fibers have a polystyrene core with acrylic cladding. Newer fibers have either a double clad or the outer part of the clad is treated to lower the index of refraction and increase the solid angle of captured photons. Double clad fibers typically capture 40% more photons. Figure 4 shows two fibers with the same K27 green fluor, made by two manufacturers. Both have polystyrene cores, but use very different cladding materials. The fiber made with a single, multi-index clad is more radiation hard. This difference in behavior is also observed in low doses of irradiation (about 100 krad). Table 2 shows the damage as a loss in light yield over a 1 meter fiber.

The most common fluor for the green WLS fibers is the K27; it is used in the Kuraray Y11 and the BICRON BCF91A fibers. Other fibers manufactured by BICRON use proprietary fluors, with the BCF92 now available commercially. One disadvantage with Y11 or BCF91A green fluors is that their decay time is a rather lengthy 10 nsec or so. Table 3 lists the decay times, as well as other characteristics of WLS fibers when used inside of a 10cm x 10cm SCSN38 or BC448 scintillator. Each fiber is 50cm long, with 35cm inside the tile.

The new proposed LHC accelerator will have a beam bunch spacing time of 25 nsecs and the long decay time of the K27 fluor presents a serious shortcoming. Unfortunately, the faster WLS fibers are too radiation soft for use with LHC detectors.

TABLE 2. Light yield loss in 1m long WLS fiber after 100 krad irradiation.

Y11 (0.83mm diameter) (Kuraray)	13%
BCF91A-MC (1mm diam) (BICRON)	2%

TABLE 3. Light yield, decay time, and radiation damage of various WLS fibers measured inside tiles.

Fiber	Light yield Relative	Decay time nsec with SCSN38	BC448	Attn length cm	D_0 radiation Mrad
BCF91A-MC	100%	10.1	8.5	160	33
Y11-250	92%	10.1	8.5	186	19
BCF99-75 SC	16%		7.4	141	8
BCF99-75 MC	20%		7.7	196	9
BCF92	49%	5.4	4.1	138	3
BCF99-29A	61%		4.0	154	2
BCF92A	21%		4.2	163	7
BCF99-29AA	64%		4.1	118	2
BCF93	60%		4.2	76	2

TABLE 4. Change in attenuation length in clear plastic fibers due to 100 krad irradiation

Type	Pre-radiation	100 krad assumed
Kuraray (0.83mm)	1,000cm	810cm
BICRON (1.0mm)	1,000cm	925cm

Clear Plastic Fibers

This type of fibers is used to carry signals to distances of up to 10 meters. A green WLS fiber with K27 fluor, for example, has a typical attenuation length of 2.5m, while the clear fiber attenuation length is usually about 10m. The measured attenuation length is a function of the quantum efficiency of the photodetector, but the ratio to first order is independent of the photodetector type. Table 4 shows the decrease of the attenuation length due to low level irradiation of 100 krad. This is a typical maximum irradiation of high energy collider detectors, in the so-called barrel portion of the calorimeters.

Quartz Fibers

In very high radiation areas, ordinary scintillators and fibers will not survive. Quartz fibers are an alternative, where the detection method is the Cerenkov light produced by charged particles. The signal is very small and in quartz fiber calorimeters, the energy resolution is substantially worse than that of the tile-fiber calorimeters. The light yield of quartz fibers has been measured as a function of length, before and after irradiation using the Cerenkov light induced by electrons from a Strontium90 source. Quartz fibers are planned to be used in very hostile environments where the radiation doses can be as high as 100 Mrad per year. Pure quartz fibers are very radiation hard, but the acrylic cladding is damaged by irradiation. An alternative method for lowering the index of refraction of the outer portion of the fiber is to treat the outside surface of the quartz core during manufacture before pulling the fiber, which allows the clad to be made of quartz as well. This type of fiber is much more radiation hard and is used in very hostile environments. Quartz fibers are very fast and are insensitive to the latent radioactivity of the surrounding detector. Typically a quartz fiber has a clad and a protective cover, called the buffer. Figure 5 shows two Polymicro fibers irradiated to 3Mrad. One fiber shows a shortened attenuation length, while the other shows no such damage. The intrinsic difference between these two fibers is in the material used for the buffer which protects, to varying degrees, the clad from radiation damage. Figure 6 shows a quartz fiber with quartz cladding irradiated to 7 Mrad with no damage. Still no damage was observed at 60Mrad (not shown). Table 5 shows 3 and 12Mrad irradiations to various Polymicro quartz fibers. The amount of damage seems to depend only on the buffer.

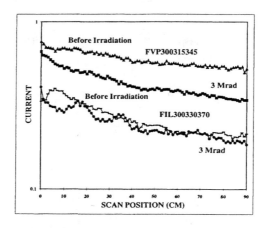

FIGURE 5. Cerenkov light yield of two quartz fibers, before and after 3Mrad radiation.

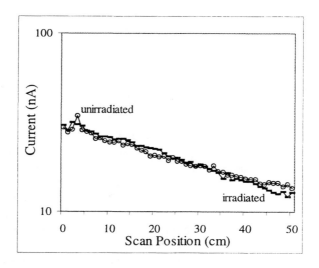

FIGURE 6. Light yield of a quartz fiber with quartz clad, before and after 7Mrad.

TABLE 5. Radiation damage of quartz fibers made by Polymicro. The damage is measured as light reduction at the center of 80cm long fiber. Excitation is by the electrons from a Strontium 90 source. The Photomultiplier used is a 12-stage bi-alkali Hamamatsu R329. Measurement errors are approximately 5%.

Fiber	Core outer Diameter Micron		Buffer	Damage 3 Mrad	Damage 12 Mrad
FVL400440520	390	520	Aluminum	0%	0%
FIA200240500	202	500	Clear	0%	10%
FSHA3003203435	297	343	Clear	20%	50%
FIP200220240	200	239	Dark	20%	40%
FIL300330370	300	428	Aluminum	0%	0%
FVP300315345	301	348	Dark	35%?	40%
FVP300330370	300	370	Dark	30%	50%

Conclusions

Radiation damage has been measured in a variety of optical fibers, with and without fluors. Some of the fibers have polystyrene cores while others have quartz cores. The radiation damage at relatively low doses of l00krad to megarad irradiation is observed as a reduction of attenuation length. For higher doses of irradiation, the fluors used in the fibers can also be damaged. As a general rule, for the same fiber and the same irradiation dose, the longer the wavelength of the transmitted light, the less is the reduction of the transmission length. In all of the fibers that were studies, the part that damages the most is the fiber clad material. In quartz fibers a buffer material is added outside of the clad. The buffer plays a role in reducing radiation damage. Plastic fibers with acrylic clad show damage with as low as l00krad of irradiation, although the magnitude of the damage is different between various manufactures of fibers.

Blue or green scintillating fibers can be used in environments up to 1 Mrad, with the green fiber showing less damage. WLS fibers can be used in radiation areas of up to several Mrads. The fibers with a single, multi index of refraction clad, show less damage than the double clad fibers. Quartz fibers also show damage down to several Mrads. This damage is due to the clad material. Quartz fibers with quartz clad have not shown any damage up to 60 Mrad. We challenge the fiber manufacturers to develop more radiation hard fibers, especially with fluors with shorter decay times.

Acknowledgments

I would like to thank my colleagues M. Bertoldi and J. Thomaston for their help in the measurements. My thanks to Prof. R. Ruchti for arranging the SCIFI97 conference and inviting me to be one of the speakers.

References

1. V. Hagopian, *Nuclear Physics B*, **61B**, 355-359 (1998).
2. BICRON Corp, Inc. 12345 Kinsman Rd. Newbury, Oh. 44065
3. Kuraray Co. Ltd. Shuwa Higashi Yaesu Bldg. 10 Fl. 9-1, 2-Chomo, Chuo-Ku, Tokyo, 104, Japan.
4. Polymicro, 18019 N. 25th Ave. Phoenix, AZ. 85023

Radiation Damage in WLS Fibres

M. David, A. Gomes, A. Maio, J. Santos, M. Varanda

LIP Av. Elias Garcia 14, 1º, 1000 Lisbon, Portugal
Univ. of Lisbon, C1 Campo Grande, 1000 Lisbon, Portugal

Abstract.   Several types of WLS fibres, candidates to be used in the TILECAL/ATLAS detector, were irradiated in a ^{60}Co γ source. Bicron, Kuraray and Pol.Hi.Tech fibres were exposed to a total dose of ~150 Krad. The degradation of light output was measured just after irradiation and followed during several days. The results are presented.

INTRODUCTION

The TILECAL hadronic barrel calorimeter of the ATLAS detector, uses scintillating tiles readout by 1 mm diameter WaveLength-Shifter (WLS) fibres, which transmit the signal to photomultipliers tubes (PMT). Due to the high radiation field predicted for the LHC detectors, degradation of the optical components is expected, so radiation damage experiments are performed in order to test the tolerance to ionising radiation. For the TILECAL at $\eta = 1.2$, the maximum expected dose due to charged particles is about 3 krad/year, and 1 krad/year due to the neutron flux $(2.0 \times 10^{12} cm^{-2} yr^{-1})$, i.e., about 40 krad for 10 years of LHC operation.

During the last 2 years, several types of new fibres were produced by Bicron, Kuraray and Pol.Hi.Tech. After Kuraray, Bicron and Pol.Hi.Tech, also developed multi-cladding fibres. Very recently Pol.Hi.Tech started the production of WLS fibres using K27 dopant, already used by Bicron and Kuraray in the BCF91A and Y11 fibres, respectively. Some of these fibres were characterized in what concerns light output and attenuation length [1,2].

The present work reports the results of radiation damage on some of these new types of fibre.

IRRADIATION AND EXPERIMENTAL SETUP

A ^{60}Co source from a radiosterilization facility, *"Unidade de Tecnologias de Radiação"* in *"Instituto Tecnológico e Nuclear"* at Lisbon, was used. At 10 cm from the ^{60}Co source, the intensity is >20 krad/h, so, in order to have a lower and

CP450, SciFi97: Workshop on Scintillating Fiber Detectors
edited by A. D. Bross, R. C. Ruchti, and M. R. Wayne
© 1998 The American Institute of Physics 1-56396-792-8/98/$15.00

rather uniform dose rate inside an area of 20x150 cm^2, several shieldings of lead and cooper were used as it is shown in fig. 1.

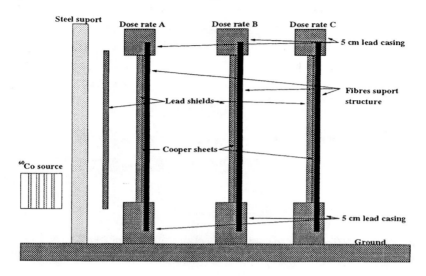

FIGURE 1. Irradiation setup.

The test bench is schematically shown in fig. 2. It consists of a X–Y MicroControl optical table, provided with two independent movements driven by two stepping motors along orthogonal directions (**X** and **Y**). It is complemented by a light read-out system. A 3 meters long table is moved in the X direction. This table supports a removable holder that contains the fibers to be tested. Each fiber is positioned in front of the light readout, and the Y motor moves the radiation source along the fiber length. At each position the light output from that fiber is recorded. The test bench is automatically controlled by a MacIntosh Personal Computer through GPIB interfaces. A dedicated software was developed for this purpose using the LabView[1] package.

A more detailed description of the irradiation facility and experimental setup can be found in [2,3].

EXPERIMENTAL CONDITIONS

In the first test, the fibres were irradiated in the central position of the irradiation setup (position B - dose rate $\sim$0.64 krad/h), shown in fig. 1:

Bicron BCF91A DC from 1995, BCF99-28 DC and BCF99-28, both from 1996; Kuraray Y11(200)MS with 600 ppm of UVA from 1997 production for the Extended

---

[1] ©National Instruments

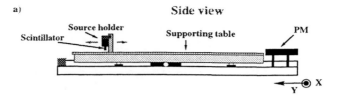

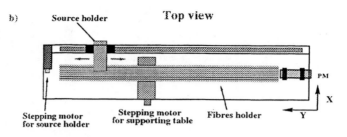

FIGURE 2. The test bench: a) side view; b) top view.

Barrel Module 0; Pol.Hi.Tech S248-100 (double cladding fibres) and S048-100 (single cladding fibres), both from the beginning of 1997. All fibres were polished one by one in a cutting machine, and the side opposite to the readout was painted black. Five fibres of each type were tested.

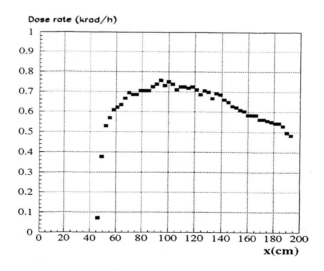

FIGURE 3. Dose profile used to irradiate the fibres.

The dose rate for this position is shown in fig. 3. The average in the plateau is about 0.64 Krad/h. The fibres were exposed to γ radiation during a total of 225 hours. It should be noted that the irradiation is not continuous (about 10 to 12 hours per day, and stopping during the weekends). The total dose delivered to the fibres was $\sim$144 Krad.

A second test was performed with new Pol.Hi.Tech double cladding fibres with K27 dopant, produced very recently (September 1997). One set of 3 S250 fibres with 200 ppm of UV Absorber (UVA), and a second set of 3 S250 fibres whithout UVA. The K27 concentration of the second set of fibres is 2 times the concentration of the first set. This (fast) irradiation was performed in position A (see fig. 1), with a dose rate about 4 times higher than position B, but the total dose was also about 150 Krad.

RESULTS FOR THE FIBRES BEFORE IRRADIATION

In order to compare the optical properties of fibres from different productions, and control the reproducibility along time, "new" fibres and "old" fibres were tested at the same time. This section summarizes the results obtained for each type of fibre before the irradiation. Table 1 shows the average for 5 fibres of the attenuation length (L_{att}) taken from one exponential fit between 70 and 170 cm, the light output at 30 cm ($I(30)$) and at 170 cm ($I(170)$). Fig. 4 shows the light output as a function of the distance to the PMT, for one typical fibre of each type.

The attenuation length of BCF99-28 fibres has lower values ($\sim$220 cm), than all the other types of fibres ($>$280 cm). Comparing the results for double cladding BCF99-28 and BCF91A fibres, the difference between these two types is in the dopant concentration. The BCF91A has half of the dopant concentration of the BCF99-28 fibres, and this is a factor that can explain the decrease in the attenuation length [4]. It can be seen that the light output at 30 cm (rather close to the readout PMT) is the same for both types of fibres. The light output at 170 cm, for the double cladding BCF99-28 DC fibres is 32% higher than for the single cladding BCF99-28 fibres.

The double cladding S248 fibres have an attenuation length of 344 cm compared with 307 cm for the single cladding S048 fibres. The light output of the double cladding S248 fibres is 16% higher than for the single cladding S048 fibres.

The Y11(200)MS fibres from the production for the Extended Barrel Module 0's (1997 production), have both attenuation length and light output similar to the BCF91A DC fibres. On the other hand, comparing the Y11(200)MS fibres from an old production (1995), with the fibres for the last production (1997), it can be seen that the Y11 fibres produced in 1997 have both light output and attenuation length lower than the ones from the 1995 production.

Both the BCF99-28 DC and S248 fibres have similar light output at 170 cm, and $\sim$11% lower than the BCF91A DC and Y11(200)MS (97) fibres.

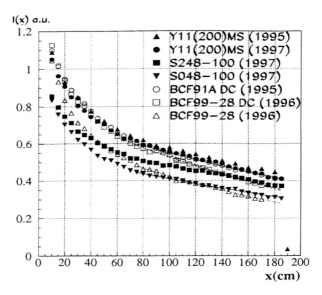

FIGURE 4. Light output of one typical fibre of each type, as a function of distance (x) to the PMT.

Fiber type	Fibre length (cm)	L_{att}(cm)	$I(30)$ a.u.	$I(170)$ a.u.
BCF99-28 DC (96)	175	223	0.850	0.381
BCF99-28 (96)	175	211	0.703	0.288
BCF91-A DC (95)	190	282	0.847	0.424
Y11(200)MS (97)	190	283	0.805	0.428
Y11(200)MS (95)	190	306	0.851	0.465
S248-100 (97)	190	344	0.669	0.385
S048-100 (97)	190	307	0.630	0.331

TABLE 1. The attenuation length taken from one exponential fit between 70 and 170 cm, the light output at 30 cm and at 170 cm are presented, as the average of 5 fibres of each type. The RMS is about 6% or less for the attenuation length and 3% or less for the light output.

RESULTS FOR IRRADIATED FIBRES

The light output and attenuation length were measured: 3 hours, 1 day and 6 days, after the of the end of the irradiation. The S250 fibres were also measured 2 weeks after irradiation. In order to quantify the degradation in the light output the following ratio was used:

$$\frac{R(x)}{R(30)} = \frac{I_{Irr}(x)}{I_{NIrr}(x)} \Big/ \frac{I_{Irr}(30)}{I_{NIrr}(30)} \tag{1}$$

$I_{Irr}(x)$ is the light output of the fibres after irradiation, $I_{NIrr}(x)$ is the light output of the fibres before irradiation and x is the distance of excitation point to the readout PMT. In the region $x = 30cm$ the dose is negligible compared with the dose in the plateau (fig 3), so this point is used to normalize the ratio.

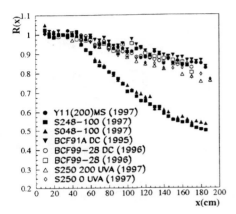 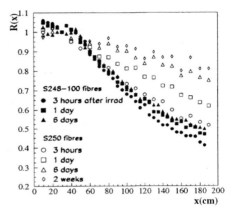

FIGURE 5. Left - Ratio $R(x)$, as a function of distance for one typical fibre of each type, for 6 days of recovery. Dose of ~150 krad and dose rate of 0.64 krad/h except the S250 fibres which were irradiated at 2.52 krad/h. Right - Ratio $R(x)$ for one S248-100 and one S250 fibre. The recovery is shown, 3 hour, 1 day, 6 days and 2 weeks after the end of irradiation.

Fig. 5-left shows $R(x)$ as a function of distance for one typical fibre of each type, with 6 days of recovery. Fig. 5-right shows the recovery between 3 hours and 2 weeks after irradiation, for the S248-100 and S250 fibres. The average of $R(170)$ is shown in table 2, for the three (or four) measurements done after the end of the irradiation. The distance $x = 170\ cm$ is a typical point of the TILECAL calorimeter for radiation damage tests.

All the fibres from Bicron, do not present any observable recovery between 3 hours and 6 days, and the degradation in light output is similar and almost independent of the fibre type. The value of $I(170)$ is about 85% of the value before irradiation.

The Y11(200)MS fibres present a slight recovery between 3 hours and 1 day, from 81% to 86% of the initial light output, and no further recovery thereafter.

The results presented in this paper are quite similar to those obtained in previous irradiations [3]. There is almost no recovery of the Bicron and Kuraray fibres, which can be explained by the irradiation being performed with a rather low dose rate (0.64 krad/h), and allowing the fibres to recover in part during the test. The RMS

67

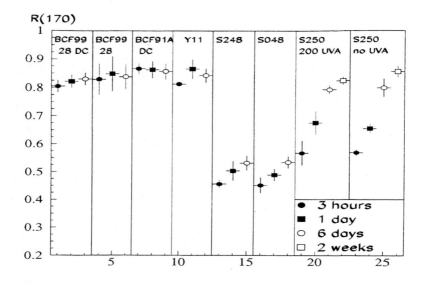

FIGURE 6. Ratio $R(170)$, corresponding to table 2.

of $R(170)$ due to fibre to fibre fluctuations and instrumental errors is less than 3.5%

The situation of Pol.Hi.Tech fibres is different. The S048 and S248 fibres show a slight recovery of 8% between 3 hours and 6 days, from 45% to 53%. Both S250 fibres present a rather strong recovery from 56% to 80% of the initial value in 6 days, and up to 82% - 85% two weeks after the end of irradiation. It should be

Fiber type	3 hours	1 day	6 days	2 weeks
BCF99-28 DC	0.805	0.822	0.830	–
BCF99-28	0.829	0.848	0.837	–
BCF91-A DC	0.866	0.862	0.856	–
Y11(200)MS	0.812	0.865	0.842	–
S248-100	0.455	0.502	0.530	–
S048-100	0.450	0.487	0.533	–
S250 200 UVA	0.566	0.674	0.792	0.823
S250 no UVA	0.568	0.654	0.798	0.855

TABLE 2. Average of $R(170)$ for 5 fibres of each type. RMS is < 3.5% for each set of fibres. Dose of ~150 krad and dose rate of 0.64 krad/h except the S250 fibres which were irradiated at 2.52 krad/h.

stressed that the dose rate during this second test is four times higher (2.52 krad/h) than in the first test.

CONCLUSIONS

Comparing the 1995 and the 1997 productions of the Y11(200)MS Kuraray fibres, it is seen that in 97, the fibres have a light output at $x = 170$ cm about 10% lower than the fibres from the older production. The BCF91A DC Bicron fibres from 95 and the Y11 Kuraray fibres from 97, present a very similar value of the light output. The DC BCF99-28 and S248 fibres, have a similar value of the light output and about 10% lower than that of the BCF91A and Y11 from 97 fibres. The DC BCF99-28 and S248 fibres have 32% and 16% higher light output than the corresponding SC BCF99-28 and S048 fibres.

For the irradiation at a dose rate 0.64 krad/h almost no recover is observed for all types of fibres, this can be due to the rather low dose rate, which can allow the fibres to recover in part during the test. The light output after irradiation is about 85% of the initial value for all types of Bicron and Kuraray fibres,and of the order of 50% for the S048 and S248 Pol.Hi.Tech. The S250 fibres, irradiated at a dose rate of 2.52 krad/h, present a fast and strong recovery due to the much higher dose rate. The light output is ~85% of the initial one, after 2 weeks of recovery.

The Bicron and Kuraray fibres tested have a good resistance to ionising radiation, similar to older productions, and in good agreement with previous results [2,3]. The Pol.Hi.Tech S250 fibres are more tolerant to ionising radiation than the S048 and S248 fibres.

ACKNOWLEDGEMENTS

The authors would like to gratefully acknowledge M. Mishina, B. Miller and J. Huston for helping us with the fiber cutting machine. To Dr. Eduarda Andrade and to Mr. Norberto Coelho for their help. This work was supported in part by Program PRAXIS XXI (Portugal).

REFERENCES

1. ATLAS Tile Calorimeter TDR, CERN/LHCC/96-42 (1996)
2. M. David Master Thesis, Fac. Cien. Univ. Lisbon 1996
3. A. Maio et al., "DOSE RATE EFFECTS IN WLS FIBERS", Sixth Topical Seminar on Experimental Apparatus for Particle Physics and Astrophysics, San Miniato, Tuscany, 21-24 May 1996
   M. David et al., proceedings of the Sixth International Conference on Calorimetry in HEP, ed. A. Antonelli, S. Bianco, A. Calcaterra, F. L. Fabbri, Frascati Physics Series vol. VI, pp 639
4. M. David et al., "Comparative Measurements of WLS Fibres", ATLAS Internal Note, TILECAL-NO-034, 28 November 1994

Scintillating Fiber Performance in High Pressure Noble Gases

S. Carabello and D. Koltick

Physics Deparment
Purdue University
West Lafayette, Indiana 47904

A. Bolozdynya and Y. Chang

Department of Medical Physics
Rush University
Rush-Presbyterian-St. Luke's Medical Center
Chicago, Illinois 60612

Abstract. The use of scintillating fibers to collect the scintillation light produced by electrons drifting in a noble gas is discussed. The imaging of photons in the 10 to 150 keV range is discussed and light yield produced by photons in this range is estimated. Results are presented from an experimental setup to measure the change in light yield of scintillating fibers as a function of time while in a 20 atmosphere Noble gas environment. An upper limit of less than $5 \cdot 10^{-5}$ per day was obtained on the change in fiber brightness as a function of time.

DETECTION OF 10-150 KeV PHOTONS

The detection of photons with energies between 10 and 150 keV is useful for many applications. Photons in this energy range are used in the medical field, for Single Photon Emission Computed Tomography (SPECT)[1], in x-ray astronomy for the detection of x-ray bursters[2], in non-destructive testing, where there is the desire to move away from film but present costs of digital system are prohibitive[3], and in particle physics for the detection of WIMPs by recoil techniques[4].

High pressure xenon electroluminescence detectors (HPXED) provide a highly efficient position-sensitive registration for gamma-radiation with energy resolution close to that of semiconductor detectors operated at room temperature. HPXED is capable of sub-millimeter position resolution with fields of view measured in square meters, yet require 100 times fewer readout channels than semiconductor detector arrays of comparable size.

The high performance of HPXED results from the effective intrinsic light-generation (electroluminescence, or secondary or proportional scintillation) which develops during the drift of primary ionization electrons through the pressurized xenon in the presence of a high electric field. In effect, this technology is a light amplifier instead of the older gas technology that relies on charge amplification. Achievable light-output is 1000 times greater than the light-output of the best standard scintillators, such as sodium iodide[5]. The Noble gases used as the working media for HPXED are inexpensive in comparison with semiconductors, and are extremely radiation hard. HPXED readout with multimode light-fiber can operate in extremely hostile environments.

Twenty years of research has been focused on this technology yet only now are real practical applications being developed. This delay was due to the necessary recent progress in gas purification methods, in methods of UV light readout with advanced photodetectors and with multimode fiber optics.

CP450, *SciFi97: Workshop on Scintillating Fiber Detectors*
edited by A. D. Bross, R. C. Ruchti, and M. R. Wayne
© 1998 The American Institute of Physics 1-56396-792-8/98/$15.00

SINGLE X-RAY AND γ-RAY IMAGING

The scale of single photon interactions with matter is set by the mass of the electron. For photons with energies greater than the electron mass, the scattered electron travels in the direction of the incident photon. The path of the electron can then be used to point to the source of the photon and an image can be formed. For photons with energies less than the mass of the electron, the scattering is more complex and to first order there is no interaction. If an interaction occurs at all, it is the total absorption of the photon. One is left only with a clump of deposited energy. Such a clump is effectively only one point on a line, so does not point to a source. The historic approach has been either to make a parallel source of x-rays in order to form a projected image onto a film or to use a pin hole camera by placing a lead collimator in front of an imaging plane.

Projection imaging, similar to the way a dental x-ray is made, has the problem that the image is not three dimensional and gives no depth of field information.

The other classical technique is to restrict the path of the photons by using a small pin hole aperture. The aperture acts as the second point needed to draw a straight line to the source. This is not a major problem when dealing with visible light because a lens can be placed in the aperture allowing for the collection of large amounts of light. However, because x-ray wavelengths are smaller than the spacing between atoms in a material, lenses are not possible. In this case the hole highly restricts the amount of light that can be collected. Typically the image contains less than 10^{-3} to 10^{-4} of the photons emitted by the source. The image resolution is inversely proportional to the size of the hole. A smaller hole means a better image but takes longer to produce and a further loss of photons.

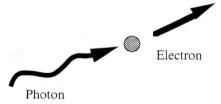

Electron

Photon

FIGURE 1. At high energies the electron moves in the direction of the photon and the source direction can be found.

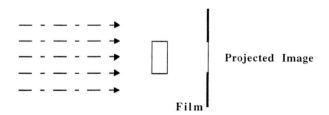

Projected Image

Film

FIGURE 2. Projection imaging. Parallel rays of x-rays give a density profile of the object of interest. The method has the problem of not being 3-dimensional nor digital.

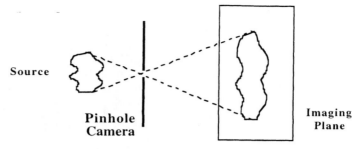

FIGURE 3. Pinhole camera imaging. The hole provides the second point allowing the image to be formed. This technique has poor photon acceptance, between 10^{-3} to 10^{-4}.

For almost 40 years the Anger camera using NaI(Tl) scintillator with a lead collimator or derivatives of this instrument, has been the major transducer used in γ-imaging[5,6]. The Anger camera has inherent limitations which restrain its performance. First, its sensitivity is greatly reduced by an extremely low collimator efficiency, typically only ~0.03% of the γ-flux is used for imaging. In medical applications count rate or imaging speed can not be increased by using larger radioactive sources because of safety limitations. Second, the system performance depends on a mechanical system that is required to precisely move a camera weighing about one ton around the source in the case of tomographic imaging.

Because of these limitations, single photon imaging has not achieved the high image quality it could. Another limitation is the available statistics (~10^6 measured γ-rays), which can not compare in quality to radiological images based on 10^8-10^{10} measured x-rays. Further desirable developments such as dynamic single photon imaging can not be considered without the development of a new highly sensitive and fast γ-imaging technology.

The idea of imaging single photon gamma-rays without the need of a collimator was first proposed by Todd et. al.,[7]. Everett et. al.,[8] presented the first simulations of a proposed camera composed of a stack of thin silicon semiconductor detectors. Sigh and Doria[9] built the first laboratory device. In this approach the incoming direction of the incident γ-radiation is not determined by the orientation of a collimator but is calculated using the measured, three dimensional positions and energies deposited at two interaction points; one where Compton scattering occurred and the other where the absorption of the scattered gamma-ray occurred.

In this technique a material is chosen that has a high probability not to absorb the photon but only to scatter the photon. A second material is chosen to surround this first material so that the Compton scattered photon is then fully absorbed. From this information the scattering angle of the photon can be found,

$$\cos\theta = 1 - m_{electron}c^2 \frac{E_{compton}}{E_{total}E_{absorption}} \tag{1}$$

Where $E_{Compton}$ is the energy deposited at the Compton scattering point, $E_{absorption}$ is the energy deposited at the photon absorption point and E_{total} is the sum of the two, which is the energy of the photon. The technique produces two interaction points that

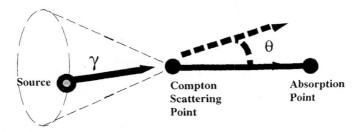

FIGURE 4. The source direction is constrained to lay on the Compton cone. Using a collection of events the intersection of the cones will form an image of the object.

allow not a line but a cone to be formed. The source of the photon is then constrained to lie on this cone. Imaging is formed by the intersection of a series of these cones. It is usual to call such a device, consisting of two detectors, a Compton camera .

Semiconductor detectors composed of light elements such as Si and Ge were the first candidates for the Compton scattering detectors[9]. Room temperature CdTe, CZT, HgI_2 semiconductor detectors are presently under discussion for the absorption detector. Unfortunately, the size of the field of view is restricted by the large number of channels required and high costs of the components.

A Compton camera based on noble gas electroluminescence has many advantages. The electroluminescence is excited by the drifting of the ionization electrons produced by the Compton scattered photon. In this process the energy taken by the drifting electrons from the applied electric field is transformed into the energy of the luminescent light. The electroluminescence light-output may consist of ~1000 photons per drifting electron per centimeter of drift path[10]. An imperical formula for the number of photons generated per drifting electron is;

$$^{Xe}N_\gamma = 70(\frac{E}{P} - 1) \, x \, p \quad .$$ (2)

Where E is the electric field in kV/cm, p is the pressure in atmospheres and x is the drift distance in cm. Typical values for an operating system are p= 20 atm, E/p=2 and x=0.5 for a 1 cm drift cell yield; $^{Xe}N_\gamma = 1400 \frac{\gamma}{cm \bullet electron}$. Such light output is hundreds of times more than the light-output of the best scintillators used in imaging devices, for a given input γ-rays energy.

Electroluminescence is a linear process and in contrast to charge multiplication processes, is capable of precise energy resolution. The best energy resolution achieved to date (2.2%FWHM at 122 keV[11]) is already close to or even better than the best measurements with room temperature semiconductor detectors. New advantages arise by combining noble gas detectors, which have intrinsic light-amplification, with a fiber optic readout system. This combination has demonstrated sub-millimeter, two dimensional, position resolution[12]. Unfortunately these first designs did not take full advantage of the fiber optic readout. The fiber array was placed on only one side of the light-production gap resulting in a 50% loss of the light. Another 25% of the light was lost because x- and y- oriented fiber arrays screened the light-production gap from one to another. This experience however gives value data for the design of future systems.

CONCEPT OF THE DETECTOR

The approach is to build a "basic" electroluminescence cell. A collection of these cells can be formed into a multi-layer structure or multi-layer electroluminescence camera (MELC) containing a stacked electrode structure placed in a high pressure gas chamber. The basic cell is then a single element in a larger detector. The cell is shown in Figure 5.

The cell will include two parallel high voltage planes made of a thin foil (aluminized Mylar). Electric fields of 20 keV/cm are required to cause the electroluminescence. An array of x-oriented fibers will be placed on one electrode while an array of y-oriented fibers will be placed on the other electrode. As soon as an ionizing event occurs, the ionization electrons begin to drift to the anode, exciting the gas and generating an electroluminescence signal. Electroluminescent UV light illuminates the fibers on both electrodes. The gas within the cell will be 99% Ar and 1% Xe at 20 atmospheres of pressure. Scintillating fibers will waveshift the 170 nm Xe scintillation light into the visible region and then transmit the light to the phototransducer outside the gas volume. The Xe luminescence will first be waveshifted to 350 nm by a coating of p-terphenyl on the outer surface of the fiber. Within the fiber, 3-HF wave shifter will shift this light to 530 nm. This final waveshift will also allow the light to transmit down the fiber to the phototransducer.

The light distribution observed on the two sets of crossed fibers give the x and y location of the γ-ray scattering point, determined by a "center of gravity" calculation.

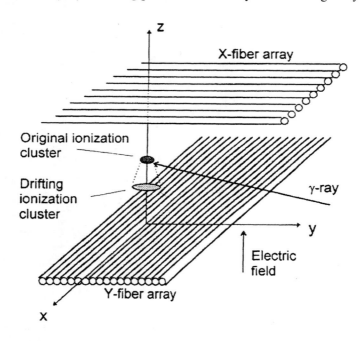

FIGURE 5. The "basic" cell . Not shown are the high voltage planes that cause the electron drift.

The amount of light that intercepts a fiber when the electron is a distance x away from the fiber is;

$$^{Xe}N_\gamma(x) = \frac{1}{\pi}\frac{dN_\gamma}{dx}\sin^{-1}\left[\frac{1}{\sqrt{1+(\frac{2x}{d})^2}}\right]. \tag{3}$$

Where d is the diameter of the fiber. The acceptance for a single photon as a function of distance from the fiber is shown in Figure 6. From this figure it might appear that only the light produced near the fiber is important. However, Figure 7 shows the relative brightness of the total scintillation signal as a function of the interaction point of the x-ray within the cell for the brightest fiber. This is just the integral of equation 2. As shown the variation in signal is only about a factor of 4 over the active region of the cell.

Because the drift velocity in the gas can easily be measured, the duration of the pulse gives the z location or distance away from the fiber planes, of the γ-ray scattering point, $z = v_d \cdot \Delta t$. The sum of all the light in the cell, divided by the duration of the pulse gives a measure of the energy deposited by the scattered photon. A second basic cell will be filled with pure Xe. The Xe cell will serve as the absorption cell while the Ar filled cell will serve as the Compton scattering cell. Collections of these two cells will form the Compton camera for γ-ray imaging.

The electrode and fiber readout structure is built from materials which have low gamma cross-sections, so that only a negligible fraction of the incident gamma-rays will interact with the electrode structure. These interactions will not generate electroluminescent signals and thus are effectively invisible.

We have studied two options for the photo-sensors. There are multi-anode photomultipliers specially developed to detect light signals from fibers, having a dynode structure with inherent low cross-talk between neighboring channels. Another possible photo-sensor array is the Visible Light Photo-Counter (VLPC) [13]. Using the measured number of scintillation photons produced within the cell, the geometric acceptance given

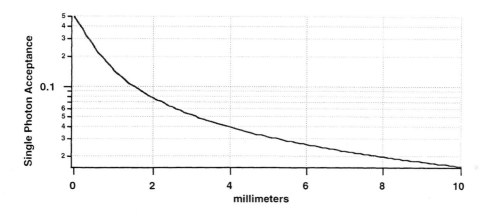

Figure 6. The single scintillating photon acceptance as a function of its creation location.

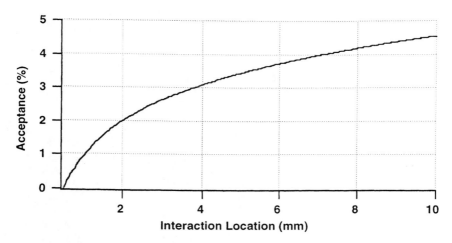

Figure 7. The total number of photons produced as a function of the x-ray interaction point within the cell for the brightest fiber when multiplied by the number of scintillating photons produced per centimeter of electron drift.

in equation 2, and the acceptances listed in Table 1 of each process required to produce a signal, our light yield calculations show that a deposition of 60 keV in a cell should yield 270 photo-electrons in the single brightest fiber if solid state photomultipliers(SSPM) are used to detect the visible light or 56 photo-electrons if photomultiplier tubes are used. The total light produced in all the fibers is calculated to be 9,000 photo-electrons if SSPMs are used or about 2,000 photo-electrons if phototubes are used. This large signal would produce excellent x-ray images. A single cell of the electroluminescent Compton camera has been tested with a photomultiplier matrix readout. It obtained light yields close to the predicted performance [14].

The number of layers and total thickness of the detector will be chosen to provide good detection efficiency. For example, a 20 cm thick detector filled with 20 bar of Xe will have an 85% photoabsorption efficiency for 140 keV gamma-rays. This is similar to the photoabsorption efficiency of a 3/8" thick NaI(Tl) crystal, normally used in Anger gamma-cameras in production today.

We have performed Monte Carlo simulations to predict the performance of the MELC with the proposed fiber optic readout system. The simulations were based on the GEANT 3.21 code for one working layer of the MELC detector. In the simulations, the fibers were 1.0 mm in diameter, the fiber planes were separated by 10 mm and thechamber pressure was 20 ATM argon (Ar) or xenon (Xe) with a 2 kV/cm-ATM electric field. The photon energies studied varied from 20 to 140 keV. The results have a three dimensional resolution with sub-millimeter position precision and energy resolution of 2-3% at 20-50 keV and 1.2% at 110 keV.

EXPERIMENTAL APPARATUS AND RESULTS

In order to determine the behavior of the scintillating fibers under high pressure, we designed and built the following apparatus shown in Figure 8. Brass tubing contained the high pressure Argon at approximately 250 PSI, in which the fiber, 1mm×3m Kuraray multi-clad, was tested. At one end, an optical glass window of 1 inch diameter

TABLE 1. Acceptance for luminescence detection

Process	Acceptance
Xe luminescence hitting single fiber	0.03
Xe luminescence converted in PTP	0.9
PTP light enters fiber (w/3HF)	0.4
PTP light converts in 3HF	.8
3HF emission propagates in Fiber (no reflection assumed)	0.05
Light not re-absorbed in fiber (attenuation length 4.6m)	0.8
Connector to photo-detector	0.8
Photo Detector Quantum Efficiency	
VLPC	0.6
Phototube	0.12
Total Acceptance	
VLPC	1.6×10^{-4}
Phototube	3.3×10^{-5}

and 0.125 inch thick, was fixed in a brass plate with a 0.5 inch diameter hole for the window. The fiber was glued to the window face using 5 minute epoxy, allowing the light from the fiber to be detected by a phototube. This window was sealed using black wax[15], and given extra stability by fixing an o-ring against the interior face of the window. During this experiment, we discovered no leakage around the window. To assure long term stability, all electrical equipment remained on during the course of the experiment. The temperature of the experiment remained constant to within $\pm 1^{\circ}C$. In order to calibrate the performance of the phototube, a blue LED[16] was also placed in this vessel, facing the window. At five points evenly spaced along the length of the fiber, brass chambers held blue LED's near the fiber.

Each LED was connected individually to the wave form generator, with a spark plug[17] serving as an electrical feedthrough. While these were quite capable of withstanding the large pressure gradient, they did leak initially. Therefore, we filled the interior section with cyano-acrylate glue, to ensure that the ceramic/metal interfaces were sealed.

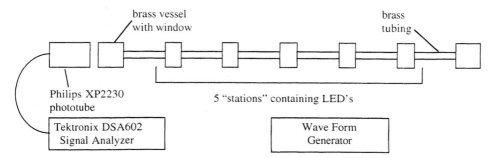

FIGURE 8. Schematic of the experimental apparatus.

77

When the LED's were pulsed, the fiber produced scintillation light which was then measured by the phototube. The data presented in the following figures represents the average integrated area under these pulses, as seen by the phototube. To ensure consistency over time, this data was normalized using the LED shining directly through the window into the phototube. Then, to allow all of the data to be assimilated into a single plot, the data for each station were normalized to 1. The results of this experiment are given in Figure 9.

The error bars on this graph are ± 1.5% of the signal level. This was the approximate range of fluctuations observed on the signal analyzer while taking data.

This plot shows that the fiber was quite stable over a 40 day period. The average slope of -2.46×10^{-4} indicates that it would take approximately 410 days for the fiber to degrade by 10 percent , assuming that the curve remained a straight line.

Fiber Feedthrough Results

Detector design, performance and fiber readout would be made much simpler and robust if the fibers themselves could be fed through the container wall, from high pressure to atmospheric pressure. The experiment was modified to determine the performance of the fibers under these conditions: the additional stress on the fiber at the interface could have unexpected effects. For this test, instead of using a glass window, a set of 3 concentric holes were drilled in a brass plate through which the fiber passed. A funnel shape connected one hole to the other. The last hole was equal to the fiber diameter. The fiber was then epoxied in place, plugging the set of holes, using Scotch-weld 2216B/A gray epoxy adhesive. The calibration LED was again placed facing the phototube, but was not under pressure.

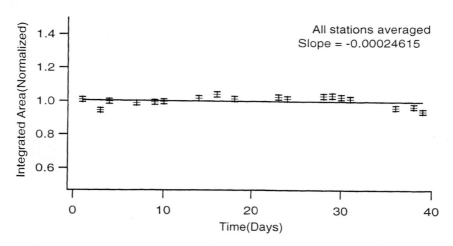

FIGURE 9. Average light output observed (normalized to 1) as a function of time. Each data point represents the sum of the data for all 5 stations, with the data for each station averaged over 256 pulses.

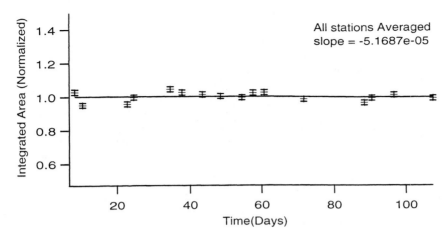

FIGURE 10. Average light output observed (normalized to 1) as a function of time. Each data point represents the sum of the data for all 5 stations, with the data for each station averaged over 512 pulses.

The results of this experiment, with the data from all five stations normalized and combined, are given in Figure 10.

Again, the performance of the fiber was quite stable, even with the additional stresses. The average slope of -5.17×10^{-5} implies that one would have to wait approximately 2000 days to observe a ten percent change in light output from the fiber. This apparent increase in fiber stability compared with the previous experiment is probably due to the fact that data was taken for a longer period of time, 110 days, making the daily fluctuations less significant in the analysis.

CONCLUSION

The calculations have shown that the HPXED Compton camera will have hundreds of times better sensitivity than a modern Anger NaI(Tl) scintillation camera because of the elimination of the collimator. It will have position resolution of 4-5 mm and precise energy resolution, which can not be achieved even with modern room temperature semiconductor detectors. The consequences of the development of such a gamma-imaging device are difficult to be predicted. But it is already clear that the radiation level required to make images will be reduced by at least factors of 10 over present single photon imaging devices. Imaging quality will also be considerably improved, and because of the greatly increased sensitivity, dynamic imaging becomes a real possibility. While the scintillating fiber have been shown to operate acceptably in a high pressure environment, further studies are required to check the effect of fiber outgassing on electron drift in the noble gas.

REFERENCES

1. Coleman, R.E. et al., *Inv Radiol* **21**(1), 1-11 (1986).
2. Meegan, C.A. et. al., *Ap. J. Suppl.* **106**, 65-110 (1996).
3. Miyahara, *J. Chemistry Today*, 29-36 (1989).
4. Cline, D., *Nucl. Instr. & Meth. A* **327**, 178-286 (1993).
5. Anger, H.O., *Rev.Sci.Instr.* **29**, 27-33 (1958).
6. Fahey, F.H., and Harkness,B.A., "SPECT imaging with rotating gamma camera systems", in *NUCLEAR MEDICINE*, Ed. By R. Henkin et al., Mosby: pp.232-246 (1996).
7. Todd,R.W., Everett,D.B., and Nighhtingale,J.M., *Nature* **251** 132-134 (1974).
8. Everett,D.B., Todd,R.W., and Nightingale,J.M., "Gamma-radiation imaging system based on the Compton effect", *Proc.IEE* **124**(11), 995-1000 (1977).
9. Singh,M., and Doria,D., *Med.Phys.* **10**(4), 428-435 (1983).
10. Belogurov,S., Bolozdynya,A., Churakov,D., Koutchenkov,A., Morgunov,V., Solovov,V., Safronov,G., Smirnov,G., Egorov,V., and Medved,S., "High pressure Gas Scintillation Drift Chamber with Photomultipliers Inside of Working Medium", *1995 IEEE Nucl. Sci. Symp. And Medical Imaging Conf. Conference Record*, October 21-28, San Francisco, **1**, pp. 519-523, 1995.
11. Bolozdynya,A., Egorov,V., Koutchenkov,A., Safronov,G., Smirnov,G., Medved,S., and Morgunov,V., "A high pressure xenon self-triggered scintillation drift chamber with 3D sensitivity in the range of 20-140 keV deposited energy," In press in *Nucl. Instr. & Meth. A*, 1997.
12. Parsons,A., Edberg,T.K., Sadolet,B., et al., *IEEE Trans.Nucl.Sci.* **37**(2), 541-546 (1990).
13. Koltick,D., "Scintillating Fiber Charged Particle Tracking at the Superconducting Super Collider", *SCIFI-93, Workshop on Scintillating Fiber Detectors*, pp.111-125, 1993.
14. Bolozdynya,A., "High pressure xenon electronically collimated camera for low energy gamma ray imaging", Report at the 1996 IEEE Nucl. Sci. Symp. And Medical Imaging Conference, November 3-9, 1996.
15. Apiezon Products Limited,4 York Rd., London S.E.1, England.
16. Panasonic # LNG992CF9; peak wavelength 450nm.
17. Champion small engine spark plugs, #843-1 CJ8.

SESSION 3: STRUCTURES

Chair: J. Huston
Scientific Secretary: R. Hooper

Fiber R and D for the CMS HCAL

H. S. Budd, A. Bodek, P. de Barbaro, D. Ruggiero, E. Skup

Department of Physics and Astronomy, University of Rochester, Rochester, NY 14627

Abstract.

This paper documents the fiber R and D for the CMS hadron barrel calorimeter (HCAL). The R and D includes measurements of fiber flexibility, splicing, mirror reflectivity, relative light yield, attenuation length, radiation effects, absolute light yield, and transverse tile uniformity. Schematics of the hardware for each measurement are shown. These studies are done for different diameters and kinds of multiclad fiber.

The CMS HCAL optical design is similar to the CDF Plug Upgrade optical design [1] [2]. A wavelength shifting (WLS) fiber, containing K27 waveshifter, embedded in the tile collects the scintillation light. Outside the tile, the WLS fiber is spliced to a clear fiber. The clear fiber takes the light to a connector at the edge of the pan. An optical cable brings the light to the optical readout box. The readout box assembles the light from layers to towers and brings the light to the photodetectors.

The fibers tested for this paper are all multiclad. The diameters of the fibers range from 0.83 mm to 1.0 mm. CMS has chosen fiber of 0.94 mm diameter. We test four types of Kuraray fiber, non-S (S-25), S-35, S-50, and S (S-70) [3] . The most flexible Kuraray fiber is S type and the least flexible fiber is non-S type. We test two batches of WLS Bicron fiber [4]. Batch 1 is an earlier version than Batch 2. We test one batch of Bicron clear fiber, which was made at the same time at Batch 1 WLS fiber. The waveshifter for all WLS fiber is K27.

We use R580-17 phototubes. This tube is a Hamamatsu 1.5 inch diameter, 10 stage tube with a green extended photocathode. The R580-17 photocathode and the photocathode for the CMS photodetectors, HPDs, are the same.

For most measurements a tile excited by a radioactive source generates the light. This insures that the spectrum of light is the same for these tests as the spectrum from the CMS HCAL calorimeter. We use both a Cs-137 γ source and a Ru-106 electron source. The Cs-137 source is collimated by a lead cone. A picoammeter reads out the phototube. The data aquisition (DAQ) program averages 20 measurements from the picoammeter and creates a pedestal subtracted data file. The absolute light yield measurement uses the Ru-106 source. Its DAQ consists of a 2249A Lecroy ADC triggered by a coincidence of two scintillators.

CP450, *SciFi97: Workshop on Scintillating Fiber Detectors*
edited by A. D. Bross, R. C. Ruchti, and M. R. Wayne
© 1998 The American Institute of Physics 1-56396-792-8/98/$15.00

TABLE 1. Flexibility of different fibers. Column 1 gives the kind of fiber. Column 2 gives the smallest bend diameter without the fiber damage. Column 3 gives the largest bend diameter with fiber damage.

Fiber type	Fiber not damaged	Fiber Damaged
Kuraray 1 mm non-S	2 1/2 in	2 in
Kuraray 0.94 mm S-35	3/4 in	5/8 in
Kuraray 0.94 mm S-50	5/8 in	-
Kuraray 0.83 mm S	5/8 in	-
Bicron 1.00 mm	2 in	1 1/2 in

FIBER FLEXIBILITY

We have studied Kuraray non-S fiber for flexibility by looking at the change in light transmission after the fibers are bent. Fibers with a change in transmission greater than 2% always have cracks or crazing in the bent portions. They have light leaking out from these cracks. Hence, we test for flexibility by looking for light leaking out of the bend portions.

We test the flexibility by wrapping the fibers around dowels and looking for cracks where light leaks out. Table 1 gives the result. The smallest fiber bend diameter for the HCAL barrel is 3 inches. Both the Kuraray S-35 fiber and Bicron fiber are flexible enough for CMS HCAL.

FIBER SPLICING

Fiber splicing is done with the semi-automated splicer developed by the CDF Plug Upgrade Group [5]. Splice transmission of WLS fibers is measured by scanning across a splice the using CDF automated UV scanner, see Figure 1. Figure 2a shows the results of splicing tests. Table 2 lists the results of the tests. Only splicing tests done at the same time should be compared. The Dec 96 splicing test shows that non-S fiber splices have higher transmission than S type fiber. The Nov 97 splicing test shows the splice transmission for non-S and S-35 fiber is the same. Splice transmission for S-50 is worse than non-S and S-35 fiber. We have chosen S-35 fiber for the HCAL preproduction prototype because of its excellent flexibility and high splice transmission. The Sept 96 splicing test shows that the splice transmission for Kuraray non-S fiber and Bicron fiber is the same.

The fibers must have a good polish for good splice transmission. We have compared splice transmission using two different polishing techniques. One polishing technique uses the Avtech polisher [6]. The Avtech polisher is a single fiber polisher which CDF used for its production. The second technique is ice polishing. Ice polishing was pioneered by Fermilab Charmonium experiment, E835 [7], and was used by Fermilab experiment HyperCP, E871 [8], and D0. The technique involves freezing fibers in water and polishing the fiber/ice combination. Many fibers can be polished at once with ice polishing. Figure 2b shows the results of splicing Kuraray

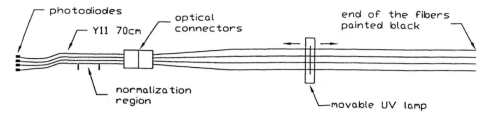

FIGURE 1. CDF automated UV scanner.

TABLE 2. Splice transmission of different kinds of fibers and different polishing techniques. Column 1 gives the kind of fiber. Column 2 gives the fluor. Column 3 gives the kind of polish. Column 4 gives the date of the test. Column 5 gives the number of fibers tested. Column 6 gives the mean of the splice distribution. Column 7 gives the RMS of the distribution. Only tests done on the same date should be compared.

Fiber type	Fluor	Polish	Date	Number	Mean	RMS
Kuraray 0.83 mm non-S	K27	Avtech	Dec 96	3	0.948	0.006
Kuraray 0.83 mm S	K27	Avtech	Dec 96	3	0.876	0.009
Kuraray 0.94 mm non-S	K27	Avtech	Nov 97	2	0.928	–
Kuraray 0.94 mm S-35	K27	Avtech	Nov 97	9	0.930	0.009
Kuraray 0.94 mm S-50	K27	Avtech	Nov 97	4	0.879	0.019
Kuraray 0.83 mm non-S	K27	Avtech	Sep 96	5	0.908	0.009
Bicron 0.83 mm	K27	Avtech	Sep 96	5	0.902	0.003
Kuraray 0.94 mm S-35	K27	Avtech	Mar 98	5	0.908	0.018
Kuraray 0.94 mm S-35	K27	Ice	Mar 98	5	0.930	0.008
Kuraray 0.94 mm non-s	clear	Avtech	Nov 97	5	0.904	0.018
Kuraray 0.94 mm S-35	clear	Avtech	Nov 97	5	0.893	0.011

0.94 mm, S-35 fibers with these two techniques. We conclude that both polishing techniques give the same transmission. CMS HCAL has chosen to ice polish their fibers.

We have tested the splice transmission of clear fibers. Figure 3 shows the setup used to measure the transmission of the splice. A connector and fiber assembly with WLS fiber at the nonconnector end is called a "pigtail". The WLS fibers in a "pigtail" are inserted into a tile excited by a radioactive source to readout the light. The pigtail for this test consists of 20 cm WLS fibers spliced to 99 cm clear fibers in a connector. The WLS fiber inserted into a tile injects a constant amount of light into the clear fiber. After the clear fiber is cut and spliced, the pigtail is remeasured. The ratio of the measurement before and after the splice is defined as the splice transmission. The results are shown in Figure 4. The test shows that the splice transmission for clear non-S and clear S-35 fiber is the same. Both clear and WLS S-35 splicing tests are done with the same splicing machine by the same operator. Table 2 lists all the fiber splicing results.

FIGURE 2. Splice transmission for WLS fibers. (a) compares different kinds of fibers and (b) compares 2 different polishing techniques

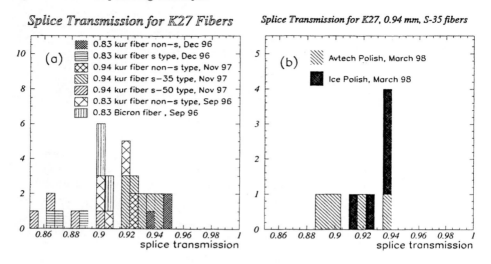

FIGURE 3. Hardware used to measure transmission of clear to clear splice. Figure 4 shows the results

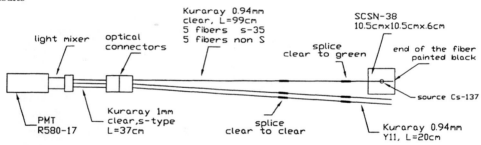

MIRROR REFLECTIVITY

The ends of the fibers are mirrored using vacuum deposition in Lab 7 at Fermilab. We have studied the mirror reflectivity for 2 types of fibers and both polishing techniques.

The reflectivity is measured using the CDF automated UV scanner (see Figure 1). A mirrored fiber is put in a connector and measured with the UV scanner near the mirror. Next, the mirror is cut off at 45° to the fiber axis. The end of the fiber is painted black to prevent any reflection from the end of the fiber. The fiber is remeasured with the CDF automated UV scanner. The reflectivity is defined as (measurement with mirror)/(measurement without mirror) - 1.

FIGURE 4. Splice transmission for clear fibers

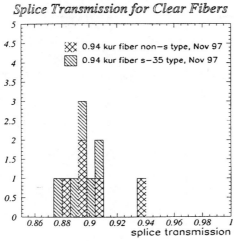

Figure 5 shows the results. We have measured the reflectivity of a 3 1/2 year old CDF pigtail, which was a spare for the CDF End Plug Hadron Calorimeter [6]. The pigtail is made of 0.83 mm, non-S fibers polished with the Avtech polisher. The mirror is dipped in Red Spot UV curable coating to protect the mirror. The measurement shows no degeneration in the mirror after 3 1/2 years. A measurement of recently mirrored non-S type fiber gives the same reflectivity. The reflectivity of S-35 fibers is roughly 5% lower. The reflectivity for ice polished fibers seems to be slightly lower than Avtech polished fibers.

We have measured the light increase from the mirror. A pigtail with 3 fibers shown in Figure 6 is made and scanned with the UV scanner. The pigtail is measured with a tile and phototubes as shown in Figure 6 Next, the mirror is cut off at 45° to the fiber axis, and the end of the fiber is painted black. The pigtail is remeasured with both setups. We get

$$\frac{\left(\dfrac{\text{light from tile with mirror}}{\text{light from tile with no mirror}}\right)}{\left(\dfrac{\text{light from UV scanner with mirror}}{\text{light from UV scanner with no mirror}}\right)} = 0.865 \quad \text{with RMS} = 0.004$$

Since the UV scanner indicates a factor of 1.77 increase for the mirror for ice polished S-35 fibers, the corresponding case for a tile is 1.53

RELATIVE LIGHT YIELD FOR DIFFERENT FIBERS

We measured the light response of different kinds of fibers. Figure 7 shows the hardware used for the relative light yield test. A clear "cable" connects the pigtail from the connector to the phototube. Table 3 gives the result for different types

FIGURE 5. Mirror reflectivity for Kuraray fiber.

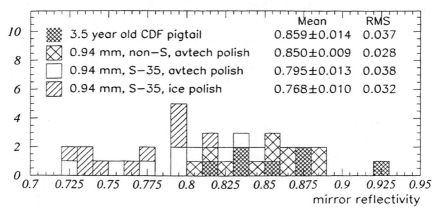

Mirror reflectivity, Y-11 Kuraray fibers

FIGURE 6. Hardware to determine the light increase in a tile from a mirror.

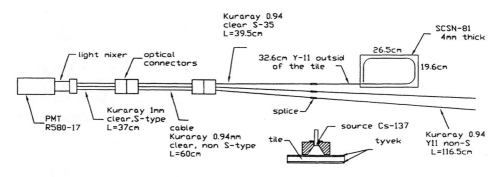

of WLS fibers using both 103 cm and 251 cm Kuraray 1 mm S type clear cables. Column 2 and 3 are separately normalized to the Kuraray 0.83 mm, 250 ppm, non-S result. For Kuraray fiber, the 0.94 mm WLS fiber yields 6% more light than a 0.83 mm WLS fiber. There is no difference in light between a 0.94 mm WLS fiber and a 1.0 mm WLS fiber to 1 mm fiber. Table 4 gives the core diameter and fiber diameter. One sample of each fiber is measured. The increase in light is smaller than the increase in either the core diameter or the fiber diameter.

FIGURE 7. Hardware for relative light yield test. 1 piece of each hardware is made except for the 251 cm cables Kuraray 1 mm S type where 2 are made. Results are shown in Table 3 and 5

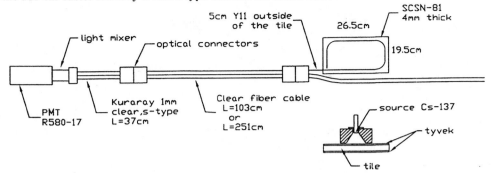

TABLE 3. Relative light yield. Column 2 gives the light using a 251 cm cable and Column 3 gives the light using a 103 cm cable.

Type of WLS fiber	251 cm Cable	103 cm Cable
Bicron 1.00 mm, 200 ppm, batch 1	0.93	0.93
Bicron 1.00 mm, 200 ppm, batch 2	1.00	1.02
Kuraray 0.83 mm, 250 ppm, non-S	1.00	1.00
Kuraray 0.94 mm, 250 ppm, S-35	1.07	1.06
Kuraray 0.94 mm, 250 ppm, S-50	1.06	1.06
Kuraray 1.00 mm, 200 ppm, non-S	1.06	1.05
Kuraray 1.00 mm, 300 ppm, S	1.08	1.07

TABLE 4. Core and fiber diameter. All units are mm. One fiber is measured for each type. Kuraray fiber has 2 claddings and Bicron fiber has one cladding. † means fiber has only one cladding. ‡ means interface between outer cladding and inner cladding was not visible enough to measure it.

Fiber type	Core diameter	Inner cladding diameter	Outside diameter
Bicron, 0.83 mm, WLS	0.789	†	0.850
Kuraray, 0.83 mm non-S, WLS	0.742	0.786	0.844
Kuraray, 0.83 mm non-S, clear	0.737	‡	0.838
Kuraray, 0.94 mm non-S, WLS	0.841	0.903	0.959
Kuraray, 0.94 mm non-S, clear	0.838	‡	0.946
Kuraray, 1.00 mm non-S, WLS	0.887	0.955	1.008
Kuraray, 1.00 mm non-S, clear	0.902	0.958	1.010

ATTENUATION LENGTH OF CLEAR FIBERS

We have looked at the relative light transmission of different clear fibers by using two different cables. Figure 7 shows the hardware. Table 5 gives the ratio of light for the two different cables. Column 2 and column 3 give the cables used. The length of the cables for column 2 are all 251 cm. Column 4 gives the ratio of the measurement of column 2 over column 3. The results are consistent with an equal attenuation length for 1 mm S, 0.94 mm S-35, and 0.94 S-50 Kuraray fiber.

89

TABLE 5. Comparison of light yields using two different clear cables.

Pigtail	Cable 1, L=251 cm	Cable 2	Cab 1/Cab 2
Kur 1.00 mm, non-S	Kur 1.00 mm S	103 cm Kur 1.00 mm S	0.79
Kur 0.94 mm, S-35	Kur 0.94 mm S-35	103 cm Kur 0.94 mm S-35	0.80
Kur 0.94 mm, S-50	Kur 0.94 mm S-50	103 cm Kur 0.94 mm S-50	0.80
Bicron 1.00 mm	Bicron 1.00 mm	251 cm Kur 1.00 mm S	0.89

FIGURE 8. Hardware to measure attenuation length of cables of Kuraray, 0.9 mm, S-type clear fibers. Results given in Figure 10a. Pigtails A and B measure different clear fibers.

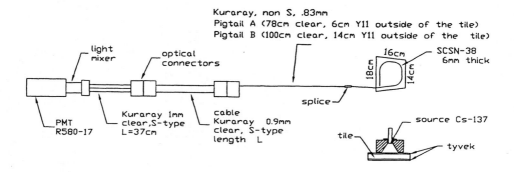

FIGURE 9. Apparatus to measure the attenuation length of 1 mm clear fiber cables. Bicron multiclad and Kuraray S-type fibers are measured. The same cables are measured with both pigtails. Results given in Figure 10b

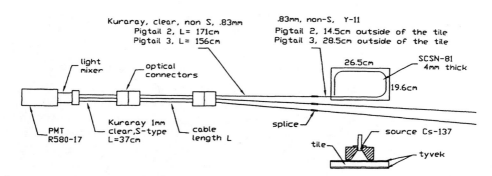

FIGURE 10. Attenuation length of clear fibers.

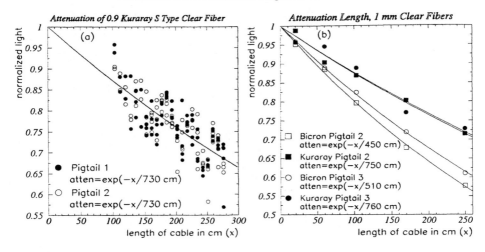

FIGURE 11. Apparatus used to measurement of attenuation length of clear fiber spliced to WLS fibers. Results given in Figure 12

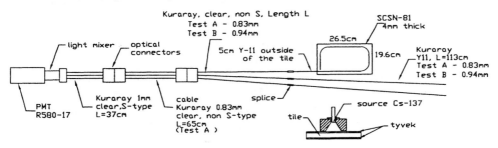

We have measured the attenuation length of the clear fiber in a cable. Figure 8 gives the apparatus for measuring Kuraray 0.9 mm S type fibers, and Figure 10a gives the result. Each point is a measurement of one fiber in a cable. Two separate pigtails are used and they give the same attenuation length of the clear fiber. The combined results of both pigtails for the attenuation length is 732 ± 13 cm. The RMS of the normalized light about the exponential curve in Figure 10a is 5.6%. The test shown in Figure 9 measures the difference in attenuation between the Kuraray 1 mm fiber and Bicron 1 mm fiber. Figure 10b gives the result. Each point is the average over the three fibers in the pigtail. The attenuation lengths for Kuraray fibers given in Figure 10a and Figure 10b agree.

We have looked at the attenuation length of the clear fiber when one end is spliced to a WLS fiber and the other is glued in a connector. Figure 11 shows the

FIGURE 12. Attenuation length of clear fiber spliced to a WLS fiber in a pigtail

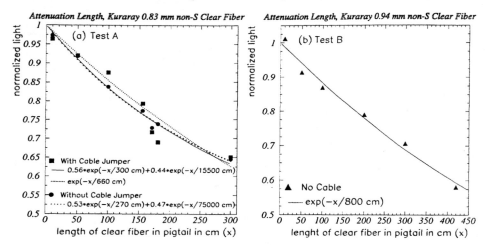

apparatus used to measure the clear attenuation length, and Figure 12 gives the results. For Test A (Figure 12a), one pigtail with 3 fibers is made for each length of clear fiber. A single exponential fit gives $\lambda = 6.4$ m with the cable and $\lambda = 6.55$ m without the cable. Hence, the attenuation length of the clear fiber does not depend on whether the cable is present. If the data are fit to a single exponential $(e^{-x/\lambda})$ with a restricted region 0.5 m $< x <$ 3.0 m, $\lambda = 6.8$ m.

For Test B (Figure 12b), one pigtail was made with three 4.2 m clear fibers. The pigtail is measured. The pigtail connector is cut off and a new connector is put on with the clear fiber reduced to 3 m. The pigtail is measured and the process continues until the pigtail has 0.1 m of clear fiber left. Test B uses the same splice between the clear and green fibers for all the clear lengths and should give a better measurement of the attenuation length.

The measurement shown in Figure 10a gives the best measurement of the attenuation length of the clear Kuraray fiber. A single exponential with $\lambda = 7.3$ m gives an adequate description of the attenuation length of all Kuraray clear fiber for lengths $<$ 4 m. The other measurements of the clear Kuraray fiber are consistent with this measurement. The results shown in Figure 12a justify using a single exponential to describe the attenuation length of clear Kuraray fiber.

ATTENUATION LENGTH OF WLS FIBERS

We have measured the attenuation lengths of various WLS fiber. The attenuation lengths are measured with a setup provided by the MINOS experiment, see Figure 14 . The MINOS setup provides an easy and quick way of measuring the attenuation length of many WLS fibers using scintillator material as a light source.

FIGURE 13. Attenuation length of different WLS fibers. The results measured with the MINOS setup are shown in (a), and the results measured with the UV scanner shown in (b). The numbers after the symbols in the plots state the number of fibers measured.

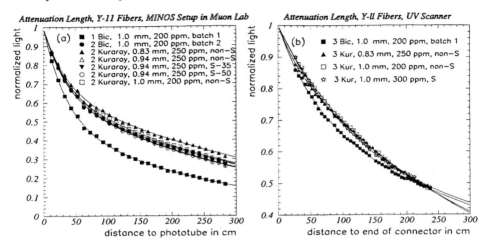

FIGURE 14. A schematic of the MINOS setup to measure fiber attenuation

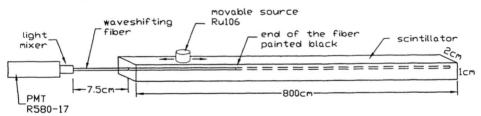

One end of the fiber is polished, and the other end is cut at 45 degrees and painted black. The fiber is inserted into a long hole in the scintillator and the polished end is pushed up against a light mixer on the phototube. The source is moved across the scintillator and the phototube current is read out with a picoammeter. The data is fit to $ae^{-x/l_1} + be^{-x/l_2}$. Figure 13a plots the distance from source to the phototube vs the normalized light. Table 6 gives the numerical results of the fits.

Almost all of our WLS fibers are 3 m long. We have some 4 m Kuraray 0.94 mm, 250 ppm, non-S fiber. To determine if the attenuation lengths using 3 m and 4 m pieces agree, we measure two 4 m pieces in the MINOS setup. We cut the 2 pieces to 3 m and remeasured them. Figure 15 plots the results. The measurement shows that an attenuation length measured with a 3 meter piece can be extrapolated to 4m.

93

FIGURE 15. Comparison of the attenuation length of Kuraray 0.94 mm, 250 ppm, non-S measured with 3 m and 4 m pieces

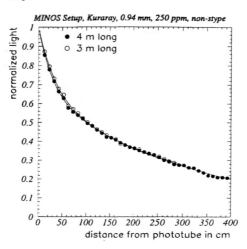

TABLE 6. Results of fit to attenuation data. The data is fit to $ae^{-x/l_1} + be^{-x/l_2}$. All units for l_1 and l_2 are cm. The column marked Setup gives the apparatus used to measure the fibers. MINOS uses the MINOS apparatus, UV used the CDF UV scanner, and SLIDE is the sliding fiber apparatus. MINOS RAD is the radiated fibers measured with the MINOS setup. UV RAD is the radiated fibers measured with the UV setup.

K27 Fiber type	Setup	a	l_1	b	l_2
Bicron 1.00 mm, 200 ppm, bat-1	MINOS	0.488	35	0.512	254
Bicron 1.00 mm, 200 ppm, bat-2	MINOS	0.357	33	0.643	343
Kuraray 0.83 mm, 250 ppm, non-S	MINOS	0.323	34	0.677	379
Kuraray 0.94 mm, 250 ppm, non-S	MINOS	0.372	39	0.628	366
Kuraray 0.94 mm, 250 ppm, S-35	MINOS	0.345	33	0.655	320
Kuraray 0.94 mm, 250 ppm, S-50	MINOS	0.348	34	0.652	317
Kuraray 1.00 mm, 200 ppm, non-S	MINOS	0.317	31	0.683	326
Bicron 1.00 mm, 200 ppm, bat-1	UV	0.304	63	0.696	611
Kuraray 0.83 mm, 250 ppm, non-S	UV	0.102	33	0.898	375
Kuraray 1.00 mm, 200 ppm, non-S	UV	0.188	79	0.812	431
Kuraray 1.00 mm, 300 ppm, S	UV	0.524	131	0.476	1407
Kuraray 0.94 mm, 250 ppm, S-35	SLIDE	0.287	31	0.713	366
Bicron 1.00 mm, 200 ppm, bat-1	SLIDE	0.381	31	0.619	303
Bicron 1.00 mm, 200 ppm, bat-1	MINOS RAD	0.484	30	0.516	259
Kuraray 1.00 mm, 200 ppm, non-S	MINOS RAD	0.343	33	0.657	250
Kuraray 1.00 mm, 300 ppm, S	MINOS RAD	0.362	31	0.638	235
Bicron 1.00 mm, 200 ppm, bat-1	UV RAD	0.305	58	0.695	358
Kuraray 1.00 mm, 200 ppm, non-S	UV RAD	0.395	148	0.605	332
Kuraray 1.00 mm, 300 ppm, S	UV RAD	0.733	152	0.267	1056

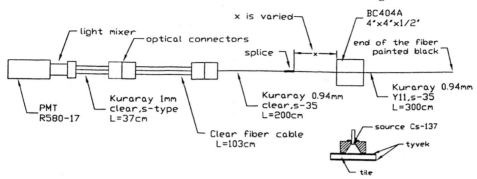

FIGURE 16. Sliding fiber setup. The results are shown in Figure 17

We compared the above measurement with a measurement using a UV light and pin diodes to read out the fibers, see Figure 1. Figure 13b plots the attenuation of the fibers vs the distance to the end of the connector. The MINOS measurement gives a greater difference between the Kuraray fiber and batch 1 Bicron fiber than the UV measurement. The greater sensitivity of pin diodes in the UV setup to long wavelengths light may be the reason. Table 6 gives the results of the fits in the rows labeled UV.

Fibers can be measured quickly with either the MINOS setup or UV scanner. Both tests are useful for comparing fibers. However, the setups may not give the correct attenuation length of the WLS fibers which are relevant for CMS HCAL design. Figure 16 shows the setup designed to give a more accurate attenuation length. Figure 17a shows the attenuation length of two kinds of fibers measured with the sliding fiber setup. We have shown the measurements of the same fibers with the MINOS measurement. Table 6 gives the results for the sliding fiber test for those entries marked SLIDE in column 2. The results of the sliding fiber setup for 0.94 mm Kuraray are used in the design of CMS HCAL.

Figure 17a shows the difference in attenuation length of a fiber measured in two different ways, but with the same photodetector and light injection. We are comparing the difference in attenuation when at clear fiber with a cable is spliced onto the fiber. Figure 17b compares the sliding tile measurement with the MINOS measurement with the origin of the x axis set 18 cm (Kuraray) and 25 cm (Bicron) into the fiber. The comparison shows 18 cm (Kuraray) of WLS fiber is acting like the clear fiber and cable.

RADIATION STUDY OF FIBERS

Some of the WLS fibers were irradiated with 127 krad at an electron source at Florida State University.

FIGURE 17. Attenuation measured using sliding fiber setup and compared with the MINOS measurement. (b) put the x origin for the MINOS measurement 18 cm (Kuraray) and 25 cm (Bicron) into the fiber.

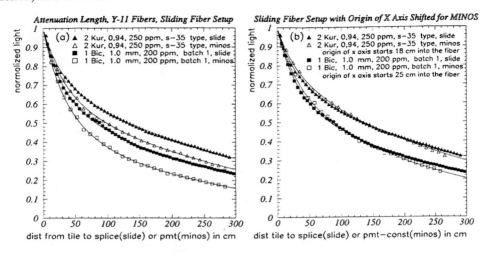

We encountered a problem in measuring the irradiated fibers. The fibers were first measured with the MINOS setup in Lab 5 at Fermilab. While the fibers were being irradiated, the MINOS setup was moved to the Muon Lab. We measured the irradiated fibers with the MINOS setup in Muon Lab.

We measured the same three Kuraray 0.83 mm fibers in both Lab 5 and the Muon Lab. The normalized light at 100 cm was .57 in Lab 5 and .52 in the Muon Lab. We have no explanation for the difference, since the hardware setups are the same. The normalized light difference between different fiber types is not affected by this problem, but the absolute normalized light is affected by this problem. The measurement of the 0.83 mm fiber in both Lab 5 and Muon Lab is used to get a correction function for the Lab 5 measurement. The normalized light yield of the fibers before irradiation is multiplied by the correction function. Figure 18 compares the normalized light of the fibers before and after radiation measured with both the MINOS setup and the CDF UV scanner. The results of the fits of the fibers after radiation are given in Table 6.

LIGHT VS TILE SIZE AND ABSOLUTE LIGHT

Figure 19 and Figure 20 show the hardware setup used to measure the light vs tile size for CMS HCAL tower 10 and 16. Tower 10 is a tower in the middle of the barrel, while tower 16 is at the high eta edge of the barrel. To measure tower 10, we make 6 tiles each for layers 1, 7 and 16. We use the same pigtail to measure the light from each of the tiles. The results are plotted as light vs perimeter/area

FIGURE 18. Comparison of attenuation length of WLS fibers before and after irradiation with 127 Krad. The fiber diameters are 1.0 mm. The 2 setups to measure the fibers are (a) MINOS setup and (b) CDF UV Scanner.

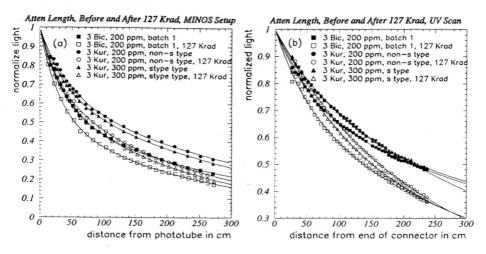

[9] [10]. The perimeter measures the length of fiber in the tile, since the distance the fiber groove is from the edge of the scintillator is kept at 0.3 cm for these measurements. Figure 21a gives the result. This result includes the additional length of WLS fiber outside the tile for layers 1 and 7. Figure 21b gives the result with the WLS fiber attenuation removed. Figure 21b gives the light vs tile size. The overall normalization is not measured by this measurement. The mean of the data for tower 10 is normalized to 1 for both Figure 21a and Figure 21b. The data on Figure 21b is fit to a straight line. The data does not measure the coefficient 3.8 in front. This coefficient is the absolute light yield. The measurement measures the coeffcient 0.077/cm, which is the intercept divided by the slope of the line.

To measure tower 16, we make 5 tiles each for layers 1, 5, and 8. Again, we use the same pigtail to measure the light from each of the tiles. The pigtails for the tower 10 measurement and tower 16 measurement are different. Hence, the normalization of the tower 10 and tower 16 measurements are independent. The mean of the data of the for Figure 21a is set to 1. For Figure 21b the normalization is set so that the mean of the perimeter/area and mean of the normalized light lie on the straight line for tower 10. This enables us to see how consistent the two measurements are.

For CMS design we used two models for the variation of light vs tile size. The first model assumes the light yield is a linear function of perimeter/area. The line is given in Figure 21b. We notice that the points for Tower 16 do not follow a straight line. In the measurement for Figure 21a the same fibers are used to measure all the tiles in a tower. The total variation of the points for Figure 21a for both Tower

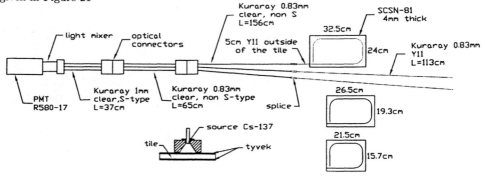

FIGURE 19. Setup used to measure relative light vs tile size for Tower 10. The results are given in Figure 21

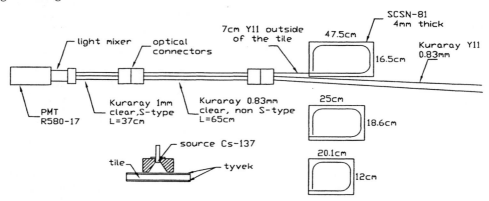

FIGURE 20. Setup used to measure relative light vs tile size for Tower 16. The results are given in Figure 21

10 and 16 is 4%. The second model assumes that the same fiber inserted into all the CMS tiles will give the same light.

Figure 21 shows the apparatus used to measure the absolute light yield of CMS tiles. A Ru-106 electron source is used instead of the Cs-137 source. Figure 23 gives the result. A CMS HCAL barrel tile gives roughly 2 photoelectrons at the photodetector with a green extended photocathode.

The light for the tiles in HCAL barrel can be predicted using the attenuation length of the green and clear fiber, the model of the light vs tower size, and the absolute light yield. From this we can get the total light of a tower. All layers of a tower, except for the first, go to the same photodetector. The longitudinal variation of light within a tower should be less than 10%. By varying the position of the splice, we can make the light uniform longitudinally in a tower.

FIGURE 21. Relative light vs tile size. (b) has the attenuation from the WLS fiber between the tile and splice removed.

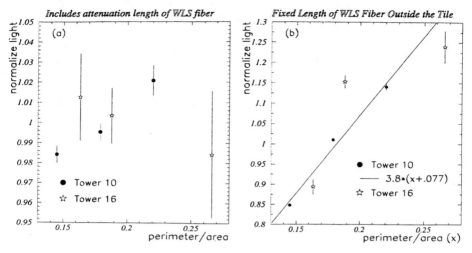

FIGURE 22. Setup to measure absolute light from a tile in CMS. Results are given in Figure 23

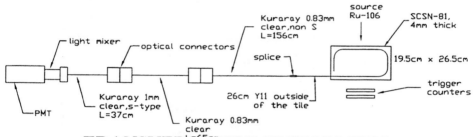

TRANSVERSE TILE UNIFORMITY

We have studied the transverse uniformity of the tiles. We constructed four tiles with the dimensions of tower 14, layer 14, which is the largest tile in the CMS HCAL Barrel. Two tiles have the fibers inserted parallel to the short side, called short side fiber insertion. The tiles with short side fiber insertion are shown in Figure 24. Two tiles have the fibers inserted parallel to the long side, called long side fiber insertion. The tiles for long side fiber insertion are shown in Figure 25. CMS HCAL tiles have long side fiber insertion. The edges of the tiles are painted with TiO_2 paint [11].

The uniformity is measured with a collimated Cs-137 γ source. The collimator constrains the radiation to a 7.5 cm diameter circle. The transverse size of a

FIGURE 23. Absolute light yield.

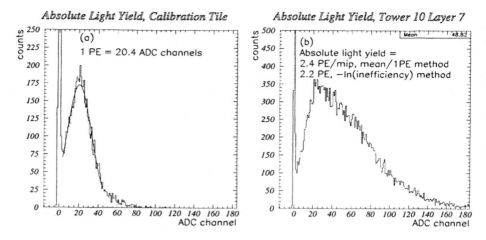

FIGURE 24. Apparatus to measure transverse tile uniformity with 1 mm air gap between the tiles. The fibers are inserted parallel to the short side. The result is shown in Figure 26

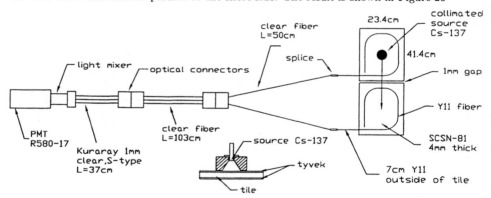

hadron shower is approximately 7.5 cm. Hence, the collimated source simulates the tranverse size of a hadron shower. Figure 26 shows the uniformity across the tile with the short side fiber insertion. The unformity was measured with both Kuraray fibers and Bicron fibers. The Kuraray measurement uses 0.94 mm S-35 fibers for both the pigtail and the cable. For the Bicron measurement, the WLS 1.0 mm bicron fiber (Batch 2) is spliced to 1.0 mm non-S Kuraray fiber. The cable for the Bicron measurement was made with S type 1.0 mm Kuraray fiber. For both kinds of fibers the tile is very uniform with a 10% increase at the boundary between the 2 tiles. The increase is due to increased light collection at the fiber. Figure 27

FIGURE 25. Apparatus to measure transverse tile uniformity with 1 mm air gap between the tiles. The fibers are inserted into the tile parallel to the scan. The measurement is shown in Figure 27

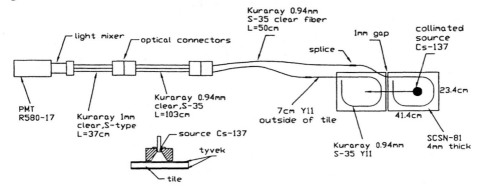

FIGURE 26. Uniformity across a tile using collimated γ source. The fiber insertion is parallel to the short side. The measurement is done with both Kuraray and Bicron fibers.

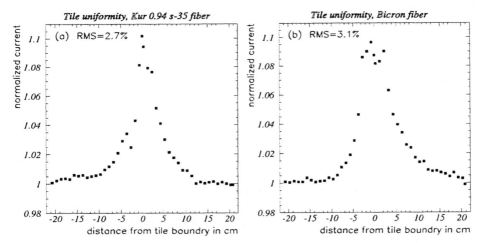

shows the uniformity with long side fiber insertion. The transverse uniformity has an RMS $\sim$ 3% regardless of the fiber type or fiber insertion point. The resolution of the CMS calorimeter is $120\%/\sqrt{E} \oplus 5\%$ [12]. The transverse uniformity across the tile should be somewhat less than the constant term, $< 5\%$, to prevent transverse uniformity from affecting the constant term in the resolution.

The HCAL CMS scintillator design has individual tiles glued together with TiO_2 loaded epoxy resin [13], to form a "megatile". The configuration at the boundary

FIGURE 27. Uniformity across a tile using collimated γ source. The fiber insertion is parallel to the long side.

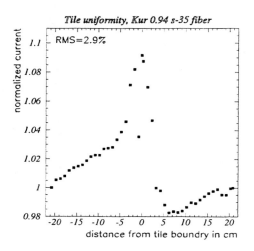

between tiles is shown in Figure 28. A 0.9 mm "separation groove" is cut to separate 2 tiles, with 1/4 mm of scintillator left uncut on the bottom of the groove. The groove is filled with TiO_2 loaded epoxy. The scintillator is marked with a black mark made with the narrow end of a "doubleshot" black marker [14]. The black mark is underneath the separation groove. The black mark is about 1.5 mm wide, slightly wider than the separation groove. The black mark reduces the light cross talk through the 1/4 mm of scintillator left at the tile boundary.

We constructed a glued megatile consisting of 2 tiles inside a piece of scintillator. The piece of scintillator was 4 mm thick, instead of 3.7 mm. The tiles for that glued megatile tile are from tower 10, layer 1. Figure 29 shows the apparatus used to measure the glued megatile. Figure 30 shows the result with and without the black mark. The light increase at the boundary is the same for an air gap and for a glue gap without the black mark. The black mark reduces the light increase at the boundary by about a factor of two. The transverse RMS is roughly 1.5% with the black mark. With the black mark, the transverse uniformity does not increase the constant term of the resolution.

We have measured the cross talk between the glued tiles. The 1/4 mm of scintillator, left uncut between the two tiles, provides a path for light to pass between 2 tiles. The cross talk is measured by first putting a fiber in one of the tiles. Next, we measure the current with the source on the following three locations: the tile with the fiber, the tile without the fiber, and just off the tile with the fiber. The last location is used to measure the source cross talk. The cross talk is 5.2% without the black mark and 1.8% with the black mark.

The black mark decreases the light output. We measured the light from glued

102

FIGURE 28. The boundary between tiles in CMS HCAL barrel.

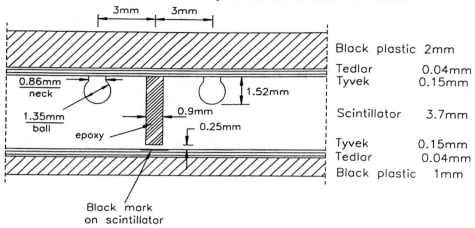

FIGURE 29. Apparatus to measure transverse tile uniformity with 1 mm glue joint between the tiles.

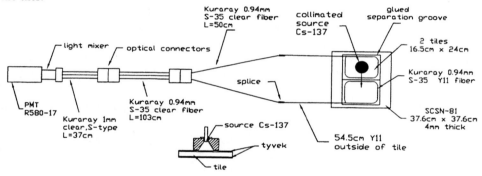

megatile. Next, the black marks are made on 3 sides of the tile, similarily to the way the HCAL CMS barrel tiles will be marked. The light goes down roughly 8%.

CONCLUSION

The CMS R and D enable us to design the optics of HCAL Barrel Calorimeter and predict its preformance. We have chosen S-35 fiber for the HCAL preproduction prototype because of its excellent flexibility, excellent mirror reflectivity, and high splice transmission. CMS HCAL has chosen to ice polish the fibers since it enables us to polish many fibers at once. We can predict the light of each tile in the barrel

FIGURE 30. Uniformity across the glued megatile using collimated γ source. The uniformity is measured with and with out the black mark.

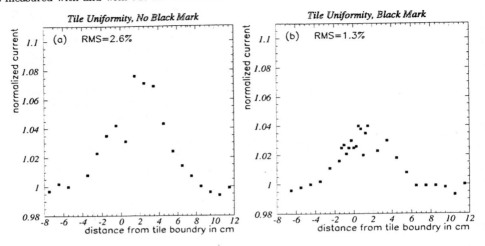

using the attenuation lengths of fibers and the absolute light vs the tile size. By varying the position of the splice for each tile, we can optimize the light distribution in a tower. CMS has chosen to have the same length WLS fiber for all layers in a tower. Measurements of the transverse uniformity shows that it does not effect the resolution of the calorimeter.

REFERENCES

1. CDF End Plug Upgrade Calorimeter Design, By P. de Barbaro et al., University of Rochester Preprint UR-1360, Jul. 1994, CDF Note 2545
2. Scintillator Tile-Fiber Calorimeters for High Energy Physics: The CDF End Plug Upgrade, Selected Articles. edited By P. de Barbaro and A. Bodek. University of Rochester preprint UR 1389, October 1994. References [1], [5], [6], [9], [10]. and [13] are in this document.
3. Kuraray Co., LTD., 8F, Maruzen Building, 3-10, 2-Chome, Nihonbashi, Chuo-ku, Tokyo, 103-0027, Japan.
4. Bicron Coroporation, 12345 Kinsman Road, Newbury, Ohio 44065-9677.
5. A Semiautomated Splicer for Plastic Optical Fibers. By J.P. Mansour, C. Bromberg, J. Huston, S. Joy, B. Miller, R. Richards, B. Tannenbaum (Michigan State U.). 1995. Notre Dame 1993, Proceedings, Scintillating fiber detectors, 534-541.
6. CDF End Plug Calorimeter Upgrade Project, by G. Apollinari, P. de Barbaro, M. Mishina. Proceedings of the IV International Conference on the Calorimetry in High Energy Physics, La Biodola Elba, Sep 19-25, 1993.
7. Performance Measurements of Histe-V VLPC Photon Detectors for E835 at FNAL, R. Mussa et al., this conference

8. The Use of WLS Fibers in a Hadronic Calorimeter for the HyperCP Experiment, by C. Durandet et al., this conference

9. Recent R&D Results on Tile/Fiber Calorimeter. By P. de Barbaro et. al., University of Rochester preprint UR-1299, SDC-93-407, January 1993.

10. Tile/Fiber Results for the Upgraded Plug Hadron Calorimeter. Present by Phillip Koehn, IEEE 1993 Nuclear Science Symposium and Medical Imaging Conference, San Francisco, October 1993.

11. Bicron BC-620 white paint, Bicron Corp, Newbury, OH.

12. Performance of a Prototype CMS Hadron Barrel Calorimeter in a Test Beam. By P. de Barbaro et. al., Proceedings of the CALOR97, VII International Conference on Calorimetry in High Energy Physics November 9 - 14, 1997 University of Arizona Tucson, Arizona - USA .

13. Techniques for Optical Isolation and Construction of Megatiles. By M. Olsson et. al., University of Rochester preprint UR-1370.

14. "Doubleshot" ink marker (No. 11120), Pentech International Inc., Edison, NJ.

Scintillator Manufacture at Fermilab

K. Mellott, A. Bross, and A. Pla-Dalmau

Fermi National Accelerator Laboratory
Batavia, IL 60510

Abstract. A decade of research into plastic scintillation materials at Fermilab is reviewed. Early work with plastic optical fiber fabrication is revisited and recent experiments with large-scale commercial methods for production of bulk scintillator are discussed. Costs for various forms of scintillator are examined and new development goals including cost reduction methods and quality improvement techniques are suggested.

INTRODUCTION

Over the past 15 years, a great deal of new work on plastic scintillators and plastic scintillating fiber (PSF) has been performed. Much of the renewed interest in scintillator detectors stems from the tremendous progress that has been made in plastic optical fibers during this period[1] and techniques involving PSF and wavelength-shifting (WLS) fiber have produced a new generation of scintillation detectors in high energy physics.[2-3] In 1988 the Scintillator Fabrication Facility (SFF) was formed at Fermilab with the purpose of extending the technology of plastic scintillator and plastic optical fiber. The original fiber R&D program at Fermilab is now continued under the auspices of the Scintillator Detector Development Laboratory (SDDL), and new work on different forms of plastic scintillator has emerged. We have developed three methods of scintillator manufacture, with the most recent method centering on a manufacturing process designed to produce large volumes of scintillator from commercially available polystyrene (PS).

The earliest method we have used, and notably the most likely to produce very high quality scintillator, produces cast shapes directly from monomer/dopant solutions which are carefully prepared and then polymerized under closely controlled cleanroom conditions. The second and somewhat more recent method is also a casting process, but begins with commercially available pre-polymerized pellets rather than monomer. By adding dopants to polystyrene pellets, we are able to quickly and inexpensively produce small scintillator samples suitable for optical and other laboratory tests. Finally, in an attempt to reduce cost and improve quality and consistency of very large ($>5 \times 10^5$ Kg) quantities of plastic scintillator, we have recently begun working on a third method involving commercial plastic processing techniques including compounding, pelletizing, and profile extrusion. There is hope this third method may reduce scintillator fabrication costs substantially. We believe the third method holds the most promise for producing very large quantities of detector grade scintillator.

PLASTIC OPTICAL FIBER-THE FACILITY AND PROCESS

In 1987 a small R&D program began at Fermilab within the Particle Detector Group. The original program intention was to investigate physical, chemical, and

CP450, *SciFi97: Workshop on Scintillating Fiber Detectors*
edited by A. D. Bross, R. C. Ruchti, and M. R. Wayne

optical properties of various fluorescent compounds and polymers that might be used in plastic scintillator. Early studies determined that commercially available monomers, polymers, and fluorescent compounds were often lacking in purity, and unsuitable for high quality scintillator use. A chemistry lab and cleanroom were subsequently constructed and utilized in a program to further understand and foster improvements in this area. Although originally envisioned as a research facility, the SFF also became a small-scale fiber preform (the scintillator rod from which fiber is drawn) production facility, initially producing over a million meters of scintillating fiber. The facility now continues as a research and development area.

The process that produces scintillating fiber preforms begins with preparation of the various chemicals used and requires substantial attention to detail if high purity preforms are to be obtained. Preform purity impacts directly on the quality of the eventual fiber and undesirable chemical impurities as well as excessive inert particulate contamination must be avoided. The facility itself must be carefully inspected and cleaned prior to each polymerization cycle, and then the styrene monomer and dopants to be used are purified. All equipment preparation, especially cleaning and assembly of the parts that come in contact with monomer and solvent, is carried out in the cleanroom. For this purpose, the cleanroom is outfitted with three large fume hoods; one serving as a distillation area, another for general cleaning and fluid transfer, and a third which houses the actual polymerization baths (Figure 1).

FIGURE 1. View of Scintillator Fabrication Facility cleanroom. This picture shows a mold set being lowered into the alcohol bath at the start of a freeze-pump-thaw cycle.

Styrene monomer preparation is the first step in the production of fiber preforms. The monomer is first deinhibited using an alumina packed column and is then vacuum distilled. Samples of deinhibited or distilled product may be taken for spectrophotometric or chromatographic analysis to determine purity. This purification step is critical to the success of the eventual scintillator. Dopants are prepared by recrystalization or other purification techniques, carefully weighed out, and are then dissolved into the monomer on the day of the polymerization run. A pressure vessel is used to accomplish the mixing step and also delivers the dopant/monomer solution through a series of filters to a set of fluoropolymer (FEP) lined cylindrical aluminum molds. Once filled the molds are placed in a low temperature bath of ethyl alcohol and evacuated in order to remove dissolved gasses from the monomer/dopant solution. The bath temperature is maintained at −90 C, just above the freezing point of the alcohol. The assembly is allowed to cool until the monomer/dopant solution solidifies and is then removed from the bath and warmed back to room temperature while pumping on the head space of the mold. This "freeze-pump-thaw" cycle is repeated several times. Next, the mold set is lowered into a heated oil bath and maintained at elevated temperature (110-140 C) for a period typically lasting several days. During this phase the monomer/dopant solution polymerizes and the molecular weight and polydispersity of the resultant polymer are fixed according to the heat profile and duration of the cycle. This polymerization cycle must be accurately controlled. Depending on initial bath temperature, the polymerizing monomer can produce significant exothermic heat about one hour into the polymerization cycle, and this heat must be removed in order to keep the mold and its contents at optimum temperature. Accomplishing this under some conditions is not trivial. A test was conducted using a mold instrumented with nine thermocouples to determine how serious a problem this might be. The experimental data indicate that the center of the preform can run 40 C hotter than the surrounding bath temperature, depending on the initial temperature of the bath and thus the initial rate of polymerization in the monomer. This large temperature difference as measured within the inner most part of the forming polymer core and the outer mold surface can yield large density variations in the finished preform which subsequently can affect fiber performance. In order to avoid this problem, the initial bath temperature is lowered thereby reducing the polymerization rate during the early part of the cycle. Temperature rampdown, especially through the glass transition ($\approx$ 110 C) of the polymer, is also critical. This step determines to a lesser but still significant extent, the eventual physical properties of the preform. Finally at the end of the heating cycle, the finished preforms are removed from their molds and samples are cut from the preform for documentation and later analysis.

When the preforms are removed, they are inspected with a 1-mW green (543nm) HeNe laser. The preform ends are diamond cut and the laser light is passed through the preform, along the axis, while looking by eye for light scattered by particles entrapped in the material. Slightly changing the entrance angle of the laser beam and scanning across the preform face gives a quick measure of the level of particulate contamination. We also use a setup that incorporates the laser, an optical chopper, reference and signal photodiodes, and a lock-in amplifier in order to measure precisely the Rayleigh scattering in the material. Preforms are next checked for roundness and then are ready to be drawn into fiber.

Drawing is an operation that must be done in a clean environment and with excellent tension and temperature control. The core/clad interface greatly influences the amount of light that can be efficiently piped in the fiber and surface contamination can be extremely detrimental to the cladding process. Lack of core uniformity over the length of the preform, including variations in molecular weight or residual stress, can be problematic in maintaining a consistent draw thereby yielding poor diameter control and possible fiber tensile strength problems.

Many manufacturers now produce high quality scintillating fiber. The present attenuation benchmark for this type fiber is 9-10 meters @ 525 nm wavelength as measured for undoped Kuraray multiclad fiber. Increasing use of plastic scintillating fiber has stimulated further research, and progress is being made in availability, quality, and cost.

LOW COST SCINTILLATOR PROGRAM-SCINTILLATING PELLETS

The advances in plastic optical fiber technology as applied to WLS fiber readout has launched a new generation of scintillator detectors, particularly in the area of calorimetry. Although cast plastic scintillator, usually polyvinyltoluene (PVT) or polymethylmethacrylate (PMMA), has been a mainstay in detector construction for decades, the cost for very large detectors using these materials becomes prohibitively expensive. WLS fiber readout has extended the applications in which plastic scintillator can be applied and in many instances has reduced assembly costs. The cost of the bulk scintillator, however, has remained high. In response to this cost problem, we began pursuing the idea that less expensive plastic scintillator could be produced using conventional plastic extrusion techniques and machinery. Discussions with vendors began in June of 1993 and the first successful attempt at producing scintillator by compounding scintillation and waveshifting dyes into commercially available polystyrene pellets using conventional extrusion equipment was accomplished in March 1994. Several different compositions have been produced using this method. The technology appears to be reasonable in cost, allows additional secondary operations, and could be widely adopted and used once standardized methods are established. The concept is promising and several groups are conducting experiments using pellets as possible detector scintillation material. The resultant material, however, is best suited for WLS fiber readout applications due to the relatively poor optical attenuation properties of the bulk polystyrene pellets as compared to conventional "cast-from-monomer" PVT, or PS scintillator. Table 1 indicates relative light yields for polystyrene scintillator samples with like geometry (2 cm cubes) and dopant systems but produced by different methods. All are referenced to Bicron BC 404, a commercially available PVT-based scintillator. The ratio between the light yield in otherwise identical PVT and PS scintillators has been reported[4] in the literature to be 1:0.8 and has been attributed to differences in the π-orbital structure of the two polymers.

TABLE 1. Relative light yield of scintillators

Sample/Method	Relative Light Yield
Bicron BC 404 (PVT)	1.0
Fermilab PS-404 (cast from monomer)	0.8
Fermilab 404-C (cast from previously compounded/extruded scintillator pellets DOW 70262)	0.8

We have tested many commercially available polystyrene pellets. Test samples are formed by casting the material of interest in test tubes and then heating and melting the contents inside a vacuum oven. Discs are cut from the resultant castings and these samples are subjected to optical, scintillation, and mechanical tests. Samples can be clear polymer or dopants can be added to produce scintillator. These test samples are used to select an initial set of materials to be used in extrusion studies. Two polymers, DOW XU 70262 and DOW Styron 663W have been found suitable for scintillator applications. Styron 663W is low-cost, readily available in commercial markets, and produces a good scintillator.

The concept of producing scintillator pellets using extrusion or compounding machinery has proven successful. This versatile material can be manufactured in small to very large quantities and yields quality high-light-output scintillator. Profile extrusion of previously compounded scintillating pellets has been accomplished, with shapes generally optimized for a particular detector geometry. Usually a hole or groove is included along one axis of the profile to contain a WLS fiber for collection of scintillation light produced in the profile. We have now used this process to successfully produce finished scintillator shapes over a wide range of production rates; from 30 to 200 pounds per hour (see Figure 2).

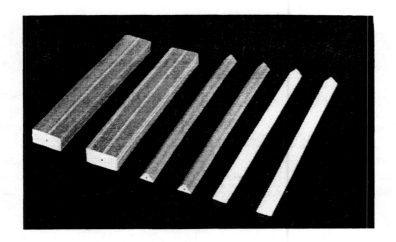

FIGURE 2. Rectangular (1 X 2 cm) and triangular profile extrusions with a hole down the axis for WLS fiber.

Production rates in the hundreds of pounds per hour range and above should markedly reduce costs, as well as provide higher quality material due to reduced residence time in the heated section of the line. Injection molding, a similar process, also uses scintillator pellets and is being used by others to produce "tiles" and various other shapes formerly machined from cast scintillator plate.[5]

This new method of scintillator manufacture uses existing equipment and technology with subtle variations from the industrial norm to make the difference between good scintillating material and just brightly colored plastic. Good quality scintillator manufacture using these methods will require some changes on the part of the processors. Companies that manufacture "clean" plastic parts, such as ophthalmic lenses, catheters, and other medical or optical devices have an advantage in that they understand contamination problems. These companies often produce parts in cleanrooms or controlled environments, providing a potential reduction in particulate loading with correspondingly less light scattering.

A recent concept that we have prototyped is the combining of the compounding step with the profile extrusion step, thereby creating a single-step process. Aside from a large cost savings brought about by the reduced handling, another potentially significant advantage of this idea would be the elimination of two heat histories and the possible loss of light yield in the scintillator. This may be critical for some polymer/dopant combinations, but would certainly be a good general practice. The process would take commercial polymer pellets and dopants and meter them through one extrusion line yielding a finished high quality scintillator shape. We have experimented with this concept and are continuing work in this area. Direct inline-doping at a concentration of approximately one percent (typical of the primary dopant in plastic scintillator) pushes the limits of ordinary extrusion practice, but products are made using similar techniques and therefore we have hope this idea may yet be achieved.

SUGGESTIONS FOR IMPROVEMENT AND FUTURE R&D

Plastic scintillator has improved in a variety of ways over the last decade. New manufacturing processes have been identified and material quality, availability, and new detector readout geometries have all significantly progressed. Lowering the cost while still producing a high-quality scintillator remains a goal. Table 2 indicates known costs for commercial PVT scintillator sheet vs. estimated and projected costs for finished profiles of PS scintillator using a similar dopant system, but produced by new methods.

TABLE 2. Cost comparison of conventional cast scintillator to extruded scintillator.

Manufacturer/Method	Cost
Bicron Cast Sheet: BC-404	~$40/Kg
Extruded Profiles from scintillator pellets	$5-6/Kg
Extruded Profiles using direct in-line doping	$3.5-4.0/Kg

Part of the challenge in continuously producing low cost quality extruded scintillator will be to understand the intuitive aspect of this technology and convert it to science. Assuming very high production rates become possible, online quality control for profile size and scintillator optical characteristics will become essential. Although very high rates have the potential to reduce scintillator costs substantially, we must be certain the parts are usable.

Finally, cultivating vendor resources, especially large polymer and chemical (dopant) manufacturers has potential for delivering advances in scintillator. Polystyrene pellets, for example, sometimes exhibit quality or consistency differences from batch to batch. Cooperation from suppliers will help us understand whether this can be resolved or needs to be factored in. Hopefully, as this new scintillator technology grows and vendor channels are established, material supply conditions will improve.

REFERENCES

1. Bross, A. D., *SPIE Plastic Optical Fibers* **1592**, 122 (1991).
2. Ruchti, R.C., *Annu. Rev. Part. Sci.* **46**, 281 (1996).
3. Bross, A.D., Ruchti, R.C., Wayne, M.R., Eds. *SCIFI 93 Workshop on Scintillating Fiber Detectors*; World Scientific: Singapore, 1995.
4. Birks, J.B., The Theory and Practice of Scintillation Counting, MacMillan, 1964.
5. Ron Richards, *these Proceedings*

Study of Light Transmission Through Optical Fiber-to-Fiber Connector Assemblies

M. Chung, M. Gutowski, M. Adams, and J. Solomon

Department of Physics, University of Illinois at Chicago
845 West Taylor Street, Chicago, Illinois 60607-7059

Abstract. Optical fiber-to-fiber connectors are now being used widely in particle tracking detectors. We describe the properties of the connectors, their production, and measurements of the light transmission through the gap of the connector assembly. We studied light transmission for various types of connectors illuminated by several different light sources. The light transmission was found to be dependent on the angular distribution of the light rays passing through a connector assembly. Two arrangements were studied, a point source and a diffuse source. A green LED with a diffuser is believed to best reproduce the angular distributions of light in the real detector applications. We also studied the transmission as a function of the index of refraction of the optical couplants. The light transmission depends on the index of refraction of an optical couplant placed in the gap, and improves as it approaches the index of refraction of the fiber core. Light transmissions of 80% ~ 88% were obtained without any optical couplant in the connector gap and transmissions of 89% ~ 99% with various optical couplants. A Monte Carlo study using measured light distributions from a fiber end produced a reasonable agreement with the transmission measurements made on a connector assembly.

1. INTRODUCTION

Optical fibers made of plastic are widely used in particle tracking detectors. Among the detectors under construction or development are the central fiber tracker (1) and the central and forward preshower detectors (2) by the D0 collaboration at Fermilab, and the hadron calorimeter of the CMS collaboration at CERN. Light passes through a series of lightguide fibers and one or more connector assemblies, and eventually reaches a photodetector. An important aspect of the work is to maximize the light transmission through the gap of a connector assembly (3, 4). For detectors that require many multichannel connectors, quality control of connector production is also very important.

We have developed, produced, and made transmission measurements for many different connectors. This paper presents our measured results and compares them to a theoretical estimate using a simple Monte Carlo program for light transmission through a connector assembly.

2. CONNECTORS

The connector bodies can be made of various materials. We have made measurements of connectors made of aluminum, epoxies, and various plastics. The plastics include Noryl, Delrin, ABS (acrylonitrile butadiene styrene), Lucite, and polyurethane.

CP450, *SciFi97: Workshop on Scintillating Fiber Detectors*
edited by A. D. Bross, R. C. Ruchti, and M. R. Wayne
© 1998 The American Institute of Physics 1-56396-792-8/98/$15.00

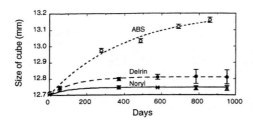

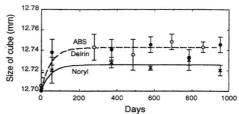

FIGURE 1. Plastics stored in ethyl alcohol.   **FIGURE 2.** Plastics stored in water.

Initially we tried machining the connectors. This was a tedious process and we also found we had difficulty getting the connectors to mate precisely. We had more success using injection molding. Our preferred material was ABS plastic because it shrunk only 0.5% when removed from the mold. Connectors, with as many as 256 fibers, could be aligned using either holes or v-grooves. The mold made of aluminum could process up to a thousand connector pairs before it had to be remade.

The long-term aging of plastic materials used in connectors was monitored. Cubes of Noryl, Delrin, and ABS plastics, 12.7 mm x 12.7 mm x 12.7 mm, were placed in five different media -- air, optical grease, mineral oil, water, and ethyl alcohol -- and their sizes were monitored. The results are shown in Figs. 1 and 2. The most significant change was observed for ABS plastic cubes stored in ethyl alcohol. In the other four media, cube sizes increased less than 0.05 mm (0.4%) over a period of three years.

Within the connector assemblies Kuraray-made multiclad fibers are being used in the Fermilab D0 upgrade detector, both in the scintillating fiber tracker and in the preshower detectors. The fibers have a core and two claddings. The core has a refractive index of 1.59, an inner cladding of 1.49, and an outer cladding of 1.42. All the fibers studied had diameters between 0.835 mm and 1 mm. Kuraray quotes the variation in the diameter of fibers to be within $\pm 1\%$ of the nominal size.

In most of our measurements, the fibers used in the connector assemblies were short, 40 to 100 cm. For such short fibers the light passing through the outer cladding is a significant portion of the total light transmitted. However, for 3-m-long fibers we found this light was negligible.

3. MEASUREMENTS OF LIGHT TRANSMISSION

In choosing a light source we attempted to find one which best reproduced the behavior in the real detector. The variables we investigated were the wavelength of the light source and the angular distribution of light produced by the source.

In the course of the study, several different light sources were tried; such as LEDs with different colors from blue (450 nm) to infrared (880 nm), and a UV source shined on a 3-hydroxyflavone (3HF) scintillating fiber from the outside. No significant dependence of light transmission on the wavelength of light was observed.

The transmission through the connector gap depends on the angular distribution of the transmitted light which depends on the type of the light source and on the distance that the light has traveled in the fiber. There are marked differences in the angular distributions between a diffuse source and a point source. For the diffuse source, we used a bargraph green LED with a diffuser in contact with the fiber end, and for the point source, a surface-mount green LED with no diffuser was placed at least 2 mm from the end of the fiber.

114

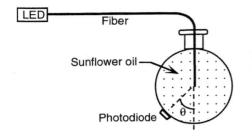

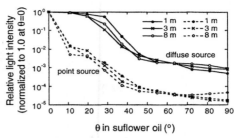

FIGURE 3. Setup to measure the light distribution.

FIGURE 4. Angular distributions measured in sunflower oil.

We measured the light distributions emerging from the end of a fiber immersed in a round beaker filled with sunflower oil, as shown in Fig. 3. The sunflower oil was used because it has the index of refraction close to that of the fiber core. Figure 4 shows the results of the measurements for the two light sources and three different fiber lengths. The two sources produced significantly different angular distributions with the light from the point source having a greater axial component than the light from the diffuse source.

In two different measurements, where light in the fiber is caused by a completely external source, we found behaviors which more closely resembled the diffuse source than the point source. 1) Using a ^{207}Bi source on a 3HF fiber and a photomultiplier tube we obtained relative light transmissions (77% ~ 87%) closer to what we measured for the diffuse source (79% ~ 89%) than what we measured for the point source (84% ~ 93%). 2) The attenuation lengths of clear fibers measured using a UV-illuminated 3HF fiber (8 m ~ 13 m) were more comparable to what we measured for the diffuse source (8 m ~ 13 m) than what we measured for the point source (13 m ~ 20 m). These agree with the conjecture that the angular distribution of light in the real detector application is broad and more closely resembles that of the diffuse source.

To study the relationship between the light transmission and the index of refraction of the optical couplant used in the connector gap, different optical couplants have been tried, and are listed in Table 1. Figure 5 is a sketch of the setup used to measure the light transmission. Transmissions of 80% ~ 88% were measured without any optical couplant in the connector gap and transmissions of 89% ~ 99% with various optical couplants. Tables 2 and 3 summarize our measurements and conclusions for the various connector assemblies.

Long-term measurements of light transmission through some of the connector assemblies with couplants are shown in Fig. 6. The variation of light transmission was less than 5% for connector assemblies which have been stored at room temperature over a period of three years.

TABLE 1. Optical couplants used in measurements.

Optical couplant	Index of refraction	Optical couplant	Index of refraction
None (air gap)	1.000	Silicone rubber index matching film	~1.4
Water	1.334	Dow Corning Q2-3067 optical grease	1.465
Ethyl alcohol	1.361	Soy bean oil	1.474
General Electric Viscasil 600M grease	1.404	Sunflower oil	1.475
Mineral oil, Osco baby oil	~1.4	Anethol[a], Good Scents Co. RW101383	1.558

[a] Anethol gave a high transmission of around 96% using the diffuse source, but it severely dissolved fiber ends in minutes. Other couplants in the table did not damage the fiber.

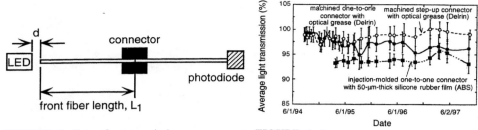

FIGURE 5. Setup for transmission measurement. **FIGURE 6.** Long-term measurements.

TABLE 2. The best values of light transmission in percent as an average of 16 or more channels, measured for connector assemblies with optical grease. The notation used indicates fiber diameter-to-diameter in microns, type of fiber (S: S type, N: non-S type) and connector body material (v: v-groove).

%	Diffuse source	%	Point source
99	• 965-to-965 clear-S, Delrin-v	99	• 965-to-965 clear-S, Delrin-v
	ø 835-to-965 3HF-N/clear-S, Delrin-v		• 965-to-965 clear-N, polyurethane
		98	• 835-to-835 clear-S, aluminum-v
96	• 1000-to-1000 clear, Delrin	96	• 835-to-835 clear-N, aluminum-v
95	ø 835-to-965 clear-S, epoxy	95	• 835-to-835 clear-S, ABS
	• 925-to-925 3HF-N, Delrin		
94	• 965-to-965 clear-N, polyurethane	94	• 835-to-835 clear-S/N, ABS
92	• 835-to-835 3HF-N, ABS-v		
91	• 835-to-835 clear-S/N, ABS		
89	• 835-to-835 3HF-N, aluminum-v		

TABLE 3. Variables that determine the connector transmission (with a few percent error).

Variable	Tried	Dependency
Fiber type	Clear, 3HF, Y11, S/non-S.	None
Connector body material	Delrin, ABS, Al, polyurethane, etc.	Δ(Delrin - others) = 0 ~ 5%
Connector type	Holes, v-grooves.	None
Type of glue used	BC600 epoxy, 5-minute epoxy, etc.	None
Wavelength of light	Blue to infrared LEDs, UV-3HF.	None
Light angular distribution	Diffuse and point.	Δ(point - diffuse) = 0 ~ 5%
Front fiber length[a]	0.2 m ~ 1 m, 3 m, 8 m.	Δ(0.2 m - 3 m) = 0 ~ 3%
Surface finishing	Diamond bit, sand paper, razor blade.	Diamond bit finishes the best.
Misalignment[b]	0 ~ 0.1 mm.	Lose 3~4%/25 μm of misalignment.
Air gap distance	0 ~ 0.5 mm.	Lose 2%/25 μm of air gap distance.
Optical couplant	With and without.	Δ(with - without) = 6% ~ 10%

[a] L_1 in Fig. 5.   [b] Step-up in fiber diameters is advantageous.

4. CONNECTOR PRODUCTION BY INJECTION MOLDING

The following describes the production of a pair of connectors designed for the central preshower detector as part of the D0 upgrade. Each connector has 16 fiber holes and is 22 mm x 4 mm x 6 mm. The production mold consists of three 130 mm x 130 mm x 65 mm aluminum blocks with 0.86-mm-diameter steel gauge pins used for the fiber holes and larger diameter steel pins, 2.1 mm and 2.4 mm, used for the alignment-pin holes. Two alignment pins have a hex head in order to keep the pins from rotating within the connector. Black ABS plastic is the material used. It took about two minutes to make one pair of connectors, and one mold was used to make a thousand pairs.

116

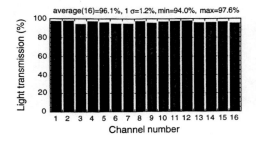

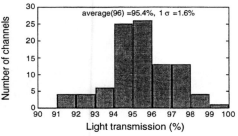

FIGURE 7. A sample of measurement with optical grease using a point source.

FIGURE 8. Distribution of the measured light transmissions.

The following procedure was used to assemble the connectors for testing the light transmission. A number of connector pairs were randomly picked from different batches. Kuraray multiclad clear fibers, 0.835 mm diameter, were cut in 19-cm-long pieces. All the fiber end faces were finished using a diamond fly-cut tool before assembly. With the connector front face secured by a piece of double-sided tape to a table, the fibers were stuffed into the holes from the glue pocket side and Bicron BC600 epoxy (5) was applied. The connectors were left overnight for curing. Optical grease, General Electric Viscasil 600M, was then applied on the face of one connector. The two mating connectors were matched with the two alignment pins and assembled using two socket-head screws.

To check the performance of the connector assemblies, a surface-mount green LED was used as a light source. The LED was placed at 1.5 cm from the front end of each fiber. The other side of the fiber channel was put into a holder of a silicon photodiode. The output of the photodiode was measured using a Keithley 485 picoammeter. The light transmission through each channel of the connector assembly was normalized to the average transmission through 10 continuous 38-cm fibers. Figure 7 shows the measured transmissions for a connector assembly with optical grease. All the measured channels yielded an average of 95%. Their distribution is shown in Fig. 8. The average value of misalignment is estimated to be 25 μm or less. The variations between fiber channels are small with a standard deviation of about 1.6%.

To simulate final detector conditions, the connectors were reassembled without the optical grease. A green LED bargraph with a diffuser at front was used to estimate a realistic value of light transmission in the preshower detector. The achieved average of transmission was 81%.

5. MONTE CARLO CALCULATION

In order to predict the light transmission through a connector assembly having an optical couplant, we wrote a simple Monte Carlo program in BASIC. A light ray exiting a fiber passes through the couplant medium and into a mating fiber. The fibers are assumed to be completely a core of polystyrene. It is assumed that there is a 1:1 ratio of transverse electric and transverse magnetic components of light. The reflection coefficients at the two boundaries are calculated. Total transmissions of individual light rays are summed up to evaluate the overall light transmission through a connection. For the distribution of light exiting the front fiber, we used the measured distribution of section 3 projected backward through the oil to the fiber core.

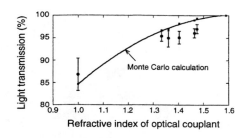

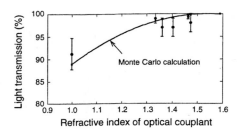

FIGURE 9. Diffuse source. **FIGURE 10.** Point source.

Figures 9 and 10 show the results of the calculations along with the measurements for a 0.835 mm-to-0.965 mm step-up connector assembly with 3-m-long fibers. Solid circles are the measurements with ± 1 σ error bars.

6. SUMMARY

We found that the light transmission depends on the angular distribution of light at the connector gap. The light distributions at the end of a fiber with the two different light sources, diffuse and point, were measured. The point source has a greater axial component. The diffuse source is believed to best reproduce the connector transmission in the real detector applications.

Optical couplants increase the light transmission by 6% ~ 10%. Light transmission using a diffuse source with a misalignment less than 25 µm is 80% ~ 88% with no optical couplant and 89% ~ 99% with an optical couplant.

By injection molding a thousand pairs of connectors were made. An average of 95% for light transmission was measured using a point source and with an optical grease in the gap.

Our Monte Carlo calculations based on the measured distributions of light agreed reasonably with the measurements made on the connector assembly.

ACKNOWLEDGMENTS

This work has been supported in part by the U.S. Department of Energy and the U.S. National Science Foundation. The authors thank all the people in the D0 collaboration who have helped the study. They also thank Carl Lindenmeyer and Masa Mishina (Fermilab) for generous advice throughout this work and Kevin Lynch (University of Illinois at Chicago) for constructing the molds.

REFERENCES

1. Baumbaugh, B. *et al.*, *IEEE Trans. Nucl. Sci.* **43** (3), 1146-1152 (1996).
2. Adams, M. *et al.*, *Nucl. Instr. and Methods in Phys. Res.* **A 378**, 131-142 (1996).
3. Aota, S. *et al.*, *Nucl. Instr. and Methods in Phys. Res.* **A 357**, 71-77 (1995).
4. Chung, M., and Margulies, S., *IEEE Trans. Nucl. Sci.* **43** (3), 1153-1156 (1996).
5. Born, M., Chung, M., and Margulies, S., D0 Note 1776, 29 June 1993.

Scintillating Fiber Ribbon Development for the D0 Upgrade

Erik J. Ramberg

Fermi National Accelerator Laboratory, Batavia, Illinois 60510

Abstract. As part of general upgrade of the D0 detector, a scintillating fiber tracker is being developed at Fermilab. This note describes the development of ribbons of scintillating fibers as the first step in production of the active element of this tracker. Methods for testing the quality of these ribbons and of mounting them on the support structure are briefly described.

THE D0 FIBER TRACKER

The D0 collaboration is developing a significantly challenging fiber tracker as part of its general upgrade of the detector. [1] The fiber tracker surrounds the silicon vertex detector and is in turn surrounded by a superconducting solenoid magnet, to provide momentum analysis. The tracker consists of 8 concentric double walled carbon fiber cylinders with scintillating fibers glued to their surface. The two innermost cylinders are 1.66 meters in length and the outer 6 cylinders are 2.52 meters in length. Half of the fibers on the cylinders will be coaxial with the beam direction and half will be mounted at a slight angle (approximately 2 degrees) to give stereo resolution. The fibers are double-clad Kuraray fibers of .835 mm diameter and contain 3HF wave shifter. [2] The wave shifter creates photons of 530 nm wavelength, which propogate through clear fiber light guides to a VLPC readout system. [3]

Because of the obvious advantages to working with collections of fibers instead of with individual fibers, the first step in creating the active detector element is to make "ribbons" of fibers, which can be grouped together into a single connector and mounted on the cylinders as a unit. This note describes the development of techniques for creating these ribbons and the impact of their design on the construction of the D0 fiber tracker.

CP450, *SciFi97: Workshop on Scintillating Fiber Detectors*
edited by A. D. Bross, R. C. Ruchti, and M. R. Wayne
© 1998 The American Institute of Physics 1-56396-792-8/98/$15.00

FIBER RIBBON CONSTRUCTION

There are a total of 76,800 fibers in the D0 fiber tracker. These fibers are formed into ribbons consisting of two layers of 128 fibers each, with the layers offset by a half fiber spacing. There are two layers of ribbons on every cylinder - the coaxial and stero views mentioned above. For triggering and analysis purposes, it is important to keep the final position of the fibers systematically controlled to the order of 50 microns. Because we expect deviations of placement of fibers to occur at every step of the construction and mounting process, a specification was made that the fibers be held within ribbons to an RMS deviation from expected position of 20 microns.

The mold to make fiber ribbons is constructed by first accurately machining 128 flat bottomed grooves in a 60 mil thick plastic Delrin sheet. This sheet is then glued to a set of accurately machined curved aluminum blocks of the appropriate radius and length. (The blocks are attached to each other to form one long mold.) This finished mold is then used in making the ribbons, testing them and in installing them on the appropriate cylinder.

Because we are reading out light from only one end of the fibers, the other end is polished and aluminized. To form the ribbon, a group of 128 aluminized fibers is hand placed within the grooved mold and checked by eye to make sure there is a single layer of fibers registered appropriately everywhere in the mold. A thin layer of glue is then spread evenly over this set of fibers. Another 128 fibers are then laid in by hand on top of this glue layer. Their registration arises because they naturally fall within the interfiber spaces of the first layer.

Through systematic testing of many types of glue, it was determined that Ciba-Geigy RP 6400 polyurethane glue gives a superior bond between fibers and dries to a relatively flexible, yet quite strong film. The glue seeps between fibers of both layers. Pull tests indicate that it takes 3-6 grams of force to liberate a fiber from the ribbon when constructed in this way. It has been determined recently that a very good way of ensuring that the final ribbon releases from the mold is to vacuum clamp a thin 1 mil layer of Teflon film into the mold grooves. After drying for at least 4 hours, the ribbon can be carefully released from the mold and can be handled with moderation. Measurements of fiber spacing on ribbons constructed in this fashion and unconstrained, yield an RMS deviation from absolute expected position of better than 15 microns, well within our tolerance.

The next step in making a finished ribbon is to attach a connector to it and polish the final product. In our case, because the fibers make a complete cylinder, it is quite difficult to spread them out and insert them into a connector with holes in it. Instead, we are developing a curved connector with grooves in it, and clamping the fiber ribbon in this set of grooves. An aluminum connector of this type has shown adequate light transmission. Currently we are trying to develop a similar connector made of Torlon, one of the strongest plastics available. The face of the connector, with fibers in place, is polished as a unit by a diamond fly cutter.

QUALITY CONTROL

After construction, it is important to test each ribbon for quality. This includes such factors as measuring the position of each fiber, light yield, attenuation length and reflectivity from some subset of fibers, and average transmission of the connector. We are developing a test machine to perform these measurements. It consists of a Co-57 X-ray source (energy of photons=136 KeV) on a platform that can scan along a circular arc at a short distance above the ribbon. This platform can also scan along the length of the fiber. The output of each fiber is multiplexed to 16 high gain phototubes for readout through a CAMAC data acquisition system. This energy of X-rays not only simulates well the average light yield of a minimum-ionizing particle, but is an optimum energy for illuminating both layers of fibers simultaneously. Furthermore, an X-ray source has the advantage that measurements can be taken with the ribbon in the same mold that it was constructed in and even with the ribbon in place on the carbon fiber support cylinders. [4]

FINAL INSTALLATION OF RIBBONS

After construction of the ribbon and installation of the curved toothed connector and its subsequent polishing, the next step is to attach ribbons to the support structures. The same polyurethane glue that was used to make the ribbons is used to glue them to the cylinders. A machine has been designed and is currently being fabricated to accurately place ribbons at any desired radius. The ribbon is placed in the mold that was used to fabricate it and the mold is then placed on top of a finely controlled vertically moving platen. One end of the appropriate carbon fiber cylinder is mounted on a rotary table and the other on a freely moving spindle. This serves as an axis for turning the cylinder so that the rising ribbon is glued accurately in place.

CONCLUSION

The D0 collaboration is developing a very sophisticated scintillating fiber tracker that is at the center of the general upgrade of the detector. We have developed methods to create coherent, stable ribbons of these fibers in order to manage their production and installation on their support structures. Machines have been designed and are currently being built to test the accuracy of placement of fibers within the ribbons and to accurately mount the ribbons to form the detector.

REFERENCES

1. For a detailed description of the central fiber tracker, see "The D0 Upgrade Central Fiber Tracker Technical Design Report".

2. Kuraray International Corporation, 200 Park Avenue, New York, NY 10166

3. For a review of scintillating fibers, see R. Ruchti, Annu. Rev. Nucl. Part. Sci. 1996; 46; 281-319

4. "Scintillating fibers position measurement in curved ribbons using an x-ray source", G. Guttierez, D0 Note 3380

SESSION 4: PHOTOSENSORS - I

Chair: R. Rusack
Scientific Secretary: H. Zheng

Advances in Vacuum Photon Detectors for High Energy Experiments

Katsushi Arisaka

Department of Physics and Astronomy
University of California at Los Angeles
Los Angeles, California 90095

Abstract. Recent progress on vacuum based photon detectors is reviewed. Even after rapid progress on various solid state detectors, vacuum based photon detectors such as photomultipliers still play a significant role in many high energy experiments where high speed detection of weak photon signals is critical. The latest development of hybrid devices such as HPD, HAPD, EBCCD and ISPA are reviewed and their advantage over conventional photon detectors is discussed in some detail.

INTRODUCTION

Primary purpose of detectors in high energy experiments is to measure the physical quantities such as energy, momentum, time and position which are carried by elementary particles. These quantities are obtained by interactions of particles in detector materials. Traditionally, scaler quantities, such as energy and time, are measured by means of fast light signals like scintillation or Cherenkov radiation, whereas vector quantities, such as momentum and position, are measured by local ionization, using gaseous detectors for example.

These detector choices are motivated by the fact that light emitted by the particle interactions is very fast (for time measurement) and very linear (for energy measurement) but hard to localize, while the ionization can be localized (for precise position and momentum vector measurement) but is a rather slow process.

In a photon detector, the light signal has to be converted into an electronic signal in order to be measured. On the other hand, local ionizations directly produce electric signals, so that photo-electric conversion is not necessary. In most of traditional applications, conventional photomultipliers for light signals and drift chambers for ionization signals are adequate.

Recent and future high energy experiments are, however, becoming more demanding in detector performance for several reasons:

1. The event rate is much higher (as high as MHz) and slow response of the drift chamber is no longer acceptable.

CP450, *SciFi97: Workshop on Scintillating Fiber Detectors*
edited by A. D. Bross, R. C. Ruchti, and M. R. Wayne
© 1998 The American Institute of Physics 1-56396-792-8/98/$15.00

2. Operation under strong magnetic field (as high as 4 Tesla) requires special photon detectors which are immune to the magnetic field.

3. For better particle identification (K/π separation up to 4 GeV/c), detection of local Cherenkov light is of critical importance.

4. For neutrino experiments, detector volume is huge (as large as 50k tons), and the number of channels is large (more than 10k channels).

As a result, various new photon detectors have recently been proposed and developed to satisfy these requirements. The purpose of this paper is to introduce and to compare these new developments in a systematic way, so that anyone in high energy experiments can decide the most suitable photon detector for their experiments in an unbiased manner.

In the following section, I first introduce various photon detectors, starting from the conventional ones. Once all the photon detectors are briefly introduced, resolutions on time measurement, energy measurement and position measurement are systematically compared for all photon detectors. Finally, as a critical ingredient of decision making, market prices of these detectors are shown.

Based on this analysis, advantages and disadvantages of various new types of photon detectors are discussed.

TYPE OF PHOTODETECTORS

A   Conventional photo detectors

Generally speaking, there are two types of photon detectors: vacuum based and solid state. A vacuum based photon detector has a photocathode which converts photons to electrons. Photo-electrons are emitted into the vacuum and hit dynode structure for fast multiplication of signals. The first process, photo-electron conversion, has rather poor efficiency (so called Quantum efficiency is typically 20%), but the rest of the process has no additional noise factor. Thus we can count the number of emitted electrons. This process is extremely fast, which is the reason why conventional photomultipliers are still widely used in high energy experiments.

On the other hand, in solid state detectors, photons are absorbed by $p - n$ junction, where electron-hole pairs are created. The quantum efficiency is very high (close to 100%), but the intrinsic noise of the device itself limits the detection of low light signals. Recent development of fine lithographic technology allows one to develop finely segmented photo detectors like CCD, which are extremely powerful for imaging purposes.

The basic properties of typical conventional photon detectors such as photomultiplier, Photo Diode and CCD etc are summarized below.

1 Photomultipliers

As mentioned already, photomultipliers are sill one of the most widely used photon detectors in high energy experiments. Here, I briefly list the latest developments.

1. Improved magnetic-field immunity: For operation under strong magnetic field (up to 1.5 Tesla) fine mesh dynode has become the standard PMT. Examples of experiments using such PMT's are KLOE at DAFNE, H1 at ZEUS, BELLE at KEK B- Factory [1].

2. Larger sensitive area for neutrino experiments: New generation of neutrino experiments require huge detector volume, viewed by large PMTs (as large as 20 inch diameter.) A well known example is the Super-Kamiokande where 11,200 of 20 inch PMTs are installed in a 50k ton water tank. The rest of neutrino experiments have adopted 8 inch PMTs so far. There are many examples: AMANDA, MILAGRO, LSND, SNO etc.

3. Pixelization for tracking: As a replacement of drift chambers, the scintillating fiber tracker is becoming a mature technology. For readout of fibers, multi-pixel photomultipliers have been developed. Philips first developed XP1700 series using foil dynodes. Later Hamamatsu developed various types; starting from mesh dynode and Venetian blind, and recently metal channel plate [2]. But the quantum efficiency remains poor. This problem can be solved by using VLPC's, a solid state device descibed later. For the preshower or shower maximum detector, multi-pixel PMTs have been adopted. The CDF plug upgrade is such an example.

4. Position sensitivity for Particle ID: For particle identification, RICH (Ring Imaging Cherenkov) is becoming standard, in particular, for the CP violation experiments in the B decays. This requires, again, well segmented, pixelized photo detectors. There is no single standard photon detector yet, so multi-pixel photomultipliers are one of the most practical options. Such an example includes HERA-B.

In summary, even though the progress in this field has not been revolutionary but rather evolutionary, contributions of a variety of newly developed photomultipliers to the new generation of high energy experiments are quite impressive. Part of the reason is their extremely reliable, maintenance-free operation, thanks to high quality control by the venders.

2 Photo Diode and APD

PIN photo diodes are also widely used in precision calorimeters, where the number of photons is large enough. Examples are BGO in L3, CsI(Tl) in CLEO-II,

BaBar and BELLE. It is shown in Section E that once the number of photons become greater than 10^6, a photo diode gives better energy resolution than a PMT. Besides, it is totally immune to the B-field.

In the future experiments, radiation hardness of crystals is another important issue. $PbWO_4$ was chosen by the CMS for the EM calorimeter for this reason. Photon yield of this crystal is quite low, so additional intrinsic gain is required for photon detection. Due to strong magnetic field (4 Tesla), conventional vacuum photon detectors are no longer allowed. Avalanche Photo Diode (APD) is currently under development [6–8]. Making a large area, yet stable APD is unfortunately non trivial, and it may take another year or two to develop reliable ones for use in the high energy community.

3   CCD and ICCD

A CCD itself has much wider applications not only in scientific but also in commercial applications as a standard imaging device of today's digital world. In high energy, it is primaly used as a tracking device by directly detecting ionization (in the SLD for example.) Once it is combined with an Image Intensifier, it can serve as a readout for high resolution scintillating fiber tracker, as long as the readout speed is not an issue. Neutrino oscillation experiments such as CHORUS and K2K adopted this approach.

B   New photo detectors

In this section, several recently developed photon detectors are reviewed.

1   VLPC/MRS

The Visible Light Photon Counter (VLPC) is an impurity conduction band device [11], which generates an avalanche like APD, But the avalanche region is very limited and due to space charge effect, the signal is saturated after $\sim 10^5$ electrons are generated. This mechanism quantizes the output pulse charge unlike APD, allowing photon counting. In other words, it has high quantum efficiency of a solid slate device as well as high gain and small excess noise of a photomultiplier. This ideal performance, however, requires low temperature ($\sim 6^o K$) operation.

Similar conditions can be realized at room temperature, if saturation can be achieved locally. The Metal Resistive Semiconductor(MRS), recently developed by Russian groups, has needle structure in PIN junction where electric field is locally realized to generate saturated avalanche [9,10] This device, however, seems to have rather large thermal noise and poor rate capability.

2 HPD/HAPD

Conventional vacuum photon detectors have metallic dynode structure where secondary electrons are emitted from the surface. Instead of metal, one can incorporate solid state devices as a dynode, where photo-electrons are directly captured internally. Such a device, HPD (hybrid Photo Diode) was originally proposed by R. Desalvo in late 80's and finally it has become a mature technology to be adopted by some of the major high energy experiments [3,4]. DEP is leading this effort and several different types are now commercially available. The current version has either single or pixelized photo diode. By pixelizing photo diode, the cost per channel can be reduced. Major advantages of such a device are as follows.

1. Immunity to magnetic field. As long as the tube axis is aligned to the B-field axis, this device is not sensitive to the field at all.

2. Very small ENF(excess noise factor). As shown later, due to large (> 1000) multiplication factor of photo electrons, ENF is almost unity.

3. Wide dynamic range. It has the same dynamic range as photo diode. Unlike PMT, there is no space charge effect which limits the linearity.

4. Uniform Gain. The gain of this device is given by the number of electron-hole pairs produced by the kinetic energy of a photo electron. Since the energy is simply given by electric potential for acceleration, gain uniformity is guaranteed.

One should however note that it still suffers from poor quantum efficiency as a vacuum device, and rather low gain makes detection of single photo-electron very difficult, unless long shaping time is allowed.

Once the photo diode is replaced by the APD, single photo-electron detection can be easily achieved, even with modest gain. Such a device is called HAPD (Hybrid APD), described below.

3 EBCCD

The electron bombarded CCD (EBCCD) is another type of the hybrid device where photo-electrons are directly bombarded into the CCD [14]. Compared to the conventional Image Intensifier where photo electrons are multiplied by the MCP and captured by the phosphor screen, there is no extra materials between photo cathode and CCD. Therefore, distortion of image and loss of linearity can be minimized.

4 ISPA

Although the EBCCD has a good image resolution, the speed is still limited due to slow readout of the CCD. This can be overcome the ISPA (Imaging Silicon Pixel

Array) which incorporates a silicon pixel detector in the vacuum instead of the CCD [13]. The Silicon Pixel detector has parallel readout chain, so this would be an ultimate photon detector for fast, precise imaging.

SYSTEMATIC COMPARISON

The primary purpose of a photon detector is to convert information carried by light to electric signals. Generally speaking, a light signal carries the following pieces of information: time, intensity (the number of photons), position, and wave length.

In most of high energy applications, wave length is not important, as it is not related to the property of particles to be measured. Therefore, I will focuse in this section on three quantities, time(T), intensity(I) and position(X).

Attention must be paid to the choice of units before we go further. A wide variety of photon detection is done in frequency domain, where signal is modulated by carrier waves. To make the analysis method simple, I will only consider time domain approach, where signal is assumed to be produced and detected as an event in time.

Then the unit of intensity can be given by the number of photons per event per unit area, where unit area is typically the sensitive area of the given photo detector. The unit of time can be the typical width of the signal, and the unit of position is the typical size of photon bucket when it arrives at the surface of photon detector.

C   Sensitivity and Dynamic Range

Let's move on to the sensitivity and the dynamic range of each photon detector in Time(T), Intensity(I) and Position(X). Table 1 summarize various photon detectors and their range of detection capabilities. Here I categolized devices into three groups; vacuum, solid state and the hybrid which is the combination of the both.

The rest of this paper is devoted to a more detailed analysis of why each detector has the sensitivity shown in Table 1. Here, as an introduction, I just summerize several key characteristics of each detector.

1. Vacuum based detectors are generally more sensitive to weak light signals and better in time resolution, thus, more suitable for detecting photons in high energy experiments.

2. Conventional solid state photon detectors are more suited for large photon signals. CCD is particularly optimized as a replacement of human eyes.

3. Hybrid devices have better dynamic range in detecting wide range of photon intensity.

Name	T(Min)	T(Max)	X(Min)	X(Max)	I(Min)	I(Max)
Human Eyes	$100msec$	$10sec$	$0.1mm$	$10cm$	10^4	10^{10}
Picture Film	$100msec$	$100sec$	$1\mu m$	$30cm$	10^2	10^5
Vacuum						
PMT	$300psec$	$1sec$	$5mm$	$50cm$	5	10^7
Pos. Sens. PMT	$300psec$	$1sec$	$1mm$	$5mm$	5	10^7
MCP PMT	$30psec$	-	$1mm$	$3cm$	10	10^4
Image Intensifier	$100psec$	$1sec$	$10\mu m$	$1cm$	10	10^4
Streak Tube	$1psec$	-	$0.1mm$	$2mm$	10	10^4
Solid State						
Photo diode	$1nsec$	$1sec$	$1mm$	$5cm$	10^4	10^{15}
APD	$1nsec$	$1sec$	$0.1mm$	$5cm$	10	10^{12}
CCD	$10msec$	10^3sec	$10\mu m$	$1mm$	10	10^5
Silicon Strip	$1nsec$	$1\mu sec$	$10\mu m$	$1mm$	100	10^6
Silicon Pixel	$1nsec$	$1\mu sec$	$10\mu m$	$0.1m$	10	10^5
Hybrid						
Photo diode	$1nsec$	$1sec$	$2mm$	$2cm$	5	10^{10}
APD	$1nsec$	$1sec$	$0.5mm$	$5mm$	5	10^8
CCD	$10msec$	10^3sec	$10\mu m$	$1mm$	5	10^4
Silicon Strip	$1nsec$	$1\mu sec$	$10\mu m$	$1mm$	5	10^5
Silicon Pixel	$1nsec$	$1\mu sec$	$10\mu m$	$0.1mm$	5	10^4

TABLE 1. Summary of some important properties of photon detectors.

Since various photon detectors and their covered ranges in three dimensional phase space in (T,I,X) is described, it is time to compare their resolution on these three dimensions. In the following sections, time, energy (i.e. intensity) and position resolutions are systematically compared.

D Time Resolution

One of the biggest advantage of the vacuum devices over solid state ones is their fast response. Thanks to the long distance between photo cathode and dynode structure, time jitter due to the intrinsic capacitance is almost negligible. The time resolution is mainly limited by the transit time spread (TTS). For large area photomultiplier, it is important to optimize photoelectron trajectory so that all electrons arrive at the dynode at the same time. In addition, time variation may be introduced by secondary electrons' spread inside of the dynode structure. This effect can be minimized by the fine dynode structure.

Such fine structure also allows to maintain position information as well. At the same time, since secondary electrons are well contained in a small area, the effect of external magnetic field can be minimized. Based on the above observations, many types of finely segmented dynode structures have been developed. Table 2 summarizes various dynode structures.

Generally speaking, the smaller the photosensitive area is, the better is the time

Name	Pitch	Max. B	Company
Box and Grid/Linear Focus	1cm	0.1Gauss	Many
Venetian Blind	2mm	1Gauss	Many
Foil Dynode	0.5mm	100Gauss	Philips
Metal Channel	0.5mm	100Gauss	Hamamatsu
Fine Mesh	20μm	1Tesla	Hamamatsu/Russia
Micro Channel Plate(MCP)	10μm	5Tesla	Many
Hybrid(PD/APD/CCD)	1cm − 10μm	> 10Tesla	DEP/Hamamatsu/API

TABLE 2. Various dynode structures.

resolution. For fair comparison of time resolution, in Figure 1, time resolution of various photon detectors are plotted as a function of sensitive area.

As one can see from Figure 1, per unit area, time resolution of a typical vacuum device is an order of magnitude better than that of a typical solid state device. That is why the old-time photomultiplier is still alive today.

Next, Figure 2 shows the effect of magnetic field on various photon detectors.

Conventional photomultipliers can be operated in magnetic fields of up to 10 Gauss or so. In stronger magnetic fields, finer dynode structure is required. The

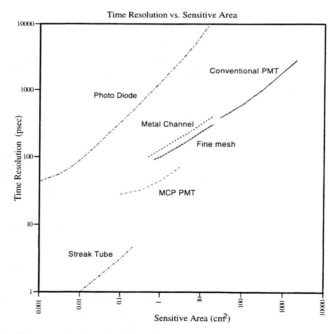

FIGURE 1. Time resolution of various photon detectors as a function of sensitive area.

finer the structure is, the less sensitive is their performance to the magnetic field. Needless to say, a solid state device is completely unaffected by a magnetic filed. A hybrid device is also insensitive to the magnetic field as long as its axis is aligned with the direction of the field.

E   Energy Resolution

In the ideal case, energy resolution is governed by the simple poisson statistics of the number of incident photons (N), given by,

$$\frac{\sigma}{E} = \sqrt{\frac{1}{N}} \tag{1}$$

In reality, several modifications to this equation are need.

1. We must take the poisson statistics of photo electrons, not photons. The number of photo electrons (N_{pe}) is given by the following formula:

$$N_{pe} = N \cdot QE \cdot \eta \tag{2}$$

Here, QE is the quantum efficiency of the device and η is the collection efficiency for photo electrons.

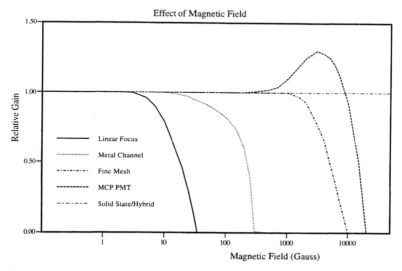

FIGURE 2. Relative gain of various photon detectors under magnetic field.

2. The poisson statistics is further modified by the Excess Noise Factor (ENF). In general, the ENF is given by the following formula:

$$ENF = 1 + \frac{1}{\delta_1} + \frac{1}{\delta_1 \cdot \delta_2} + \cdots + \frac{1}{\delta_1 \cdot \delta_2 \cdots \delta_N} \tag{3}$$

Here δ_N stands for the the multiplication factor of the N'th dynode. For typical PMTs, δ_N is $5 \sim 10$, While it is about two for the fine mesh and MCP. As for solid state device, the photo diode is one, but the APD has two or greater than two.

3. Lastly, there is an additional contribution from the Equivalent Noise Charge (ENC). A typical amplifier has about 1000 e^- of ENC. This factor must be normalized by the effective output signal level, given by $N_{pe} \cdot G$, where G is the gain of the photo detector,

Taking all these factors into account, the energy resolution is given by,

$$\frac{\sigma}{E} = \sqrt{\frac{ENF}{N_{pe}} + \left(\frac{ENC}{N_{pe} \cdot G}\right)^2} \tag{4}$$

Table 3 summarizes the energy resolution calculated by this formula, assuming the typical ENC level. The results are given as a function of the number of photons (not photoelectrons).

Type	Name	Q.E.	δ_i	ENF	Gain	Energy Resolution
Vacuum	Conventional PMT	0.3	10	1.2	10^6	$\sqrt{\frac{4}{N}}$
	Fine Mesh PMT	0.3	2	2.0	10^6	$\sqrt{\frac{7}{N}}$
	MCP PMT	0.2	-	1.5	10^6	$\sqrt{\frac{7}{N}}$
Solid State	PIN Photo Diode	0.8	-	1	1	$\sqrt{\frac{1.4}{N} + (\frac{1000}{N})^2}$
	APD	0.8	2	2	100	$\sqrt{\frac{3}{N} + (\frac{14}{N})^2}$
	MRS	0.2	-	1.1	10^6	$\sqrt{\frac{5}{N}}$
	VLPC	0.8	-	1.1	10^5	$\sqrt{\frac{1.6}{N}}$
Hybrid	HPD	0.3	1000	1.0	10^3	$\sqrt{\frac{3}{N} + (\frac{3}{N})^2}$
	HAPD	0.3	1000	1.0	10^5	$\sqrt{\frac{3}{N}}$

TABLE 3. Comparison of Energy resolution.

The energy resolution of various devices is plotted in Figure 3 as a function of number of incident photons.

Based on this analysis, I would like to list several important conclusions.

1. Thanks to its high gain($> 10^4$), the vacuum devices can reduce the electric noise contribution to completely negligible level. This makes photon counting possible.

2. On the other hand, the photo diode (PD) achieves the best energy resolution, if the number of photons exceeds 10^6, thanks to its high QE and low ENF.

3. The charecteristics of APD fall between the PMT and the Photo diode. It has better energy resolution than the PMT and the PD in the range of 100 and 10^6 photons. This is because the APD has higher QE than PMT and higher gain than the PD, but lower gain than PMT and worse ENF than PMT and PD.

4. The HPD is somewhat similar to the APD. It's poor QE can be compensated by the excellent ENF. The gain is higher than APD, which makes the resolution better than the APD in the region of $10 \sim 100$ photons. The HAPD can cover even fewer photon numbers, down to single photon because of its high gain. However both HPD and HAPD still suffer from poor QE, which is an intrinsic property of the vacuum device.

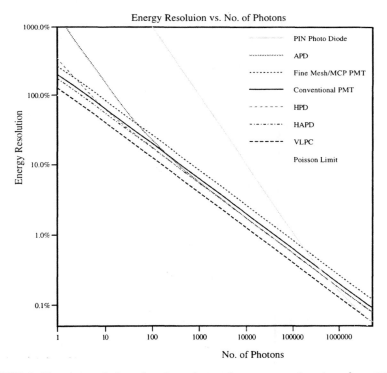

FIGURE 3. Energy resolution of various photon detectors as a function of sensitive area.

5. The VLPC is the ideal device for photon counting with high QE (as a solid state device), high gain and low ENF (similar to the PMT).

As one can see from this analysis, the major advantage of solid state devices at the large number of photons comes from their high quantum efficiency (QE). In order to further improve the energy resolution of the vacuum devices, increased QE is essential. To achieve this, solid state photo cathodes such as GaAs and GaAsP have been under development [12]. Figure 4 shows the quantum efficiency of various photo cathodes as a function of wave length. As shown here, the newly developed GaAsP photo cathode by Intevac has achieved the QE as high as 50%.

F   Image Resolution

Traditionally, Image resolution is given by the Signal to Noise Ratio (SNR). However, by taking the inverse of the SNR, one can define the image resolution as a natural extension of the energy resolution as shown below.

$$(\text{Image Resolution}) = (\text{Energy Resolution}) \cdot (\text{Position Resolution}) \qquad (5)$$

$$= \frac{\sigma}{E} \cdot \frac{1}{MTF} \qquad (6)$$

$$= \frac{1}{SNR} \qquad (7)$$

Here, MTF is so-called Modulation Transfer Function at the given spatial frequency.

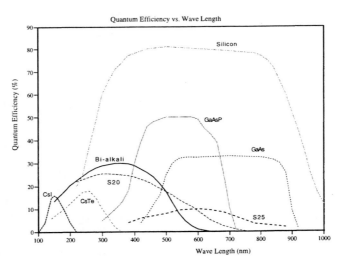

FIGURE 4. the quantum efficiency of various photo cathode as a function of wave length.

Table 4 summarizes the image resolution calculated by this formula. For simplicity, the MTF at $10lp/mm$ is given here. The image resolution is given as a function of the number of photons (not photoelectrons).

Type	Meaning	Q.E.	ENF	Gain	MTF (at $10lp/mm$)	Image Resolution
CCD	Charge Coupled Device	0.3	1	1	0.8	$\sqrt{\frac{5}{N} + (\frac{200}{N})^2}$
BCCD	Back Illuminated	0.8	1	1	0.8	$\sqrt{\frac{2}{N} + (\frac{80}{N})^2}$
Cooled BCCD		0.8	1	1	0.8	$\sqrt{\frac{2}{N} + (\frac{10}{N})^2}$
ICCD	Intensified by MCP	0.2	2	600	0.3	$\sqrt{\frac{100}{N} + (\frac{1}{N})^2}$
EBCCD	Electron Bombarded	0.2	1.1	3000	0.6	$\sqrt{\frac{15}{N} + < (\frac{1}{N})^2}$
ISPA	Imaging Silicon Pixel Array	0.2	1.1	5000	0.4	$\sqrt{\frac{30}{N} + < (\frac{1}{N})^2}$

TABLE 4. Comparison of Image resolutions.

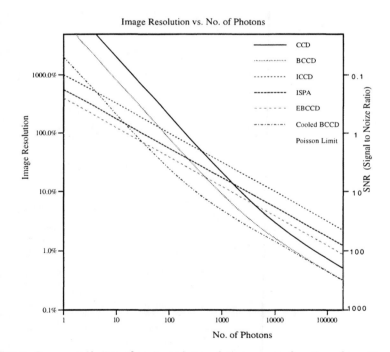

FIGURE 5. Image resolution of various photon detectors as a function of sensitive area.

The image resolution of various devices is plotted in Figure 5 as a function of the number of incident photons. Based on this analysis, I can conclude the following:

137

1. In general, by using vacuum devices, one can reduce the electric noise to negligible level. This is the common feature of ICCD, EBCCD and ISPA.

2. Among these three, EBCCD has the best image resolution, thanks to its excellent ENF and MTF. ICCD is the worst due to its poor ENF and MTF. ISPA is not as good as EBCCD, simply because of larger pixel size. But the main advantage of ISPA over EBCCD is readout speed.

3. Once the number of photons exceeds several hundreds, Back illuminated CCD (BCCD) shows the best image resolution, thanks to its high QE. Its electric noise level can be reduced by cooling the device to the level of less than 10 electrons. This is the reason why astronomical observations are performed by the cooled, back-illuminated CCD.

In summary, newly developed EBCCD or ISPA is a powerful device for the detection of photons in a range of < 10 photons. However, for most of scientific applications where one can collect more than 10 photons be integrating over time, the cooled BCCD is better suited and less expensive.

I MARKET PRICE AND FUTURE PROSPECT

A Comparison of Market Price

The market price is one of the most important factors to make a final decision on the choice of the photon detector to be used. So I would like to make some comments on this issue. Figure 6 shows the market price of various photon detectors as a function of dimension (diameter) of the sensitive area. The prices were taken from the large scale high energy experiments.

I can conclude the following from this figure.

1. Above 10mm diameter, the PMT is still the least expensive. Any other devices cost more. Therefore, unless there is a specific reason why PMT's can not be used, motivation for using other devices is not compelling. Fine mesh PMTs have been used when the magnetic field is somewhere 0.1 - 1.5 Tesla, and the APD and HPD are under consideration by the CMS because of 4 Tesla environment. Otherwise, even though APD and HPD have slightly better performance in terms of energy resolution and linearity than the PMT, additional cost is hard to justify under today's market prices.

2. For small pixel size (< 0.1mm). CCD is the only practical solution, and it is cheap. For high energy applications, however, its slow readout speed and electric noise level make it difficult to use.

3. Between 0.1mm and 1mm, there is no single solution. ICCD is OK if readout speed is not an issue. But this is not the case for most of high energy applications.

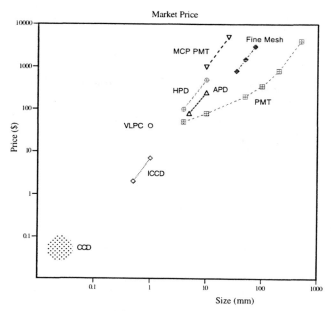

FIGURE 6. The market price of various photon detectors which have been used in large scale high energy experiments.

4. Between $1mm$ and $5mm$, again, there is no easy solution. Idealy the price should be an order of $10 at 1 mm. In reality, the VLPC costs $50 (including associated cryogenic and preamp), and the position sensitive PMT costs $\sim$ $20 per channel. This is the area where future development of inexpensive photon detectors can contribute the most.

The market price is one of the most important parameters for the final decision. Developing a new detector with 10% better energy resolution or time resolution is an interesting R & D project. However, if it costs twice than the conventional one at the end, it is hard to justify such a new device for large quantities. Considering the fact that the cost of the photon detectors is a significant part of budget for today's large scale detectors, further competition and price reduction are more important than anything else in my opinion.

B Future prospect

In the past 10 years or so, many new types of photon detectors have been proposed and developed. From time to time, people thought that these new devices, such as APD and HPD would eventually replace all the PMTs in the world. As we know, this has not yet happened. In this paper, I tried to address this question by taking a systematic approach.

We have more variety of photon detectors than ever. However, as I demonstrated, there is no single device for general purposes, which makes our research more interesting.

As a conclusion to my paper, I would like to propose my dream photon detector for the scintillating fiber read out. It is a hybrid APD with GaAsP photo cathode. My specifications is given in Table 5.

Item	Specification
Photo cathode	GaAsP with 50% QE
Inside	APD array (64 pixels)
Gain	10^5
ENF	1.0
Cost	$10/pixel

TABLE 5. Katsushi's dream detector.

REFERENCES

1. T. Iijima *et al.*, Nucl. Instr. and Meth. A **379** (1996) 457.
2. Y. Yoshizawa *et al.*, Nucl. Instr. and Meth. in Phys. Res. A **387** (1997) 33.
3. R. DeSalvo *et al.*, Nucl. Instr. and Meth. A **342** (1994) 558.
4. R. DeSalvo, Nucl. Instr. and Meth. in Phys. Res. A **387** (1997) 92.
5. P. Cushman *et al.*, Nucl. Instr. and Meth. in Phys. Res. A **387** (1997) 107.
6. J. P. Pansart, Nucl. Instr. and Meth. in Phys. Res. A **387** (1997) 186.
7. Th. Kirn *et al.*, Nucl. Instr. and Meth. in Phys. Res. A **387** (1997) 199, 202.
8. A. Karar *et al.*, Nucl. Instr. and Meth. in Phys. Res. A **387** (1997) 205.
9. N. Bacchetta, Nucl. Instr. and Meth. in Phys. Res. A **387** (1997) 225.
10. A. V. Akindinov, Nucl. Instr. and Meth. in Phys. Res. A **387** (1997) 231.
11. M. R. Wange, Nucl. Instr. and Meth. in Phys. Res. A **387** (1997) 278.
12. S. M Bradbury, Nucl. Instr. and Meth. in Phys. Res. A **387** (1997) 45.
13. T. Gys *et al.*, Nucl. Instr. and Meth. in Phys. Res. A **387** (1997) 131.
14. M Gruwa, Nucl. Instr. and Meth. in Phys. Res. A **387** (1997) 282.

The Latest Vacuum Photodetectors for Scintillating Fiber Read Out

Y. Yoshizawa, I. Ohtsu, N. Ota, T. Watanabe, J. Takeuchi

Hamamatsu Photonics K.K., Electron Tube Center
314-5, Shimokanzo, Toyooka Village, Iwata-gun, Shizuoka Pref.,438-01, JAPAN

Abstract. In applications such as radiation measurement and medical instrumentation, development for equipment which utilize scintillating fibers or combination of plastic scintillators and wave length shifters has been continually progressed. This has led to a strong demand for multi anode photodetectors. Additionally, in the recent experiments for high energy physics (HEP), luminosity and energy of beams have been increased to achieve the precise measurement. As a result, the size of the detector as well as number of read out channels have been also increased. In such case, multi anode photodetectors with reasonable cost performance is required. Hamamatsu is continuously developing new types of photodetectors to keep up with those trends. The performance and the prime features of the latest vacuum photodetectors for scintillating fiber read out are discussed in this paper.

INTRODUCTION

Hamamatsu is one of the leading manufacturers of vacuum photodetectors in the world for several fields of application. Requirements from those fields and our investigations of the demands motivate us to develop new types of detectors.

In this paper, following 2 detectors are discussed as the latest vacuum photodetectors for scintillating fibers read out.

A. Metal Package photomultiplier tube (PMT) series;

B. Image Cain Unit.

A. Metal Package PMT series

R5900 series

CP450, *SciFi97: Workshop on Scintillating Fiber Detectors*
edited by A. D. Bross, R. C. Ruchti, and M. R. Wayne
© 1998 The American Institute of Physics 1-56396-792-8/98/$15.00

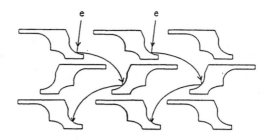

FIGURE 1. Cross section of the metal channel dynodes

In radiation measurement and medical instrumentation, the development of more compact and portable equipment has continuously progressed. This has led to a strong demand for miniaturization of highly sensitive photodetectors like PMTs. However, it's difficult to miniaturize conventional PMTs with glass envelopes and sophisticated electrode structures.

To meet the increasing needs for compact photodetectors with high sensitivity, R5900 series (1,2) using a metal package in place of the traditional glass envelope, has been developed. R5900 incorporates an 10-stage electron multiplier constructed with stacked thin electrodes, it's called Metal Channel Dynode, into a metal package of 28 x 28 mm square and 20 mm in height. The development of this metal package and its metal channel dynode have made the fabrication of a compact PMT possible. Figure 1 shows cross section of the metal channel dynodes. The structure of inner electrode was designed by advanced analysis of electron trajectory. Furthermore, our long experience with micromachining technology has achieved a closed proximity assembly of these thin electrodes. The dimensional outline of R5900U (3) is shown in Figure 2.

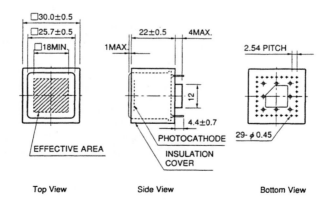

FIGURE 2. Dimensional outline of R5900U

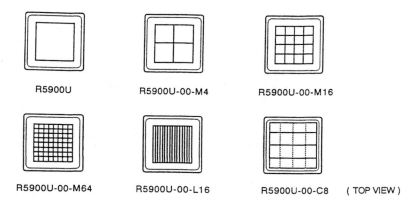

FIGURE 3. Anode variation of R5900 series

In this figure, "U" means a tube having insulation cover. It's necessary to prevent electrical shock, because a metal package has a cathode potential voltage.

As the metal channel dynode is a sort of an array of small linear focused dynodes, secondary electrons hardly go to adjacent dynode channel in a process of multiplication. It's possible to make multi anode PMTs utilizing its feature. R5900 series is offering 5 types of multi anode shapes as well as single channel. These types are 4 [2x2], 16 [4x4] and 64 [8x8] matrix channels, 16 [1x16] linear channels and crossed-wire [4X + 4Y] anode. Each type has its own dynode pattern which matches to its anode shape. Figure 3 shows the anode variation of R5900 series. It has bialkali photocathode as its standards. R5900U is for general scintillation counting. R5900U-00-M4 (4), -M16 (5) and -M64 (6) are suitable for scintillating fiber readout as well as

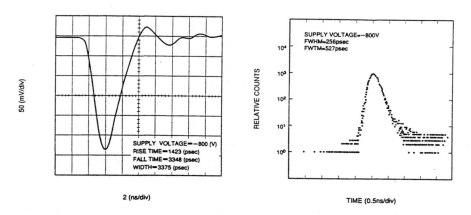

FIGURE 4. Anode output waveform of R5900U   **FIGURE 5.** TTS (Transit Time Spread) of R5900U

143

Anode Uniformity

100	97	97	92
89	82	66	70
66	72	66	76
70	68	69	96

(unit=%)

< Condition >
Applied Voltage = 800V
Light Source = W lamp
    (uniform DC light)
Full illumination

FIGURE 6. Anode uniformity of R5900-00-M16

Cross-talk

0.1	0.8	0.1	---
0.5	100	0.3	---
0.1	0.3	0.1	---
---	---	---	---

(unit=%)

< Condition >
Applied Voltage = 800V
Light Source = W lamp
    (uniform DC light)
Spot illumination (4.0 x 4.0 mm)

FIGURE 7. Cross talk of R5900-00-M16

for RICH (ring image cherenkov counter), R5900U-00-L16 (7) is suitable for coupling with slit shaped scintillators and scintillating fiber ribbons. R5900U-00-C8 (8,9) is a tube having 4 + 4 crossed plate, it's possible to get position information by using center of gravity method. It is suitable for PET or X ray imaging detection.

One of remarkable feature of R5900 series is fast time response. Figure 4 and Figure 5 show anode output waveform and TTS (Transit Time Spread) of R5900U. The typical rise time and TTS are 1.5 ns and 260 ps, respectively. On the other hand, the anode uniformity and the cross talk are important feature for scintillating fiber read out. As examples, those of R5900-00-M16 are shown in Figure 6 and Figure 7. In those data, each square corresponds to each anode, W (tungsten) lamp was used as uniform DC light source. In the uniformity test, all of useful area of the cathode were illuminated, and signal from each anode was measured. The numbers in squares show relative values of anode output signals. The variation of anode sensitivity is within factor 2 in this data. In the cross-talk test, aperture of 4.0 x 4.0 mm was used to illuminate only one channel. Anode signal of the channel as well as its adjacent channels were measured. Cross-talk is less than 1% in this condition.

R5900-00-M64

Recently, R5900-00-M64 has been developed as a new multi anode PMT for scintillating fiber read out. It uses the same metal package and has 64 [8x8] matrix channels, but its pixel size is 2 x 2 mm and its pitch is 2.3 mm. This small size is still suitable for read out of 1 mm diameter scintillating fibers as well as optical fibers. The anode uniformity of R5900-00-M64 is shown in Figure 8. In this uniformity test, W (tungsten) lamp was used as uniform DC light source, and all of useful area of the cathode were illuminated.   Signal from each anode was measured,   and is shown as a

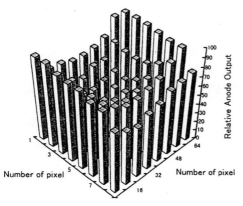

Number of pixel

Relative Anode Output

Number of pixel

0.3	1.3	0.3
0.8	100	0.8
0.2	1	0.2

Light Source: SCSF78(1.0mm)
Applied Voltage: 800(V)

FIGURE 8. Anode uniformity of R5900-00-M64

FIGURE 9. Cross talk of R5900-00-M64

height of a pole in the Figure. As the pixel size is smaller, the variation between pixel to pixel becomes larger compared with those of R5900-00-M16, but its variation ratio is approximately 1:3 in this data. The cross talk of R5900-M64 is shown in Figure 9. In this test, scintillating fiber, which is double cladding SCSF78 with 1 mm in diameter and is cortesy from Kuraray, was used as DC spot light source. It was set on the face plate where corresponds to a center of one pixel. UV light was used to excite the scintillating fiber of which peak wave length is around 440 nm. Anode signal of the channel as well as its adjacent channels were measured. As you see, the cross talk is less than 2 % in this condition. At the beginning of the development, the thickness of the face plate was 1.3 mm, but finally it became 0.8 mm to minimize optical cross talk inside it.

R5600-00-M4

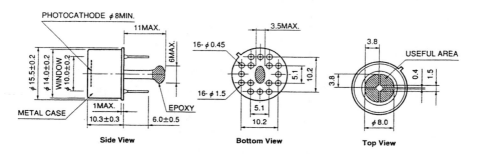

FIGURE 10. Dimensional outline of R5600-00-M4

145

Anode Uniformity

85	100
90	92

(unit=%)

< Condition >
Applied Voltage = 800V
Light Source = W lamp
(uniform DC light)
Full illumination

FIGURE 11. Anode uniformity of R5600-00-M4

Cross-talk

100	0.85
1.88	0.06

(unit=%)

< Condition >
Applied Voltage = 800V
Light Source = W lamp
(uniform DC light)
Spot illumination (3.8 x 3.8 mm)

FIGURE 12. Cross talk of R5600-00-M4

R5600-00-M4 (10) is one of R5600 series (11,12), which is another version of the metal package PMT. It has 4 channel multi anode, of which shape is fan-shaped. It incorporates an 8-stage metal channel dynodes into a TO-8 type metal package of 15 mm in diameter and 10 mm in height. Its dimensional outline is shown in Figure 10. The anode uniformity and the cross talk are shown in Figure 11 and Figure 12. In those data, each square corresponds to each anode, W (tungsten) lamp was used as uniform DC light source. In the uniformity test, all of useful area of the cathode were illuminated, and signal from each anode was measured. The numbers in squares show relative values of anode output signals. The variation of anode sensitivity is within 85 % in this data. In the cross-talk test, aperture of 3.8 x 3.8 mm was used to illuminate only one channel. Anode signal of the channel as well as its adjacent channels were measured. Cross-talk is less than 2% in this condition.

B. Image Chain Unit (Large Diameter Image Intensifier Chain Unit)

The image chain unit (the large diameter image intensifier chain unit) (13) is a device for low light level imaging for particle tracking with scintillating fibers, which are widely used in HEP experiments. It consists of a large diameter image intensifier followed by 3-stage conventional image intensifiers and CCD camera coupled with fiber optical plate.

Figure 13 shows the cross section of the image chain unit. The large diameter image intensifier is used as the first stage of light amplifier. It consists of input window, photocathode, focusing electrodes and phosphor screen target. A fiber optical plate (FOP) is used as the input window, but it doesn't include EMA (Extra Mural Absorption). In this case, the transmittance of the FOP is better than that of FOP with EMA. Although EMA is usually used to get good position resolution for image intensifiers, but this application doesn't need such a high position resolution. The large

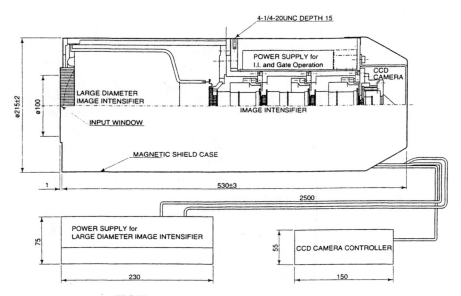

FIGURE 13. Cross section of the image chain unit

diameter image intensifier has bialkali photocathode which match a spectrum of a scintillation light. The following 3-stage image intensifier consist of the same components, but they have multialkali photocathode, which match spectrum of the emission light of the phosphor screen. When particle hit a scintillating fiber bundle, which is coupled to the input window, it generates scintillation light as the signal. It's converted into photoelectrons at the photocathode. These photoelectrons are electrostatically focused and are accelerated onto the phosphor screen, with keeping the position information, by the focusing electrodes. As a result, the signal light is

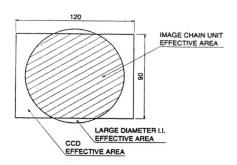

FIGURE 14. Effective image area of the image chain unit

amplified. The amplification process is repeated in the later 3-stage image intensifiers, then finally the signal light is converted into video signal at CCD camera. The total light amplification is around 1×10^5 times, but it could be adjusted by the applied voltage for the image chain unit. It means that signal light with single photon level can be detected.

The effective imaging area on the input window is shown in Figure 14. If scintillating fibers with diameter of 0.5 mm are used, more than 20k fibers can be coupled to one image chain unit. If the timing as well as the energy information of the signal aren't so important and the count rate of signal is low, the image chain unit could be the cheapest solution for the read out of a lot of fibers.

Due to the decay time of the phosphor screen of the image intensifiers, the image signal can be delayed by approximately 20 u sec. As the one of the later image intensifiers works with gating pulse, it's possible to detect only interesting signal with high signal to noise ratio.

CONCLUSION

Development of photodetectors are required and progressed daily for the applications and their operating conditions which are also moving day by day. Hamamatsu will continue its effort to the research and the development as one of the leading manufacturers of photodetectors.

REFERENCE

1. Kyushima, H., et al, "Photomultiplier Tube of New Dynode Configuration", *IEEE Trans. Nucl. Sci.,* Vol.41, 1994, pp.725.
2. Yoshizawa, Y., et al, "The latest vacuum photodetector", *Nucl. Instr. and Meth.vol.A387,* 1997, pp. 33-37.
3. Hamamatsu Catalog, "R5900U", Cat. No. TPMH1109E.
4. Hamamatsu Catalog, "R5900U-00-M4", Cat. No. TPMH1126E.
5. Hamamatsu Catalog, "H6568, H6568-10", Cat. No. TPMH1137E.
6. Hamamatsu Catalog, "R5900-00-M64", Cat. No. TPMH1192E.
7. Hamamatsu Catalog, "R5900U-L16 series", Cat. No. TPMH1146E.
8. Hamamatsu Catalog, "R5900-00-C8", Cat. No. TPMH1139E.
9. Watanabe, M., et al, "A compact position-sensitive detector for PET", *IEEE Trans. Uncle. Sci., Vol.42,* 1995, p.1090-1094.
10. Hamamatsu Technical Information, "Metal Package Photomultiplier Tubes R5600 Series and Photosensor Modules", Cat. No. TPMH 9001E.
11. Hamamatsu Catalog, "R5600 series", Cat. No. TPMH1066E.
12. Hamamatsu Catalog, "Photosensor Module", Cat. No. TPMH1168E.
13. Hamamatsu Catalog, "H6600 series", Cat. No. TPMH1159E.
14. Hamamatsu Catalog, "Large Diameter Image Intensifier, Image Chain Unit", Cat. No. T I I 1040E.

Recent Developments in Avalanche Photodiodes for Scintillating Fiber Applications

Richard Farrell, Kofi Vanderpuye, Stefan Vasile,
Jeffrey S. Gordon, and Prakash Gothoskar

Radiation Monitoring Devices, Inc.
44 Hunt Street, Watertown, Massachusetts 02172

Abstract. Research is ongoing to tailor proportional mode avalanche photodiodes (APDs) for use in nuclear radiation environments and for scintillating fiber readout. We report progress on APD design modifications directed toward minimizing signals from ionizing particle interactions and also toward producing APD arrays for scintillating fiber readout. In addition, we present results for very high gain APDs fabricated using a new planar process which shows great promise for lower production costs of both APD arrays and discrete devices.

INTRODUCTION

Radiation Monitoring Devices, Inc. has been producing and continuing development of beveled edge silicon APDs for almost 15 years (1) with a primary emphasis on large area detectors. These devices are in principle a relatively simple structure with a deep diffusion of p-type dopant into a very uniformly doped n-type substrate. The finished device has a depleted or space sharge region on either side of the p-n junction and a region of undepleted material that we call the drift region between the space charge region and the sensing surface on the p-type side of the diode. The beveled edges allow the structure to achieve bulk breakdown voltages without the edges breaking down first (2,3,4,5).

We have three areas of research which are relevant to the topic of scintillating fiber readout. All are outgrowths of previous work which led to the development of very high gain proportional mode APD structures (6). First, work is underway to design and build 64 element arrays of 1 mm^2 high gain APD pixels intended specifically for scintillating fiber readout. Secondly, we are involved in a continuing effort to modify our APD device structure in order to minimize the signal generated from direct interaction of ionizing particles with the detector. This is important in situations where the device must be located in a particle flux environment. Finally, we

CP450, SciFi97: Workshop on Scintillating Fiber Detectors
edited by A. D. Bross, R. C. Ruchti, and M. R. Wayne
© 1998 The American Institute of Physics 1-56396-792-8/98/$15.00

are developing and have successfully demonstrated a planar process for producing APDs and APD arrays which we expect to lead to availability of such detectors at lower costs.

ARRAYS FOR SCINTILLATING FIBERS

In the recent past we have found techniques for fabricating arrays (7) of independent APD pixels by etching deep grooves which essentially form independent bevels around each pixel. Such arrays perform well but are mechanically fragile and labor intensive to fabricate. We are presently developing a planar process, successful initial results of which will later be presented in this paper, in order to allow the fabrication of larger area arrays at affordable costs. The earlier arrays were 4 x 4 element units with 1mm x 1mm pixels. With the new process we are aiming at 8 x 8 element units with the same pixel size. We expect that the performance of each pixel in the new structure to be the same as for the old structure, based on results from our first planar processed APDs.

Reviewing performance results for a 1mm x 1mm pixel (7), at room temperature the noise minimum occurred at an APD gain between 1000 and 2000 and was equivalent to an energy of 100 eV FWHM at the input of the APD, calibrated relative to detected low energy X-rays. Low numbers of optical photons could be detected by moderately cooling the array. A 90% detection efficiency was measured for 50 photon pulses at room temperature, 20 photon pulses at 20C, and about 10 photon pulses at –43C.

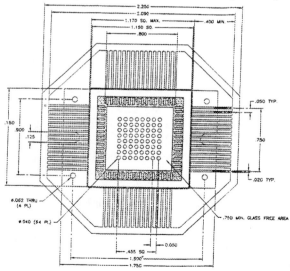

FIGURE 1. Package for 8 x 8 element array, with independent windows spaced 0.050 " apart.

The planned hermetic package (Fig. 1) for the 8 x 8 element array has an independent glass-in-kovar window for each pixel and the silicon array will be permanently bonded optically in alignment with the window array. A fiberoptic bundle of 1mm diameter fibers, with matching pitch of 0.050", is intended for coupling to the window surface on the exterior surface of the package.

MINIMIZING THE NUCLEAR COUNTER EFFECT

Another topic of recent APD research has been directed toward the development of a modified APD structure designed to minimize the nuclear counter effect which results from ionizing radiation directly interacting with the APD and degrading the spectra being sought from scintillation signals being simultaneously detected. This work has been targeted for the demanding requirements of the Compact Muon Solenoid (CMS) calorimeter, but is also relevant for situations where scintillating fiber signals are being detected in a particle flux environment.

The basic idea in this work has been to take our existing APD technology for very high gain detectors, but to perform our deep diffusions into much lower resistivity silicon. Among the consequences of making this change are a reduction of operating voltage from near 1700V to less than 500V with no increase in the total noise level of the detector. Though significantly less than for the higher voltage structure, APD gains of several hundred are routinely achieved.

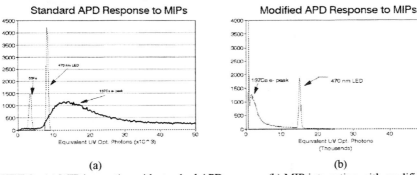

(a)   (b)

FIGURE 2. (a) MIP interaction with standard APD.   (b) MIP interaction with modified APD.

The nuclear counter effect has been dramatically diminished due to a significantly reduced thickness of silicon on the sensing side of the p-n junction for the modified structure. A comparison (Fig. 2) of the magnitude of the signals from a minimun ionizing particle (MIP) interaction with a standard APD is shown next to that of the modified structure. Electrons from ^{137}Cs were used as a MIP source, a stable 470 nm pulsed LED was used to simulate a lead tungstate scintillator input, and a low energy (5.9 keV) X-ray source was used for calibration. For the standard APD the 3dB width of the MIP peak is equivalent to a 1.9 GeV gamma interaction with lead tungstate. For the modified structure it has been reduced to 114 MeV. The improvement is mostly due to a reduction in the thickness of the silicon into which the MIP can deposit energy, but it is also in part due to an improvement in detector quantum efficiency (Fig. 3) which results from a modified detector surface treatment.

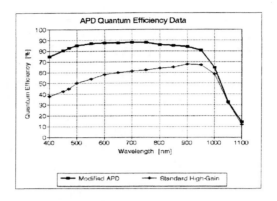

FIGURE 3. Quantum efficiency comparison

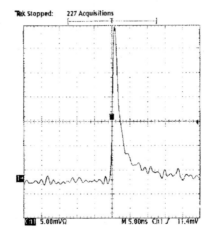

FIGURE 4. Speed of response for modified APD.

The speed of response (Fig. 4) was recorded as the pulse generated by the direct interaction of an alpha particle with the modified APD. The APD output was fed into the 50 ohm termination of a fast oscilloscope. The observed pulse width of 7ns is less than required for the 25 microsecond reset time of the CMS experiment.

The stability of gain with temperature and bias voltage are a concern to systems designers using APDs. The temperature stability target for CMS is 2% / °C at an APD gain of 50. Fig. 5 shows slightly more than 2% at a gain of 70 for the modified APD. CMS also requires less than 5% per volt for gain stability with voltage and less than 150 MeV MIP sensitivity. At APD gains above 200 (Fig.6), both of these requirements are satisfied. The active areas of the devices tested were all 8mm x 8mm.

Work is continuing on this project with focus on further reduction of the thickness of the undepleted(drift) region of the APD in order to further lessen MIP sensitivity while also lessening the damage induced (8) by high levels of neutron irradiation. In addition we are working to address issues involved with manufacturing large numbers of such detectors for applications such as CMS.

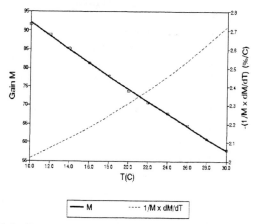

FIGURE 5. Gain stability vs. temperature for modified APD structure.

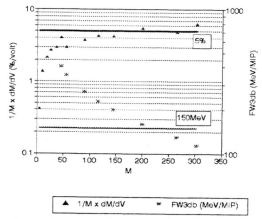

FIGURE 6. Gain stability and MIP sensitivity for modified APD structure.

A PLANAR PROCESS FOR MANUFACTURING

Bevel formation is the most difficult aspect of producing beveled edge APDs. This is true of the formation of the fragile edge around the active area of discrete devices as well as of the deep grooves that form bevels around the elements of arrays. In order to address this problem we have been working to develop a planar process by which we can bring the electric field spreading function of the bevel into the back plane of the silicon wafer from which the APDs are being fabricated. This allows the APDs to be cut out of the wafer by traditional dicing methods as a final step before packaging, consistent with many other production processes for semiconductor devices. Since this approach allows bevel formation on a wafer scale, many devices can be brought to completion at once rather than needing to be handled as individual chips for much of the processing.

We have very recently demonstrated such a process by fabricating planar arrays of 3mm x 3mm elements on two inch diameter silicon wafers. Devices were tested on the whole wafer as well as after dicing them into discrete 3mm x 3mm devices. The performance was unchanged from before to after the dicing operation.

A pulse height spectrum of ^{55}Fe (Fig. 7) taken using one such element is shown along with the pulse from an electronic pulser injected into the test circuit to characterize the device noise. The noise FWHM from the pulser peak, calibrated relative to the 5.9keV X-ray peak, is 370 eV at a test temperature of 25C.

With the noise calibrated relative to low energy X-rays, as above, and using a 700nm pulsed LED for a signal, a noise and gain versus bias voltage curve (Fig. 8) was generated. Maximum gains well in excess of 10,000 have been measured using these planar processed devices.

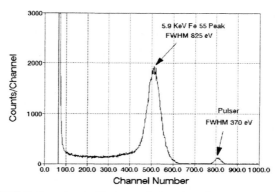

FIGURE 7. Low energy X-ray spectrum using planar processed APD.

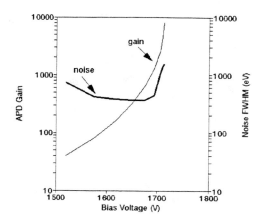

FIGURE 8. Gain and Noise vs. Bias Voltage for planar processed APD.

In another spectrum (Fig. 9) a 2mm x 2mm x 10mm LSO (lutetium orthosilicate) scintillator is coupled to a planar fabricated APD element. The 11% resolution of a 662 keV gamma peak is as narrow as any resolution previously reported for a scintillator of this size.

Continuing work in this area is directed toward the fabrication of planar arrays for scintillating fiber and other array based imaging applications. In addition planar processing will be applied to the fabrication of large area planar discrete APDs for applications such as CMS.

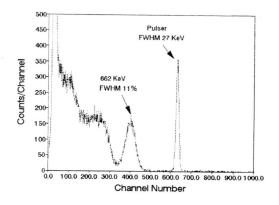

FIGURE 9. Planar APD with 2mm X 2mm X 10mm LSO scintillator. ^{137}Cs Spectrum.

155

REFERENCES

1. Squillante, M.R., Reiff, G., and Entine, G., IEEE Trans Nuc. Sci., **NS-32,** 563 (1985).
2. Huth, G. C., Trans. Nuc. Sci., **NS-13,** 36 (1966).
3. Locker, R. J., and Huth, G. C., Appl. Phys. Lett., **9,** 227 (1966).
4. Farrell, R., Olschner, F., Frederick, E., McConchie, L., Vanderpuye, K., Squillante, M. R., and Entine, G., Nucl. Inst. and Meth. in Phys. Res., **A288,** 137 (1990).
5. Farrell, R., Vanderpuye, K., Squillante, M.R., and Entine, G., IEEE Trans. Nucl. Sci., **38,** 2, 144 (1991).
6. Farrell, R., Vanderpuye, K., Cirignano, L., Squillante, M.R., and Entine, G., Nucl. Inst. and Meth. in Phys. Res., **A353,** 176 (1994).
7. Farrell, R., Redus, R., Gordon, J.S., Gothoskar, P. , "High Gain APD Array for Photon Detection" submitted to *Proceedings of SPIE's Symp. on Opt. Sci., Eng., Instrum.* (1995).
8. Ruuska, D., "Study of APDs Exposed to Neutron Fluences", Presented at the SCIFI97: Workshop on Scintillating Fiber Detectors, South Bend, Indiana, Nov. 2 - 6, 1997.

Temperature Dependence of Avalanche Photodiodes for Scintillating Fiber Readout

T. Yoshida, T. Okusawa, Y. Sasayama, and M. Tsuji

Department of Physics, Osaka City University
3-3-138 Sugimoto, Sumiyoshi-ku, Osaka 558-8585, Japan

Abstract. We have evaluated the performance of avalanche photodiodes (APDs) as photosensors for scintillating-fiber charged particle tracking detectors, putting special emphasis on their temperature dependence. For this purpose, we built a prototype of a scintillating fiber hodoscope with 3HF scintillating fibers 0.5 mm in diameter. The fiber array structure was so designed that each APD could receive photons from two fibers along a particle trajectory penetrating the array. By cooling the APDs to -40^0C, we could achieve the signal-to-noise ratio larger than 25. The detection efficiency reached more than 98 % with reducing the noise count rate to 0.5 %.

INTRODUCTION

Charged particle tracking detectors designed with scintillating fibers for recent high energy physics experiments typically consist of 0.5−1.0 mm diameter scintillating fibers a few meters long spliced with clear fibers several meters long [1,2]. When a charged particle traverses a fiber in those detectors, the number of photons expected at the photosensor coupled to the fiber can be as small as 10−20 [3]. Thus, the photosensor is essentially required to have high quantum efficiency in order to realize efficient readout of each individual fiber. The visible light photon counter (VLPC) [4,5] is one of the most elegant solutions because of its high quantum efficiency of over 65 % and its high gain of over 20,000. However, one often hesitates to use it owing to its optimum operating temperature around 6.5 K.

Instead of the VLPC, we have been testing avalanche photodiodes (APDs) counting on their quantum efficiency as high as that of the VLPC. APDs can be operated at room temperature or lower temperatures easily achievable with conventional cooling devices. However, the signal-to-noise ratio of an APD is not good due to their low gain around 100. In order to make practical use of APDs for scintillating fiber readout, we have to overcome this problem. In this paper we show how much we can improve the signal-to-noise ratio and the detection efficiency by lowering the APD temperature.

CP450, *SciFi97: Workshop on Scintillating Fiber Detectors*
edited by A. D. Bross, R. C. Ruchti, and M. R. Wayne
© 1998 The American Institute of Physics 1-56396-792-8/98/$15.00

TABLE 1. Specifications of some commercial APDs potentially suitable for scintillating fiber readout.

APD Type	197-70-72-520	S5343	C30626E
Manufacturer	Advanced Photonix	Hamamatsu	EG&G
Sensitive area	ϕ5 mm	ϕ1 mm	5×5 mm^2
Quantum efficiency at 530 nm	70 %	80 %	70 %
Typical bias	2350 V	150 V	460 V
Gain at the bias	100	100	100
Dark current at the bias	50 nA	1 nA	5 nA
Junction capacitance	15 pF	15 pF	25 pF
Window	None	Glass	Glass

Table 1 summarizes the specifications of some commercial APDs potentially suitable for scintillating fiber readout [6–8]. The decisive factors in selecting these APDs are : (1) the high quantum efficiency at the wavelength of 530 nm where the 3HF-doped green scintillating fiber have its emission maximum [9], (2) the small dark current, and (3) the small junction capacitance. The intrinsic APD noise is dominated by two sources; one is the shot noise caused by the statistical fluctuation of the APD dark current, and another is the noise charge stored in the APD junction capacitance by the Johnson noise of the resistance connected to the APD such as the input resistance of the preamplifier [10,11]. Therefore the small dark current and the small junction capacitance are important factors for the low-noise characteristic of the APD. Among the candidates in Table 1, we eventually chose the Advanced Photonix 197-70-72-520, because we could put fiber ends directly on the silicon surface of this APD.

Throughout this work, the APDs were operated with bias voltages slightly smaller than their breakdown voltages, where the gains at room temperarure were expected to be 50–100 according to the data sheets that the manufacturer supplied. The problematic Geiger mode operation with a bias exceeding the breakdown voltage was avoided.

TESTS AT ROOM TEMPERATURE

The first step of our study was to measure the detection efficiency of the scintillating fiber detector coupled to an APD operated at room temperature. The details of our measurements are described elsewhere [12].

The measurements were made with several different thicknesses of fiber bundles penetrated by ^{90}Sr β-rays. Each bundle was made by stacking a number of fiber layers, and all the fibers in the bundle were directly coupled on the silicon surface of a 197-70-72-520 APD with optical grease. Another end of the bundle was terminated by a sheet of aluminum foil light reflector, which reflectivity was measured to be 70 %. The length of each bundle was 20 cm. The fiber that we used was a 3HF-doped multiclad type (Kuraray SCSF-3HF). The outer diameter was 0.5 mm, and the core diameter was 0.44 mm. A couple of trigger counters were placed underneath the fiber bundle. The signals from the APD were amplified by a low-noise and high-gain (30 mV/fC) charge sensitive preamplifier (Digitex HIC-1576) which had an RC pulse shaping circuit with a time constant of 150 ns at its output stage. The output signals of the preamplifier were further amplified by a main amplifier (Phillips Model 777) and digitized by a peak sensing ADC (LeCroy 2259B).

As a result, an APD placed at room temperature had to receive photons from at least four layers of fibers along a particle trajectory in order to realize reasonably good separation of signals from noises. With four layers, the detection efficiency reached 98 % with a threshold by which the noise count rate was reduced to 0.5 %.

TESTS AT LOWER TEMPERATURES

The second step of our study was to lower the APD temperature to improve the signal-to-noise ratio with fewer layers of fibers, for instance, with two layers. Lower temperatures reduce the number of thermal electrons diffusing from the valence band to the conduction band in an APD, and then reduce the dark current. Consequently, the shot noise is reduced. Lower temperatures also calm down the crystal lattice vibration which interrupts the avalanche electrons in the APD. Then, the gain of the APD increases as the APD temperature is lowered. By these effects, we can expect that the signal-to-noise ratio is improved at lower temperatures.

We made a scintillating fiber array as shown in Figure 1 with multiclad 3HF scintillating fibers 0.5 mm in diameter. The fiber length was 54 cm. The fiber array was divided into six segments so that it could work as a hodoscope equipped with six APDs. A segment was 1.6 mm wide. One end of each segment was coupled to a 197-70-72-520 APD. A sheet of aluminum foil reflector terminated another end of the segment. The structure of this fiber array is what we call "two-layers", where an APD receives photons from two fibers along a particle trajectory traversing the array in the normal direction. The temperature of the six APDs was lowered by thermoelectric cooling modules in a vacuum vessel as shown in Figure 2. The vacuum was necessary to protect the APDs from frost and to improve the thermal efficiency. The scintillating fibers were led into the vacuum vessel through the holes punched in an acrylic plate, and then the opening spaces at those holes were hermetically sealed with epoxy glue. The signals from each APD were read out in the same way as in the tests at room temperature mentioned above [12]. With ^{90}Sr β-rays penetrating the scintillating fiber array in the normal direction, we measured the signal-to-noise ratio and the detection efficiency as a function of APD temperature. The distance from the β-source to the APDs was 46 cm.

Temperature dependence of the APD dark current is shown in Figure 3a. By cooling the APD from room temperature to -40°C, the dark current is reduced by one order of magnitude ; -40°C is the lowest temperature allowed for this type of APD in its specifications. Figure 3b shows the breakdown voltage at -40°C for each of the six APDs that we used. The gain of an APD increases rapidly as the bias

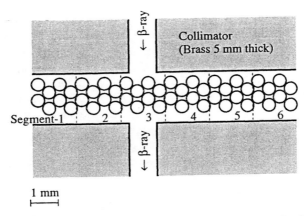

FIGURE 1. The "two-layer" structure of the scintillating fiber array. Dashed lines indicate the segment boundaries.

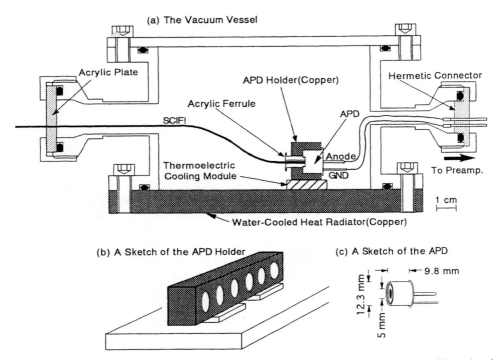

FIGURE 2. (a) The vacuum vessel in which the APD temperature was lowered. (b) A sketch of the APD holder made of copper. (c) A sketch of the 197-70-72-520 APD.

160

voltage approaches its breakdown voltage. Since we biased those six APDs with one common high voltage power supply when we used them in the scintillating fiber hodoscope, the bias voltage that we could apply was restricted by the Segment-3 APD which had the smallest breakdown voltage, and each of the other APDs had its gain diminished because of the larger margin to its breakdown voltage. When we applied a bias of 2280 V at an APD temperature of -40^0C, the pulse height of the signal from Segment-2, where its APD had the highest breakdown voltage, was about 1/3 of that from Segment-3, even though the difference in breakdown voltage was only a few percent of the bias voltage. In this paper, we show below the results obtained with β-rays collimated on Segment-2 and -3.

Figure 4 shows typical preamplifier output signals obtained with β-rays penetrating Segment-3 at APD temperatures of $+23^0$C and -40^0C. The oscilloscope was triggered by coincidences of two trigger counters placed underneath the fiber array. The pulse height spectra of the preamplifier output signals measured at various APD temperatures are shown in Figure 5 together with the noise spectrum measured at random timing. Those spectra were obtained in the off-line analysis by transformation from the digitized pulse heights of the main amplifier output signals. The pulse height increases typically by one order of magnitude by lowering the APD temperature from $+20^0$C to -40^0C. At -40^0C, clear separation of signals from noises is realized, and the detection efficiency reaches 99 % with reducing the noise count rate to 0.5 %.

All the results to characterize the temperature dependence of the APDs are summarized in Figure 6. The temperature dependence of the noise is shown in Figure 6a, where the noise is defined as a standard deviation of the gaussian fitted to the noise spectrum at the stage of preamplifier output. The noise has no apparent temperature dependence, even though the dark current is reduced so much by cooling the APD. This means that the shot noise is not a dominant noise source in this type of APD. Our noise estimation suggests that the noise in this APD is dominated by the noise charge stored in the APD junction capacitance [12]. In Figure 6b is shown the temperature dependence of the signal defiend as the avarage pulse height of the preamplifier output signals obtained with β-rays penetrating the segment. The signal increases exponentially as the APD temperature is lowered. Consequently, the signal-to-noise ratio increases by one order of magnitude as the APD is cooled from $+10^0$C to -40^0C as shown in Figure 6c. Even at Segment-2 the signal-to-noise ratio reaches 26 at -40^0C. Figure 6d shows temperature dependence of the detection efficiency with a threshold by which the noise count rate is reduced to 0.5 %. At -40^0C, the efficiency is more than 98 % both at Segment-2 and at Segment-3.

SUMMARY AND DISCUSSION

By cooling the APD from room temperature to -40^0C, we observed that the avalanche gain increases by one order of magnitude without increasing noises. Tak-

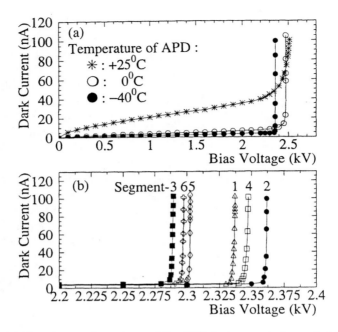

FIGURE 3. (a) Temperature dependence of the APD dark current observed at Segment-2. (b) The dark current of each APD at -40^0C vs. bias voltage near the breakdown voltage.

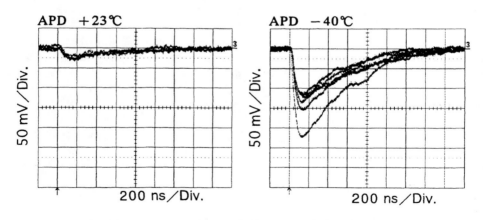

FIGURE 4. Typical preamplifier output signals obtained with β-rays penetrating Segment-3 at the APD temperatures as indicated.

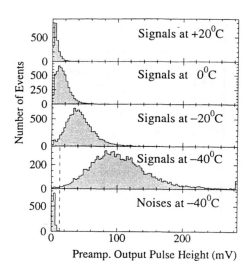

FIGURE 5. Temperature dependence of the preamplifier output pulse height spectrum measured with β-rays penetrating Segment-3. The APD bias voltage was 2280 V. The noise spectrum at -40^0C is also given. The dashed line indicates the threshold which reduces the noise count rate to 0.5 %.

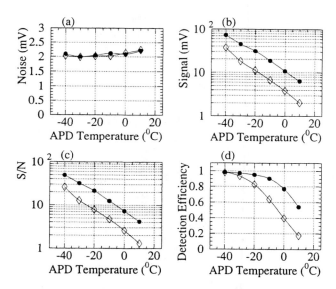

FIGURE 6. Results of our measurements at Segment-2 with a bias of 2270 V (open diamonds) and at Segment-3 with a bias of 2250 V (closed circles). Temperature dependence of : (a) the noise, (b) the signal obtained with β-rays, (c) the signal-to-noise ratio, (d) the detection efficiency with a threshold by which the noise count rate is reduced to 0.5 %. See the text for details.

163

ing advantage of this effect, we have demonstrated experimentally that an APD can detect with reasonably high efficiency the minimum ionizing particles traversing two layers of scintillating fibers 0.5 mm in diameter.

Our next step is to test with one layer of fibers and with longer fibers up to a few meters. New types of APDs such as the Advanced Photonix 197-70-74-520 which has higher quantum efficiency (90 %) at the emission maximum of the 3HF scintillator may bring us additional improvement.

The APD that we employed for our tests is a single channel type where one APD much lager than the fiber diameter is contained in a sizable package of about 1 cm. In reality, scintillating fiber tracking detectors used in real high energy physics experiments consist of more than thousands of fibers. If we use APDs as photosensors for such detectors, the APDs must be as compact as possible. It is technically possible to make an APD array in which a number of small APDs are contained compactly in a small package with keeping the same performance as we demonstrated above. We can even have a prospect that reduction of the APD size helps to reduce the APD junction capacitance and to reduce the noise charge stored there which is a dominant noise source for the moment. Thus, our goal is to realize a scintillating fiber tracking detector with compact APD arrays.

ACKNOWLEDGEMENTS

We thank Y. Shimizu from Riken-sya Co., Ltd. for his useful suggestions about the design of the vacuum vessel. This work was supported by the Sumitomo Foundation.

REFERENCES

1. Warchol, J., "UPGRADE OF THE D0 DETECTOR" in *SCIFI93 Workshop on Scintillating Detectors*, World Scientific Publishing, 1995, pp.151-167.
2. The CDF Collaboration, "Proposal for Run II Tracking System Upgrades for CDF", CDF/DOC/TRACKING/PUBLIC/3079, unpublished.
3. Baumbaugh, B. et al., *Nucl. Instr. and Meth.* **A345**, 271 (1994).
4. Atac, M. et al., *Nucl. Instr. and Meth.* **A314**, 56 (1992).
5. Abbott, B. et al., *Nucl. Instr. and Meth.* **A339**, 439 (1994).
6. Advanced Photonix, Inc., 1240 Avenida Acaso, Camarillo, CA 93012, USA.
7. Hamamatsu Photonics K.K., 1126-1 Ichino-cho, Hamamatsu City, 435-91 Japan.
8. EG&G Optoelectronics Canada Ltd., 22001 Dumberry Road, Vaudreuil, Quebec, J7V 8P7.
9. Renschler, C.L. and Harrah, L.H., *Nucl. Instr. and Meth.* **A235**, 41 (1985).
10. Lorenz, E. et al., *Nucl. Instr. and Meth.* **A344**, 64 (1994).
11. Webb, P.P. and McIntyre, R.J., *IEEE Trans. Nucl. Sci.* **NS-23**, 138 (1976).
12. Okumura, S., Okusawa, T., Yoshida, T., *Nucl. Instr. and Meth.* **A388**, 235 (1997).

Irradiation Damage of APDs for CMS Using Neutrons From [252]Cf

Y.Musienko[1], S.Reucroft[1], R.Rusack[2], D.Ruuska[1], J.Swain[1]

1. Northeastern University, Boston, MA 02115 USA
2. University of Minnesota, Minneapolis, MN 55455 USA

Abstract. We report the results of exposing three APDs (avalanche photodiodes) from two manufacturers, EG&G and Hamamatsu, to a total fluence of $2 \cdot 10^{13}$ neutrons/cm^2 in a fast neutron field produced by [252]Cf at ORNL (Oak Ridge National Laboratory). The effects of this type of radiation on many parameters such as QE (quantum efficiency), voltage coefficient of the gain, wavelength dependence of the gain, intrinsic dark current, and voltage dependent capacitance for these devices are shown and discussed.

INTRODUCTION

APDs are the baseline readout device for the approximately 100,000 PbWO$_4$ crystals of the barrel segment of the ECAL (electromagnetic calorimeter) of the CMS (Compact Muon Solenoid) detector of the anticipated Large Hadron Collider at CERN in Geneva, Switzerland. APDs were chosen for their high QE, compact size, insensitivity to magnetic fields, (4Tesla in the ECAL barrel region), and internal gain. These devices are to be exposed to a fluence of $2 \cdot 10^{13}$ neutrons/cm^2 over the proposed ten year operation of the CMS detector (1). We are participating in an extensive study of the APDs resistance to neutron irradiation along with a number of other groups in the CMS ECAL collaboration, see for example Refs. (2) and (3). Our devices were irradiated for eighteen days with neutrons produced by [252]Cf to obtain a ten year CMS equivalent fluence at the REDC's Californium Users Facility[1] at ORNL. The APDs were kept at a constant temperature of 20C and within +/-1.5 volts of the bias for a gain of 50 by means of a feedback loop in the bias control chain. Thermal neutrons were absorbed by a 1mm thick Cadmium shell surrounding each group of two APDs. The 1-MeV equivalent neutron fluence is calculated by convoluting the known [252]Cf neutron emission energy spectrum with the ASTM ,KERMA function (4). Performing this calculation yields an effective fluence equal to 1.06 times the same number of mono-energetic, 1-MeV neutrons. A more detailed description of the Californium User Facility and our setup can be found in (5).

DEVICE DESCRIPTION

The three devices described here were developed by Hamamatsu and EG&G under the auspices of an ongoing R&D program for the CMS ECAL detector. These devices are characterized primarily by the avalanche multiplication region being located near the light entry surface, and a p-n junction depth of

---

[1] Radiochemical Engineering Development Center   http:\\redc.ct.ornl.gov\cuf.htm

CP450, SciFi97: Workshop on Scintillating Fiber Detectors
edited by A. D. Bross, R. C. Ruchti, and M. R. Wayne
© 1998 The American Institute of Physics 1-56396-792-8/98/$15.00

only 4-5μm. This minimizes both response to wavelengths greater than 800nm and MIP (minimum ionizing particle) interaction signal offsets, the so-called nuclear counter effect.

The Hamamatsu device (p⁺-p-n-n⁻-n⁺ structure) is produced by epitaxial growth on a low resistivity silicon substrate. The surface is a thin SiO₂ anti-reflective layer protected by 0.5mm of a transparent silicone rubber. The device is mounted in a ceramic package with cathode and anode contacts. The sensitive area is round with a 5mm diameter. The depletion region of this device is only 25-30μm wide.

The EG&G device (p⁺-p-n-π-n⁺ structure) is produced by ion implantation and deep diffusion on a high resistivity, 3kΩ·cm wafer. The Si₃N₄ surface is a λ/4, antireflective layer optimized for blue light. The device is mounted in a ceramic package with cathode, anode and guard ring contacts. The sensitive area is a 5x5mm square. This device fully depletes before reaching the operating bias.

PARAMETER COMPARISON BEFORE & AFTER IRRADIATION

The three different devices we will discuss are the Hamamatsu BC-16 device and two EG&G devices: 3 (243 μm thick) and 15 (196 μm thick). The following characteristics were measured before and after neutron irradiation: QE dependence on wavelength, gain dependence on bias voltage, gain dependence on wavelength at constant voltage, and dark current and capacitance dependence on bias voltage. In addition, the device dark currents were recorded during irradiation.

The QE loss is greater than 25% at 450nm for the Hamamatsu device, Fig.1, whereas we detected no significant change for those of EG&G, Fig.2. This effect on the Hamamatsu device may be due to charge carriers being trapped in surface states produced by neutron damage at the Si-SiO₂ interface. As for the EG&G devices, the Si₃N₄ surface does not appear to have this problem. Likely as well is a thin (~0.2-0.4μm) undepleted region near the surface where produced charge carriers may only enter the depleted region through much slower diffusion processes.

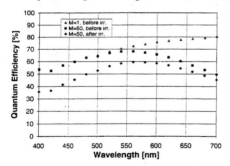 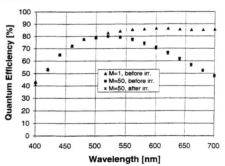

FIGURE 1. QE before and after irradiation for Hamamatsu BC-16 device.

FIGURE 2. QE before and after irradiation for EG&G device 15.

At low biases both device types show a marked effect on the gain profile due to neutron damage. Before irradiation there existed a wide 'plateau' region where the gain was assumed to be nominally one. However, after irradiation this 'plateau' no longer exists for any of the devices and determination of the gain one bias becomes difficult. This is not totally unexpected as neutrons create trapping centers in the silicon with different time constants and effectively these trapped charge carriers become 'lost' to the measurement at hand. We also see that this effect can be minimized by allowing sufficient time between measurements for the devices to settle to an equilibrium state. Irradiated devices have been seen to have 20-40 min settling times after bias changes (5).

For higher gains the Hamamatsu device does not appear to be affected by the neutrons, Fig.3. The EG&G devices demonstrate a 'shift' in the bias for a given gain, Fig.4. After taking this shift into account, the gain dependence on the bias voltage for these devices remains stable. This shift is proportional to the thickness of the device and hence is dependent on the high resistivity, intrinsic π-type wafer material. Our models have shown that this can be explained by an effective doping change in this region of the device from the creation of acceptor like states. Support for this is that the gain dependence is not altered as much as it is shifted. Shifting the curves by 9 and 30 volts, for devices 3 and 15 respectively, demonstrates a very good match for gain curves before and after irradiation. At the highest biases there is considerable self-heating of the devices due to power dissipation of 30-40μA currents. A temperature coefficient of the gain of -2.8% for the EG&G devices can cause a gain drop from self-heating effects at these currents.

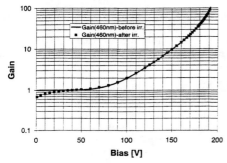

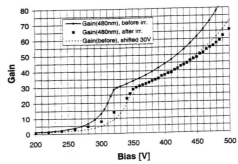

FIGURE 3. Gain before and after irradiation for Hamamatsu BC-16 device. λ=460nm, T=20C.

FIGURE 4. Gain before and after irradiation for EG&G device 3 showing result of 30V gain shift. λ=480nm, T=20C.

Gain as a function of wavelength is seen to drop off by at least a factor of 2 in all devices for wavelengths longer than 750nm. This is due to the rather thin avalanche region being located very near the light entry surface of the devices. Longer wavelengths penetrate further into the device and produce electron-hole pairs behind the high field region. Holes passing back through this region undergo a smaller amplification due to a lower ionization coefficient, resulting in a lower net gain. The fall off of the EG&G devices at shorter wavelengths than the Hamamatsu device illustrates the fact that their high field region is contained much closer to the front surface.

Dark current dependence on bias voltage is seen to increase linearly with integrated neutron flux as expected from what we have seen during previous irradiations (5). This holds for the total current read out from the Hamamatsu device and the separate bulk and surface currents of the EG&G devices.

The capacitance dependence on the bias for the Hamamatsu, Fig.5, does not change overall, but there are some structures that appear to have shifted with bias in the device after irradiation. These can be shown clearly by a plot of $dV/d(1/C^2)$ vs. bias. This is something that we do not have a satisfactory explanation for as of yet. As for the EG&G devices, Fig.6, we see more marked changes, especially in the region where full device depletion occurs. We attribute this to the 'effective' doping change in the thick, depleted π-region which causes the shift in the gain curves. This supports our hypothesis that the gain mechanism has not been affected, but rather a higher voltage is needed after irradiation to fully deplete the device.

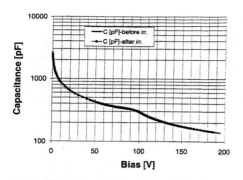

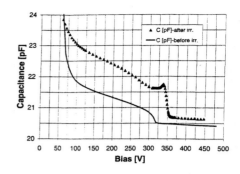

FIGURE 5. Capacitance before and after irradiation for Hamamatsu device BC-16. f=1MHz.

FIGURE 6. Capacitance before and after irradiation for EG&G device 3 showing capacitance change relative to 30V gain shift. f=1MHz.

CONCLUSIONS

We have described the neutron irradiation effects on three devices from two manufacturers up to a fluence of $2 \cdot 10^{13}$ neutrons/cm^2. The most marked effects seen are the QE loss at all wavelengths for the Hamamatsu device and an increased full depletion bias causing a gain 'shift'in the two EG&G devices. Most importantly, the gain mechanism for all of the devices appears to remain basically unaltered. The dark currents of both device types increase in a linear, predictable manner. The overall capacitance of the devices do not change enough to be of concern for the preamplifier designers. As anticipated the devices show some degradation, however they can clearly still be effective in the context of use in the CMS ECAL detector. Some problems still need to be overcome, but we are working in close cooperation with the manufacturers to optimize the next generation of these devices for testing.

ACKNOWLEDGMENTS

The authors wish to thank the following:

At ORNL: the members of the ORNL REDC Californium User's Facility for all of their professionalism and courtesy in helping us with the use and handling of the ^{252}Cf neutron sources.

At Northeastern University: the Physics Department Machine Shop personnel for test setup fabrication, and Matthew Marcus, a physics undergraduate who helped greatly with the development and running of the irradiation test.

REFERENCES

1. CMS ECAL Technical Design Report, **CERN/LHCC 97-33**, (1997)
2. Karar, A., et al., Ecole Polytechnique, IN2P3-CNRS, **X-LPNHE/95-10**, (1995)
3. Bateman, J.E., Stephenson, R., Rutherford Appleton Laboratories, **RAL-TR-96-009**, (1996)
4. Annual Book of ASTM Standards, E722-94, **12.02**, 318-326 (1996)
5. Reucroft, S., Rusack, R., Ruuska, D., Swain, J., *Nucl. Instr. Meth.* **A394**, 199-210 (1997)

SESSION 5: TRACKING - I

Chair: S. Gruenendahl
Scientific Secretary: H. Zheng

THE SCINTILLATING FIBER TRACKER AND THE OPTO-ELECTRONIC READOUT SYSTEM OF THE CHORUS NEUTRINO OSCILLATION EXPERIMENT

Dirk Rondeshagen

Institut für Kernphysik
University of Münster
Wilhelm-Klemm-Strasse 9
D-48149 Münster
Germany

E-mail:rondesh@uni-muenster.de

Representing the CHORUS Collaboration

Abstract. A scintillating fiber tracker system consisting of about 1.1 million fibers has been constructed successfully and is operational in the CHORUS neutrino oscillation experiment at CERN. The tracker system composes of a target tracker and a magnetic spectrometer. The target tracker consists of 8 tracker planes with 4 projections each, and the magnetic spectrometer consists of an air-core magnet and 3 tracker planes. The design, the performance and the opto-electronic readout system are described.

I INTRODUCTION

CHORUS[1] [1] is a short baseline neutrino oscillation experiment. It is located in the CERN wide band neutrino beam and is searching for the transition $\nu_\mu \rightarrow \nu_\tau$. It is using a hybrid detector which consists of an emulsion target followed by a real-time electronic detector (Figure 1). Data taking started in April 1994 and was completed in November 1997. With the complete data of 4 years, CHORUS will reach a sensitivity in the ν_μ/ν_τ mixing angle down to $\sin^2(2\theta_{\mu\tau}) \leq 2 \cdot 10^{-4}$ for large Δm^2. This is an interesting parameter space for the hypothesis that ν_τ is a Dark Matter candidate [2].

---

[1] Cern Hybrid Oscillation Research apparatUS

CP450, *SciFi97: Workshop on Scintillating Fiber Detectors*
edited by A. D. Bross, R. C. Ruchti, and M. R. Wayne
© 1998 The American Institute of Physics 1-56396-792-8/98/$15.00

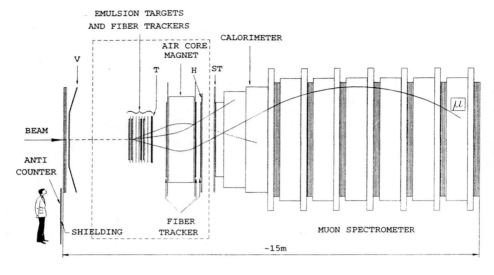

FIGURE 1. The CHORUS detector.

The CHORUS experiment searches for charged current interactions of ν_τ's from a $\nu_\mu \to \nu_\tau$ oscillation producing a negative τ-lepton

$$\nu_\tau + N \to \tau^- + X$$

whereby it is sensitive to the following decay channels of the τ^-:

$$
\begin{aligned}
\tau^- &\to \mu^- \bar{\nu}_\mu \nu_\tau & (17.3\ \%) \\
&\to h^-(n\pi^0)\nu_\tau & (49.8\ \%) \\
&\to (\pi^+\pi^-\pi^-)(n\pi^0)\nu_\tau & (14.9\ \%).
\end{aligned}
$$

To identify the short flight path of the τ^--lepton and its decay topology, CHORUS uses ~ 800 kg nuclear emulsion as an active target. Nuclear emulsion offers a spatial resolution of ~ 1 μm with a hit density of ~ 300 grains/mm and is therefore ideal for the detection of short lived particles. The emulsion target is divided into 4 stacks with an area of (1.44×1.44) m^2 and a thickness of 2.75 cm each. Each stack is subdivided into 35 plates of 800 μm thickness each. Behind each stack, there are three emulsion sheets, which act as an interface between the trackers and the emulsion stacks. The first, called special sheet (SS), is mounted directly behind the emulsion target. The other two, called changeable sheets (CS, named because of their frequent exchange to avoid excessive track densities) are close to the target trackers. These large arrays of high resolution scintillating fiber trackers are placed between the four emulsions stacks. Measured track data from the trackers are used to first project the tracks coordinates onto the changeable sheets. A high precision in the prediction of exiting tracks is required to get unambiguous and efficient track finding and to reduce the scanning load. Through such

172

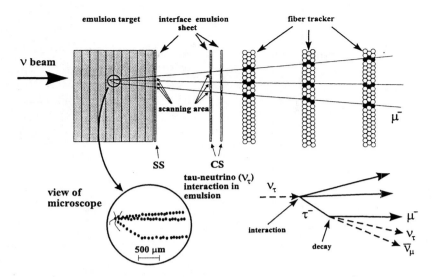

FIGURE 2. The target region with an emulsion target, three interface emulsion sheets (CS and SS) and three fiber trackers is shown schematically. A ν_τ-CC interaction with a "kink" structure as it would be visible under the microscope is shown.

an arrangement (Figure 2), the track prediction accuracy progressively improves from the fiber tracker to the CS and further down to the SS, right in front of the emulsion target.

II SCINTILLATING FIBER

The CHORUS tracker system [3,7,8], consisting of target trackers and trackers associated to an air core magnet, contains $\sim$ 1.1 million (about 2500 km) plastic scintillating fibers[2] of 500 μm diameter. The fibers consist of a polystyrene core (refractive index 1.59) doped with 1 % butyl-PBD[3] and 0.1 % BDB[4], surrounded by a 3 μm PMMA[5] cladding (refractive index 1.42). The emission spectrum peaks at $\sim$ 420 nm.

Fiber ribbons with 7 layers in a "staggered" geometry (Figure 3) were constructed as basic elements of the tracker planes. These ribbons were produced with a special fiber-winding machine [4]. A single fiber taken from a bobbin was wound around a drum. A spiral grove on the drum leads the fiber and guarantees the coherence of the fibers in the ribbon. Each layer was painted with a TiO_2-based white paint

---

[2] KURARAY SCSF-38/Japan
[3] butyl-2-phenyl-5(4-biphenyl)-1-3-5-oxadiazole
[4] 4,4'-bis-(2,5-dimethylstyryl)-diphenyl
[5] polymethyl methacrylate

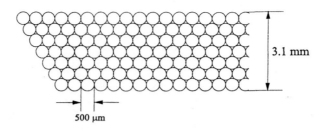

3.1 mm

500 μm

FIGURE 3. Cross section of a scintillating fiber ribbon, showing its "staggered" geometry.

which acts as glue and as extra-mural absorber (EMA) to prevent cross talk. One end of each ribbon was polished and sputtered with aluminium, to act as a mirror ($\sim$ 80 % reflectivity). The procedures and quality control of the mass production of the fiber ribbons have been reported in [5].

The tracker inefficiency per ribbon (the probability of a minimum ionizing particle passing through without producing any signal recorded by the opto-electronic readout chain) was evaluated using beam muon data, and was measured to be $\sim 2 \times 10^{-3}$.

The attenuation length of the scintillating fibers was monitored over 4 years using beam muons. The ageing [7,8] is negligible with less than 2.5 % loss of performance per year. This result can be attributed to special attention paid during construction and operation of the tracker system - e.g. minimized exposure to ultra-violet light, handling with clean gloves, storage and operation at 5°C ambient temperature.

III TARGET TRACKER

The main purpose of the target tracker is to reconstruct the position of the neutrino interaction vertices, and to provide accurate predictions for tracks at the interface emulsion sheet (shown as CS in Figure 2), which are essential for track scanning in the emulsion. Therefore large arrays of 8 scintillating fiber trackers, which provide information in 4 projections each, are placed downstream of each emulsion stack (Figure 4).

The basic elements of the target tracker are scintillating fiber ribbons of 2.3 m length, 40 cm width and an active length of 1.6 m. Four of these ribbons were combined to form a subplane with an active area of (1.6×1.6) m^2 (Figure 5). Each tracker plane consists of 4 subplanes (Y,Z, and Y$^\pm$, Z$^\pm$ inclined by $\pm 8°$ relative to Y,Z). Combining the information from the four projections, a unambiguous three-dimensional trajectory can be determined for each traversing particle. For the entire target tracker system $\sim$ 720,000 fibers are used.

The target tracker, which consists of 8 tracker planes, is compact to avoid loss of accuracy by extrapolating tracks over large distances (Figure 4). The distance of

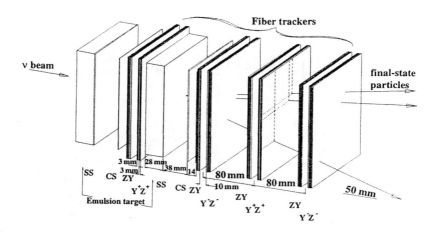

FIGURE 4. The geometry of one half of the module of the target region, showing the emulsion targets and interface sheets with the target tracker. The second half is identical.

52 mm between the target emulsion and the closest downstream tracker plane was chosen as a compromise between the two-track separation and the track prediction accuracy.

At the readout end, each ribbon was split into 5 strips of 8 cm width. 16 of these strips were bundled together, and coupled to an opto-electronic readout system. In total, 40 of these readout bundles were produced.

IV MAGNETIC SPECTROMETER

The target region is followed by a magnetic spectrometer [7,8] (Figure 1). It consists of an air-core magnet [6] and 3 scintillating fiber trackers, one in front of the magnet and two behind. The spectrometer allows determination of charge and momentum of low energy hadrons which are mainly produced in tau decays. The magnet is operated in pulsed mode in synchronisation with the beam extraction of the SPS and provides a momentum resolution of $\Delta p/p \approx 0.3$ at 5 GeV.

The tracker of the magnetic spectrometer consists, like the target tracker, of 7-layer ribbons with a fiber length of ~ 2.6 m. It consists of $\sim 330,000$ fibers. The readout ends of the fibers are prepared in the same way as for target tracker section. They are read out with 18 opto-electronic chains. More details about this magnetic spectrometer are given in [8].

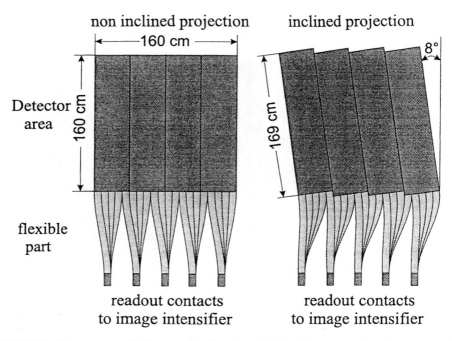

FIGURE 5. The geometry of a "non-inclined" and an "inclined" projection for the target tracker planes.

V CALIBRATION SYSTEM

Because of image distortion in the image intensifier chain, a calibration system is necessary to obtain an accurate mapping between the input window and the CCD[6]-pixel coordinates. The 16 fiber strips, which are bundled together, are separated by spacers (Figure 6). Five of these spacers ("flappers") contain 9 clear light fibers each (127 μm diameter). These "fiducial" fibers are grouped together and coupled with a fiber optic cable to a LED pulser module (Figure 6). Details of the construction and tests of the calibration system can be found in [9]. The exact dimensions of the fiber bundles and the position of the fiducial fibers were measured with photographic contact prints before the opto-electronic readout chains were mounted. The geometry at the input window is shown in Figure 7a. The image distortion due to the image intensifiers turns this pattern into the one shown in Figure 7b.

The fiducial fiber system is also used to optimize the operation of the opto-electronic readout. The LED intensity is adjusted to produce single photo-electrons at the first photo-cathode of the readout chain. The optimum parameters, such

---

[6] Charge Coupled Device

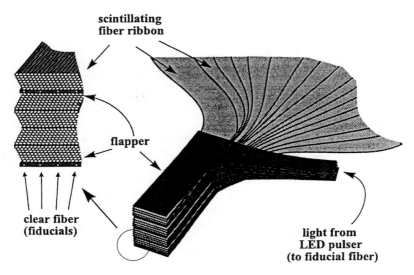

scintillating fiber ribbon

flapper

clear fiber (fiducials)

light from LED pulser (to fiducial fiber)

FIGURE 6. Schematic diagram of the construction of the fiber bundle and calibration fiducial fibers to be coupled to the opto-electronic readout chain

as the image intensifier focus, the zooming voltage, the MCP voltage and the gate time, can be determined by studying the response of the readout system (i.e. the quantum effiencies, gain, spatial resolution, and signal-to-noise ratio).

VI OPTO-ELECTRONIC READOUT

Since only few photons are produced by a charged particle passing through the fiber ribbon, an opto-electronic readout chain (Figure 8) is used to amplify the signal. It consists of 4 image intensifiers: a large surface demagnifying stage (100 mm to 25 mm diameter) with low gain but high quantum efficiency, is followed by a second electrostatic image intensifier to increase the gain, a high gain gatable Micro Channel Plate (MCP) image intensifier and a final electrostatic demagnifying stage (25 mm to 11 mm diameter). The chain is readout by a CCD camera.

The photon detection efficiency is determined by the first image intensifier. It was measured to be about 18 %, which is a convolution of a photo-cathode (plus fiber optical window) QE of 22 % and a phosphor QE of 80 %. A slow P11 phosphor (50/90 % light collection in 50/500 μs) was chosen for the first stage to avoid large light losses due to trigger delay during the first 0.5 μs. The other image intensifiers use a fast P46 phosphor. On average 5-7 photo-electrons per traversed ribbon are produced at the photo-cathode of the first image intensifier by minimum ionizing particles. The net gain of the entire chain is about 10^4 to 10^5.

There are two aspects which characterize the spatial resolution of the readout

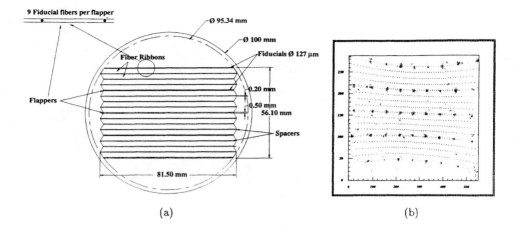

FIGURE 7. The fiber ribbon bundle as measured in window of the opto-electronic chains [Figure a)], and the CCD readout [Figure b)], showing the effects due to electrostatic distortion from the image intensifiers. The units are CCD pixel numbers, translating to X=145 μm and Y=209 μm per unit in detector space. Small black dots denote the parametrized ribbon boundary.

system. The "spot size" (width of the charge distribution induced on the CCD from a single photo-electron in the first image intensifier) is due to the intrinsic resolution of the opto-electronic readout chain, whereas the "spot displacement" (deviation of the center of gravity of a cluster from its mean position) is caused mainly by the focusing of the first image intensifier. To obtain good spatial resolution as well as good gain, the LED calibration system is used to tune each image intensifier chain to its optimum. The standard deviations of the spot size and its displacement, based on the measured fiducial width, are 136 μm and 89 μm at the input window, respectively.

The CCD sensor contains an image zone and a memory zone, with an anti-blooming gate in the image zone to provide a 1 μs fast clear. Each zone consists of 550 × 288 pixels with dimension (16 × 23) μm^2. Each pixel corresponds to (145 × 209) μm^2 in detector space. The transfer from the image to the memory zone takes ~ 125 ms. Details of the image intensifier and CCD readout system are given in [10,11].

In total 58 opto-electronic readout chains are used to read out about 1.1 million fibers.

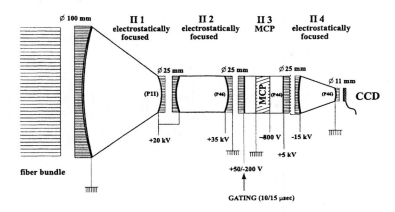

FIGURE 8. The configuration of the opto-electronic readout chain.

VII PERFORMANCE

Beam muons and cosmics were used to measure the properties of the fiber tracker and to achieve precise relative alignment of the various tracker planes.

The obtained hit residuals for beam muon tracks (deviation of hits from the best fitted trajectory) is $\sim$ 330 μm, while the "track-element" (center of gravity of all hits in one ribbon) residual is $\sim$ 180 μm. The two track resolution is $\sim$1 mm. The prediction accuracies (deviation of the target tracker prediction on the emulsion interface sheets from the found tracks) have a σ of $\sim$ 150 μm in position and $\sim$ 2.5 mrad in angle. Folding in an expected intrinsic resolution of 1.5 mrad for the emulsion sheets, the angular resolution of the fiber tracker is $\sim$ 2 mrad.

VIII CONCLUSION

The CHORUS scintillating fiber tracker system has been constructed successfully and has taken data for more than four years. The expected hardware performance in terms of hit density, fiber attenuation length and readout spatial resolution has been achieved.

IX ACKNOWLEDGEMENTS

We gratefully acknowledge the financial support of the different funding agencies: in particular, the Institut Interuniversitaire des Sciences Nucleaires and the Interuniversitair Instituut voor Kernwetenschappen (Belgium), the German Bundesministerium für Bildung und Forschung under contract numbers 05 6BU11P for Berlin and 05 7MS12P for Münster (Germany), the Japan Private School Promotion Foundation and Japan Society for the Promotion of Science (Japan), the

Foundation for Fundamental Research on Matter FOM and the National Scientific Research Organisation NWO (the Netherlands).

REFERENCES

1. N. Armenise et al., CHORUS Proposal, CERN-SPSC/90-42, (1990);
   M. de Jong et al., CERN-PPE 93-131 (1993);
   E. Eskut et al., Nucl. Instrum. Methods **A 401** (1997) 7-44
2. Ya. B. Zel'dovich and I. D. Novikov, Relativistic Astrophysics, Nauka, Moscow, 1967;
   H. Harari, Phys. Lett. **B 216** (1989) 413;
   J. Ellis, J. L. Lopez and D. V. Nanopoulos, Phys. Lett **B 292** (1992) 189;
   H. Fritzsch and D. Holtmannspötter, Phys. Lett **B 338** (1994) 290
3. S. Aoki et al., Nucl. Instrum. Methods **A 344** (1994) 143
4. T. Nakano et al., Proc. IEEE 1991 Nuclear Science Symp., Santa Fe (1991)
5. T. Nakano et al., Proc. of Scintillating Fibers Symposium 1993, University of Notre Dame (1993)
6. F. Bergsma et al., Nucl. Instrum. Methods **A 357** (1995) 243
7. P. Annis et al., Nucl. Instrum. Methods **A 367** (1995) 367
8. P. Annis et al., Nucl. Instrum. Methods, to be published; CERN-PPE/97-100
9. D. Rondeshagen, Dipl. Thesis (1994), Westfälische Wilhelms-Universität, Münster, Germany
10. M. Gruwé, Ph.D. Thesis (1994), Université Libre de Bruxelles,Belgium
11. C. Mommaert, Ph.D. Thesis (1995), Vrije Universiteit Brussel,Belgium

Results from the E835 Cylindrical Scintillating-Fiber Tracker

M. Ambrogiani, W. Baldini, D. Bettoni, R. Calabrese,
E. Luppi, R. Mussa and G. Stancari

Dipartimento di Fisica, Università di Ferrara, 44100 Ferrara (FE), Italy
Istituto Nazionale di Fisica Nucleare, Sezione di Ferrara, 44100 Ferrara (FE), Italy

Abstract. A cylindrical 860-channel scintillating-fiber tracker for the measurement of the polar coordinate θ has been built for experiment 835 at Fermi National Accelerator Laboratory; it was used during the data-taking period, from October 1996 through September 1997.

The high granularity, flexibility and fast response of the scintillating fibers are combined with the high quantum efficiency of the Visible-Light Photon Counters. From this combination, a new kind of detector arises, capable of high-resolution tracking, fast first-level kinematical triggering and single/double-track discrimination.

Results of the tracker performance are presented.

INTRODUCTION

As often happens in high-energy physics, a detector needed to be developed to extract rare reactions from a huge background. A constraint usually exists on the amount and rate of data that an acquisition system can manage. Also, fast high-luminosity data-taking is sought, but it requires radiation-hard detectors [1]. For its physics goals, the E835 collaboration decided to build a scintillating-fiber tracker (SFT) read out by Visible-Light Photon Counters (VLPCs).

The thinness of the fibers allows a precise position determination; their flexibility makes possible a high-precision measurement of the angle θ relative to the beam axis. Due to its fast response, a detector made of scintillating fibers can also be used in a first-level trigger logic. The VLPCs compensate for the small amount of light produced by the thin fibers, because of their high quantum efficiency. The homogeneity of their response makes the analog readout of each channel useful for pulse-height analysis. Moreover, both the attenuation length of the fibers and the gain of the VLPCs have been shown to be nearly independent of the data-acquisition rate, at E835 luminosities ($\approx 2 \cdot 10^{31}$ cm$^{-2} \cdot$ s^{-1}) [2].

CP450, *SciFi97: Workshop on Scintillating Fiber Detectors*
edited by A. D. Bross, R. C. Ruchti, and M. R. Wayne

THE EXPERIMENT

Experiment 835 studies the direct formation of charmonium mesons in anti-proton-proton annihilations ([3], [4], [5], [6]). By scanning an interval of center-of-momentum (CM) energy and counting the number of events that satisfy the charmonium-formation hypotheses, we obtain an excitation curve from which we infer masses, widths and decay branching fractions of these ($c\bar{c}$) states. About 150 pb^{-1} worth of data in the CM energy region between 2.9 GeV and 4.2 GeV were collected during the data-taking period, which lasted from October 1996 through September 1997.

The antiproton beam is stored and stochastically cooled in the Antiproton Accumulator, where it intersects a gaseous hydrogen jet-target, defining an interaction region of about $5 \times 5 \times 5$ mm^3. The detector is designed to select the final states of the charmonium mesons that include electrons, positrons or photons. It is symmetrical around the beam axis and covers the polar angle from 15° (downstream region) to 65° (upstream). Its main components are an inner tracking system, a threshold Cherenkov counter and a lead-glass electromagnetic calorimeter, as shown in fig. 1.

The signature for charmonium formation is given by an electron and a positron with invariant mass around 3.1 GeV/c^2, indicating the inclusive production of the

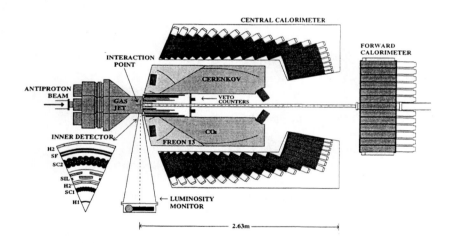

FIGURE 1. The E835 Detector. The inner tracking system consists of three hodoscopes (H1, H2', H2), two sets of straw tubes (SC1, SC2), silicon strips (SIL) and the two layers of scintillating fibers (SF). The central calorimeter is segmented in both θ (20 elements, shown) and ϕ (64 elements, not visible in the picture). The z axis of the lab frame (centered on the interaction point) is defined by the antiproton momentum, while y is the vertical axis.

J/ψ meson and its decay into an e^+e^- pair. Also, the decay of a charmonium state into two photons is directly detectable.

A complementary approach to the identification of charmonium uses hadronic final states. For the study of the $\eta_c(1^1S_0)$ state and the search for the $\eta_c'(2^1S_0)$ state, the processes $\bar{p}p \rightarrow \bar{p}p$ and $\bar{p}p \rightarrow \phi\phi \rightarrow K^+K^-K^+K^-$ were chosen. In this context, the SFT represented the ideal choice: high tracking resolution for the topological identification of those reactions and fast first-level trigger capabilities to improve the trigger performance.

THE SCINTILLATING-FIBER TRACKER

Detector Design

This cylindrical detector is used to measure the polar coordinate θ in the region between $15°$ and $65°$. The 860 fibers are laid on two staggered layers. Following is a brief description of the layout of the tracker and of its readout. Details regarding the fibers (Kuraray SCSF-3HF-1500), the cylindrical supports, the VLPCs (HISTE-V, EOC-low), the cryostat and the readout electronics can be found in [2], [7] and [8].

LIGHT COLLECTION. Each scintillating fiber (core diameter = 740 μm) is wound around one of the two support cylinders (radii 144.0 mm and 150.6 mm), on the surface of which a set of U-shaped grooves has been machined (pitches 1.10 mm and 1.15 mm). The depth of the grooves varies linearly with the azimuthal coordinate ϕ,

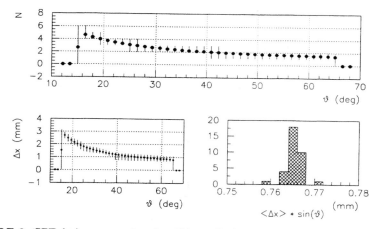

FIGURE 2. SFT design parameters from Monte Carlo calculations. The number N of hit fibers and the active thickness Δx of the detector are shown as a function of the polar angle θ. The dots represent the average over an arbitrary θ slice, while the error bars indicate the minimum and maximum possible values. The distribution of the average active thickness, multiplied by $\sin(\theta)$, is also plotted (lower right).

so that the fiber overlaps itself after one turn. On one end, the fibers are aluminized, to increase the light yield and improve its homogeneity; on the other end, they are thermally spliced to clear fibers, which, after a four-meter path, bring the light to the surface of the VLPCs, housed in a cryostat and kept at a temperature of 6.5 K.

SIGNAL READOUT. The electronic signal generated by the VLPCs is amplified by QPA02 cards, designed at Fermilab. After the amplification stage, the signal is sent to discriminator-OR-splitter modules, custom made in Ferrara. These modules provide an analog and a digital output for each input channel, together with the digital OR of all inputs. The analog signal is then sent to ADCs, while the digital output is read out by latches. The digital OR of a set ('bundle') of adjacent channels is sent to TDCs and to the first-level trigger logic of the experiment.

GEOMETRICAL MONTE CARLO. Fig. 2 shows the number N of fibers (summed over both layers) whose core is intersected by a straight track originating from the interaction vertex; in order to maximize its geometrical acceptance, the detector has been designed so that, in the region $15° < \theta < 65°$, there is always at least one fiber on the track path. The active thickness Δx is the sum, over all hit fibers, of the track path lengths inside the traversed fiber cores; even though its fluctuations are not negligible, its mean $\langle \Delta x \rangle$ is proportional to $1/\sin(\theta)$; this yields to an extrapolated normal-incidence active thickness of 0.765 ± 0.001 mm.

The total thickness of the tracker, dominated by the acrylic support cylinders, corresponds to $\approx 10^{-2}$ radiation lengths per layer at normal incidence.

Detector Performance

RELIABILITY OF THE EQUIPMENT. Throughout the run, the cryostat worked very efficiently. The liquid helium flux was constant and the temperature stable $(6.500 \pm 0.004$ K$)$. Four inevitable warm-ups occurred, due to scheduled shutdowns or to power outages; the temperature of the cryostat rose typically up to ≈ 70 K in a few hours. That was the main cause of loss of channels, attributed to the poor

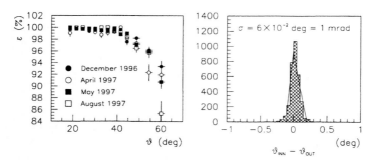

FIGURE 3. Tracking performance. On the right, the detection efficiency as a function of the polar angle and time; on the left, the distribution of θ residuals.

quality of the micro-welding joining the VLPC substrates to the electrical cables. Four channels out of 860 were not working at the beginning of the run; at the end, we had almost eighty dead channels [2].

SIGNAL. The pulse generated by a track crossing one fiber, as seen at the input of the discriminator-OR-splitter modules, is typically 180 mV high and 80 ns wide, corresponding to a collected charge $\approx$ 0.2 nC. The discriminator thresholds are set to about 40 mV.

DETECTION EFFICIENCY AND TRACKING RESOLUTION. We measured the detection efficiency with J/ψ-inclusive and ψ' events, when the J/ψ or ψ' decay into an $e^+ e^-$ pair ($\approx 4 \cdot 10^4$ events over the whole data-taking). For each track, we look for an associated hit in the SFT above a given software threshold (typically, 0.2 mip), within a polar window of ± 50 mrad; the results are shown in fig. 3. The seeming detection inefficiency at $\theta > 40°$ is actually accounted for by a loss of geometrical acceptance, caused by an improper mechanical alignment of the two layers (the relative shift, measured independently in two-body reactions, is 0.8 mm larger than the design value). Variations of the efficiency over time are due to different run conditions (gate width of the ADCs) and to loss of channels.

The SFT is by far the detector with the best spatial resolution in our apparatus. For this reason, we can only measure what we call 'intrinsic' tracking resolution, i. e. the standard deviation (divided by $\sqrt{2}$) of the distribution of the differences $\theta_{INN} - \theta_{OUT}$, where, for a given track, θ_{INN} (θ_{OUT}) is the polar angle measured by the inner (outer) layer. The resolution, averaged over all polar angles, comes out to be (0.7 ± 0.1) mrad, as shown in fig. 3.

SINGLE-FIBER AND DETECTOR CALIBRATION. In order to get, for each VLPC channel, the one-photoelectron (phe) equivalent in ADC counts (1 ADC count = 0.25 pC), we performed an LED test on all channels with the setup actually used

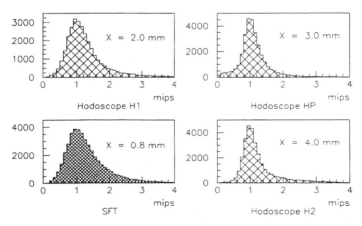

FIGURE 4. Comparison between the energy deposit distribution in three hodoscopes and in the SFT. X is the normal-incidence thickness.

in the experiment (VLPC cassettes, cryostat and readout electronics). The pulse charge in ADC counts generated by a minimum-ionizing particle, instead, is obtained studying a high-statistics hadronic sample of punch-through tracks in the electromagnetic calorimeter ($\approx 10^3$ events/fiber). The ADC/mip variable reflects fluctuations from many sources: the response of the scintillating fibers; the light transmission of the connections; the coupling between clear fibers and VLPC surfaces; the gains of the VLPCs and of the amplifiers; and the readout electronics. Still, the distribution of this variable shows an excellent homogeneity (mean ≈ 400 ADC/mip, standard deviation $\approx 12\%$). The mean number of photoelectrons per minimum-ionizing particle turns out to be distributed around 14 phe/mip. More details can be found in [2], [8] and [7].

PULSE-HEIGHT ANALYSIS AND SINGLE/DOUBLE-TRACK DISCRIMINATION. Fig. 4 shows a comparison between the fluctuations of the energy deposit in the scintillating-fiber tracker and those in a traditional hodoscope/phototube system. The normal-incidence thicknesses X are very different (≥ 2 mm for the hodoscopes, < 0.8 mm for the SFT) and so are the light-guide lengths L (typically 50–100 cm between a hodoscope and a phototube, while the clear fibers are more than four meters long); also, the thickness oscillations of the SFT are significant. Still, the fluctuations of the energy deposit in the SFT (hence the amount of light collected) are comparable with those in hodoscope H1.

These relatively small fluctuations, together with the good spatial resolution, can be helpful when applied to suppress a kind of background particularly important for E835, and common to many high-energy-physics experiments: e^+e^- pairs with small opening angle, generated by photon conversions or by Dalitz decays of neutral pions, simulating a single track.

The tools that we use off-line to reject this background are the pulse height in

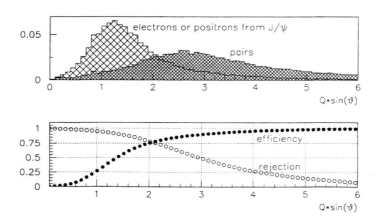

FIGURE 5. Single/double-track discrimination power using charge information only. A clean sample of electrons or positrons is compared with a sample of pairs.

186

the hodoscopes, the signal in the threshold Cherenkov counter and the energy and shape of the showers in the central calorimeter. The SFT can add two tools: pulse height and granularity. When the opening angle of the e^+e^- pair is so small that just one cluster (defined as a set of adjacent hit fibers) is generated, the energy deposit is likely to be big; whereas when the pair's separation is big enough, an extra cluster in the detector will appear. In fig. 5 one can see the discrimination, based on charge deposition only, between a clean sample of single tracks (electrons or positrons) and a sample of pairs, defined as charged tracks which, combined with a neutral deposit in the central calorimeter, form a π^0 invariant mass. Q is the sum of the charges deposited in all SFT clusters in a $\pm 2°$ window centered around the track. For instance, a 50% rejection factor is achieved with 90% efficiency, cutting at $Q \cdot \sin(\theta) = 3$.

TIMING RESOLUTION. In order to evaluate the intrinsic time resolution of the detector, we select those tracks that hit at least two fibers belonging to adjacent bundles, and are therefore read out by different TDC channels. We then define the time resolution as the standard deviation, divided by $\sqrt{2}$, of the distribution of the variable $t_i - t_{i+1}$, when the same track crosses both a fiber of bundle i and a fiber of bundle $i + 1$ ($i = 1, 18$). This intrinsic time resolution comes out to be around 3.5 ns, as can be seen in fig. 6; it is mostly due to the decay time of the scintillator.

FIRST-LEVEL TRIGGER. Given the excellent timing performance of the detector in terms of both fast response and intrinsic resolution, a fast first-level trigger logic based on the SFT was designed. The goal is to observe the decay of the $\eta_c(1^1S_0)$ and $\eta_c'(2^1S_0)$ mesons into $\phi\phi \rightarrow K^+K^-K^+K^-$ and the interference between resonant and elastic $\bar{p}p$ final states near 90° in the CM frame. Coincidences between sets of fiber bundles can select the right polar kinematics, whereas the hodoscopes provide azimuthal information. This first-level selection yields affordable trigger rates — 20–110 Hz for the $\bar{p}p$ trigger and 600–1000 Hz for the $\phi\phi$ trigger, as the CM energy varies between 4.2 GeV and 2.9 GeV.

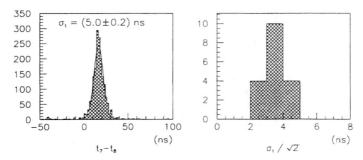

FIGURE 6. Intrinsic time resolution. The first plot shows the time difference distribution for a couple of adjacent bundles, while the one on the right is the standard-deviation distribution for all eighteen couples of adjacent bundles.

187

CONCLUSIONS

We presented the results of the first detector used in a high-energy-physics experiment which exploits scintillating fibers and Visible-Light Photon Counters.

It is a unique detector, because it combines a high tracking performance (resolution < 1 mrad, detection efficiency ≈ 99.5%), fast first-level trigger characteristics (time spread < 5 ns) and single/double-track discrimination power.

The technique is ready for future large-scale applications.

ACKNOWLEDGEMENTS

The realization of this project would not have been possible without the valuable support of the Fermilab Particle Physics Division; the Fermilab Cryogenics Group; the D0 and CDF Collaborations at Fermilab; the University of Ferrara and INFN; and, of course, the E835 collaborators. In particular, we wish to thank Alan Bross (Fermilab and D0 Collaboration) and Stephen Pordes (Fermilab, Particle Physics Division).

Discussions with M. Atac, C. Lindenmeyer, M. Mishina and R. Richards represented a source of helpful suggestions and effective ideas.

REFERENCES

1. Ruchti R. C., *Ann. Rev. Nucl. Sci.* **46**, 281–319 (1996).
2. Mussa R., "Performance Measurements on Histe-V VLPC Photon Detectors for E835 at FNAL", in these proceedings.
3. Pastrone N. (for the E835 Collaboration), "New Results from E835 at Fermilab Antiproton Accumulator — Study of ($c\bar{c}$) States formed in Antiproton-Proton Annihilations", presented at the 4th International Workshop on Progress in Heavy-Quark Physics, Rostock, Germany, September 20–22, 1997.
4. Zioulas G. (for the E835 Collaboration), "First Results on Charmonium Spectroscopy from Fermilab E835", presented at the 7th International Conference on Hadron Spectroscopy, Brookhaven National Laboratory, Upton, New York, August 25–30, 1997.
5. Cester R. and Rapidis P. A., *Ann. Rev. Nucl. Sci.* **44**, 329–371 (1994).
6. Armstrong T. et al. (E760 Collaboration), Fermilab Proposal P-835-REV (1992).
7. Luppi E., "The E835 Scintillating-Fiber Tracking Detector", presented at the 5th International Conference on Advanced Technologies and Particle Physics, Villa Olmo (CO), Italy, October 1996.
8. Ambrogiani M. et al., *IEEE Trans. Nucl. Sci.* **44**, 460–463 (1997).

The Scintillating Fiber Detectors of the H1 Forward Proton Spectrometer

J. Bähr, U. Harder, K. Hiller, H. Lüdecke, R. Nahnhauer

DESY–Zeuthen, 15738 Zeuthen, Germany

Abstract. Since 1995 the H1 experiment at HERA is operating a Forward Proton Spectrometer (FPS) employing the HERA machine magnets adjacent to the interaction zone as spectrometer magnets. The FPS consists of four stations, two vertical stations and two horizontal stations. Scattered protons are detected in pairs of stations behind the interaction point with scintillating fiber hodoscopes. The scintillating fibers are readout by Position-Sensitive Photomultipliers (PSPM) in the case of the vertical stations and by Micro-Channel Photomultipliers (MCPM) for the horizontal stations.

INTRODUCTION

At the HERA collider at DESY a class of events is observed in deep inelastic electron proton scattering and photoproduction, which is characterized by the absence of secondary particles in a region of phase space between the outgoing proton debris and the target jet. These so called rapidity gap events have all features of diffractive interactions known from hadron–hadron scattering. One can expect, that in a fraction of these events the proton stays intact and leaves the interaction point unobserved in the central H1 detector through the beam pipe. In order to detect them the HERA machine magnets adjacent to the interaction region are used in a proton spectrometer [1]. After about 80 m protons, which are emitted at angles less than 0.5 mrad with respect to the beamline, appear at a distance of several millimeters from the nominal orbit so that they can be registered in detectors close to the circulating proton beam.

A permanent installation of counters a few millimeters from the proton beam is impossible, because they would cut into the machine aperture needed for injection and acceleration. The detector elements are thus mounted inside movable vacuum sections, so called Roman Pots , which are retracted during injection and are brought close to the beam after stable conditions have been reached. Fig. 1 shows a sketch of such a Roman Pot device with fiber detectors and housing for front-end electronics.

CP450, *SciFi97: Workshop on Scintillating Fiber Detectors*
edited by A. D. Bross, R. C. Ruchti, and M. R. Wayne
© 1998 The American Institute of Physics 1-56396-792-8/98/$15.00

In the 1995 until 1997 running periods the H1 group operated two Roman Pots [2]. From 1997 four stations [3] with scintillating fiber hodoscopes were operated to measure the position and direction of scattered protons.

PRINCIPLE OF THE FPS

The HERA beam line is sketched in Figure 2. It provides a horizontal bend of the proton beam by dipoles around 20 m and a vertical deflection by three dipole magnets between 60 and 80 m. Also indicated are the positions of the two detector stations at 81 and 90 m, which serve to reconstruct trajectories of scattered protons.

Due to the fact that the HERA beam line introduces a dispersion in two planes, longitudinal momenta can be measured twice by making use of the horizontal and vertical deflections. The two independent measurements which have to agree within errors, if they belong to one track coming from the interaction point, provide efficient means to reject background tracks. The detectors inside the Roman Pot plunger vessels thus have to record two coordinates at two positions behind the interaction region such that a straight line through the measured space points determines a trajectory which can be traced back to the origin by employing the known transfer properties of the magnetic channel.

The detector is based on scintillating fibers and makes use of their main advantages, which are fruitfully used also in other experiments of High Energy Physics e.g. in the UA2-Experiment [4], in CHORUS [5] and D0 [6] as

- time resolution (matching to the HERA cycle of 96 ns)

- spatial resolution

- robustness

- relative radiation hardness

and especially also the possibility , that by use of light guide fibers , a distance in space can be realized between the fiber detector and the front-end electronics, which is a clear advantage compared to other detector techniques.

The opto-electronic readout is realized by multi-channel photomultipliers (PSPMs) for the vertical stations at 81 m and 90 m and by multipixle microchannel devices (MCPMs) at the horizontal stations at 64 m and 80 m.

The trigger consists of four scintillator planes per station and gives a fast trigger signal within a few nanoseconds.

THE SCINTILLATING FIBER DETECTORS

The H1 Forward Proton Spectrometer (FPS) consists of four stations. Two vertical stations are installed at 81 m and 90 m since 1995, two horizontal stations at 64 m and 80 m were brought into operation in 1997. The distance from the interaction point and orientation of the detectors were chosen basing on beam-optical and Monte Carlo calculations which optimized the acceptance for different physical processes. Figure 3 shows the scheme of a detector insert of the horizontal stations. The detector insert consists of the main parts: Fiber detector support and electronic box. The fiber detectors including trigger planes and the PMs for the trigger readout of the horizontal stations are mounted on the fiber detector support. The multi-channel opto-electronic devices and the readout electronics are distributed in the electronic box.

The scintillating fiber detector of one station consists of two subdetectors, which are separated by 60 mm. This detector geometry allows background reduction and a rough measurement of track angles. In figure 4 a sketch of a fiber detector and the trigger tiles of a vertical station is shown. Methodical investigations using radioactive sources, cosmic particles and electrons of 3 GeV were performed during the development of the fiber detectors [7].

Each subdetector consists of two coordinate detectors u,v. The fibers are oriented under 45 degrees relative to the horizontal and vertical coordinate axes (x,y). The coordinate detectors consist of five layers of scintillating fibers of 1 mm diameter. To reach a spatial resolution of about 100 μm the fiber layers are staggered to each other by pitch/5 = 0.21 mm, while the pitch amounts to 1.05 mm. Figure 5 shows the design of a coordinate detector of a horizontal station.

There are 48 fibers per layer in the detectors of the vertical stations and 24 fibers per layer in the horizontal stations. The scintillating fibers and the light guide fibers are produced by pol.hi.tech. [1].

The optical contact of high transmission is realized by thermal splicing of the scintillating fibers to the light guide fibers of about 50 cm length. Therefore the damage of the opto-electronic devices and the front-end electronics in the period of beam filling and acceleration is excluded with high probability. To splice scintillating fibers to light guide fibers one fiber of both types are put in a precise quartz tube. Under slight vacuum a resistive wire wound around the quartz tube produces heat by a certain electric current of well defined duration. The splice point is mechanically stable , the optical transmission is about 95 percent. The fiber detectors were produced by GMS Berlin [2].

The endface of the fiber detectors is polished to optical quality. The far ends of the light guide fibers are glued in masks, which couple to the opto-electronic readout devices. The basic characteristics of the fiber detectors are summarized in table 1.

---

[1] pol.hi.tech., s.r.l., S.P. Turanense, 67061 Carsoli(AQ), Italy

[2] GMS- Gesellschaft für Mess- und Systemtechnik mbH, Rudower Chaussee 5, 12489 Berlin, Germany

TABLE 1. Parameter list of the fiber detectors for the FPS

Fiber hodoscopes	
Fibers per coordinate and subdetector	5 layers with 48 fibers (vertical stations)
	5 layers with 24 fibers (horizontal stations)
Fiber diameter	1.00 mm
Fiber type	POLHITECH 042–100
Staggering	0.21 mm

OPTO-ELECTRONIC READOUT

The demand to the opto-electronic readout consists in a high time resolution matched to the HERA beam cycle to keep the possibility of deriving a fast trigger signal from the fibers and in the registration of the amplitude of each fiber in such a way, that the spatial resolution of the detector is fully used.

Vertical station

In preparation of the construction of the vertical stations a careful comparison of different techniques of opto-electronic devices was performed [8]. The result was a decision to use the device H4139-20 from Hamamatsu [3] basing on fine-mesh dynodes with improved shielding against cross-talk. The multi-pixle PM can be operated in magnetic fields up to 1 T. Table 2 shows basic characteristics of the PSPM H4139-20 and of the MCPM used in the horizontal stations. To minimize the number of pixels needed, the fiber detector readout is multiplexed: Four fibers out of four different regions of each coordinate detector are readout by one pixel. The demultiplexing is realized by use of trigger tiles, see below. The devices work essentially stable since 1995. One device had to be exchanged in the 1996/97 shutdown because of internal discharges. The analog data are readout from the anode contacts by preamplifiers. After digitization by FADCs they are transmitted in the event pipeline of H1.

Horizontal stations

The field of opto-electronic readout devices is rapidly developing. Therefore, in preparation of the construction of the horizontal station, again a review of multi-

---

[3] Hamamatsu Photonics K.K., electron tube division, 314-5, Shimikanzo, Tokyooka Village. Iwatagun, Shizuoka-ken, Japan

TABLE 2. Parameter list of the opto-electronic multi-channel devices

Multi-channel devices		
	Vertical stations	**Horizontal stations**
Principle	position sensitive PM fine mesh dynode	micro-channel plates
Type	Hamamatsu 4139–20	MCP-2FEU-124
Number of pixels	64	124
Pixel size	4 mm	1.5×1.5 mm^2
Amplification	at 2 kV $3 \cdot 10^6$	$3 \cdot 10^5$
Quantum efficiency at 420 nm	20 %	15 %
Cathode	Bialkaline	Multialkaline
Front plate	glass window	fiber optical plate
Pitch of pixels	5.08 mm	2.7 mm

pixle opto-electronic devices was performed. Kapishin et al. from Dubna proposed the application of the micro-channel device technique. The investigation of russian devices resulted in the principial applicability, if the quantum efficiency and gain would be enlarged [9] . In 1995 this further development took place and resulted in a quantum efficiency of 15 percent and a gain of about $3 \cdot 10^5$. Figure 6 shows the sketch of a MCPM[4] . A fiber-optical plate in front is used to reduce optical cross-talk. The photoelectrons are focused by an electrostatic lens onto two micro-channel plates.

Because of the small pixel size, multiplexing is not possible. Each fiber of the fiber detectors couples to one pixel of the MCPM by means of masks. The preamplifier were adapted to the lower gain of the MCPM compared to the PSPM of the vertical stations. Tests with cosmic particles gave efficiencies of about 65 percent.

THE TRIGGER SETUP

There are two aims for the use of the trigger setup in the Roman pots of the H1 FPS:

- Formation of a trigger element for the central H1 trigger processor

- Demultiplexing of the multiplexed readout in the vertical stations

The trigger is realized by plastic scintillator tiles readout via bundels of clear optical fibers (diameter 0.5 mm) and photomultipliers. One trigger plane of 5 mm

---

[4] Moscow radio tube factory, Moscow, Russia

thickness for the vertical stations and of 3 mm for the horizontal station is combined with one coordinate plane of the fiber detector, i.e. there are four trigger planes per station. In the vertical stations PMs of type XP1911 from Philips [5] are used. PMs of type R5600U from Hamamatsu are applied in the horizontal detectors. The trigger efficiency is about 99 percent.

RUNNING AND RESULTS

The vertical stations were under operation in three luminosity periods (1995 - 1997). The horizontal stations were installed in January 1997. Concerning the horizontal stations most of the time in 1997 was used for methodical investigations. So, for example different stations had to be shielded against several disturbing influences like magnetic stray fields, synchroton radiation, high frequency background and high temperatures.

When stable beam conditions are reached in a luminosity run the detectors are moved into the measuring positions about a few millimeters away from the circulating proton beam. The approach is controlled by monitoring a set of counting rates.

The efficiencies of the scintillating fiber hodoscopes are inferred from the number of hits belonging to a reconstructed track. In 1997 on average 6.6 hits per coordinate are observed in a layer of 10 scintillating fibers resulting in an average efficiency of 66%, which compares well with the geometrical efficiency, whereby four weak readout channels were not considered. Figure 7 shows the efficiency for 10 fiber layers of the v projection of the vertical station of 81 m for one particular run in autumn 1997. The observed internal spatial resolution of $\sigma_{xy} = 130 \ \mu$m is close to the design value. Fig. 8 show the internal resolution of the same run as above and projection v as well.

SUMMARY

A forward proton spectrometer at the H1 experiment working at the HERA collider was successful and stable operated since 1995.

The FPS improves the physics scope of the experiment H1 by providing a momentum measurement of forward going protons. The HERA beamline magnets are employed as a spectrometer which records scattered protons in four stations. Detectors with scintillating fiber hodoscopes for measuring two space points on a trajectory have proven to be robust instruments with a spatial resolution of about 130 μm.

The detector is based on the Roman Pot technique and scintillating fiber detectors. The readout of the scintillating fibers is realized by multi-channel PMs

---

[5] Philips Photonique, Avenue Roger Roncier, B.P.520, 19106 Brive la Gaillarde, Cedex-France

and micro-channel photomultipliers respectively. The FPS consists of two pairs of detector stations which cover complementary kinematical ranges.

ACKNOWLEDGMENTS

We are grateful to the HERA machine group whose outstanding efforts have made and continue to make this experiment possible. We thank the engineers and technicians for their work in constructing and now maintaining the H1 detector, our funding agencies for financial support, the DESY technical staff for continual assistance.
The work of M.Pohl and S.Ruzin for the development of electronic subsystems is acknowledged.
The succesful operation of the FPS is based to large extent on the skillful work of the mechanical workshop of our institute.

REFERENCES

1. Proposal for a Forward Proton Spectrometer of H1, H1-10/94-381; PRC 94/03, Oct.94.
2. H1-Collaboration, Proceedings of the 28th Intern. Conf. on High Energy Physics, Warsaw, Poland, 1996, eds. Z.Ajduk, A.K.Wroblewski V.II, p.1759.
3. Upgrade of the H1 Forward Proton Spectrometer, H1 - 12/95 - 467; PRC 96/01.
4. Ansorge, R. et al., *NIM* **265**, 33 (1988).
5. Annies, P. et al., *NIM* **A367**, 367 (1995).
6. Bross, A.D., *Nucl. Phys.B (Proc. Suppl.)* **44**, 12 (1995).
   Adams, D. et al., *Nucl. Phys.B (Proc. Suppl.)* **44**, 332 (1995).
7. Bähr, J. et al., *DESY 93-200* (1993) and Proceedings of the "International workshop on scintillating fiber detectors SCIFI-93", Notre Dame 1993, USA.
8. Bähr, J. et al., *DESY 93-201* (1993) and Proceedings of the "International workshop on scintillating fiber detectors SCIFI-93", Notre Dame 1993, USA.
9. Bähr, J. et al., Internal Report, DESY-Zeuthen 95-01 (1995).

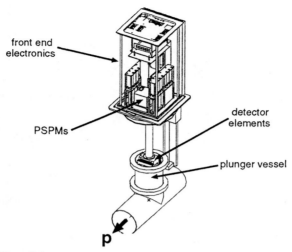

FIGURE 1. Schematics of a Roman Pot device for a scintillating fiber hodoscope detector.

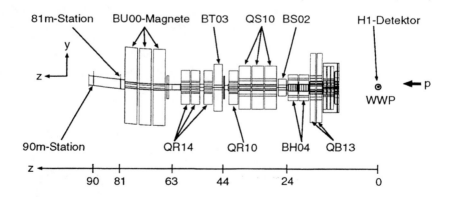

FIGURE 2. HERA–beam line adjacent to the H1 interaction region. The magnets BH04 introduce a horizontal bend, BU00 are vertical dipoles. Quadrupoles are QB, QB, and QS, correction coils are BT and BS, also shown is the position of the detector stations.

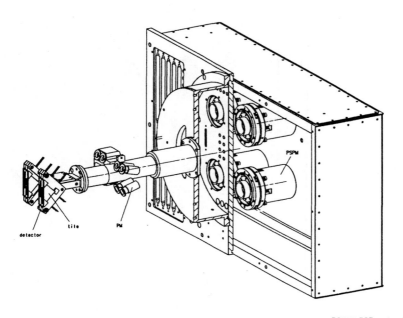

FIGURE 3. Detector insert for a horizontal station.

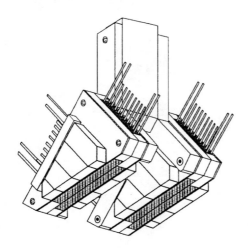

FIGURE 4. Scheme of a fiber detector of a vertical station.

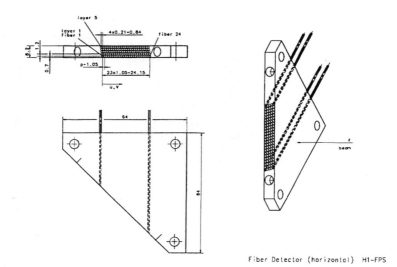

Fiber Detector (horizontal) H1-FPS

FIGURE 5. Scheme of a coordinate detector of a horizontal station.

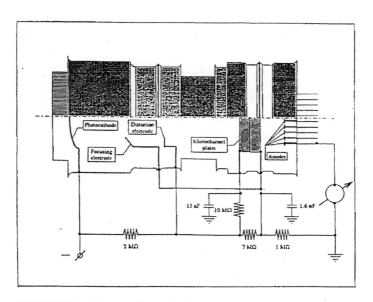

FIGURE 6. Scheme of a multichannel tube FEU-2MCP-124.

198

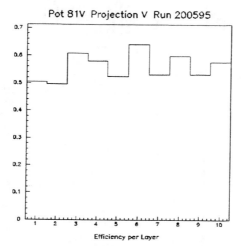

FIGURE 7. Efficiency per layer of the v projection of the vertical station at 81 m.

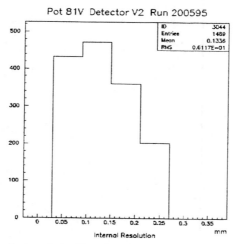

FIGURE 8. Internal resolution of the fiber detectors of the v projection of the vertical station at 81 m.

199

High Resolution Tracking Devices Based on Capillaries filled with Liquid Scintillator

P. Annis[c], A. Bay[g], D. Bonekämper[h], S. Buontempo[i], C. Currat[g],
R. v. Dantzig[a], A. Ereditato[i], J.P. Fabre[e], D. Frekers[h], A. Frenkel[k],
F. Galeazzi[k], F. Garufi[i], J. Goldberg[f], S.V. Golovkin[j], K. Hoepfner[f1],
K. Holtz[h], J. Konijn[a], E.N. Kozarenko[d], I.E. Kreslo[d2], B. Liberti[k],
M. Litmaath[e], G. Martellotti[k], A.M. Medvedkov[j], P. Migliozzi[i],
C. Mommaert[c3], J. Panman[e], G. Penso[k], Yu.R. Petukhov[d],
D. Rondeshagen[h], V.E. Tyukov[d], V.G. Vasil'chenko[j], P. Vilain[c],
J.L. Visschers[a], G. Wilquet[c], K. Winter[b], T. Wolff[h4], H. Wong[e5],
H.J. Wörtche[h]

presented by Thomas Wolff

[a] *NIKHEF, Amsterdam, The Netherlands;* [b] *Humboldt-Universität, Berlin, Germany;*
[c] *IIHE(ULB-VUB), Bruxelles, Belgium;* [d] *JINR, Dubna, Russia;* [e] *CERN, Genève, Switzerland;*
[f] *Technion, Haifa, Israel;* [g] *Université de Lausanne, Switzerland;* [h] *Westfälische Wilhelms-Universität Münster, Germany;* [i] *Unversità "Federico II" and INFN, Napoli, Italy;*
[j] *IHEP, Protvino, Russia;* [k] *Unversità di Roma "La Sapienza" and INFN, Roma, Italy.*

Abstract. The aim of this project is to develop high resolution tracking devices based on thin glass capillary arrays filled with liquid scintillator. This technique provides high hit densities and a position resolution better than 20 μm. The radiation hardness of more than 100 Mrad makes capillary arrays superior to other types of tracking devices with comparable performance. The technique is attractive for inner tracking in collider experiments, micro-vertex devices, or active targets for short-lived particle detection in high-energy physics experiments. During 1996 and 1997, two prototype capillary bundles (1.8 m and 0.9 m in length) were put into operation. One of them was equipped with a conventional opto-electronic read-out chain and the other, for the first time, with a new device, the Electron-Bombarded CCD (EBCCD) imaging tube. The development of planar capillary layers is another aspect of the R&D activity. Details of the design and the performance are given.

---

[1] now at DESY, Hamburg, Germany
[2] now at INFN, Roma, Italy
[3] now at CERN, Genève, Switzerland
[4] Member of the Graduiertenkolleg of elementary particle physics at Humboldt-Universität, Berlin, Germany
[5] now at Institute of Physics, Academia Sinica, Taiwan

CP450, *SciFi97: Workshop on Scintillating Fiber Detectors*
edited by A. D. Bross, R. C. Ruchti, and M. R. Wayne
© 1998 The American Institute of Physics 1-56396-792-8/98/$15.00

I INTRODUCTION

The aim of the CERN RD46 Collaboration [1] is to build high resolution vertex or tracking detectors based on the scintillating fiber technique to be used for different high energy physics application. Among the figure of merit of the detector, we mention a high radiation hardness and a good spatial resolution and track identification in events with high multiplicity.

The Collaboration tries to reach these goals by using glass capillaries filled with liquid organic scintillator. They combine the low cross-talk of glass-fibres, achieved by using an Extra Mural Absorber (EMA), with a high attenuation length and a fast response of plastic fibres [2]. They are read out by an opto-electronic chain with a Megapixel CCD at its end. The chain provides the light intensification to detect photons emitted by a minimum ionizing particle.

II GLASS CAPILLARIES

The glass capillaries are being developed in collaboration with industry[6]. Starting from a single macro-capillary of borosilicate glass with a diameter of the order of 1 mm, a multistructure of several hundred capillaries (Figure 1) is formed through continuous drawing under high temperature, resulting in a coherent array of microcapillaries. The lateral deviations from a straight line of a single capillary in these arrays was measured to be less than 2 μm over 10 cm [3]. Several of these multistructures are assembled together and drawn again to form a bundle. Capillaries can be built with a length up to a few meters. They can be produced with

---

[6] Schott Fiber Optics Inc., Southbridge, MA, USA.

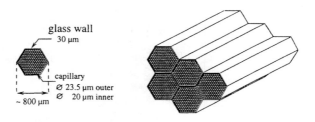

| single multistructure | multistructures forming a bundle |

Schott glass 8250	
density	2.28 g/cm³
radiation length	12.4 cm
refractive index	1.49
thermal expansion	5.10⁻⁶ K⁻¹
transformation temperature	492°C
softening point	715°C

FIGURE 1. Example of a multistructure. Several of these multistructures form a bundle.

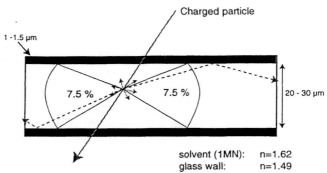

FIGURE 2. The liquid scintillator is excited by a traversing charged particle and deexcites by isotropic light emission, 15 % of which being trapped inside the capillary. The angle of total reflection at the liquid-glass interface is 66.9°.

inner diameters between 2 and 1000 μm and with a typical capillary wall thickness around 5 to 10 % of the inner capillary diameter. Typically, the most recent capillary arrays have an active scintillation volume of 68 - 78 % of the total volume.

To reduce cross-talk, EMA is inserted. The EMA consists of small black borosilicate slabs stacked in the interstice between the capillary tubes. They absorb the light that is not trapped inside the capillary.

III   LIQUID SCINTILLATOR

The liquid organic scintillator most frequently used consists of 1-Methyl-Naphthalene (1MN) as solvent to which one adds 3 g/l of a pyrazolin derivate (e.g. 3M15 or R45) as dye[7]. The scintillator has a large Stokes-shift, to minimize the self-absorption in the scintillator.

In Figure 2, a cut through a filled capillary is shown. The borosilicate glass has a refractive index of n = 1.49 and the solvent one of n = 1.62. This results in a trapping efficency of 15 % for the photons generated isotropically inside the capillary by a traversing charged particle. An excellent liquid-glass reflectivity makes the attenuation length sensibly independent of capillary diameter to first order, even for diameters as low as 20 to 30 μm diameter.

For the presently used scintillators, the attenuation length and the radiation hardness was studied. An attenuation length of more than 3 m was measured for a 500 μm capillary using a movable low energy beta source to excite the scintillator [4]. To measure the radiation hardness, the capillaries filled with liquid scintillator were first irradiated with a ^{60}Co source. After that, the scintillation efficency was measured with a ^{90}Sr source. A degradation of only 30 % of the light collection was found after a dose of 180 Mrad [5].

The Collaboration also studied the variation of the light yield of the liquid scintillator when exposed to various gases or in vacuum. This is important, since the

---

[7] The used scintillators are trademarks of Geosphaera Inc., Moscow, Russia

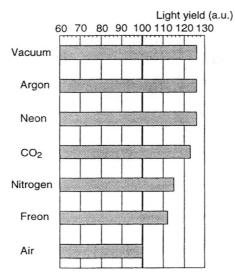

FIGURE 3. Observed scintillation light yield using different atmospheres during filling.

light yield decreases due to quenching effects, in particular from oxygen, when the scintillator is pushed into the capillaries using a gas under slight overpressure. On the other hand, using suction by evacuating the capillaries, gives, of course, the best results. However, vacuum is difficult to establish in the capillaries. We therefore resorted to the first method using purified neon or argon. These noble gases leave the light yields essentially unaffected (Figure 3). The highest quenching effect was observed when using air where the light yield decreased by 27 % [6].

IV OPTO-ELECTRONIC CHAINS

For the read-out of the capillary bundle, the following has to be taken into account:

- Low light level imaging: Only few photons exit the capillaries to make up the image that is to be processed. This requires high amplification factors for any type of light sensing device to be used.

- Large number of read-out channels: There are about 10^6 capillaries forming a prototype bundle requiring a Megapixel CCD for image storage.

- Trigger: Typically, a CCD is read out with a clock frequency of 10 to 20 MHz, which corresponds to 50 to 100 ms read-out time for a Megapixel CCD. Therefore only low rate events can be acquired. For high rate events a trigger must be used which allows to select "interesting" events, and to reduce the acquisition rate down to an acceptable value. In order to wait for the trigger decision, we must either store the image before it is projected on the CCD, or use a CCD with special features , like buffering, fast clear or fats memory

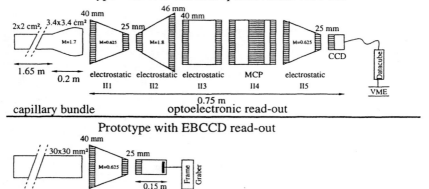

FIGURE 4. The setup of the two different opto-electronic chains.

dump. The first solution can be achieved either by taking advantage of the decay time of the image intensifier phosphor(s) which act as an optical memory, or by using a high speed gateable image pipeline which provides an electronic delay of the image [7].

Two different types of image intensifiers chains have been tested in situ:

- Conventional chain

  The chain (Figure 4 (top)) consists of several conventional image intensifiers. To detect the few photons exiting the capillaries, a high quantum efficiency of the first image intensifier is needed, which is about 23 % in the present setup. The main intensification and gating is accomplished by a proximity focused image intensifier with a Micro-Channel Plate (MCP) inside. The image is stored in a Megapixel CCD with a pixel size of (19×19) μm^2.

- EBCCD chain

  This chain consists of a first generation image intensifier and an EBCCD imaging tube[8] (Figure 4). The image intensifier with a slow phosphor decay time needed to be placed in front of a EBCCD imaging tube, in order to achieve an optical memory for a trigger delay of about 1 μs. The EBCCD imaging tube itself (Figure 4 (bottom)) is a electrostatically focused image intensifier. The novel feature of this imaging tube is the CCD chip incorporated inside the tube at the location of the phosphor screen. The CCD chip (pixel size

---

[8] supplied by Geosphera Inc., Moscow, Russia.

13.4 × 13.4 μm^2) is thinned down to 8 μm and backside bombarded by photoelectrons accelerated to 15 KeV. Using frontside bombardment, the gate
structures on the chip would be quickly destroyed. However, when an electron
hits the backside of the CCD, it produces a cloud of free secondary electrons.
In order to reduce recombination of free electrons with holes in the silicon and
to obtain an acceptable yield of electrons in the potential wells of the CCD,
the CCD chip must be thinned.

There are various advantages of the EBCCD imaging tube over a conventional
opto-electronic chain. The EBCCD imaging tube has a high gain of about 4000,
a signal-to-noise ratio of about 15, is gateable and features a spatial resolution
better than 10 μm [8]. The EBCCD imaging tube is also capable of resolving
single photoelectrons [8]. Obviously, the use of an EBCCD imaging tube sensibly
reduces the overall length of the read-out chain, correspondingly the price.

V PROTOTYPES & RESULTS

Two prototypes of those devices (capillary bundle + image intensifier chain) were
installed in the CERN wide-band neutrino beam in front of the CHORUS detector
[9,10]. In such a setup one can take advantage of the good tracking capability
and particle identification of the CHORUS detector, in order to identify events
occurring in each of the capillary bundles.

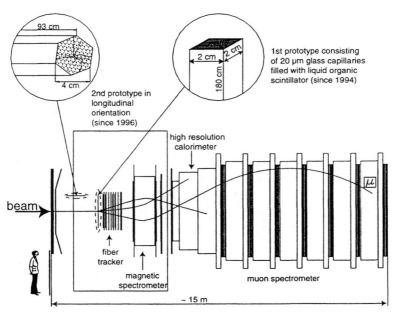

FIGURE 5. The two prototypes are installed in front of the CHORUS detector in the CERN
wide-band neutrino beam.

The first conventional prototype was installed in 1994. It consisted of a set of five image intensifiers as shown in Figure 4 and a $(2 \times 2 \times 180)$ cm^3 capillary bundle. The capillaries were filled with 1MN and 3M15 as scintillating dye. The inner diameter of the capillaries was 20 μm. The bundle had an active scintillation volume of 55 %. The prototype was placed vertically to the neutrino beam just in front of the fiber tracker section of the CHORUS detector [11] (Figure 5).

In Figure 6, a neutrino interaction recorded with this prototype is shown. Six tracks can be identified. The highly ionizing track belongs to a nuclear break up. Three tracks were reconstructed in the CHORUS detector. Two of them are oppositely charged muons. Analyzing the tracks in the bundle, the track to which the μ^+ is attached shows a kink from a decaying particle coming from the vertex. Also the hit density of that track differs before and after the kink. Taking into account additional information from the CHORUS calorimeter, which indicates a signal from a hadronic non-charged particle, this track can be identified as a D$^+$ ($c\tau=317$ μm) which decays after 200 μm into $\bar{K}^0\mu^+\nu_\mu$.

For this prototype the following results have been achieved by analysing muon- and neutrino-events [4,13]:

- A vertex resolution better than 30 μm.

- A track residual (spread of hits around the straight line fit of the track) better than 15 μm.

- An angular resolution (uncertainty of the slope in a straight line fit) of about 0.14 mrad.

- 5-8 hits per mm along a track of a minimum ionising particle, which corresponds to about 2000 hits per radiation length.

- A two-track resolution of about 33 μm.

A second prototype was installed in 1996, using the above described EBCCD chain (Figure 4). The target consisted of a hexagonal bundle with a diameter of 4 cm and a length of 93 cm. It was filled with 1MN and 3M15 as scintillating dye. Here, the inner diameter of the capillaries was 30 μm and the active scintillation volume was about 70 %. The prototype was placed longitudinally to the neutrino beam (Figure 5). A typical event from this prototype is shown in Figure 7. The figure shows a clearly resolved image of the interaction vertex, which is a vast improvement over an earlier result using plastic fibres without EMA [12]. Also imperfections in the bundle are indicated in the picture. These imperfections are likely due to non-uniform capillary arrays and dead regions in between the multistructures. Despite of this non perfect bundle, this configuration shows the capability of using such a device as an active target for the detection of short-lived particles.

The Collaboration also tested planar capillary bundles[9], so called capillary layers. Such structures may eventually be used as micro-vertex detectors [3]. A test setup

---

[9] manufactured by Geosphaera Inc., Moscow, Russia

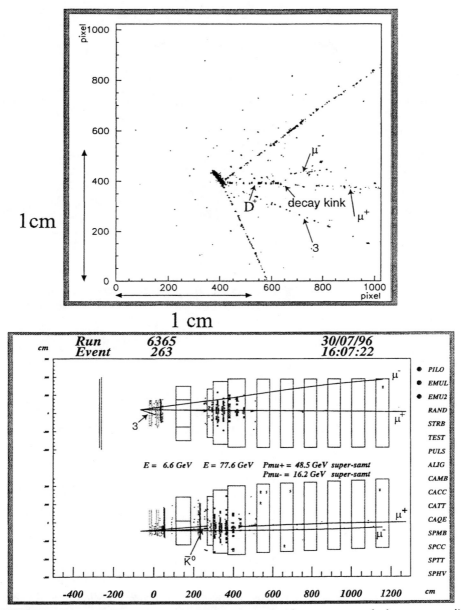

FIGURE 6. A neutrino interaction in the conventional prototype and the corresponding CHORUS event. The μ^+ track indicates a D^+ decaying into $\bar{K}^0 \mu^+ \nu_\mu$. The neutrino beam enters from the left side.

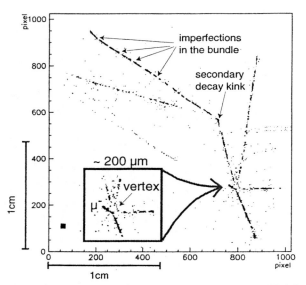

FIGURE 7. A neutrino interaction recorded by the prototype with EBCCD read-out. Note, that the vertex is clearly visible. The neutrino beam points out of the plane. The short track in the enlarged window corresponds to a produced μ^- identified by the CHORUS detector. Zooming into the vertex region, one can appreciate the vertex resolution.

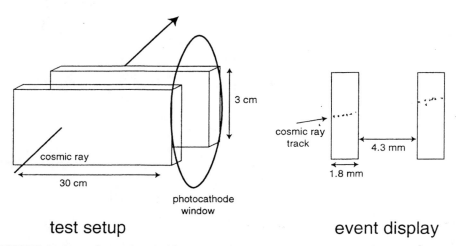

FIGURE 8. A cosmic ray detected by a test setup using 2 capillary layers (3×30 cm^2, thickness 1.8 mm and 4.3 mm apart) and EBCCD read-out. The track can be identified from the hits in the capillaries of both layers and a track vector can be assigned indicated by the arrow. The right side shows the projected image of the cosmic ray track as it appears in the event display.

(Figure 8) using cosmic rays consisted of 2 capillary layers ((3×30) cm^2 and a thickness of 1.8 mm placed 4.3 mm apart) and an opto-electronic read-out like the above described EBCCD chain. A typical event is shown. The track of the cosmic ray can be identified in each layer and a track vector can be easily resolved (indicated by the arrow in the event display in Figure 8).

VI CONCLUSION

A new detector has been developed based on coherent arrays of capillaries, which have an inner diameter of 20 to 30 μm and are filled with liquid organic scintillator. This device can be used as tracking detector or active target to detect short-lived particles in future high-energy physics experiments. The capillary production allows to build capillary arrays in various geometries which are suitable for different applications. The possibility of using capillary layers as a micro-vertex detector has been demonstrated.

Two prototypes of the detector have been tested in the CERN wide-band neutrino beam in conjunction with the CHORUS detector. Various neutrino interactions have been recorded and the quality of the images demonstrates the potential of the detector for future applications in high-energy physics.

REFERENCES

1. R. v. Dantzig et al. RD46 Proposal, CERN/LHCC 95-7, P60/LDRB (1995).
2. C. Angelini et al., NIM **A295** (1990) 299.
3. J. Konijn et al., proceedings VERTEX97.
   G. Martellotti et al., NIM **A384** (1996) 179.
4. S. Buontempo et al., NIM **A360** (1995) 7, CERN-PPE/94-142.
5. S.V. Golovkin et al., NIM **A346** (1994) 163.
6. S.V. Golovkin et al., IHEP 97-12, 1997.
7. A.G. Berkovski et al., NIM **A380** (1996) 537.
8. Talk given by I.E. Kreslo, these proceedings.
9. M. de Jong et al., CERN-PPE/93-131.
10. E. Eskut et al., NIM **A401** (1997) 7.
11. Talk given by D. Rondeshagen, these proceedings.
12. C. Angelini et al., NIM **A289** (1990) 342.
13. P. Annis et al., NIM **A367** (1995) 377.
    P. Annis et al., NIM **A386** (1997) 72.
    P. Annis et al., Nucl. Phys. **B54** (Proc. Supp.) (1997) 86.
    J. Konijn et al., RD46 Status Report, CERN/LHCC 97-38
    P. Annis et al., to be submitted to NIM.

Fast Scintillation Counters
with WLS Bars

V.Bezzubov[1],D.Denisov[2],S.Denisov[1],H.T.Diehl[2],
A.Dyshkant[1],V.Evdokimov[1],A.Galyaev[1],P.Goncharov[1],S.Gurzhiev[1],
A.S.Ito[2],K.Johns[3],A.Kostritsky[1],A.Kozelov[1],D.Stoianova[1]

[1] *Institute for High Energy Physics, Protvino 142284, Russia*
[2] *Fermi National Accelerator Laboratory, Batavia, Illinois 60510*
[3] *University of Arizona, Tucson, Arizona 85721*

Abstract. The DØ collaboration is building 4608 scintillation counters to upgrade forward muon system for the next Fermilab Collider run. Each counter consists of 12.7mm thick scintillator plate with two WLS bars along two sides for the light collection. With average 10^2 photoelectrons from *mip* particle the counters provide time resolution below $1ns$ and have good energy resolution. Results of Bicron 404A scintillator and Kumarin 30 WLS aging under irradiation up to $3Mrad$ are presented. With specially designed magnetic shielding counters can operate in magnetic filed up to $500G$.

INTRODUCTION

The goal of the DØ Upgrade [1] is to exploit the physics potential to be presented by the Fermilab Main Injector and Tevatron. An integrated luminosity of 2 fb^{-1} is expected with instantaneous luminosity of up to $2 \cdot 10^{32} cm^{-2} s^{-1}$ accompanied by a reduction in a bunch spacing. This factor of 10 increase in integrated luminosity over previous run will provide an opportunity for significant improvements sweeping the wide range of physics studied at the DØ experiment. Capability to identify and trigger on muons is one of the key features necessary to exploit these possibilities. The upgraded DØ muon system features full coverage for $|\eta| < 2$.

The proposed layout of the forward $(1.0 < |\eta| < 2.0)$ muon system includes three layers of scintillation counters for triggering on events with muons on both sides of the detector [2]. Counters are arranged in $R - \phi$ geometry to match central tracking detector trigger segmentation and muon bending direction in the toroidal magnets.

Advantage of the scintillation counters is their ability to count substantially less background hits than other types of detectors [3,4]. Decrease of the gate width from $100ns$ to $20ns$ reduces the number of background hits per plane by ten

CP450, *SciFi97: Workshop on Scintillating Fiber Detectors*
edited by A. D. Bross, R. C. Ruchti, and M. R. Wayne

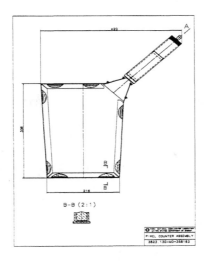

FIGURE 1. Design of the scintillation counter.

times (keeping 100% efficiency of muon detection). The *mip* energy deposition in 12.7mm scintillator is 2.5MeV. Setting detection threshold at 0.5MeV will reduce counting rate due to neutrons by a factor of 3 [4] in comparison with gas detectors. Scintillation counters can be easily made of any specific size matching required $R - \phi$ geometry.

In this article we present design and results of studies of the scintillation counters for the DØ forward muon system upgrade.

COUNTERS DESIGN

The size of the muon trigger planes is from 7×7 m^2 for counters closest to interaction region to 12×12 m^2 for counters situated near walls of the collision hall. The counters are arranged in $R - \phi$ geometry with 0.1 *eta* and 4.5^o *phi* segmentation. The total number of counters is 4608. The counters sizes vary from 10×15 cm^2 to 60×106 cm^2. The counter design is optimized to provide good time resolution and uniformity for background rejection, high efficiency of muon detection and reasonable cost for production of 4608 counters. The selected design is shown in Fig. 1 for counter with sizes 216×338 mm^2. It consists of 12.7mm thick Bicron 404A scintillator plate cut to trapezoidal shape with two Kumarin 30 WLS bars [5] for light collection. The bars are 4.2mm thick and 12.7mm wide. They are installed along two sides of the counter and bended at 45° to collect light on the phototube. Use of WLS bars provides higher efficiency of light collection in comparison with optical fibers. Parameters of scintillator and WLS are presented in Table 1.

TABLE 1. Characteristics of scintillating materials.

Material	Decay time, ns	Peak emission, nm	Attenuation length, m
Bicron 404A scintillator	2.0	408	1.7
Kumarin 30 WLS	2.7	460	1.4

Green extended phototube 115M from MELZ [6] is used for light detection (Table 2). The scintillator and WLS bars are wrapped with Tyvek material [7] (for better light collection) and black paper (for light tightness). The wrapped counter is placed into 1 mm thick aluminum container with stainless steel part for connection of phototube assembly.

TABLE 2. 115M phototube characteristics.

Diameter of photocathode, mm	20
Quantum efficiency at 500 nm, %	$15^{+5}_{-2.5}$
Non-uniformity of response of photocathode, %	< 10
Operating voltage for anode sensitivity of 100 A/Lm, kV	1.8 ± 0.2
Gain at operating voltage with anode sensitivity of 100 A/Lm	$2 \cdot 10^6$
Rise time, ns	3

PROTOTYPE COUNTERS PERFORMANCE

Counters performance has been studied at Fermilab 125 GeV/c pion test beam. On Fig. 2 the dependence of counter efficiency and time resolution vs high voltage is presented for counters of 3 different sizes: "large" - $60 \times 106 \ cm^2$, "typical" - $24 \times 34 \ cm^2$ and "small" - $17 \times 24 \ cm^2$. Single threshold electronics has been used with detection threshold set at $10mV$. Time resolution achieved is 0.5-$1.0ns$. At high phototube gain time resolution is limited by photoelectron statistics. Small counters produce more photoelectrons and their light collection time is faster what improves time resolution [8]. Counters detection efficiency on platou is above 99.9%.

In order to make efficient energy deposition cut number of photoelectrons should be large to reduce statistical fluctuations and non-uniformity should be reasonably low. Non-uniformity was measured by irradiating counters in different points by Sr^{90} radioactive source and cross checked by cosmic ray studies. For all counters sizes measured *rms* non-uniformity is less than 10%.

In Table 3 average number of photoelectrons detected for cosmic ray muons passing counter perpendicular to its surface is presented. Calibration of ADC scale in number of photoelectrons has been done by LED producing single photoelectron pulses. The number of photoelectrons is large and *mip* amplitude distribution is mainly determined by Landau fluctuations.

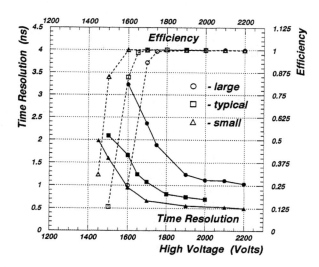

FIGURE 2. Time resolution and detection efficiency of scintillation counters.

TABLE 3. Average number of photoelectrons.

Counter size, cm^2	14×14	46×32	106×60
Average number of photoelectrons	184	115	61

Now we turn to specific studies performed to ensure reliable long term operation of the scintillation counters.

SCINTILLATOR AND WLS RADIATION AGING

An important characteristic for detectors operating under heavy radiation is stability of their parameters during long run. An estimation of the total radiation dose for integrated luminosity of 2 fb^{-1} for the hottest region gives $\sim 1krad$ for the scintillation counters. Aging of scintillation counters under radiation depends mainly upon radiation hardness of the scintillation materials. We performed detailed studies of scintillator and WLS aging under irradiation to ensure reliable operation of counters over long term run. The studies have been done using set of Cs^{137} radioactive sources producing close to uniform irradiation in the sample volume with the rate of $5.3rad \cdot s^{-1}$. On Fig. 3 Bicron 404A scintillator light output as a function of accumulated dose is shown. Up to $1Mrad$ dose radiation damage is below 20%.

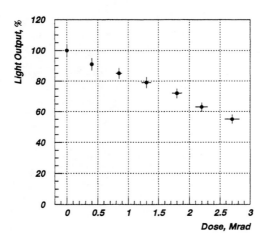

FIGURE 3. Bicron 404A scintillator light output vs accumulated dose.

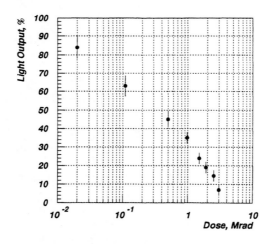

FIGURE 4. Light output for pair: non-irradiated Bicron 404A scintillator and irradiated Kumarin 30 WLS.

On Fig. 4 the light output for pair of non-irradiated Bicron 404A scintillator and irradiated Kumarin 30 WLS bar is presented. WLS demonstrates considerably faster radiation aging than scintillator.

In order to study aging at closer to real conditions time scale and irradiation flux prototype counter with sizes 33×22 cm^2 was placed for 9 months in the DØ collision hall during fixed target run operation. The total accumulated neutron dose was $49krad$ and gamma dose was $30krad$. The decrease of the signal before/after irradiation is $11 \pm 3\%$ and can be attributed to aging of WLS.

PHOTOTUBE MAGNETIC SHIELDING

The fringe magnetic field of the DØ detector superconducting solenoid and toroidal magnet reaches $350G$ in the area where phototubes are situated [9]. In order to reduce magnetic field in the phototube area down to $\sim1G$ necessary for normal operation shielding made of $1.2mm$ thick mu-metal and soft steel tube with 3 or $6mm$ thick wall has been designed [2]. The soft steel flange mounted on the counter case (see Fig. 1) serves as a part of the shield also. For the field perpendicular to the tube axis no reduction in output pulse has been observed up to $700G$. For the field parallel to the tube axis the reduction in pulse amplitude vs magnetic field is presented in Fig. 5. The effect of magnetic field on phototube gain is less than 10% up to $350G$ for $48mm$ external shield diameter.

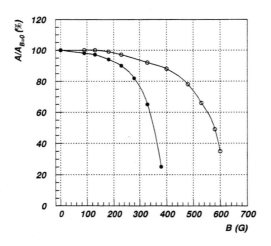

FIGURE 5. Relative gain change of 115M photomultiplier in magnetic field parallel to the tube axis: solid circles - $3mm$ steel tube wall ($42mm$ external diameter shield), empty circles - $6mm$ steel tube wall ($48mm$ external diameter shield).

PHOTOTUBE AGING

Phototubes gain decreases during operation due to the interaction of accelerated electrons with surface of dinodes and anode. This aging is commonly represented in terms of accumulated anode charge [10]. For integrated luminosity of $2fb^{-1}$ the maximum accumulated anode charge will be $\sim 20C$. In Fig. 6 typical 115M phototube aging curve is presented. The drop of phototube gain is less than 5% for accumulated charge of $20C$. This drop in gain could be compensated by adjusting electronics threshold or phototube operating voltage.

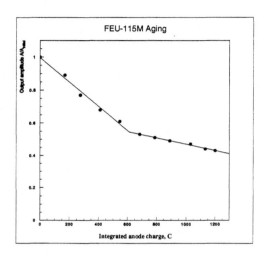

FIGURE 6. Phototube gain as a function of integrated anode charge.

CONCLUSIONS

Scintillation counter design based on light collection by two WLS bars has been developed for the forward DØ muon system upgrade. Average number of photo-electrons for the counter with sizes $1.3 \times 60 \times 106 \; cm^2$ is 61 providing high *mip* detection efficiency, good amplitude resolution and time resolution of $1ns$. Use of fast and transparent scintillator and WLS determines counters response uniformity of 10% and high time resolution. Proposed magnetic shielding design provides counters operation in the filed up to $500G$. Degradation of counters performance starts at accumulated doses of $0.1Mrad$ and mainly determined by radiation aging of Kumarin 30 WLS bars.

REFERENCES

1. The DØ Collaboration, "The DØ Upgrade", Preprint Fermilab-FN-641, 1995.
2. V. Abramov et al., "Technical Design Report for the DØ Forward Trigger Scintillation Counters", DØ Note 3237, Fermilab, 1997.
3. B. Baldin et al., "Upgrade of the DØ forward muon system with pixel scintillation counters", DØ Note 2358, Fermilab, 1994.
4. Yu. Nikolaev et al., "The efficiency of plastic scintillators to neutrons and gammas", Preprint SDC-93-512, SSC Laboratory, 1993.
5. S. Belikov et al., "Physical Characteristics of the SOFZ-105 Polymethyl Methacrylate Secondary Emitter", Instruments and Experimental Technique, 36, p.390, 1993.
6. S. Belikov et al., "Characteristics of Phototube 115M for Electromagnetic Calorimeter of PHENIX Experiment", Preprint IHEP 96-42, Protvino, 1996.
7. DuPont Nonwovens, Tyvek type 1056D.
8. Chi Peng Cheng et al., Nucl. Instr. and Meth. A252(1986) 67.
9. R.Yamada et al., "2-D and 3-D display and plotting of 3-D magnetic field: Calculation for upgraded DØ detector", DØ Note 2023, Fermilab, 1994.
10. "Photomultiplier tube: Principle to Application", Hamamatsu, 1994.

SESSION 6: TRACKING - II

Chair: R. Mussa
Scientific Secretary: E. Ivanov

The DØ Scintillating Fiber Tracker

A. Bross[a], S. Choi[c], G. Gutierrez[a], S. Grunendahl[a], D. Lincoln[a], E. Ramberg[a], R. Ray[a], R. Ruchti[b], J. Warchol[b], M. Wayne[b].

[a]Fermi National Accelerator Laboratory, [b]University of Notre Dame, [c]Seoul National University.

Abstract. The DØ detector is being upgraded in preparation for the next collider run at Fermilab. The Central Fiber Tracker discussed in this report is a major component of the DØ upgrade. The expected Tevatron luminosity of $2 \times 10^{32} \mathrm{cm}^{-2}\mathrm{sec}^{-1}$, the 132ns bunch crossing time, and the DØ detector constraints of a 2 Tesla solenoid and a 52 cm lever arm, make a scintillating fiber based tracker an optimal choice for the upgrade of the DØ detector.

INTRODUCTION

The Central Fiber Tracker (CFT) discussed in this report is an important part of the upgrade of the DØ detector. From b tagging to improve the top physics capabilities, to a more accurate W mass measurement to constrain the Higgs mass, from improving the ability to look for new physics, to providing a better calorimeter calibration, the Central Fiber Tracker will provide a substantial improvement in the physics capabilities of DØ.

Several constraints coming from the accelerator or the DØ detector make a scintillating fiber based tracker an optimal choice for the DØ upgrade. For Run II the Tevatron luminosity is expected to reach 2×10^{32} cm^{-2}sec^{-1}, with a bunch crossing time of 132 ns. This requires small cell sizes and fast detector response, which is satisfied by fibers with a 775 µm active diameter and few nanosecond response time. The small lever arm (52 cm) and the 2 Tesla B-field of the DØ detector require a very good spatial resolution. The expected resolution for the CFT is less than 100 µm per doublet layer. The first level trigger capability requires not only speed, but also high efficiency to simplify the trigger hardware. The expected efficiency of the CFT is better than 99.5% per doublet layer.

This paper provides an overview of the CFT design. For more detailed information see Reference (1).

OVERALL DESIGN

A schematic view of the Central Fiber Tracker is shown in Figure 1. The baseline design of the CFT consists of scintillating fibers completely covering 8 concentric support cylinders occupying the radial space from 20 to 50 cm. The detector will

CP450, *SciFi97: Workshop on Scintillating Fiber Detectors*
edited by A. D. Bross, R. C. Ruchti, and M. R. Wayne

contain 76,800 Kuraray multiclad S-type scintillating fibers. A fiber doublet layer oriented with the fibers in the axial direction (i.e. parallel to the beam line) will be mounted on each of the eight support cylinders. An additional doublet layer oriented in either -2° or +2° stereo angle will be mounted on successive cylinders. The stereo angles will alternate from the smallest to the largest cylinder.

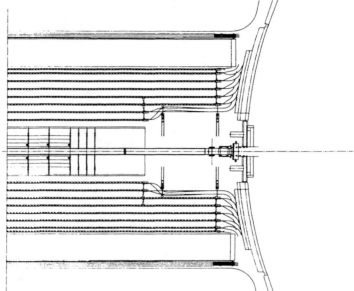

FIGURE 1. Schematic view of the DØ upgrade tracking system

TABLE 1. Design parameters of the Central Fiber Tracker

Layer	Radius (cm)	# of Fibers per sector	# of Fibers per layer	# of Fiber Ribbons	Fiber pitch in microns	Active length (m)
1	19.99	16	1280	10.0	979.3	1.66
1U	20.15	16	1280	10.0	987.2	1.66
2	24.90	20	1600	12.5	975.8	1.66
2V	25.60	20	1600	12.5	982.1	1.66
3	29.80	24	1920	15.0	973.4	2.52
3U	29.97	24	1920	15.0	978.6	2.52
4	34.71	28	2240	17.5	971.7	2.52
4V	34.87	28	2240	17.5	976.2	2.52
5	39.62	32	2560	20.0	970.4	2.52
5U	39.78	32	2560	20.0	974.4	2.52
6	44.53	36	2880	22.5	969.5	2.52
6V	44.69	36	2880	22.5	972.9	2.52
7	49.43	40	3200	25.0	968.7	2.52
7U	49.59	40	3200	25.0	971.8	2.52
8	51.43	44	3520	27.5	916.1	2.52
8V	51.59	44	3520	27.5	919.0	2.52

The diameter of the scintillating fibers will be 835 microns (775 micron active diameter) and the lengths of the fibers will range from 166 to 252 cm. Each scintillating fiber will be mated, through an optical connector, to a clear fiber waveguide, which will pipe the scintillation light to a Visible Light Photon Counter (VLPC). The clear fiber waveguides will vary in length between approximately 8 to 11 meters. Additional details of the tracker design are given in Table 1.

The small (835 µm) fiber diameter, together with the approximate 100 µm gap between fibers, gives an inherent doublet layer resolution for perpendicular tracks of about 70 µm. In order to preserve this resolution capability, the location of the individual fibers must be known to an accuracy better than 30 µm in the r-phi direction. This will be achieved by building up accurate ribbons of fibers and placing them precisely onto the support cylinders. Doublet layer resolution of 90 µm has been achieved with cosmic rays in a test of a large-scale fiber tracker prototype (2-4).

The other significant factor in the momentum resolution of the upgrade tracking system is the amount of material contributing to multiple scattering of charged particles. The amount of material per cylinder is about 0.9% of a radiation length, of this 0.5% corresponds to the scintillating fibers, 0.25% to the support cylinders, and 0.15% to the glue.

The small fiber diameter (and large channel count) also gives the tracker sufficient granularity both to find tracks and to trigger in the complex event environments expected in Run II.

The expected transverse momentum resolution for the DØ upgrade tracking system is shown in Figure 2a. This calculation of the resolution was done with the following parameters: the resolution of the scintillating fiber doublet is 100 µm, the resolution of the silicon barrels is 10 µm, the non-active material in the silicon detectors has a radial distribution as given in Figure 1.7 of the DØ Silicon Tracker Design Report (5), the thickness of the barrels supporting the scintillating fibers is 0.086 g/cm^2 for barrels 3 and 4 and 0.065 g/cm^2 for all other barrels, and the interaction vertex is inside the beam envelope, which has an rms of 35 µm. The transverse momentum resolution in the region covered by cylinder 8 ($\eta < 1.7$) may be parametrized as:

$$\sigma_{p_t} / p_t = \sqrt{(0.015\sqrt{ch\eta})^2 + (0.0014p_t)^2} \, .$$

The expected impact parameter resolution is shown in Figure 2b. The solid line shows the resolution of the combined silicon and scintillating fiber tracker while the dashed line shows the resolution of the silicon detector only. The scintillating fiber tracker contributes significantly to the quality of the measurement of the impact parameter. The reason for this is that to extrapolate to the vertex the momentum of the particle is needed, and the silicon vertex detector alone provides a very poor measurement of the momentum.

To evaluate the effect of *b* tagging on the identification of top quark events, a Monte Carlo sample of top quark events plus background was generated. These events were then simulated and reconstructed. A tagging scheme that requires at least three reconstructed tracks, each with impact parameter significance (the ratio of impact parameter to its error) greater than three, retains more than 50% of top events and only 2% of background events.

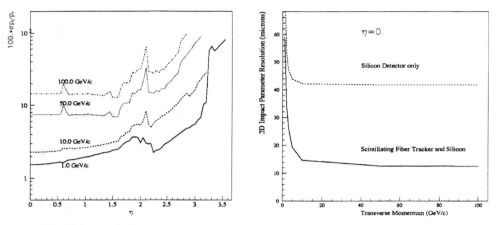

FIGURE 2. a) (left) Transverse momentum resolution as a function of η . b) (right) 2-dimensional impact parameter resolution at $\eta=0$ vs. transverse momentum.

A high doublet layer efficiency is essential to the fiber tracker performance, and studies have shown that to maintain this efficiency the mean number of detected photoelectrons per fiber must exceed 2.5 for a minimum ionizing particle. This number is a function of the intrinsic photoyield of the scintillator, the light transmission properties of the fiber and connectors, and the quantum efficiency of the VLPC. In the aforementioned cosmic ray test, a mean of about nine photoelectrons per fiber was obtained, leading to doublet layer efficiencies of better than 99.5%. To guarantee good efficiency in the final detector, the roughly 77,000 channels of VLPCs delivered from the manufacturer (Boeing North American) must have less than 0.1% dead channels and each channel must satisfy a series of performance requirements, including quantum efficiency, gain and noise count.

Only the fibers themselves will be susceptible to any radiation damage. Studies indicate that no more than a 30% reduction in light yield is expected for the innermost fiber cylinder over the entire Run II; with correspondingly less damage to fibers at larger radii (6).

The central fiber tracker is an important element of the DØ trigger system. Due to the fiber tracker's fast response time, the total time of the collection of signals from the central fiber tracker from one interaction is considerably shorter than the 132 ns bunch spacing expected in Run II. This enables the fiber tracker to participate in the

DØ Level 1 trigger without contributing any dead time. The trigger is implemented using field programmable gate arrays, FPGA's. First the signals from singlet axial layers are combined into hits. Coincidences between eight hits form tracks. The tracks are combined with central preshower clusters to form an electron trigger, and with muon detectors to form a muon trigger. However in order to perform this operation in the 4 μs time allowed for Level 1 processing, the tracker has to be divided into 80 equal azimuthal sectors for parallel processing.

MECHANICAL CONSTRUCTION

As was indicated in the previous section, the ideal resolution for tracks perpendicular to a fiber doublet is 70 μm. Tests done with a 3000 channel prototype of the fiber tracker indicate that a hit resolution of approximately 80 μm perpendicular to the principal fiber axis is attainable (1-3). This sets the precision required for overall fiber placement in the detector: the goal is to build a detector where each fiber is at its nominal position within an uncertainty of better than 30 μm rms on the surface of each cylinder, and 150 μm in the radial direction.

The process of "mounting" a scintillating fiber in the detector will be broken down into four separate steps:
1. Alignment of individual scintillating fibers into a precise ribbon.
2. Mounting of the fiber ribbons precisely onto support cylinders.
3. Alignment of the cylinders relative to each other and to the silicon tracker within the mechanical support frame.
4. Alignment of the tracker in the detector relative to the nominal beam line axis.

For the trigger the fibers on different cylinders have to line up in Φ. Since no alignment corrections can be made at the trigger level, in step 3 above the cylinders will have to line up relative to each other with an accuracy of 50 μm.

Ribbons

Fiber ribbons consist of 256 fibers in two layers of 128 fibers each. The top layer is shifted by half a fiber diameter so that it registers on the bottom layer. Each layer has a precise fiber spacing between 912.0 and 987.2 μm, depending on the tracking layer radius in the detector. Ribbons are fabricated by placing one layer of 128 fibers into a grooved ribbon mold, applying a layer of polyurethane resin, adding the second layer of fibers and compressing the assembly while the resin cures. A 30 μm thick Teflon film is held with vacuum over the grooves to assure that the fibers will easily release from the mold. A single scintillating fiber is placed into an edge groove under the Teflon film. This serves as a support for positioning the overhanging fiber on the upper layer of a molded ribbon.

A ribbon mold consists of a flexible sheet of 1.5mm thick Delrin plastic attached to a curved aluminum support that has been machined to the correct radius. Groves with the appropriate pitch are machined in the flat Delrin plastic, which is then glued to the aluminum support at the appropriate angle for axial or stereo ribbons. Pitch uniformity on molds fabricated with this process is about 20 μm rms, which is better than the overall specified circumferential accuracy of fiber placement on a cylinder. The basic idea is illustrated in Figure 3a. After all the ribbons are made, these molds will be used to mount ribbons on the support cylinders.

Grooved curved connectors will be glued to the fibers at one end of the ribbon. These connectors provide the coupling to the clear fibers that will carry the light to the VLPCs. After glueing and embedding the fibers into the connector, the connector's front surface will be diamond-polished. The alignment with the mating clear fiber connector will be done by means of pins inserted into precision holes in the connecting surfaces. Studies with aluminum connectors have shown that the light transmission efficiency through the connector coupling is better than 95% with a 1.5% rms (7). Studies are under way to replace the aluminum connectors with machined Torlon plastic ones. This will reduce the mass at the ends of the CFT.

At the end opposite to the connectors, the fibers will be aluminized to improve the light collection efficiency. The aluminum sputtering planned for the aluminization is expected to provide reflectivities on the order of 80%.

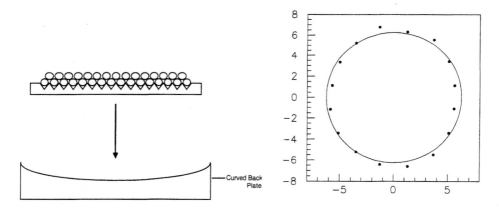

FIGURE 3. a) (left) Technique for curved scintillating fiber ribbon fabrication. b) (right) Average deviations from a cylindrical fit for the 12" diameter, 80" long double-wall R&D cylinder fabricated at Fermilab. One division equals 0.00625 inches.

Support cylinders

In order to minimize the mass, the support cylinders will be built using a sandwich of high module carbon fibers and a rohacell core. The cylinders will be constructed using linear carbon fibers impregnated with about 40% epoxy. According to ANSIS calculations the lay-up that minimizes the sagging of the cylinders is: 0°/60°/-60°/core/-60°/60°/0°. The numbers are the angles in degrees of the carbon fibers with respect to the axis of the cylinder. The required modulus of the carbon fibers is determined by the sagging or beam deflection. Under the load of the scintillating fibers, plus their own weight, the cylinders deflect by stretching the bottom and compressing the top. Therefore for a given carbon fiber modulus the deflection increases as the radius decreases. To keep the beam deflection to the required 30 μm, and to minimize the mass, the highest modulus carbon fibers (105 mpsi) will be used for cylinders 3 and 4. As the radius increases, the fiber modulus can be relaxed. To reduce the cost, cylinders 5 to 8 will be built using 62 mpsi carbon fibers. The thickness of each layer of carbon fibers and epoxy will be 0.003 inches, and the thickness of the rohacell-31 core will be 0.250 inches. This amounts to a total of 0.25% of a radiation length for each of the support cylinders.

R&D support cylinders have been built and measured in a CMM at Fermilab. The measurement for the largest of them (12" diameter by 80" long) gives a deviation from a perfect cylinder of 0.0031 inches rms, well within the required tolerance. Figure 3b shows the average of the measurements made along the axis as a function of the angle around the circumference. The (slightly) oval shape may be due to the mandrel being out of round at high temperature. This is now under investigation.

Waveguide and VLPC

The clear waveguide fibers route the light from the scintillating fiber connector to the VLPCs. These bundles typically consist of 256 undoped fibers of the same construction and diameter as the scintillating fibers. The bundles vary in length from about 7-11 meters, depending upon location. At the detector end the clear fibers are terminated in a v-groove connector which mates to the connector on the scintillating fiber ribbon. The connectorization at the other end of the waveguide bundles is under design. To map the 80 trigger sectors of the CFT to the correct VLPC cassettes requires segmenting the 256 fibers in a waveguide bundle into 32 separate, 8-fiber units. Two options to solve this problem are under study: a) a patch panel, and b) a mixer box.

VLPCs operate at a temperature of 7-10K. The VLPCs will be housed in 1024 channels cassettes, which in turn will be housed in two 51 cassette cryostats. Details of the VLPC cassette design can be found in Reference (8). Prototype cryosystems used in VLPC testing maintained the VLPC set temperature to within 50 mK, which is the specification for the final Central Fiber Tracker cryostats.

READOUT ELECTRONICS

Figure 4a shows a sketch of one front end electronics trigger channel. The VLPC output (superimposed on the 5 to 8 volts bias return line) is capacitively coupled to the SIFT chip. Each SIFT chip has 16 input channels and a common threshold. The chip first deamplifies the signal and then buffers it. For each input the SIFT chip outputs two signals: 1) for those channels above threshold a 3.3V single ended output for the level 1 trigger, and 2) for every channel the SIFT chip outputs an analog signal to the

SVX for digitization. Only the axial fibers in the CFT send information to the level 1 trigger.

The SVXII chip (9) and its readout support were designed for the silicon tracker but are well suited to the fiber readout as well. Detailed information on the silicon tracker read out system can be found in References (5,10). The SVXII stores the signal in a 32 deep pipeline analog buffer. Upon receiving a level 1 accept, all the SVX chips stop accruing data and digitize the corresponding data stored in the analog pipeline. The Port Card Board (PCB) reads out the digitized and zero suppressed values from the SVX chains and transmits the data via fast optical link to the VME Readout Buffers (VRB). A sketch of this readout chain is shown in Figure 4b.

The SIFT and SVXII chips sit on the Stereo/Trigger boards that are mounted directly on the VLPV cassettes. The PCBs (also known as Sequencer) are located on the center platform of the detector in the collision hall, and the VRBs are located in the moving counting house.

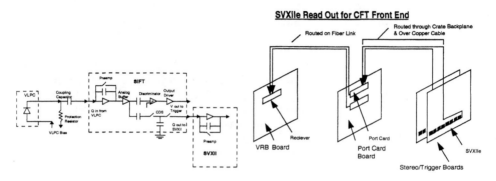

FIGURE 4. a) (left) Schematic for one trigger board channel showing the VLPC, the SIFT-2b chip and the first part of the SVXIIe chip. b) (right) Overall layout of the read out system. The digitized and zero suppressed data from the SVXIIe chips are read out by the PCB via a copper link. The data are then sent to the VRB board over a fast serial link.

REFERENCES

1. "The DØ Upgrade Central Fiber Tracker. Technical design Report."
2. D. Adams et al., IEEE Trans. on Nucl. Sci. **42** (1995), 401.
3. B. Baumbaugh et al., IEEE Trans on Nucl. Sci. **43** (1996), 1146.
4. B. Baumbaugh et al., *Proceedings of DPF 96*, Minneapolis, MN (to be published).
5. "DØ Silicon Tracker Design Report", DØ Note **2169** (1994).
6. S. Margulis et al., SciFi 93, *World Scientific*, A. Bross, R. Ruchti and M. Wayne editors.
7. M. Chung and S. Margulis, *Proceedings of SPIE* **2281** (1994), 26; and **2551** (1995), 10.
8. "DØ Fiber Tracker 1024 Channel VLPC Cassette Technical Design Report and Specification." A. Bross et al, "VLPC Characterization", these Proceedings.
9. R. Yarema et al., "A Beginners Guide to the SVXIIE", Fermilab-TM-1892, June 1994 (revised October 1996)
10. M. Utes, "SVX Sequencer Board", DØ Engineering Note 3823.110-EN-480, November 1997.

R&D Results for the Proposed Intermediate Fiber Tracker of CDF

Juan A. Valls

Rutgers, The State University of New Jersey

for the CDF collaboration

Abstract. In this paper, R&D results for the proposed Intermediate Fiber Tracker detector for the CDF upgrade are presented. We show a detailed study of the performance of the tracker under the high radiation levels expected during the Tevatron Run II operation. The system is proved to be highly efficient even after the expected 2 fb^{-1} of delivered luminosity. A rate capability study has also been conducted on readout SSPM to address the issue of high rate operation. High efficiency is well maintained up to several MHz expected for Run II.

INTRODUCTION

Fermilab is currently facing a major challenge with the operation of the Run II Tevatron collider. The accelerator structure will be enhanced by the fall of 1999 with the inclusion of the Main Inyector. Anticipated instantaneous luminosities of up to 2×10^{32} cm^{-2} s^{-1} and bunch crossings of 132 ns will require a major upgrade for the CDF and DØ detectors. This represents a major change with respect to the past Tevatron operation where maximum instantaneous luminosities of 1.7×10^{31} cm^{-2} s^{-1} with 3.5 μs bunch interval times were reached.

CDF is in the proccess of a tracking system upgrade to accomodate the higher luminosities and shorter bunch crossing times planned for Run II. The silicon vertex detetor (SVX) will be replaced with a 5-layer, double-sided system (SVX II) much longer along the beam axis and with a pipeline readout. The original Central Tracking Chamber (CTC) will be replaced with a new Central Outer Tracker (COT) with a similar open cell structure but four times smaller drift distances. For the space between the SVX II and the COT (approximately between 19 and 40 cm) a fine granularity scintillating fiber detector was originally proposed, referred to as the Intermediate Fiber Tracker (IFT). The IFT detector consists of 12 doublet layers (6 axial and 6 stereo) of 800 μm diameter scintillating fibers. The length of the fibers vary from 180 to 300 cm and cover the pseudorapidity range $|\eta| < 2$. Each of the fibers is connected to a 0.8 mm clear fiber which is routed through the gap between the central structure and the end-plug calorimeter to the outer

CP450, *SciFi97: Workshop on Scintillating Fiber Detectors*
edited by A. D. Bross, R. C. Ruchti, and M. R. Wayne

surface of the detector structures, where Visible Light Photon Counters (VLPC's) are housed in liquid helium cryostats.

LIGHT YIELD

Several factors affect the light yield transmission of scintillating fibers. These are:

- the scintillating fiber length, L_{sc}, and its attenuation length, λ_{sc},

- the clear fiber length, L_{cl}, and its attenuation length, λ_{cl},

- the mirror reflectivity, R, if the far end of the fiber is mirrored,

- the connector efficiency, ϵ_{conn}, between scintillating and clear fibers, if they are spliced together,

- the cassette transmission efficiency, ϵ_{cass}, from the clear fibers to the readout VLPC (Visible Light Photon Counter), and

- the quantum efficiency of the VLPC, QE.

The light transmission may then be written as:

$$ T = \left[exp\left(-\frac{z}{\lambda_{sc}}\right) + R\, exp\left(-\frac{2L_{sc} - z}{\lambda_{sc}}\right) \right] \epsilon_{conn}\, exp\left(-\frac{L_{cl}}{\lambda_{cl}}\right) \epsilon_{conn}\, \epsilon_{cass}\, QE \qquad (1) $$

where z is the distance between the particle track and the connector to the clear fiber (near end).

Attenuation Length

A systematic measurement of the attenuation length of both scintillating and clear fibers of various diameters and types has been performed at Fermilab within the CDF group [1]. The type of fibers included standard 3HF (1% PTP, 1500 ppm) and fibers with slightly different 3HF derivatives. Fibers from both Kuraray and Bicron were used. 3HF and clear fibers were spliced together and placed in a dark box. Scintillating fibers were then illuminated with a UV lamp and light yield was measured by a photodiode at the end of the clear fibers. The measured attenuation curves were well fitted by single exponentials for both 3HF type fibers and clear fibers. The results are shown in Fig. 1 and Fig. 2. In the case of 3HF fibers there is no diameter dependence among the fibers. The clear fibers showed a mild dependence on the diameter.

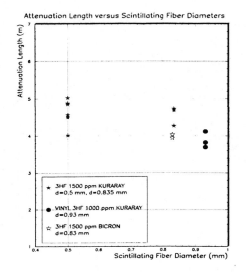

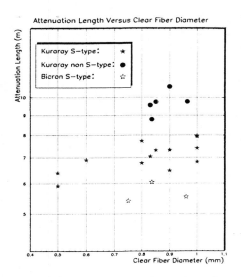

FIGURE 1. Attenuation length vs fiber diameter for scintillating fibers.

FIGURE 2. Attenuation length vs fiber diameter for clear fibers.

RADIATION DAMAGE

Light transmission in plastic optical fibers is known to suffer from radiation exposure. The radiation hardness of plastic fibers has been reported by the the DØ group [2] in the range of 100 krad and above. More recently, the University of Tsukuba group at CDF extended these measurements to a lower range, from 700 krad to 0.4 krad for both 3HF (1% PTP, 1500 ppm) scintillating and clear fibers [3]. The fibers used in this study were Kuraray multiclad s-type fibers of 0.5-0.75 mm diameter for 3HF and 0.8 mm for clear fibers. A ^{60}Co γ-source was utilized to irradiate the fibers at different radius ($R=12$, 35 and 75 cm). More irradiation results have also been obtained when using a ^{137}Cs γ-source and fast neutrons at the JINR pulsed reactor in Dubna [4].

The light output degradation of 3HF fibers can be attributed to the degradation of the attenuation length and not to the scintillating mechanism itself. The radiation hardness of both scintillating and clear fibers is identical if expressed in terms of the ratio of attenuation lengths before and after irradiation, and it is well described as a linear function of $\log_{10}$ of dose. The results are shown in Fig. 3.

Expected Degradation and Radiation Dose Profile

The radiation level in the IFT can be estimated using the radial dose profile measured with the CDF SVX' detector during the 1992-93 run [5]. The dose mea-

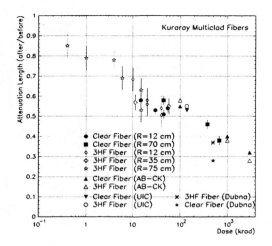

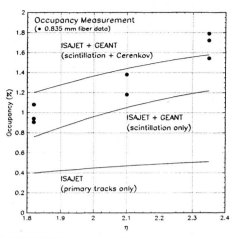

FIGURE 3. Ratio of attenuation lengths before and after irradiation. Measurements AB-CK (A. Bross, C. Kim) and UIC are taken from [].

FIGURE 4. Comparison of the occupancy measured by 0.8 mm scintillating fibers and Monte Carlo simulation results.

sured with a dosimeter was 300 rad/pb^{-1} for the SVX layer 0 (r=3 cm). A radial dependence of $r^{-1.7}$ was found from the increase in the leakage current measured for layers 0 to 3 ($3 < r < 8$ cm). Extrapolating the dose profile to the inner (r=19 cm) and outer (r=40 cm) layers of the proposed system, radiation doses of 24 and 7.3 krad are expected, respectively, for 2 fb^{-1} delivered luminosity during Run II.

Another way to evaluate the light yield drop due to radiation damage is to estimate the dose profile using Monte Carlo. The dose profile in the CDF Run II environment has been evaluated using a complete ISAJET + full GEANT simulation [6]. The energy deposition in the fibers was translated into dose using the measured inelastic $p\bar{p}$ cross section and the charged multiplicity distribution ($dN/d\eta = 4.5$). A realistic calibration of the simulation was provided by the measurement made with 3HF fibers inserted into the CDF detector in the last Run 1C in 1996 [7]. A bundle of 16 scintillating fibers of different lengths and diameter were place at a radius of 27 cm and readout by 8 m long clear fibers with Solid State Photomultipliers (SSPM's). The measured occupancy is compared with the predictions of the simulation in Fig. 4. Considering the uncertainty in the Cerenkov contribution produced in the clear fibers strung through the 10° hole of the end-plug calorimeter, there is good agreement between the data and the simulation results.

Fig. 5 shows the simulation results for the dose profiles along the fibers [6]. The distribution is essentially flat along the z-direction. The dose over the length of the cylindrical part of the innermost layer at $r = 21$ cm is ~ 22 krad for 2.5 fb^{-1}

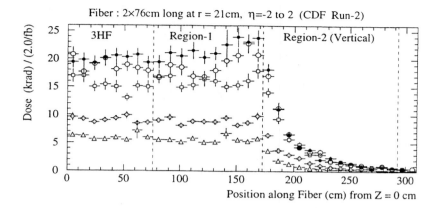

FIGURE 5. Dose profile evaluated for 2 fb^{-1}. The abcissa corresponds to the fiber length from $z = 0$. The fiber is parallel to the beam ($L < 1.6$ m), routed between the COT and the end-plug calorimeter ($1.6 < L < 2.5$ m).

reasonably consistent with the value extrapolated from the SVX measurement. A significant enhancement of the background due to low energy particles looping in the magnetic field, electron pairs converted from photons and back splash from the end-plug calorimeters is seen.

Taken this expected behavior to model the light yield transmission along the fibers, Fig. 6 shows the overall transmission of a signal generated at the middle of the tracker ($\eta = 0$). Two curves ($R=0.9$ and $R=0.8$) are shown for the two conditions considered before and after 2.5 fb^{-1} of integrated luminosity. The right-handed side scale indicates the expected number of photoelectrons. This number has been estimated from several measurements done within the CDF group [8] and found in agreement, within uncertainties, with the observed number by the DØ group with their cosmic ray test stand [9] as the realistic basis.

Fig. 7 shows the efficiency of each doublet layer as a function of the average number of photoelectrons per single fiber. Doublet layer efficiencies above 99.5% for ≥ 6 PE indicate that the proposed IFT has an operational margin greater than 50% even at the weakest point at $\eta = 0$.

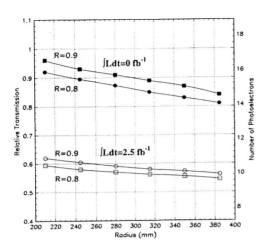

FIGURE 6. Relative transmission and expected number of photoelectrons for signals generated at the middle point of the tracker before and after integrating 2.5 fb^{-1} of delivered luminosity.

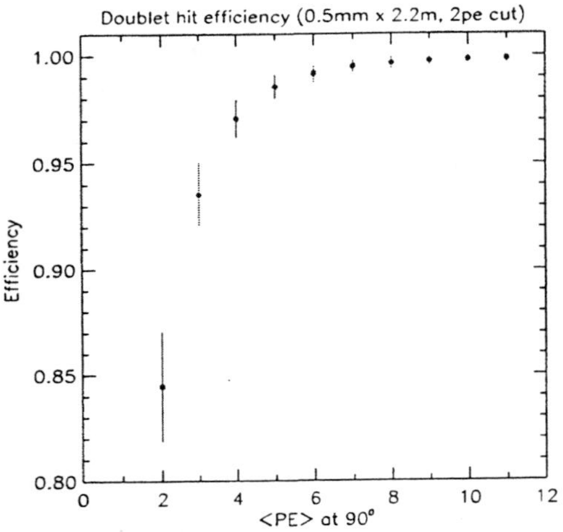

FIGURE 7. Doublet layer hit efficiency as a function of the average number of photoelectrons per single fiber.

OCCUPANCY AND PHOTOELECTRON RATE

A study of the rate capability of SSPM [10] and VLPC [11] has been performed within the CDF collaboration to address the issue of the performance of these readout devices under high rate operation. The setup used is similar to the originally developed by DØ [12]. For the case of the SSPM measurement, light from two laser diodes[1] illuminated the input connector of a cassette holding four 8-channel SSPM cells. The 32-channel cassette was suspended in a liquid helium cryostat. The light emission spectrum from the diodes peaks at 635 nm, well within the broad peak of the SSPM's. One of the diodes, referred to as "background", was excited by a pulse train with frequencies between 100 kHz and 6 MHz. This pulse height generated about 2 photoelectrons and was intermittent with a frequency of 100 Hz, as shown in Fig. 8. A 2 ms pause was imposed after every 8 ms. During this pause, the second diode, referred as "signal" was fired with a variable "quiet time" after the last pulse of the background pulse train. In such arrangement, the amount

Data was taken for the set of parameters shown in Table 1. For each set of parameters the pulse height spectra of the SSPM output was fit with the following formula:

---

[1] Hitachi HL6314MG.

TABLE 1. Parameter sets for the rate capability measurement.

TABLE 1. Parameter sets for the rate capability measurement.

Temperature	Bias voltage, V_{bias}
7.5 K (72 kΩ)	−6.0 V
6.9 K (86 kΩ)	−6.5 V
6.5 K (96 kΩ)	−6.5 V, −7.0 V
6.0 K (112 kΩ)	−7.0 V, −7.5 V

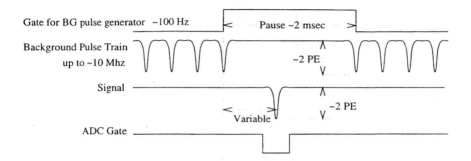

FIGURE 8. Timing diagram for the rate capability test.

$$S(q) = \sum_{n=0}^{\infty} S_n(q) = N \sum_{n=0}^{\infty} G_n(q) \frac{(\overline{N_{pe}})^n}{n!} e^{-\overline{N_{pe}}}$$

$$G_n(q) = \frac{1}{\sqrt{2\pi}\sigma_n} \exp\left[-\frac{1}{2}\left(\frac{q - p - n \cdot g}{\sigma_n}\right)^2\right] \quad ; \quad \sigma_n = \sqrt{\sigma_p^2 + n \cdot \sigma_{gain}^2}$$

where p is the pedestal, σ_p its Gaussian spread, g the average gain of avalanches, σ_{gain} its spread, and $\overline{N_{pe}}$ the average number of PE's. The Gaussian spread of the peaks and the gain, which is the incremental distance between the peaks, are assumed to be the same among all peaks. The gain is linear and the fit is excellent within the measured pulse height range, as shown in Fig. 9 and Fig. 10 for single PE rates of 220 kHz and 11 MHz, respectively.

The number of photoelectrons, which is interpreted as a quantum efficiency in relative scale, increases with both the temperature and the bias voltage and, in both cases, this behavior is more significant at high rates. No sizable dependence of the pedestal position is observed on the temperature and bias voltage at low rates. However, at higher frequencies there is a striking decrease of the pedestal position towards higher values of the temperature and bias voltage.

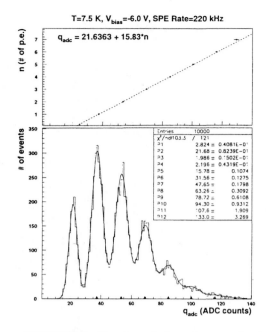

FIGURE 9. Number of photoelectrons (top) and pulse height distribution (bottom) for a single PE rate of 220 kHz.

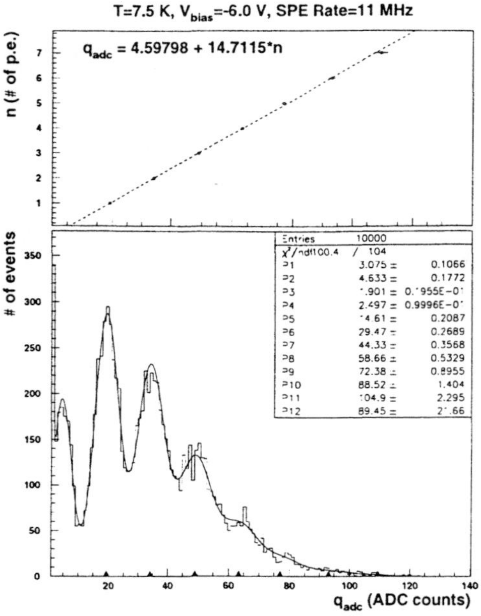

FIGURE 10. Number of photoelectrons (top) and pulse height distribution (bottom) for a single PE rate of 11 MHz.

Within the tested range of the operating conditions, Figure 11 shows a gradual decrease of the gain with the single PE rate above 1 MHz. This decrease in gain at high frequecies is slightly improved by increasing the temperature. In these ideal conditions the gain reduction is only a few percent. The quantum efficiency seems to be even less affected by the increase of rate and no visible effect can be noticed up to 4 MHz SPE rate, as shown in Figure 11. Above this rate, an overall trend of decrease is observed below several percent level up to 12 MHz.

CONCLUSIONS

A fine granularity scintillating fiber tracking system was proposed for the intermediate radial range of the CDF detector. A detailed study of its performance, specially under the high rate conditions of Run II, has proved the detector to be highly efficient for integrated luminosities of up to 2 fb^{-1}. The rate capability of SSPM has also been studied for various combinations of temperature and bias voltage. Using a laser diode flasher, measurements were made up to a single photoelectron rate of 12 MHz. Only a slight decrease, less than several per cent, of the

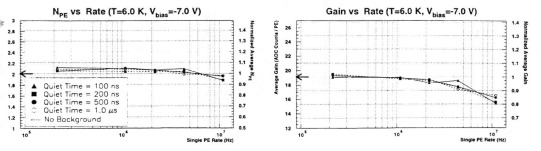

FIGURE 11. $\overline{N_{pe}}$ and gain vs single PE rate for T=6.0 K and V_{bias}=-7.0 V.

quantum efficiency was observed at this rate whereas a 10% drop of the gain was observed.

ACKNOWLEDGEMENTS

I would like to thank the organizers of the conference for the excellent work. My sincere thanks also for prof. M. Atac, M. Mishina and F. Bedeschi for let me participate in this project and the large experience gained by working with them.

REFERENCES

1. D. Cassettari *et al*, Measurement of Attenuation Length of Scintillating and Clear Fibers, CDFnote 3560, February, 1996.
2. S. Margulies *et al*, Proceedings of SCIFI 93 Workshop on Scintillating Fiber Detectors, World Scientific, pp. 421-430, October 1993.
3. K. Hara *et al*, Radiation Hardness of Kuraray Optical Fibers, CDFnote 3727, June 25, 1996.
4. F. Bedeschi *et al*, Radiation Hardness of New Kuraray Double Cladded Optical Fibers, CDFnote 3583, March, 1996.
5. P. Azzi *et al*, Noise and Radiation Effects on SVX' and Other Silicon Sensors, CDFnote 3278, August 1995.
   M. A. Frautschi, Radiation Damaga Issues for the SVXII Detector, CDFnote 2368, January, 1994.
6. T. Yoshida, private communication.
7. G. Bolla, R. Demina, Measurement of the Intermediate Fiber Tracker Occupancy, CDFnote 3623, June 1996.
8. M. Atac *et al.*, Measurement of the Light Yield from 0.5 mm Scintillating Fibers with the IFT prototype, CDFnote 3569, April 1996.

R. Demina *et al.*, Measurement of the Relative Photoelectron Yield of 0.5 mm and 0.835 mm Scintillating Fibers, CDFnote 3559, February, 1996.

9. D. Adams *et al.*, Cosmic Ray Test Results of the DØ Prototype Scintillating Fiber Tracker, Fermilab-Conf-95/012-E, January, 1995.

10. J. A. Valls *et al.*, Study of the Performance of SSPM's at High Rate, CDFnote 3790, July, 1996.

    M. Gubinelli *et al.*, Measurement of the Rate Capability of Solid State Photomultipliers Using Visible Light, Fermilab-PUB-97-104, April, 1997, submitted to Nucl. Instr. and Meth.

11. F. Bedeschi *et al.*, Measurement of VLPC Gain and Q.E. as a Function of Pulsing Rate, CDFnote 3752, January, 1997.

12. J. Warchol, *these proceedings*.

Development of a Scintillating Fiber Tracking Detector for the K2K Neutrino Oscillation Experiment

Presented by Atsumu Suzuki for the KEK E362(K2K) Collaboration

Physics Department, Kobe University, 1-1, Rokkodai-cho, Nada-ku, Kobe, 657, Japan

Abstract. We are preparing a scintillating fiber tracking detector as a part of the near fine-grained detector in the K2K long baseline neutrino oscillation experiment between KEK and Super-Kamiokande. We use Kuraray SCSF-78, 0.7 mm diameter fiber with Hamamatsu IIT-CCD camera read out system. The choice of the fiber is based on a series of measurements of the light yield and aging of the candidate fibers under various conditions. It was found that SCSF-78 has enough light yield and lifetime for our purposes. We have also checked the performance of the SCIFI sheet-IIT-CCD system by source(^{90}Sr) and cosmic rays. The detection efficiency was found to be more than 99%. The full SCIFI detector construction is current under way.

K2K EXPERIMENT

The results of the former Kamiokande and the recent Super-Kamiokande show the strong possibility of neutrino oscillation. The parameter region for ν_μ to ν_τ oscillation which they show is $\sin^2 2\theta \sim 1$ and Δm^2 between 10^{-3} and 10^{-2} eV^2([1,2]). The K2K experiment is a long baseline neutrino oscillation experiment between KEK and Kamioka(K2K means KEK 'to(2)' Kamioka, [3]). The distance is 250km and the mean neutrino energy is about 1 GeV. We can survey the above parameter region.

We will use Super-Kamiokande detector as a rear detector and construct a new detector as a front detector in KEK site(Figure 1). The front detector consists of the 1kt water Cherenkov detector used in KEK PS E261A [4] and a fine-grained detector(FGD). The FGD is composed of the SCIFI detector, the lead glass counter, and the muon range counter. We compare the ν_μ flux and energy spectrum between the front and rear detectors to search for neutrino oscillations between ν_μ and any other neutrino species(the ν_μ disappearance analysis).

CP450, *SciFi97: Workshop on Scintillating Fiber Detectors*
edited by A. D. Bross, R. C. Ruchti, and M. R. Wayne
© 1998 The American Institute of Physics 1-56396-792-8/98/$15.00

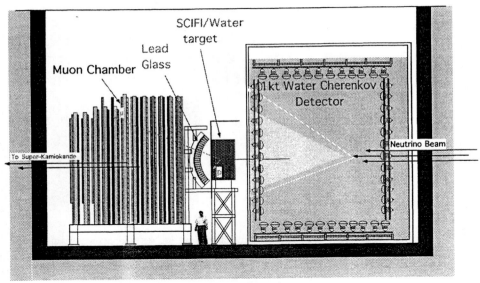

FIGURE 1. K2K front detector.

SCIFI TRACKING DETECTOR

We need a tracking detector for the FGD. It is required to have good position resolution, to be thin to increase the acceptance of the down stream muon range counter as much as possible, and to have less material compared with the water target. We selected the SCIFI tracking detector as such a detector. Its schematic view is shown in Figure 2. The main purpose of the SCIFI detector is charged track reconstruction. From the charged track reconstruction, we can determine the fiducial volume precisely ($\Delta V/V \sim$ a few %) to measure the neutrino flux. We will use quasi–elastic scattering $\nu_\mu + n \to \mu^- + p$ to get the ν_μ energy. It is calculated by the following formula;

$$E_{\nu_\mu} = \frac{m_N E_\mu - m_\mu^2/2}{m_N - E_\mu + p_\mu \cos\theta_\mu} \tag{1}$$

E_{ν_μ} : ν_μ energy
E_μ : muon energy
m_N : nucleon mass
m_μ : muon mass
p_μ : muon momentum
θ_μ : muon scattering angle.

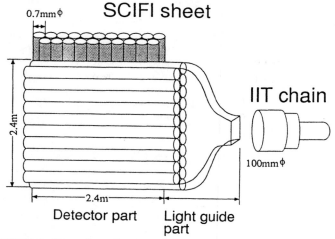

FIGURE 2. The schematic view of the SCIFI tracking detector of K2K.

Our scintillating fiber tracking detector consists of 20 of 2.4m × 2.4m modules placed 6cm apart. Each module contains 2 double-layered SCIFI sheets in horizontal and vertical directions, respectively. The readout device is Hamamatsu IIT-CCD chain, which consists of V5502PX, V1366GX(MCP-IIT), 1/3 lens, and C3077(CCD). We need 24 of this device for the all 20 SCIFI modules.

FIBER TYPE CHOICE

We measured the light yield and attenuation length of various types of scintillating fibers using a photo-multiplier tube(Hamamatsu R329-058) by exciting the fiber with a β ray source ^{90}Sr [5]. The light yield after the propagation of distance x in the fiber is expressed as follows;

$$Y(x) = Y_0 \, e^{-x/\lambda}, \tag{2}$$

where Y_0 is an initial light yield($Y(0)$) and λ is an attenuation length. We estimated Y_0 from the extrapolation of the measurements at $x = 1.5$, 2.0, 2.5, and 3.0m. The results are summarized in Table 1. Kuraray SCSF-78M is a multi cladding fiber. The other Kuraray's SCSF fibers and Bicron BCF-12 are single cladding ones. When we require more than 99% detection efficiency, there are two candidates left. To save the readout area, we chose thinner one, SCSF-78M 0.7 mmϕ.

TABLE 1. Light yield Y_0, attenuation length λ, and detection efficiency ε at $x = 4$m.

SCIFI type	Diameter[mmϕ]	Y_0[p.e.]	λ[m]	Efficiency[%]
SCSF-78M	0.5	4.35 ± 0.15	3.46 ± 0.18	97.2 ± 0.7
	0.7	6.49 ± 0.20	3.57 ± 0.17	99.6 ± 0.1
SCSF-78	0.7	5.16 ± 0.13	3.48 ± 0.13	98.6 ± 0.3
	0.9	7.24 ± 0.15	3.86 ± 0.13	99.9 ± 0.03
SCSF-77	0.7	4.22 ± 0.14	4.39 ± 0.27	98.9 ± 0.3
BCF-12	0.7	3.60 ± 0.13	3.36 ± 0.18	94.2 ± 1.2

AGEING MEASUREMENT

Scintillating fibers are known to age. They have the finite lifetime. We measured the time variation of the light yield of the fibers at $x = 2.6$m ($Y(2.6m)$)to know how long they would maintain their performance. The tested fibers are Kuraray SCSF-78 with single and multi cladding of which diameters are 0.7mm. The conditions under which we put the fibers are shown in Table 2. We ordinary wind the fibers and put them into the vessels to save the storage space. To study the effect of this mechanical stress, we also prepared straight fibers. The time variation of the light yield of each fiber at 70°C is shown in Figure 3. We evaluated the lifetimes of fibers as the time for the light yield to become 90% of the initial state. The lifetimes of single and multi cladding SCSF-78 were $4.0^{+1.4}_{-0.9}$ days and $1.8^{+0.4}_{-0.3}$ days, respectively. We could not see any ageing effect in nitrogen gas. This means that oxidation mainly causes the ageing. This is also confirmed from the fact that SCSF-78M coated with aluminum has more than 3 times longer lifetime ($5.8^{+2.5}_{-1.5}$ days). From these results, Kuraray developed SCSF-78M version 11. It is an anti-O_2 type and its lifetime was 8.95 ± 0.87 days. It is about 5 times longer than that of the original SCSF-78M. The straight SCSF-78M has a longer lifetime($5.0^{+1.2}_{-0.9}$ days) than that of wound one. This fact shows that the mechanical stress also causes the ageing.

Temperature dependence of the fiber lifetime is known as Arrhenius plot. From the extrapolation of the lifetimes in the air at 41C°, 55C°, 67C° and 70C°, we estimated the lifetimes of SCSF-78M version 11 at 15C° and 20C° (Figure 4). They are 8160 and 3970 days, respectively. We can permit 10% light yield loss and keep the temperature of the experimental hall less than 20C°. So these estimated lifetimes are long enough for our experiment.

TABLE 2. The conditions for the ageing measurement of SCSF-78

Winding	SCSF-78(single cladding) @70C°
	SCSF-78M(multi cladding original type) @70C°
	in the air
	in $N_2(O_2 < 1\%)$
	Al coated
	SCSF-78M version 11(anti-O_2 type) @41, 53, 67, and 70C°
Straight	SCSF-78M(original type) @70C°

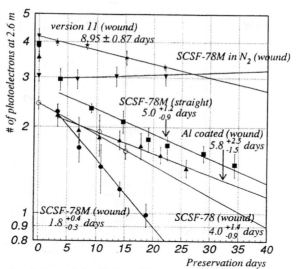

FIGURE 3. Time variation of the light yield of each fiber at 70C°. The 90% lifetimes are shown in the figure.

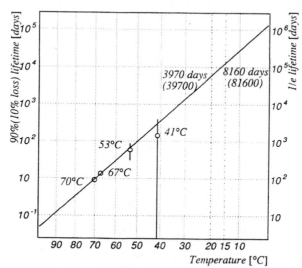

FIGURE 4. The Arrhenius plot of SCSF-78M version 11(anti-O_2 type). The numbers in the parentheses are the 1/e lifetime.

PERFORMANCE OF SCINTILLATING FIBER TRACKER

We measured the detection efficiency and position resolution of the scintillating fiber tracker by radio isotope source(^{90}Sr) and cosmic ray. Figure 5 shows N_{hit} distribution by source test, where N_{hit} means the number of hit pixels on a CCD camera. When $N_{hit} > 5$(it corresponds to 0.5 photoelectrons)is required, the detection efficiency is 99.2% and we can suppress the noise contamination less than 0.5%.

We used five layer of doublet SCIFI sheets for the cosmic ray test. The interval between them is 6cm. We set trigger counters above and under the SCIFI sheets. We reconstructed a cosmic ray track with the upper two and lower two layers after the clustering algorithm was applied to CCD image, and took a residual between the reconstructed track and the real hit position at the center layer(Figure 6). From the quick analysis, we preliminary got the position resolution of 275μm. This is enough position resolution for our experiment because we need at least 1mm position resolution. However we will be able to get better resolution by further analysis.

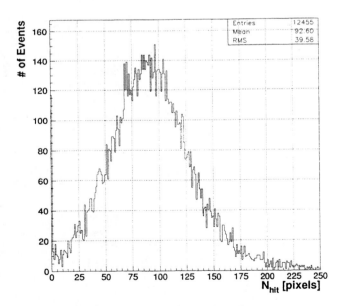

FIGURE 5. N_{hit} distribution by RI source test.

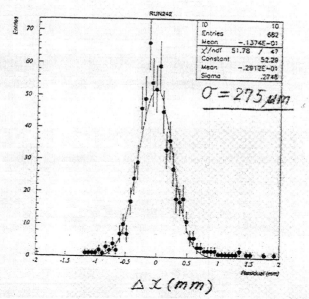

$\sigma = 275 \mu m$

$\Delta \chi \, (mm)$

FIGURE 6. Position resolution of the prototype SCIFI tracker by cosmic ray test.

CONSTRUCTION STATUS

The schematic view of our SCIFI sheet is shown in Figure 7. We will make 240 of such a sheet.

Our sheet production method is same as that of CHORUS group [6,7]. It is shown in Figure 8. The fiber is spooled by the drum with a spiral groove. The first layer fiber is guided to the groove. After the first layer spooling, we fix the position of the fiber by gluing with paint. The second layer fiber is guided between the first layer fibers. To make the SCIFI sheet flat after removing from the drum, we add a suitable tension to the second layer fiber and extend its length during spooling.

We set the SCIFI sheet production machine at Kobe University (Figure 9). The diameter of the drum is 1.2m. The mass production of the SCIFI sheets using this machine started in July 1997 and it is going very well.

The structure of our SCIFI detector is now under preparation at KEK and Boston University. The schematic view is shown in the Figure 10. We glue 12 SCIFI sheets on a honeycomb board, 6 of them horizontally on one side and the other 6 of them vertically on the other side. Our SCIFI detector consists of 20 layers of such a module.

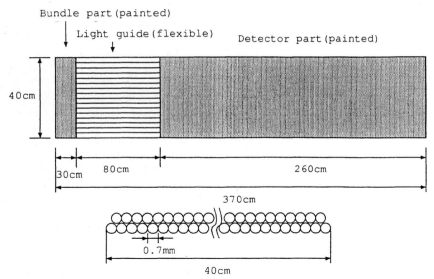

FIGURE 7. Schematic view of our SCIFI sheet. The upper figure shows the top view and lower one shows the cross sectional view.

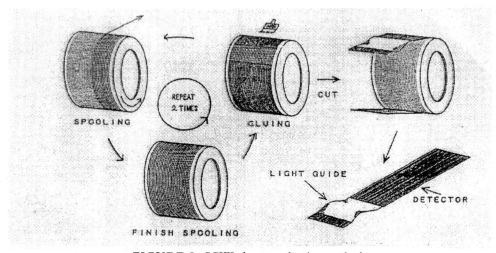

FIGURE 8. SCIFI sheet production method.

246

FIGURE 9. Scintillating fiber spooling machine for the K2K experiment. It is in Kobe University.

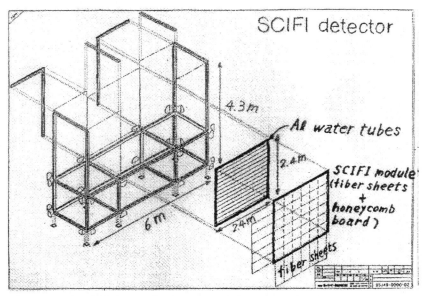

FIGURE 10. The structure of the K2K SCIFI detector.

SUMMARY AND SCHEDULE

We will use scintillating fiber Kuraray SCSF-78M version 11(anti-O_2 type) for the K2K long baseline neutrino oscillation experiment. It has enough light yield and long lifetime for the experiment. The SCIFI detector made of this fiber has good detection efficiency($> 99\%$) and we preliminary got the position resolution of $\sim 280\mu m$ by quick analysis.

The SCIFI sheet mass production started at the end of July 1997. It is going very well. It will be finished at the end of July 1998. Assembling of the SCIFI detector will begin in January 1998 and end in July 1998. Its installation will begin in October 1998 and end in December 1998. The K2K experiment will begin in January 1999.

REFERENCES

1. Fukuda, Y., et al(KamiokandeIII Collaboration), *Phys. Lett.* **B335**, 237–245(1994).
2. Totsuka, Y., presented at the Lepton Photon Symposium, Hamburg, July, 1997.
3. Nishikawa, K., et al, *Proposal for a Long Baseline Neutrino Oscillation Experiment, using KEK-PS and Super-Kamiokande*, April 1995.
4. Kasuga, S., et al(E261A Collaboration), *Phys. Lett.* **B374**, 238–242(1996).
5. Etoh, M., Master Thesis, Tokai University, March 1997(in Japanese).
6. Nakano, T., et al, *IEEE Trans.*, **39**, No.4, 680–684(1992).
7. Ishikawa, S., et al, *"Hoshasen"*,**21**, No.3, 85–119(1995)(in Japanese).

Tests of a fiber detector concept for high rate particle tracking

E.C. Aschenauer, J. Bähr, V. Gapienko[1], B. Hoffmann[2],
H. Lüdecke, A. Menchikov[3], C. Mertens[4],
R. Nahnhauer, R. Shanidze[5]

DESY-Zeuthen, 15738 Zeuthen, Germany

Abstract. A fiber detector concept is suggested allowing to registrate particles within less than 100 nsec with a space point precision of about 100 μm at low occuppancy. The fibers should be radiation hard for 1 Mrad/year. Corresponding prototypes have been build and tested at a 3 GeV electron beam at DESY. Preliminary results of these tests indicate that the design goal for the detector is reached.

INTRODUCTION

The advantageous use of fiber detectors for particle tracking has been demonstrated for very different conditions e.g. in the UA2-Experiment [1], CHORUS [2], for D0 [3] and the H1-Forward Proton Spectrometer [4]. Due to the different experimental situation in this applications three types of optoelectronic read out techniques are applied – Image Intensifier plus CCD's, Visible Light Photon Counters and Position Sensitive Photomultipliers. However, all have in common that the precision of space point measurements is given by hits of overlapping fibers of several staggered fiber layers. For high rate experiments demanding online tracking of several hundred particles per 100 nsec bunch crossing such a concept may not work due to too high occupancy of single fiber channels.

We propose in the following to use overlapping fiber roads reading out several thin scintillating fibers with one clear optical fiber. The demands and the solutions presented below match to a possible application of the detector as the inner tracker in the HERA-B project at DESY [5]. Similar ideas have been used by others [6] to build a fiber detector for the DIRAC experiment at CERN.

---

[1] on leave from IHEP Protvino, Russia
[2] now at Esser Networks GmbH, Berlin
[3] on leave from JINR Dubna
[4] Summerstudent from University of Clausthal Zellerfeld
[5] on leave from High Energy Physics Institute, Tbilisi State University

CP450, *SciFi97: Workshop on Scintillating Fiber Detectors*
edited by A. D. Bross, R. C. Ruchti, and M. R. Wayne
© 1998 The American Institute of Physics 1-56396-792-8/98/$15.00

DETECTOR PRINCIPLE

The fiber detector under discussion is aimed to detect throughgoing particles with more than 90 % efficiency within less than 100 nsec and a precision of better than 100 μm. The fibers should not change their characteristics significantly after an irradiation of 1 – 2 Mrad. The sensitive detector part should have a size of 25 x 25 cm^2. The scintillating fibers should be coupled to clear optical fibers of about 3m length guiding the light to photosensors placed outside the experimental area.

It is assumed that most particles of interest hit the detector perpendicular, i.e. with angles less than five degrees with respect to the beam axis. In this case low occupancy and high light yield are guaranteed by using overlapping fiber roads like schematically drawn in fig. 1. One fiber road consists of several thin scintillating fibers arranged precisely behind each other and coupled to one thick light guide fiber. The scintillating fiber diameter determines the space point resolution of the detector. The number of fibers per road is fixed by the scattering angle of particles and the allowed amount of multiple scattering. It will also influence the factor of background suppression for tracks with larger inclination or curvature. The pitch between fiber roads is defined by demanding a homogeneous amount of fiber material across the detector width.

Keeping in mind the conditions at HERA-B, we made the following choices:

$$
\begin{array}{ll}
\Phi_{fib} = 480 \mu m & N_{fib/road} = 7 \\
L_{fib} = 30 \text{ cm} & p_{road} = 340 \ \mu m \\
\Phi_{lg} = 1.7 \text{ mm} & N_{road} = 640 \\
L_{lg} = 300 \text{ cm} & W_{det} = 217.6 \text{ mm}
\end{array}
$$

with Φ and L: diameter and length of scintillating and light guide fibers, $N_{fib/road}$: number of fibers per road, p_{road}: distance between neighboured road centers, N_{road}: number of roads per detector, W_{det}: detector width.

The light guide fibers are read out with the new Hamamatsu[6] 64 channel PSPM R5900–M64 with a pixel size of 2 x 2 mm^2 [7]. To diminish optical cross talk the thickness of the entrance window of the device was decreased to 0.8 mm.

The coupling between scintillating and light guide fibers is done by loose plastic connectors. The light guides are coupled to the PSPM using a plastic mask fitting the corresponding pixel pattern.

MATERIAL STUDIES

Double clad fibers of three different producers[7][8][9] were tested concerning light output, light attenuation and radiation hardness for several fiber diameters and

---

[6] Hamamatsu Photonics K.K., Electron tube division, 314–5, Shimokanzo, Tokyooka Village. Iwatagun, Shizuoka-ken, Japan

[7] BICRON, 12345 Kinsman Road, Newbury, Ohio, USA

[8] Pol. Hi. Tech., s.r.l., S.P. Turanense, 67061 Carsoli(AQ), Italy

[9] KURARAY Co. LTD., Nikonbashi, Chuo-ku, Tokyo 103, Japan

wavelengths of the emitted light. Details of these measurements are given in [8]. A few results are summarized below.

The light output of fibers of 500 μm diameter is shown in fig. 2. Generally it can be seen, that the light yield decreases with increasing scintillator emission wavelength because the PM sensitivity curve is not unfolded. There is no remarkable difference between the best materials of the three producers. A mirror at the end of the fiber increases the light output by a factor 1.7.

Several tests were performed to couple scintillating and light guide fibers. Finally the coupling efficiency became better than 95 %, independent of the medium between both parts (air,glue,grease).

The light attenuation of clear fibers was measured coupling them to single scintillating fiber roads excited by a Ruthenium source. The clear fibers were cutted back piece by piece to the length under investigation. Results for two producers are given in fig. 3.

Radiation hardness tests of fibers were made using an intense 70 MeV proton beam at the Hahn–Meitner–Institute Berlin. 1 Mrad radiation was deposited within a few minutes. For all materials investigated we observed a damage of the scintillator and the transparency of the fiber which was followed by a long time recovery of up to 600 h. An example is shown in fig 4. More detailed studies using glued and nonglued fibers and irradiate them in air and nitrogen atmosphere are still ongoing.

Summarizing all results of our material studies we decided to use the KURARAY fibers SCSF-78M with a diameter of 480 μm for the scintillating part of our detector prototypes. For clear fibers still two choices seem to be possible: 1.7 mm fibers from KURARAY or Pol. Hi. Tech..

DETECTOR PRODUCTION

Using winding technology as developed for the CHORUS experiment [9] we built a detector production chain at our institute. A drum of 80 cm diameter allows to produce five detectors at once. The production time for winding one drum is about 14 h. Sticking the fibers to the connector holes is still done by hand and rather time consuming. A part of the polished end face of one of our detectors is shown in fig. 5.

Two other detector prototypes are ordered from industry. GMS-Berlin[10] followed a technology proposed by the university of Heidelberg [10] mounting single layers on top of each other using epoxy glue. Each layer is prepared on a v-grooved vacuum table. One layer per day can be produced in this case. The connector is here also added by hand. The production technology used by KURARAY is unknown to us.

To get the precision of the detector geometry quantified we measured the coordinates of all fibers of the polished end face of the three detectors. In fig. 6 the deviation from the ideal position is given per fiber road. Some stronger local effects

---

[10] GMS - Gesellschaft für Mess- und Systemtechnik mbH, Rudower Chaussee 5, 12489 Berlin, Germany

are visible. Averaging these results characteristic accuracies of 20 μm, 50 μm and 10 μm are calculated for the Zeuthen, GMS and KURARAY detectors respectively.

TESTRUN RESULTS

Two testruns were performed to measure the properties of the produced fiber detectors in a 3 GeV electron beam at DESY. The setup used in both cases was very similar and is schematically drawn in fig. 7. Four silicon microstrip detectors are used together with two fiber reference detectors and an external trigger system to predict the position of a throughgoing particle at the detector to be tested. A precision of 50 μm and 80 μm was reached for that prediction using the geometrical arrangements of testrun 1 and 2. The fiber detector signals were registrated after 3m of light guide in the first case using a 16 channel PM R5900–M16 read out with a charge sensitive ADC. In the later run the 64 channel R5900–M64 was used and the signals were transfered via a special multiplexer to a flash ADC.

In April 1997 first small eight road detectors were investigated to measure the light profile across a fiber road. The result is shown in fig. 8. The data can be described simply by taking into account fiber geometry seen by a throughgoing particle. They allow to calculate the detector efficiency for any particular pitch between the fiber roads.

During the testrun in October 1997 the three full size detector prototypes described in section 4 were investigated in detail. Up to now only preliminary results are derived from about 4 Gbyte of data.

A relation of 0.9/1.0/0.8 was found for the average light output of the Zeuthen, GMS and KURARAY detectors. It seems to be due to the different quality of the end face polishing rather than to the mechanical detector precision.

The detector efficiency and resolution is dependent on the hit selection method used. With a maximum amplitude search for all PSPM pixels we calculated rough values of 97 $\pm$ 3 % for the efficiency and about 140 μm for the resolution of the three detectors. (see also figs. 9 and 10). Taking into account the finite resolution of our track prediction of 80 μm and the total mechanical alignment not better than 50 μm this points to a fiber detector resolution of better than 100 μm.

Work is in progress to qualify these results. In addition the detector noise has still to be studied in detail. Optical and electrical cross talk will influence the choice of cuts and the hit selection methods and in this way also efficiencies and resolution.

SUMMARY

Three fiber detector prototypes have been tested. They are made out of 640 overlapping roads of seven 480 μm diameter fibers coupled to 1.7 mm diameter light guides of 3 m length read out with 64 channel photomultipliers. For all three detectors a preliminary analysis gives an efficiency of about 97 % and a resolution of about 100 μm. These results together with radiation hardness studies of the used fiber material seem to make it possible to use a corresponding detector in a high rate experiment like HERA-B. In such case special care has to be taken to keep noise from optical and electrical cross talk at an acceptable level.

Acknowledgement

Part of this work was done in close collaboration with groups from the universities of Heidelberg and Siegen. We want to thank our colleagues for their good cooperation and many fruitful discussions.

The fiber irradiation tests were possible only due to the kind support of the Hahn-Meitner-Institute Berlin. We are deeply indebted to the ISL accelerator team and want to thank in particular Dr. D. Fink, Dr. K. Maier and Dr. M. Müller from HMI and Prof. Klose from GMS for a lot of practical help.

We acknowledge the benefit from the DESY II accelerator crew and the test area maintainance group.

REFERENCES

1. Ansorge, R., et al., *NIM* **265**, 33 (1988)
2. Annies, P., et al., *NIM* **A367**, 367 (1995)
3. Bross, A.D., *Nucl. Phys. B (Proc.Suppl.)* **44**, 12 (1995)
   Adams,D.,et al., *Nucl. Phys. B (Proc.Suppl.)* **44**, 332 (1995)
4. Bähr, J., et a., *Proceedings of the 28th Intern. Conf. on High Energy Physics, Warsaw, Poland, 1996*, eds. Z.Ajduk,A.K.Wroblewski V. II, p. 1759
5. Lohse, T., et al., *HERA-B Technical Proposal, DESY-PRC* 94/02 (1994)
6. Ferro–Luzzi, M., et al., contribution presented by A.Gorin to this workshop
7. Yoshizawa, Y., contribution to this workshop
8. Aschenauer, E.C., et al., preprint *DESY* **97-174** (1997)
9. Nakano, T., et al., *Proceedings of the workshop SCIFI93, Notre Dame, USA, 1993*, eds. A.Bross, R.Ruchti, M.Wayne, p. 525
10. Eisele, F.,et al., private communication

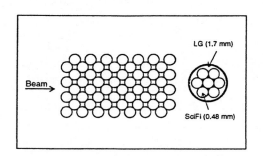

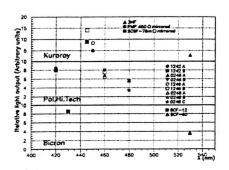

FIGURE 1. Schematic view of the proposed fiber detector cross section and coupling principle (LG: light guide fiber)

FIGURE 2. Light output from 500 μm diameter fibers for several fiber materials of three producers

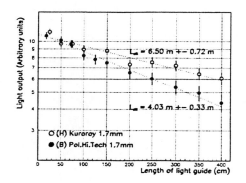

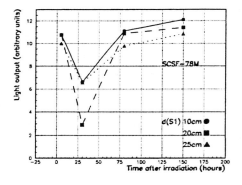

FIGURE 3. Light attenuation in clear fibers of 1.7 mm diameter produced by Kuraray and Pol.Hi.Tech.

FIGURE 4. Time evolution of light output for KURARAY SCSF 78M fibers irradiated with 0.2 and 1.0 Mrad at 10 and 20 cm respectively. The solid, dashed and dotted curves correspond to measurements with a source placed at 10,20 and 25 cm.

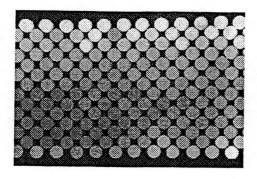

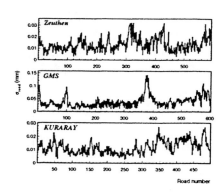

FIGURE 5. Photograph of part of the polished end face of a Zeuthen prototype detector.

FIGURE 6. Deviation of fibers from ideal position per fiber road for three prototype detectors from Zeuthen, GMS and KURARAY.

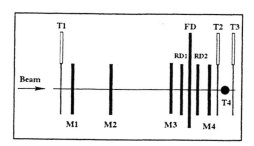

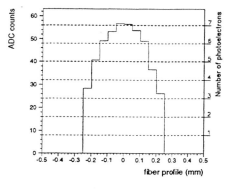

FIGURE 7. Set up for testruns 1 and 2. M1-M4: silicon microstrip detectors, RD1 and RD2: fiber reference detectors, T1-T4: trigger paddels, FD: detector to be tested.

FIGURE 8. Light output across a fiber road of seven 500 μm KURARAY fibers coupled to a 3 m long light guide of 1.7 mm diameter.

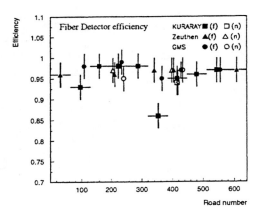

FIGURE 9. Efficiency at different positions of three fiber detector prototypes, averaged for 64 channels. Particles hit 5cm from the near (n) or far (f) end of the ordered detector part.

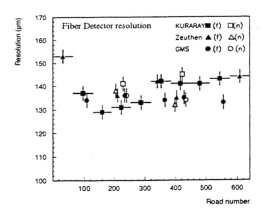

FIGURE 10. Resolution at different positions of three fiber detector prototypes, averaged for 64 channels. Particles hit 5cm from the near (n) or far (f) end of the ordered detector part.

Fiber technology applications for a future e^+e^- linear collider detector

H. Leich, R. Nahnhauer, R. Shanidze[1]

DESY-Zeuthen, 15738 Zeuthen, Germany

Abstract. The advantages and possibilities of fiber technology for the detection of particles in 500 GeV e^+e^- reactions are considered. It is suggested to build a fast trigger which could be used also for intermediate tracking. A fiber preshower in front of the electromagnetic calorimeter would allow to identify electrons and photons with a space precision better than 100 μm.

INTRODUCTION

e^+e^- reactions have particular advantages for the study of particle properties and interaction pecularities. That is the reason for the worldwide discussion about problems and possibilities of future e^+e^- linear colliders in the energy range between 500 – 1500 GeV [1].

Detectors at such colliders have to handle very short bunch crossing times and/or tremendous background rates originating from new sources due to the strongly collimated beams. All these questions were studied in detail during an Joint ECFA/DESY workshop in 1996 [2]. The following considerations arised from discussions during this workshop.

REACTIONS STUDIED

To investigate the capabilities of special detector configurations we simulated benchmark e^+e^- reactions at 500 GeV :

$$
\begin{aligned}
e^+e^- &\rightarrow W^+W^- & (1)\\
e^+e^- &\rightarrow t\bar{t} & (2)\\
e^+e^- &\rightarrow HX & (3)\\
\text{beamstrahlung} & & (4)
\end{aligned}
$$

---

[1] on leave from High Energy Physics Institute, Tbilisi State University

CP450, *SciFi97: Workshop on Scintillating Fiber Detectors*
edited by A. D. Bross, R. C. Ruchti, and M. R. Wayne
© 1998 The American Institute of Physics 1-56396-792-8/98/$15.00

and traced the resulting secondary particles through the detector, using the program systems PYTHIA [3], ABEL [4] and GEANT [3].

The number of particles to be detected varies between 2 (pure leptonic W^+W^- decays) and 200. Averages are given in table 1.

TABLE 1. Average number of all particles, charged particles and leptons for W, top and higgs production. The event rates correspond to a luminosity L= $2 \cdot 10^{33}$ cm^{-2} sec^{-1}

	$< N >$	$< N_{ch} >$	$< N_{lep} >$	ev. rate
W^+W^-	57	26	0.9	1/min
tt	115	55	2.1	1/10min
HX	78	36	1.5	~1/h

The kinematics is different for all processes considered (see fig. 1). W^+W^- pairs are produced strongly in forward and backward direction in contrast to top and higgs production. Particles from beamstrahlung are mostly low energetic and bounded to the inner part of the detector near to the beam pipe, if a magnetic field of 3 T is assumed.

FAST INNER TRIGGER

To suppress background and select events with high transverse momentum jets, a fast inner trigger could be useful in particular for collider operations with very short bunch crossing times. We propose a layout as schematically drawn in fig. 2.

In a magnetic field of 3 Tesla two cylinders of 1 m length surround the interaction point at r = 18 cm and r = 28 cm. Each cylinder consists of 4 layers of scintillating fibers of 1 mm diameter parallel to the z–axis. The fibers are combined to form Φ–slices for the inner and outer cylinder. For a one degree resolution one would have to handle therefore $2 \cdot 360$ channels. The light is read out by light guides coupled to normal photomultipliers or by hybrid devices withstanding magnetic fields [5].

The simplest trigger condition is to demand signals above a certain threshold from the same Φ–segment of both cylinders. Trigger times below 10 nsec seem to be in reach with the above arrangement. A possible hardware scheme is given in fig. 3. It would allow even to form clusters between neighboured slice signals. The trigger efficiency is limited only by geometry. For t$\bar{t}$ and Higgs production it is nearly 100 %. The number of fake triggers may be decreased by smaller cell sizes. With the configuration described, the average occupancy is 5 %(W^+W^-), 14 %(t$\bar{t}$) and 7 %(HX). In fig. 4 the clear correlation between high p_t tracks and trigger clusters observed is demonstrated. No trigger is found using simulated tracks from beamstrahlung. A typical particle distribution for this process in one bunch crossing is shown in the rΦ–plane in fig. 5.

CENTRAL INNER TRACKER

It is now easy to extend the mechanical trigger layout to build a central fiber tracker. In addition to the fibers parallel to the z-axis four layers of staggered fibers inclined by 11 degrees will be added to both cylinders. With a pitch of 1.08 mm between adjacent fibers one would have 1 050 and 1 650 fibers per layer for the inner and outer cylinder respectively, adding up to a total number of 21 600 channels to be read out e.g. via light guides and multianode photomultipliers [6] or VLPC's [7].

The space point resolution of such a configuration would be $\Delta\ r\Phi = 80\ \mu m$ and $\Delta z = 410\ \mu m$. The geometrical acceptance and average occupancy per layer is given in tab.2.

TABLE 2. Geometrical acceptance and average occupancy for reactions 1-3 at the inner and outer cylinder of a central inner fiber tracker.

cylinder	geometr. acceptance			aver. occupancy/layer		
	W^+W^-	$t\bar{t}$	HX	W^+W^-	$t\bar{t}$	HX
inner	0.67	0.93	0.92	0.021	0.060	0.031
outer	0.51	0.86	0.84	0.011	0.038	0.019

Due to multiple scattering and energy loss in the fiber material the described detector will naturally influence the precission in reach for a following outer tracker. That seems to be important only for particles with energies below 2 GeV, as can be seen from fig. 6.

FIBER PRESHOWER

A fiber tracker and preshower was first successfully used in the UA2–experiment [8]. We will closely follow that concept using however only fibers in z–direction because no precise tracking in two coordinates is necessary.

We suggest to build a fiber–lead sandwich cylinder with a radius of 1m and a total length of 7 m splitted in two parts in the middle. Using fibers of 1mm diameter and 3.5 m length arranged with a pitch of 1.05 mm this results in 6 000 fibers per layer and cylinder half. Six staggered inner layers parallel to z would allow a resolution of $\Delta\ r\Phi = 50\ \mu m$. After 9 mm of lead corresponding to 1.6 radiation lengths, four staggered outer layers still would give $\Delta\ r\Phi = 80\ \mu m$.

The very good two track resolution and electron identification of such a device allows excellent $\gamma/e/\pi$ separation important for the precise measurement of many physical variables. Adding up the signals of hitted fibers for the inner and outer preshower layers the corresponding difference of the number of detected photoelectrons is shown in fig 7 for photons, electrons and pions. Correcting for the different number of layers one finds integral values of ΔN_{pe} of $\gamma/e/\pi = 33/105/0$.

A typical $t\bar{t}$–production event as seen by the preshower is shown in fig. 8. In fig. 9 the position of tracks weighted with light amplitudes from the hitted fibers is plotted for sector 12 of fig 8. One clearly can see the resolving power of the detector.

An open question is how to read out the 120000 channels of the preshower. Keeping in mind the low rate of interesting events there is no need to do it fast. Therefore image intensifier chains and CCD already applied for UA2 and now used for the 10^6 channel tracker of the CHORUS experiment [9] may be a todays solution also here.

SUMMARY

Fiber technology provides various interesting detector applications for a 500 GeV e^+e^- collider.

A very fast inner trigger could select e.g. top and higgs production events with high efficiency within about 10 nsec. Extended to a two coordinate measuring inner tracker a space point resolution of $\Delta r\Phi = 80$ μm and $\Delta z = 410$ μm could be reached. A nearly complete suppression of background of beamstrahlung seems to be possible.

A one coordinate high resolution fiber preshower provides excellent $\gamma/e/\pi$ separation within $\Delta r\Phi$ below 100 μm.

REFERENCES

1. Loew, G.A., (ed.), International Collider Technical Review Committee Report, SLAC-R-95-471, 1995
2. Brinkmann, R., et al.,(eds.), Conceptual Design of a 500 GeV e^+e^- Linear Collider with Integrated X-ray Laser Facility, DESY 1997-048/ECFA 1997-182
3. CERN Program Library Long Writeups W5035 and W5013
4. Yokoya, K., KEK report 85-9 (1985), see also: Schulte, D., preprint DESY-TESLA 97-08 (1997)
5. Arisaka, K., contribution to this workshop
6. Bähr, J., et al., Proceedings of the workshop SCIFI93, eds. A. Bross, R. Ruchty, M. Wayne, Notre Dame, USA, 1993, p. 183
7. Adams, D., et al., *Nucl. Phys.* **B** (Proc.Suppl.) 44, 340, (1997)
8. Ansorge, R., et al., *NIM* **265**, 33, (1988)
9. Annies, P., et al., *NIM* **A367**, 367, (1995)

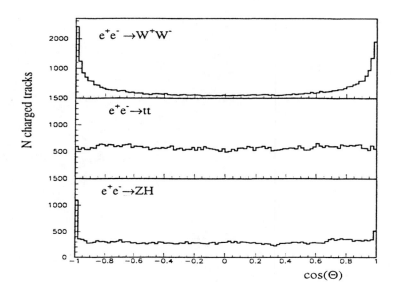

FIGURE 1. Angular distribution of secondary particles from W^+W^-, $t\bar{t}$ and higgs production reactions in 500 GeV e^+e^- scattering

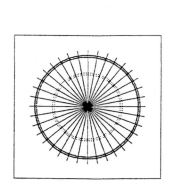

FIGURE 2. Schematic rϕ-view of a possible central trigger structure

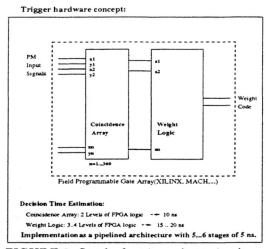

FIGURE 3. Scetch of a trigger electronic scheme

$e^+e^- \rightarrow t\bar{t}$ at 500 GeV, fast trigger

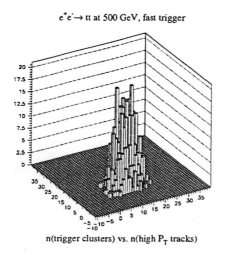

n(trigger clusters) vs. n(high P_T tracks)

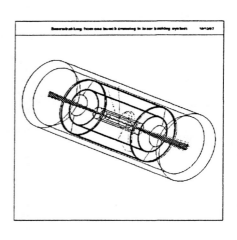

FIGURE 4. Number of high P_t tracks vs. number of detected trigger clusters for 500 GeV $t\bar{t}$-production

FIGURE 5. Distribution of tracks from beamstrahlung produced in one bunch crossing in the inner tacking system in a 3T magnetic field

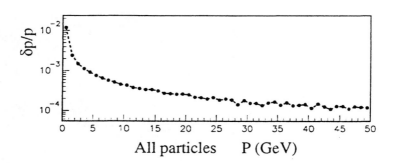

All particles   P (GeV)

FIGURE 6. Relative precision of particle momenta after having passed the central inner fiber tracker

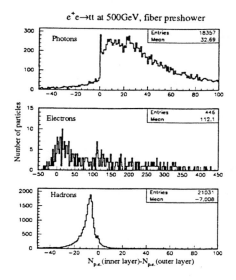

$e^+e^- \rightarrow t\bar{t}$ at 500GeV, fiber preshower

FIGURE 7. Difference of number of photo-electrons for inner and outer preshower layers normalized to equal numbers of them for photons, electrons and pions

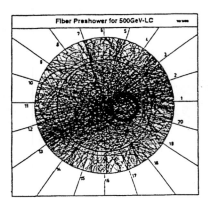

FIGURE 8. Typical event from $t\bar{t}$ production at 500 GeV as seen by the preshower

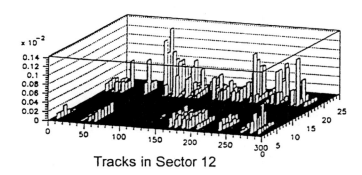

Tracks in Sector 12

FIGURE 9. Particle tracks from preshower fibers weighted with the corresponding light amplitudes for sector 12 of fig. 8

263

SESSION 7: APPLICATIONS - I

Chair: Y. Onel
Scientific Secretary: R. Hooper

Thin scintillating tiles (with fiber readout and high light-yield) for the OPAL endcaps

Austin H. Ball

Department of Physics,
University of Maryland,
College Park, Maryland 20742

Abstract. Scintillating tiles with embedded wavelength-shifting fiber readout have recently been installed in the OPAL endcaps to improve trigger performance, time resolution and hermeticity for experiments at LEP II.

The design is constrained to provide hermetic coverage of the available area with high single particle efficiency, uniform response and good time resolution, notwithstanding the limited space for the detector and its long readout cables, and despite the strong endcap magnetic field. A high light yield per embedded fiber is required.

This paper motivates and describes the design, and demonstrates that the performance meets the required targets. A light yield of 14 photoelectrons/MIP and a time resolution of 3ns have been obtained during 1997 LEP operation.

INTRODUCTION AND MOTIVATION

The OPAL detector [1] at LEP has been improved for experiments at higher collision energies, by installing scintillating tile layers (denoted TE and MIP plug) in the endcaps, as shown in fig. 1. The outer radius of the largest tile sector is 1750mm.

In contrast with earlier LEP studies at the Z^0 resonance, the ongoing program at LEP II is characterised by low cross-sections, an emphasis on new particle searches (particularly using missing energy signatures) and accelerator operation with high currents, often using a multi-bunch structure. Many sources of machine-related background increase with both current and collision energy.

The TE scintillators, complementing an existing system in the barrel region, detect charged particles or converted photons and give precise timing information. This is used in triggering [2], for better background rejection, and in interpreting the signals from slower sub-detectors, particularly the endcap electromagnetic calorimeter, where isolated clusters with no associated charged particle track (photon candidates) may be faked by out-of-time tracks, such as cosmic rays. The MIP

CP450, *SciFi97: Workshop on Scintillating Fiber Detectors*
edited by A. D. Bross, R. C. Ruchti, and M. R. Wayne

plug scintillators around the beam-pipe catch high momentum tracks which otherwise would have escaped detection. The better hermeticity improves the sensitivity of searches using missing momentum signatures.

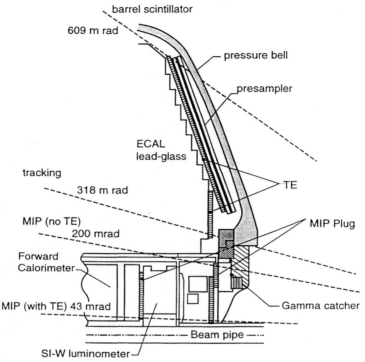

FIGURE 1. Position of the tile endcap scintillators (TE and MIP plug) relative to the beam pipe, pressure bell, presampler and leadglass electromagnetic calorimeter (ECAL)

CONSTRAINTS

The essential requirements for all the intended functions were high and uniform light yield, low noise and good time resolution. Low construction cost, low maintenance and high reliability were also important, the latter because of infrequent access opportunities. The physical constraints were the limited ($\sim$20mm) thickness obtainable by displacing existing detectors, the restricted volume of cables which can exit the endcap, the 0.4T magnetic field and the required solid angle coverage. In the angular range 318 to 609 mrad, the TE tile system (see fig 1) has to overlap with the existing barrel scintillators and extend down to the limit of tracking. Maximum hermeticity is needed in the MIP plug region below 200mrad.

Thus no azimuthal dead-space and very little radial dead-space could be left within the total available volume, ruling out the use of conventional light guides or local readout. The only practical solution was to use tiles with embedded wavelength-shifting (WLS) fiber, coupled by clear fibers to remote photo-transducers outside the endcap. Here the magnetic field is low enough for conventional photomultipliers to operate (with magnetic shielding). The restricted total volume of clear fibers means that a high light yield per embedded fiber is needed.

OPTIMIZATION OF LIGHT-YIELD

In evolving a workable design, much use was made of recent experience elsewhere [4–7]. In addition, to better understand some of the factors affecting light-yield, a number of tile-fiber configurations were studied using a combination of prototype measurements and simulation.

Using a cosmic ray track reconstruction telescope and a collimated Ru source mounted on an x-y scanner, tests were made on identical $300 \times 300 \times 10$mm tiles of Bicron BC408, prepared using different grooving, polishing and wrapping techniques and with a 1.8m total length of 1mm diameter Kuraray Y11 200 non-S WLS fiber embedded in either straight, 'sigma'-shaped or more complex grooves.

The WLS fiber ends were coupled either directly, or via connectors and clear fiber, to a photomultiplier. Light yield was measured by fitting the observed cosmic ray charge distributions (corrected for normal incidence) for the mean number of photoelectrons n_{pe}, using a Poisson distribution convoluted with a Gaussian single photon resolution function. The photoelectron count was checked by probing the single photoelectron response of the photomultiplier using a progressively attenuated LED.

The tile with the best light yield for the fewest grooves had diamond-milled grooves on both sides of the tile, with complementary groove patterns and two fibers per groove. For tiles of this size, with diamond-milled edges, and wrapped in Dupont Tyvek 1073D, the best uniformity was obtained from grooves evenly spaced over the tile surface. It was also noted that coupling both ends of each fiber to the photomultiplier improves time resolution, avoids mirroring and allows easy and thorough pre-testing of the WLS/connector assemblies before gluing into the scintillator tiles. In this way, expensive tile wastage due to faulty WLS fibers can be avoided.

Full Monte Carlo simulation of the light collection from the complex geometry of a tile-fiber detector requires a large amount of computer time. Analytical calculations are very fast and work well for simple geometries, but are not practical for complex ones. To understand the main parameters affecting light yield and uniformity, a simple technique [3] was developed which combines features from both the Monte Carlo and analytic approaches and which can simulate arbitrarily complicated geometries without requiring large amounts of computer time.

The light yield calculation was factorised into three components viz. the transport of scintillation photons to an absorption point in a WLS fiber, the transport of re-emitted (green) photons through WLS fiber, connector and clear fiber, and finally the characteristics of photoelectron emission and amplification within the photomultiplier tube.

After tuning to results from the $300 \times 300 \times 10$mm test tiles, this simulation was able to correctly predict the light-yield, uniformity and time resolution of more complex geometries, enabling the final design to be optimised without building multiple prototypes.

DESIGN DETAIL

The 152 scintillator tiles in each endcap are arranged in five rings composed of identical sectors. The typical structure is illustrated by fig 2, which shows the two (double layer) rings around the beampipe (labelled 'MIP Plug' in fig 1).

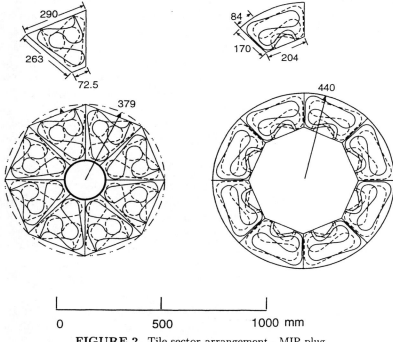

FIGURE 2. Tile sector arrangement - MIP plug

Single loop grooves, 2.5mm deep, are diamond-milled on both sides of the tile, following a complementary toolpath. The tile areas vary from 470 to 900 cm^2, the

270

exact sizes being dictated by trigger and time resolution requirements, mechanical constraints from adjacent detectors, and manufacturing feasibility. The complex groove paths allow long lengths of fiber (2 fibers per groove) to be embedded, whilst respecting a 40 mm minimum bending radius and distributing fiber uniformly over the tile surface. The total embedded fiber length divided by the tile area is typically ~ 0.01 mm^{-1}, with detailed adjustment for each of the five tile types needed, to compensate the light yield for the incident angle of tracks from the interaction point. The fiber ends exit the tile close together on an accessible edge, allowing termination in a 4-fiber connector very close to the tile. Bicron BC 408 scintillator was chosen because it gave acceptable light yield (when matched with all the other components), combined with reasonable durability, availability and cost. In addition, the manufacturer was willing to develop the milling and quality control techniques required to reliably produce complex groove paths, with a well controlled depth.

The double-clad WLS fibers are Kuraray Y11-200 non-S, and Bicron BCF91a mc. Both have an inner core of polystyrene, with $n_{wlscore} = 1.6$ and successive cladding layers of polymethyl methacrylate (PMMA) and fluorinated PMMA ($n_{wlscladding}$ = 1.49 and 1.42 respectively). The absorption spectra of these WLS fibers are a reasonable match to the BC408 scintillator output. In addition, the emitted wavelength spectrum is well transmitted by the chosen clear fiber, within the sensitive range covered by several commercially available phototubes with 'green-extended' cathodes.

WLS 'looms' were constructed by gluing the WLS fibers into precision optical connectors, using Nuclear Enterprises NE581 or Bicron BC 600 optical cement, on plastic jigs replicating exactly the scintillator groove pattern. The ends of the fibers were flycut and polished using Novus type 2 polishing compound to produce optically flat surfaces. Pre-assembly of these looms allowed the fibers to be tested in their connectors before being glued into the scintillator.

The 1mm diameter clear fibers are Hoechst Infolite ER51, a singly clad fiber with $n_{clearcore} = 1.49$ and $n_{clearcladding} = 1.42$. This was selected for low attenuation (120db/km at 500nm) and a thin, light-tight polyethylene outer jacket, giving an overall diameter of 1.5mm, small enough to pack sufficient fibers in the available space for exit from the endcaps. For the tiles around the beampipe, Mitsubishi SH4001-16, with overall diameter 1.6mm and otherwise identical properties, was substituted.

The 4-fiber, dowelled optical connectors are machined from Dupont Delrin plastic and provide for stable concentric alignment of the fibers to 10 μm with a surface flatness of a few microns. Losses due to the poor match between the core refractive indices of the WLS and clear fibers are more than compensated by the much lower attenuation in the clear fiber of wavelengths near the peak sensitivity of the photomultiplier tube.

Simulation results showed that much of the light emerging from the WLS fibers is carried in the outer 20% of their diameter, in helical propagation modes, suggesting transmission characteristics strongly dependent on connector tolerances. However, such modes are subsequently suppressed by attenuation in the long clear fiber, and the connector introduces a loss of only $\sim$ 15% in the intensity of light arriving at the photomultiplier tube. The clear fibers carry the light up to 15m, terminating in cylindrical Delrin couplers which mate the polished fiber ends onto the window of a photomultiplier tube.

Electron Tubes Ltd. 9902SKA photomultipliers were chosen for their quantum efficiency at 500nm (Rb-Cs photocathode), gain ($\sim$ 2×10^7), low dark current, cost, size and uniformity of photocathode response. They are mounted in accessible housings, on the outside of the endcaps, where the residual non-uniform magnetic field is $\sim$ 100 G. Special tubular housings (QL38) were developed in collaboration with the manufacturer, featuring magnetic, electrostatic and rf shielding, spring loading and positive registration of the Delrin optical coupler with respect to the photocathode.

For reasons of economy, fibers from up to four different tiles arrive at the cathode of a given photomultiplier. Differences in response between tiles induced by photocathode non-uniformities were limited to 6%, without the need for a mixer box, by designing the coupler with a suitable surface distribution of the clear fiber end positions. This was derived from detailed scans (2mm grid) of the photocathode response of the first 100 tubes accepted, made using a computer-controlled 500nm light-spot.

The gain of each photomultiplier and associated readout, as a function of operating voltage, was determined after installation, using LEDs installed in each tile (see below). For low LED intensities, a central sampling time much shorter than the LED on-time per pulse, and for high first stage phototube gain, the gain and n_{pe} values, for a particular LED drive voltage and photomultiplier high voltage, are well estimated by fits to the distribution of output charge.

Although all LEDs were pre-calibrated for light output against applied voltage, gain equalisation using this technique does not require knowledge of the geometry or absolute output of the LED light sources, only that the light output per pulse be stable throughout the duration of each test.

ASSEMBLY AND TESTING

To validate the overall design, a prototype tile with production optical connectors, clear fibers and photomultiplier, was exposed to beams of electrons, muons and pions. By fitting the charge distribution observed for a 100 GeV muon beam, the number of photoelectrons n_{pe} was estimated to be 14.5 ± 1.3, and a uniform efficiency of $\geq$ 99% was recorded over the tile surface for a 2 photoelectron equivalent threshold.

WLS fibers were first tested by injecting, at various points, light from a DC tungsten halogen lamp, passed through a Wratten 47B filter to simulate scintillator light, and thence through a splitter to provide reference and test signals. By comparing light which emerged from the fiber with the reference, defects were readily identified. The same test was used to check the fibers in completed looms.

A particularly simple visual test proved equally effective in finding most of the flaws in the WLS looms, and was used for subsequent checks during tile assembly. The looms were covered with an opaque sheet such that only the ends were open to room light. Significant defects showed up as a reduction in the light transmitted from one end to the other and an increase in that backscattered to the same end it entered. This was identifiable by eye through the change in brightness of one end when the other was covered. In total, approximately 30% of the WLS fiber was rejected as sub-standard.

The clear fibers were tested in groups of four by coupling through standard connectors to WLS fibers exiting from a block of clear plastic, excited from the side by pulsed blue LEDs. The other ends of the fibers were fly-cut and coupled to a photomultiplier tube. Once again, good correlation was seen between low light output and failure of a simple visual test for reflections. Accepted fibers show a response uniformity of approximately 7% (standard deviation).

Each scintillator tile was checked for dimensions, surface finish, groove quality and groove depth before gluing. The two fiber loops in a loom were glued one above the other in a groove while a precision jig correctly located the tile and connectors. BC 600 or NE581 scintillator cement was used. Final assembly of the sectors, which is illustrated in fig. 3, involved precision placement of the scintillator-fiber units

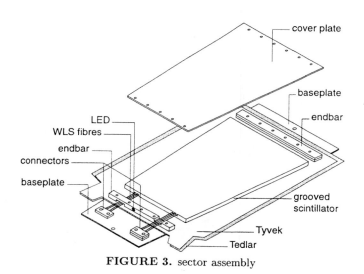

FIGURE 3. sector assembly

onto backplates, to which reflective 150μm Dupont Tyvek 1073D and light-tight

273

Dupont 50μm Tedlar wrapping materials had been prebonded using 3M 465 2mil high tack adhesive transfer tape. The endbars, bolted through the wrappings onto to the back and cover plates at the inner and outer radii, are slotted to allow the fibers to exit. The scintillators are held in place only by the endbars and wrapping materials, while the connectors are bolted to the back plate through oversize holes, allowing some tolerance and avoiding straining the fibers. A pre-calibrated blue LED (Cree Research C470-5D36), press-fit into a hole in the centre of one end bar, illuminates the scintillator for calibration and monitoring. Before applying sealants, the WLS fiber loops were checked by eye as described above, as was the excitation of the WLS fibers by the LED. Finally, both ends of the package were light sealed in a two stage process, with Dow Corning 3145 and Sylgard 150 sealants.

Each WLS fiber end was then coupled, through standard connectors, and 10m of clear fiber, to a separate photomultiplier tube, to check first for light leaks and then for uniformity of response between fibers (by fitting the charge distribution induced by LED pulses). With all fibers connected to a reference photomultiplier operating at a known gain, uniformity was checked by a source scan and light yield by fitting the charge distribution produced by near vertical cosmic rays. The distribution of the average number of photoelectrons for the first 196 assemblies is shown in fig. 4.

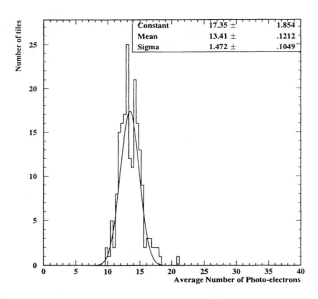

FIGURE 4. Distribution of mean n_{pe} from cosmic ray events for 196 sectors

PERFORMANCE DURING DATA-TAKING WITH LEP

Event time distributions for electron-positron collisions at the Z^0, observed during bunch train operation with two bunchlets per train, are shown in fig.5 for small and large tiles. Corrections for differing fiber paths have been applied. The observed inter-bunchlet spacing is 335.2ns $\pm$ 1.0ns (334ns expected) and the peaks have $\sigma = 3.0ns$ compared with the design target of 5ns.

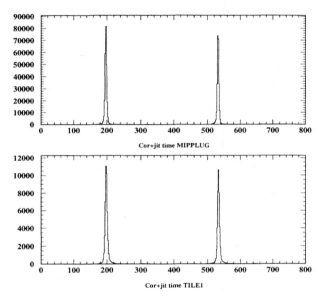

FIGURE 5. Bunchlets from Z^0 run as seen by the tiles in the MIP plug (upper histogram) and by TE (lower histogram). the horizontal scale is in ns and the data have been corrected for differences in fiber path lengths.

The good uniformity, calibration, efficiency and light yield of the tile-endcap are demonstrated by its response to single tracks from a selection of low multiplicity Z^0 decays. Isolated charged particle tracks, with momenta greater than 5 GeV/c, were selected if they projected through any single tile-endcap readout segment. The charge (0.25 pC counts) recorded for all such single tracks, corrected for angle of incidence, and summed over all segments, is shown in fig. 6. The minimum ionizing peak is well separated from pedestal (see inset figure), while the broader distribution at higher charge corresponds to electrons which have interacted in the two radiation lengths of material encountered before intersecting the scintillator. The minimum ionizing peak is easily isolated by vetoing tracks with an associated electromagnetic calorimeter energy greater than 3 GeV/c. A gaussian fit to the minimum ionizing peak gives an estimated light yield of 14 ± 2 photoelectrons (a lower limit since any variation in gain for different channels broadens the MIP peak). The same procedure applied to a single channel (with looser selection to increase statistics) yields 15 ± 3 photoelectrons.

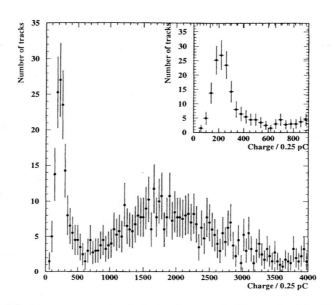

FIGURE 6. TE charge response to single tracks corrected for angle of incidence and summed over all segments

Observed sector hits are 99% correlated with those expected from track extrapolation, indicating that the tile-endcap has the high efficiency and low noise characteristics required for use in triggering. This is confirmed by a study of randomly triggered beam crossings (where no physics hits are expected), from which the probability of a noise hit above the two photoelectron threshold was determined to be $\sim 0.01\%$ per photomultiplier.

CONCLUSION

The OPAL tile-endcap is a successful application of tile-fiber technology, using thin tiles, to the detection of minimum ionizing particles with high efficiency. A time resolution $\sigma(t) = 3ns$ has been achieved and experience so far indicates that the detector system is reliable and can be readily calibrated to give uniform response.

ACKNOWLEDGEMENTS

I would like to thank the organisers for the invitation to an exceptionally well-run and stimulating conference, characterised by much cross-fertilisation between different research and application disciplines.

The tile-endcap was designed, constructed and installed by teams from the universities of Alberta and Maryland, CERN, and University College, London. Individuals from the universities of Bologna, Budapest, Heidelberg and Montreal, Rutherford Appleton Lab and Brunel University also made important contributions. It was a pleasure to work with all these collaborators. The facilities provided by the team institutes were especially appreciated and the consistent support of R.D. Heuer, D.J. Miller, D.E. Plane, A.M.Smith and A. Skuja were vital to the successful completion of the project. For initial concepts and the benefit of much accumulated experience, we are indebted to A. Bamberger, A. Benvenuti, C. Hearty, J. Freeman, H. Grabosch, P. le Du, P. Melese and H.Tiecke. The productive partnerships with Bicron Corporation and Electron Tubes Ltd., to customize products for our applications, contributed much to the successful completion of the detector system. We gratefully acknowledge the financial support of the following: the Department of Energy, USA, under grant DEFG05-91ER40670, The Natural Sciences and Engineering Research Council of Canada, the Particle Physics and Astronomy Research Council, UK and the Bundesministerium fur Forschung und Technologie.

REFERENCES

1. K. Ahmet et al., *Nucl. Instr. and Meth.* **A305** (1991) 275-319.
2. M. Arignon et al., *Nucl. Instr. and Meth.* **A313** (1992) 103.
3. TE group, *OPAL Technical Note* **TN524** 1997.
4. P. de Barbaro and A. Bodek, *University of Rochester preprint*, **UR 1398**, (1994).
5. ZEUS collaboration, *DESY PRC*, **93-07**, and *ZEUS-Note*, **94-001**, 1994.
6. R. Wojcik et al, *IEEE Trans Nucl.Sci.*, **40** (1993)4.
7. T. Asakawa et al., *Nucl. Inst. and Meth.*, **A340** (1994) 458.

Scintillation Fiber Hodoscope
for Topological Triggering

M.Ferro-Luzzi[1], A.Gorin[2], M.Kobayashi[3], V.Korolev[2], A.Kuznetsov[2],
T.Maki[4], I.Manuilov[2], K.-I.Kuroda[5], A.Penzo[6], A.Riazantsev[2],
A.Sidorov[2], F.Takeutchi[7], K.Okada[7], Y.Yoshimura[3]

1. CERN, Geneva, Switzerland;
2. IHEP, Protvino, 142284, Russia;
3. KEK, Tsukuba, Japan;
4. UOEH, Kitakyushu, Japan;
5. Waseda University, Tokyo, Japan;
6. INFN, Trieste, Italy;
7. Kyoto-Sangyo University, Kyoto, Japan.

Abstract. A fast and precise scintillating fiber (SciFi) hodoscope based on KURARAY SciFi and HAMAMATSU multianode H6568 tubes was constructed to be used in DIRAC experiment (PS-212) at CERN for tracking and topological triggering information. The front-end electronics was custom developed to allow on-line discrimination after built-in dynamic cross-talk rejection. Preliminary tests of the detector using more than 300 read-out channels have shown an excellent stability and optimum performance for this system: a uniform detection efficiency of 98%, a 125 μm single hit and ~0.5 mm two-hit space resolution, 0.6 ns time resolution.

1. INTRODUCTION

Our work consists in the development of the traditional approach to realize scintillating fiber (SciFi) detectors using position-sensitive photomultipliers (PSPM). Our group is concerned with the designing of a *topological trigger* device [1], utilizing simultaneously **both** trigger and tracking functions in high intensity experiments.

To satisfy stringent conditions required by new DIRAC experiment [2], such as real time identification of double tracks (π^+, π^-) at small (< 9 mm) distances amongst more than 10^7 incoming charged particles per second, **parallel readout** of the PSPM was chosen.

Special care was paid to decrease the observed multiplicity of the detector caused by non-ideal parameters of the fibers and tubes (cross-talk effects) by designing multichannel discriminators with **dynamic cross-talk rejection**.

Tests of the first prototype of this detector are described in [3].

The proposed DIRAC experiment, a magnetic double arm spectrometer, is

CP450, *SciFi97: Workshop on Scintillating Fiber Detectors*
edited by A. D. Bross, R. C. Ruchti, and M. R. Wayne

aimed to measure the lifetime of exotic $\pi^+\pi^-$ atoms in the ground state with 10% precision which would offer the understanding of chiral symmetry breaking of QCD.

The main function of the SciFi hodoscope placed near the target in the DIRAC set up is the measurement of the particles coordinates to find out a relative momentum of a pion pair. A suitable cut on the opening angle (≤3 mrad) of $\pi^+\pi^-$-pairs allows one to reject efficiently the background coming from free pairs (≤50 mrad) by two orders of magnitude; and this function can be realized with a topological triggering.

2. DETECTOR STRUCTURE

The SciFi hodoscope consists of two independent units to measure projections of the X- and Y- coordinates of an incident particle whose trajectory is normal to the SciFi plane. The fiber array covers 105×105 mm^2 area and consists of five layers of KURARAY SCSF38 single cladding SciFi with a diameter of 0.5 mm. The total amount of read out channels is 240 for each array and every fiber column is connected to the corresponding PSPM channel. The real length of SciFi is 120 mm and each fiber is connected to a clear fiber light guide ~300 mm long with an optical epoxy. The 5-fiber columns were glued with white paint with a pitch of 0.43 mm, and this is the deliberate compromise between column's gap and overlapping. The fiber edges are glued (and polished) into a square black plate to fix the position on the PSPM photocathode, no optical grease was used. The optical contact was also checked and we got rather a small increase in the light yield (15%). The far end of the SciFi array was tightly pressed to Al mirror, vacuum evaporated onto organic glass plate. We did not find any significant light attenuation along the fiber.

As a photosensor a new metal dynode PSPM [4] of the H6568 type was selected. The level of cross-talk of 1% (or less) was experimentally confirmed (a light spot diameter was about 1.6 mm). The measured linear range of this tube (for the linear bleeder) was up to 15 photoelectrons at 950 V. In contrast to the HAMAMATSU data we obtained much smaller gain non-uniformity as shown in Fig.1. This type of tube has surprisingly good single electron spectrum (Fig.1) and the noise level (exponential tail of the pedestal) is almost negligible. These both effects allow one to improve significantly the parameters of the hodoscope.

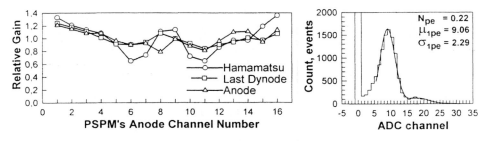

FIGURE 1. H6568 channel's gain non-uniformity and single photoelectron spectrum.

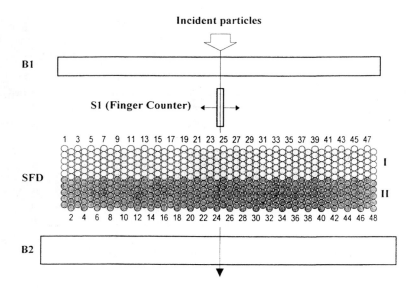

FIGURE 2. Beam test experimental set-up.

3. EXPERIMENTAL SET-UP AND READ-OUT ELECTRONICS

Experimental set-up of the hodoscope (SFD) and trigger counters for the beam test are shown in Fig.2. The main part of the measurements was done with a «wide» trigger (B1*B2). The thin (0.5 mm) scintillating counter S1 was used in some cases in coincidence with B1*B2 to enrich the statistics for desired region of the hodoscope. The beam measurements were performed in T10 PS beam line at CERN.

For the test purpose two 48 - column superimposed layers of SFD were used to get position information for tracking. Three H6568 PSPMs in each layer were used for the light collection from the fiber columns. The PSPM signals from the first layer (**I**) passing through the dedicated front-end electronics were detected with TDC and from the second layer (**II**) - directly with ADC. Layer **II** gives coordinate information to check efficiency and multiplicity of layer **I**. The maximum amplitude criterion using ADC information, supplying very good efficiency [3], is used to select a particle track for off-line analysis. The definition of the detection efficiency ε for the coordinate X is the ratio of the number of events detected in layer **I** (for $X-1$, X or $X+1$) to those obtained for layer **II** (for X). The multiplicity M is defined as a number of tracks detected by layer **I** for the required region of the selected single tracks in layer **II**.

Conventional LeCroy CAMAC TDC (2228) and ADC (2249A) modules were used to test the detector performance. As a front-end electronics we used peak-sensing circuit (PSC) NIM modules, developed in the collaboration. A detailed description of this module [5] will be published soon. Peak-sensing algorithm is expressed as

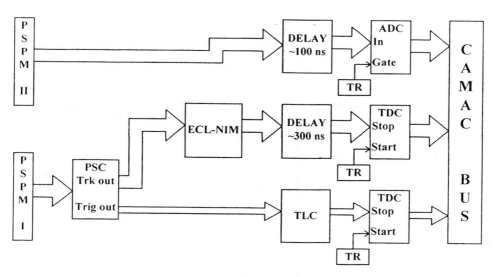

FIGURE 3. Block diagram of the read-out electronics.

$$A_i > \frac{A_{i-1} + A_{i+1}}{2} + U_{th},$$

where A_i is the signal amplitude in the i-th channel and U_{th} - the threshold value. The real meaning of this function is a dynamic rejection of any nature cross-talk in adjacent channels. The module has 32 input/output channels for *Tracking* and 16 output channels for topological *Triggering*, which are the logical sum of 2 adjacent channels (1+2, 3+4, etc.).

To realize the topological triggering, the dedicated outputs of PSC are used. Several prototypes of the Trigger Logic Circuit (TLC) were tested for a restricted amount of channels. The obtained level of the TLC efficiency is near 99 ± 03% for the hit distances from 1 to 15 doubled columns, and time resolution varies from 1.1 to 2.2 ns in σ. The latest version of TLC was proposed to be realized in FASTBUS standard to cover all the 120 channels. The TLC should measure the distance between the nearest hits in the range from 0 to 15 channels. These output signals will be used to make a final trigger signal.

A simplified block diagram of the read-out electronics is shown in Fig.3.

4. EXPERIMENTAL RESULTS

The optimal threshold of the PSC can be found by efficiency scanning in a wide range of high voltages at different thresholds. In our case the detection efficiency was measured for three values of threshold (20, 50 and 120 mV) at different high voltages. The probability to have one and only one hit in PSC was also calculated. The

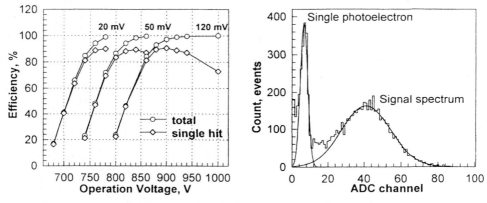

FIGURE 4. PSC detection efficiency and raw amplitude spectrum.

results are presented in Fig.4. The curves show clearly the maximum value of the single hit probability (above 90%), corresponding to the optimal operation conditions.

A «raw» amplitude spectrum of the hodoscope (single anode) with subtracted pedestal is also presented in Fig.4. The single photoelectron peak is clearly visible and well separated from the signal spectrum. The main origin of the single photoelectrons is a low optical cross-talk (inside the SciFi bundle and in the photocathode window). A good separation of the spectra allows one to tune the threshold level in the optimum region (around 15th ADC channel) to get high efficiency at low multiplicity.

Typical multiplicity and hit distributions for the layer **I** obtained at the optimal conditions are shown in Fig.5. The PSC multiplicity distribution was obtained for single particle tracks detected in column range 5 - 44 of layer **II** to avoid the edge effects. For the track distribution in this case only the events which give a track in column 24 of layer **II** were taken. The slightly asymmetric tails of the distribution are due to a small inclination of the hodoscope plane from the normal to beam direction.

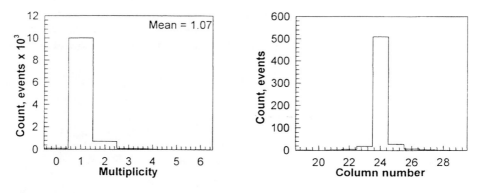

FIGURE 5. Examples of PSC multiplicity and track distributions.

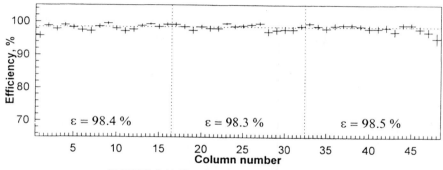

FIGURE 6. Uniformity of the detection efficiency.

The obtained space resolution (r.m.s.) is very close to the theoretical limit of $0.43/\sqrt{12} = 124\ \mu m$.

The detection efficiency uniformity is shown in Fig.6. The three regions separated by the vertical dashed lines correspond to the different PSPMs. One can see that there is no boundary effects between regions observed by separate phototubes. The mean detection efficiency is 98.3±0.5% with r.m.s.≤1%.

The ability of the hodoscope to detect double-hit events was checked with high multiplicity beam. The reconstruction efficiency dependence as function of the distance between hits is shown in Fig.7 and one gets the double-track detection efficiency of 50% at 0.5 mm distance; for larger distances the mean value of the efficiency is ~92% (for 96% single-hit detection efficiency).

Typical time resolution of the SciFi hodoscope obtained with PSC is also shown in Fig.7. Taking into account the resolution of trigger counters, the intrinsic time resolution of the hodoscope with PSC is ~0.6 ns.

The behavior of the hodoscope at high intensity beam was checked for the prototype. The measured values were the amplitude of the pulse, detection efficiency,

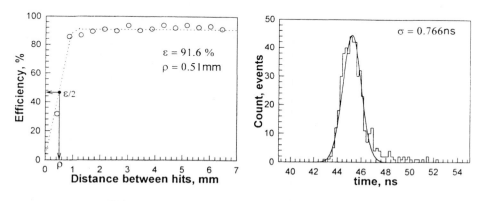

FIGURE 7. Double track detection efficiency and time resolution.

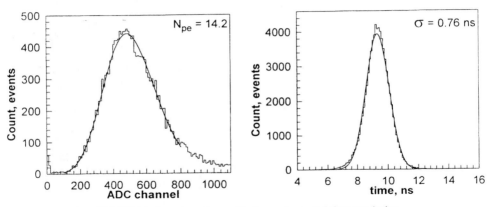

FIGURE 8. Last dynode amplitude spectrum and time resolution.

mean multiplicity. No dependence of these values from the beam intensity was found. The ability of the 100×100 mm^2 hodoscope to work with beam intensity up to 3.6×10^7 particle/s was revealed.

The PSPM tubes were modified by our request to have a last dynode signal that can be used as a monitor or/and in the trigger scheme. The last dynode amplitude spectrum is shown in Fig.8. A simple analysis of the last dynode spectrum with individual anode «tracking» signals allows one to get the amplitude spectrum of a separate PSPM channel to check the quality of the detector and, for example, SciFi aging effects. Good correlation between these signals is shown in Fig.1. The small amplitude of the 8th anode is explained by the high attenuation in a coaxial cable, which was revealed after this measurement.

The time distribution dependence was measured for the difference (to exclude the contribution of the trigger counter timing) of two PSPM's last dynodes. The measured value of σ is 0.76 ns (Fig.8); in the assumption of the equal resolution for both tubes, we get $0.76/\sqrt{2} = 0.54$ ns; and this value is close to the limit of time resolution for our hodoscope.

The full size hodoscope was tested during the August 97 run at the relatively low beam rate.

The measurements started from the parallel positioning of two hodoscope planes (in X-direction). It was important to check the detection efficiency for the well separated detectors and compare the value with our «artificial» definition, used for calibration. The mean efficiency is a little bit smaller - of the order of $0.3\div0.5\%$. The multiplicity distribution practically did not change in comparison with the results obtained with the 48 - channel prototype. It means that the multiple hit events come mainly from the optical cross-talk in the fiber array and in the photocathode, and there is practically no contribution from the PSPM noises.

5. CONCLUSION

The obtained parameters of 105×105 mm^2 SciFi hodoscope with 480 read-out channels are:

- high and uniform detection efficiency - 98%,
- track multiplicity - below 1.1,
- space resolution - 125 µm,
- time resolution - 0.6 ns.

These allow one to construct universal tracking detector which can be used not only in physics experiments, but also in other fields of science and technology. Our current estimation for the detector size limit is 30×30 cm^2.

The main problem of the present solution is a relatively high cost of the detector, consisting mainly of the PSPM (50%) and front-end electronics (PSC - 25%).

ACKNOWLEDGMENTS

We are grateful to the stuff of the CERN PS group and PS/SPS Coordinator, who offered the beam facilities under the best conditions. We are much indebted to C.Detraz for his permanent technical assistance. Special thanks are to V.Agoritsas for his invaluable assistance.

REFERENCES

1. V.Agoritsas et al., *RD-17 proposal*, CERN/DRDC 91-8, DRDC/P-25.
2. DIRAC; *SPSLC Proposal*, CERN/SPSLC, SPSLC/P28.
3. V.Agoritsas et al., preprint CERN-PPE/96-143, 2.12.96.
4. *H6568 Data Sheet*, HAMAMATSU PHOTONICS K.K., 14-5, Shimokanzo, Toyooka-village, Iwata-gun, Shizuoka-ken, 438-01, Japan.
5. A.Gorin et al., «*Peak-Sensing Discriminator...*», will be published in NIM.

A Beam Hodoscope Based on a Scintillating-Fiber Array with Multianode Photomultiplier Readout

D. Dreossi[*], N. Akchurin[§], A. Bravar[*,a], R. Giacomich[*],

M. Gregori[*], E. Gulmez[×], M. Iori[+], M. Kaya[§], A. Lamberto[+],

C. Newsom[§], D. Northacker[§], Y. Onel[§], E. Ozel[§],

S. Ozkorucuklu[§], A. Penzo[*], P. Pogodin[§], G.F. Rappazzo[*],

S. Reia[*], P. Schiavon[*], G. Venier[*], R. Winsor[§]

[*] *Dipartimento di Fisica e Sezione INFN, Universita' di Trieste, Trieste - Italy*
[§] *Department of Physics and Astronomy, University of Iowa, Iowa City - USA*
[+] *Dipartimento di Fisica e Sezione INFN, Universita' di Roma I, Roma - Italy*
[×] *Physics Department, Bogazici University, Istanbul - Turkey*
[a] *Presently at Mainz University, Mainz - Germany*

Abstract. A hodoscope system constructed with a scintillating fiber array and a multi-anode photomultiplier is described. The results of laboratory and beam tests on the hodoscope response, efficiency and spatial resolution are discussed.

INTRODUCTION

Many high-energy experiments demand precise multi-track reconstruction in a high-rate environment; good space resolution ($\sigma_x \approx 100\mu m$) and fast time response ($\sigma_t \approx 1ns$) are however seldom combined in the same detector.

The advent of scintillating fiber (Scifi) detectors yields the potential of a relatively simple, inexpensive and reliable technology for large-area applications with sub-millimetric precision and a fast response time. Readout devices have to match these properties, in particular the fast response of scintillators ($t_r \approx 1 - 5ns$).

Recent evolution in photomultiplier construction, including fine-grid (or channel) dynodes, has added novel features to this mature technique (still among the ones with best time response): position sensitivity and immunity to magnetic field.

Extensive R&D efforts [1, 2] have advanced the development of systems incorporating position-sensitive photomultipliers (PSPM) coupled to Scifi arrays, affording elegant solutions to the problem of high density of particles in space and time.

An upshot of this course of progress, the recently built Scifi hodoscope described here, embodies new features to improve the already high standards of this type of detectors. Its performance was subjected to close scrutiny, both on a global scale and locally (within the single fiber size), by measuring the properties of the device with a 600 GeV/c beam [3] at the Tevatron.

CP450, *SciFi97: Workshop on Scintillating Fiber Detectors*
edited by A. D. Bross, R. C. Ruchti, and M. R. Wayne
© 1998 The American Institute of Physics 1-56396-792-8/98/$15.00

DESIGN CRITERIA

The performance of individual scintillating fibers with PSPM readout, for precise and efficient charged-particle tracking, might be adversely affected by

- low light yield of individual fibers,
- non uniformity of response due to
  - variations in fiber thickness across the sensitive area,
  - sensitivity differences on the photocathode surface,
- sizable cross-talk between pixels.

The approach developed here addresses the above difficulties with the following design criteria:

⋄ multi-layer (staggered) Scifi arrays act as a nearly uniform medium to:
  - provide a local thickness (number of hit fibers) close to the average,
  - achieve high efficiency ($\approx 100\%$) with a sufficient effective thickness,
⋄ the position of hits is assigned by means of center-of-gravity criteria to:
  - localize tracks with precision better than the size of the fibers,
  - handle the cross-talk (to advantage of spatial resolution, if possible),
⋄ each fiber is individually positioned with high precision, thus allowing to:
  - avoid gradual or random shifts of fiber alignement,
  - implement complex fiber patterns, with a controlled amount of overlap;

both requirements are awkward, for usual tight-packed arrangements of fibers.

CONSTRUCTION AND BENCH TESTS

The mechanical structure of the fiber array is illustrated in Figure 1. A black Delrin frame holds the fibers rigidly in place. Reference holes are drilled with $\pm 5\mu m$ precision on the frame using a CNC machine on the two opposite sides defining the position of each fiber. The fibers are pulled through the corresponding holes on either side and glued in place. With this construction technique, each fiber can be precisely located, independent of their relative positions and diameter variations.

The resulting Scifi array consists of 16 layers (rows) of 20 fibers (0.5 mm Ø SCSF38 - Kuraray) each, grouped in 80 columns, each with 4 fibers parallel to each other. Each column (segment) is shifted from the adjacent one by a fiber radius, 0.25 mm. The active area consisting of the 80 segments, is 20 mm wide and 30 mm long. Each fiber in a segment extends for about 25 cm from the active area to the PSPM window, where it is grouped with the other fibers belonging to the same segment, such that they are viewed together by one of the anode pads, separated by 2.54 mm. This arrangement is achieved by locating the readout end of the 4 fibers inside precision holes drilled in a black Delrin square template which ensures proper positioning on the photocathode. The groups of fibers are glued in the template holes and the fiber ends are polished. Four reference pads at the corners of the template are coupled with clear fibers to LED's. For finer alignment, a positioning jig is mounted on the PSPM assembly box and used to adjust the template position on the photocathode window, in order to simultaneously maximize the signals from the illuminated reference pads; at this point the template is fixed in place and the jig is removed.

The PSPM used for this hodoscope is the Hamamatsu H4140-20, a high density device ($16 \times 16 = 256$ pads) with low cross-talk ($< 2\%$) and high gain ($> 10^6$). The 80 groups of fibers are located on the central part of the photocathode, where gain variations between channels are smaller ($\pm 15\%$), following a zigzag pattern (Figure 1.C), a configuration suitable for handling cross-talk between adjacent rows [2].

The anode pad signals are transmitted individually with coaxial cables, amplified (LRS 612A) and read-out via FERA ADC (LRS 4300) for the laboratory tests and via FASTBUS ADC (LRS 1881 M) during the beam measurements. The hit position is determined by the center-of-gravity of the charge in active pixels.

The PSPM was bench-tested prior to beam measurements, in a light-tight box using a green LED coupled to a 2 mm diameter fiber, facing the PSPM window. The position of the fiber with respect to the PSPM could be remotely controlled micrometrically and the light spot on the PSPM was collimated to a diameter of 0.5 mm. With this setup the local properties of the PSPM were measured. The profiles of adjacent channels were measured individually and are shown in Figure 2. The response curve follows a gaussian distribution, with $\sigma = 1.107mm$. The cross-talk is close to the 2% level specified by the manufacturer. The measured map of the PSPM response for all channels, also compatible with the manufacturer data sheet, was used to normalize the different channels.

BEAM MEASUREMENT RESULTS

The hodoscope was installed on the beam in between microstrip silicon detectors, measuring the incident beam trajectory to a precision of about 5 μm in the x and y directions. In this way, it was possible to investigate the properties of the hodoscope as a function of the particle impact position over the whole hodoscope active area and across an individual fiber segment at a few microns scale. The fibers were positioned horizontally (along the x-axis), measuring the vertical y coordinate.

The thickness profile of the scintillating material in the hodoscope, as a function of the y-coordinate, is shown in Figure 3. The average thickness traversed by a beam particle is 2.75 mm. For this thickness, the expected average number of photoelectrons [4] is $< N_{phe} >^{exp.} \approx 19.0$ for a minimum-ionizing particle (MIP). The amplitude distribution of signals from the last dynode (LD) of the PSPM, viewing the whole active area, is shown in Figure 4. From this distribution, the number of photoelectrons can be estimated, as usually done, by

$$< N_{phe} >= k \left(< A > /\Delta A \right)^2 \qquad (1)$$

where $< A >$ is the mean value of the amplitude distribution and ΔA is the RMS deviation. The factor k is equal to 1 in the case of pure Poisson statistics for the photoelectrons; however the PSPM gain spread and first dynode collection efficiency affect the photoelectron distribution [5] in such a way that, in general, $k \geq 1$; the precise value of k depends on the specific device (and can reach values of $k \approx 1.3$ or higher). An estimate based on Poisson statistics ($k = 1$) leads to $< N_{phe} >^{est.} = 14.5 \pm 0.8$; for $k = 1.3$, $< N_{phe} >^{est.}$ comes very close to the expected value.

The amplitude distribution of signals from one anode channel of the PSPM is shown in Figure 5 (full line). As a function of the impact point, determined with the silicon detectors, the shaded area corresponds to hits that span the selected segment; the sharp peak at small amplitudes (dash line) is produced by hits on adjacent segments and results from the overlapping parts of the fibers and the respective cross-talk, whose main source can be traced to the spread of the fiber image through the photocathode window.

Typical y- coordinate distributions of the impact point on the hodoscope, reconstructed with the silicon microstrip telescope, for events where there is a signal from one anode channel [ADC(A_j) > 0], are shown in Figure 6: in (A) a well localized peak appears at the position corresponding to the active segment, with a small background extending over the hodoscope area; in (B) spurious peaks accompany the main peak at different positions, corresponding to cross-talk from segments far away from the active segment on the fiber array, but close to it on the photocathode.

By performing this procedure for all the readout channels, the entire map of the hodoscope is determined (Figure 7), where the main peak position for each channel is represented by (+), while the spurious peaks are represented by (×). This map is improved substantially when a modest threshold cut is applied to the anode amplitude distributions, with a substantial reduction of the crosstalk ($\approx 2\%$).

The amplitude distributions for particles through the central part (A) of a fiber segment ($\pm 125\mu m$) or the two edges (B), are shown in Figure 8. Estimates of corresponding numbers of photoelectrons are given in Table I.

Table I

Hodoscope Element	Av. Thickness (mm)	Expected	$< N_{phe} >$ Estimated	Corrected
Last Dynode (LD)	2.75	19.0	14.5 ± 0.8	19.3 ± 1.2
Segment (A+B)	1.35	9.4	5.1 ± 1.2	6.8 ± 1.5
Center section (A)	1.80	12.4	6.2 ± 1.3	8.2 ± 1.6
Edge sections (B)	0.95	6.5	3.2 ± 1.3	4.3 ± 1.6

It turns out that LD provides as many photoelectrons as expected, but the single-segment yield of photoelectrons is lower by about 30% with respect to the expected one. This discrepancy is mainly accounted by the dispersion at the level of the photocathode window, which would allow to collect on the pixel only about 64% of the photons from the corresponding fibers.

The response of a single segment consisting of 4 aligned fibers was studied locally within the size of the fiber by selecting beam tracks in Y intervals of 25 μm width within a $\pm 250\,\mu m$ range, where Y is the transverse coordinate relative to the axis of the fiber. Figure 9 shows the distribution of estimated $< N_{phe} >$ measured for beam tracks in each interval. This distribution appears to be slightly broader than the fiber thickness profile (full line; see also Figure 3), for two reasons: an increased trapping efficiency in the fibers for photons emitted close to the fiber periphery and cross-talk from adjacent and overlapping fibers.

The efficiency was also studied locally for single segments and for triplets of adjacent segments, in order to take into account the relative overlap. Figure 10 shows the single-segment efficiency (shaded area) reaching 94% in the center of the segment and decreasing to 55% at the edges. Conversely the efficiency of the adjacent segments (shifted by 250 μm) is maximal there and decreases towards the central position (dash lines). The overall efficiency for the triplet is represented by the upper (full) line, which averages at 98% (dash-dot line). All the previous results were obtained at a PSPM high voltage $HV = 2350V$; the LD high voltage curve (Figure 11) gives also an overall efficiency approaching 99% on plateau.

The ADC information from individual segments was used to determine the center-of-gravity of the hit channels for each event, as

$$\mathcal{Y} = (\sum_j y_j w_j A_j)/(\sum_j w_j A_j) \qquad (2)$$

where y_j and $w_j A_j$ are the positions of the hit segments and the corresponding normalized amplitude. The displacement of $\mathcal{Y}$ with respect to the intercept Y_S from the silicon telescope (Figure 12.a) is fitted with a gaussian distribution ($\sigma_Y = 132\pm12\mu m$), overlapping with a small background ($\lesssim 5\%$). Similar resolutions result from using, instead of Y, the position of the segment with maximum amplitude. The background is almost totally suppressed and the resolution slightly improved ($\sigma_Y = 111\pm10\mu m$), for events with no more than two active ADC channels (Figure 12.b), as ideally implied by the array configuration for a single particle through the hodoscope. This condition can be easily implemented in hardware (Figure 13) with an accuracy $\lesssim 150\mu m$ and a timing spread $\approx 1 - 5ns$, the jitter of the hodoscope channels.

SUMMARY AND CONCLUSIONS

The measured performance of the hodoscope can be summarized as follows:

Hodoscope Parameters		Reconstruction Methods		
		C-G	M-A	I-A
Space resolution, σ_x	(μm)	132 ± 12	146 ± 18	111 ± 10
Efficiency, η	(%)	98 ± 0.8	98 ± 1.3	89 ± 1.6
Linearity	(%)	±0.5	±1.5	±0.8
Time window, Δt	(ns)	$40 - 70$	$40 - 70$	$1 - 5$

where C-G refers to the center-of-gravity algorithm, M-A to the maximum amplitude criterion and I-A to the interpolation of adjacent channels. The first two methods use ADC's and need a sufficient time range for the strobe; the hardware implementation of the third method relies on much faster coincidence or analog-sum circuits.

The results on the response from fractional parts of the fibers bear out the concept that precision patterns of partially overlapping fibers provide resolution much better than the fiber size, when the information from adjacent channels is fully and properly used.

Acknowledgments: This work has benefited from encouragement, suggestions and help by a number of members of the FAROS (RD-17) and SELEX (E-781) Collaborations, that are herewith gratefully acknowledged.

REFERENCES

1. FAROS Collaboration, *RD − 17 Status Report*, CERN/DRDC 93/47 (1993)
2. Agoritsas. V. et al., *Nucl.Phys.B(Proc.Suppl.)* **44**, 323-331 (1995)
3. J. Russ (for E-781), *SELEX − Hadroproduction of Charm Baryons at large x_F*, Talk at the Workshop on Production and Decay of Hyperon, Charm and Beauty Hadrons, Strasbourg, Sept. 1995.
4. Bähr J. et al., *Nucl.Instr.Meth.Phys.Res.* **A371**, 380-387 (1996)
5. J. P. O'Callaghan et al., *Nucl.Instr.Meth.Phys.Res.* **225**, 153-163 (1984)

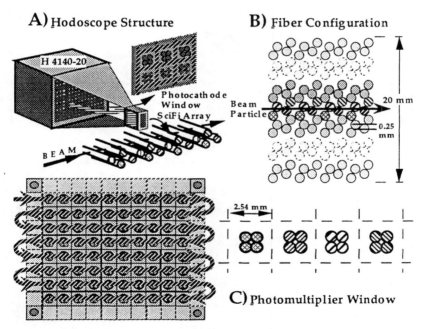

FIGURE 1. A) the hodoscope structure; B) the fiber array configuration and C) the organization of the fiber readout on the PSPM photocathode; the zigzag line shows the readout order of the segments. Four pixels at the corners of the template are illuminated with LED's for alignment of the fibers on the PSPM.

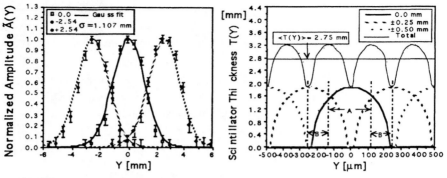

FIGURE 2. Photocathode pixel scan.   **FIGURE 3.** Thickness profile of the array.

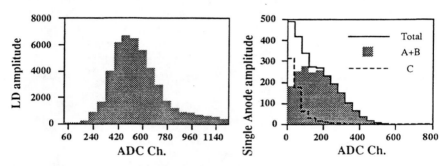

FIGURE 4. Amplitude distribution of LD.   **FIGURE 5.** Anode amplitude spectra.

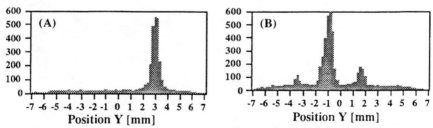

FIGURE 6. Position of hits from the silicon telescope with one anode on (ADC > 0).

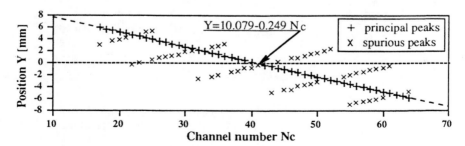

FIGURE 7. Map of the hodoscope response: position sensitivity and linearity.

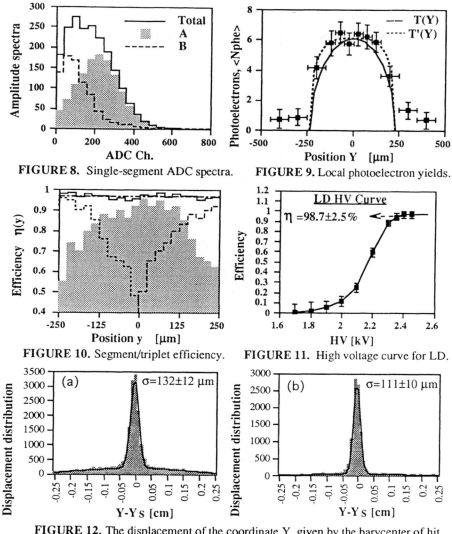

FIGURE 8. Single-segment ADC spectra.

FIGURE 9. Local photoelectron yields.

FIGURE 10. Segment/triplet efficiency.

FIGURE 11. High voltage curve for LD.

FIGURE 12. The displacement of the coordinate Y, given by the barycenter of hit segments, with respect to the track position Y_S from the silicon beam telescope.

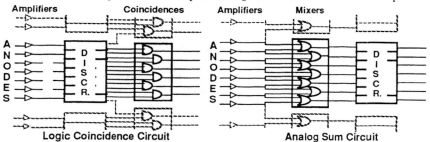

FIGURE 13. Schemes for interpolating between adjacent channels with overlap.

Photon Veto Counters at KTeV/KAMI with Blue WLS Fibers

Kazunori Hanagaki

Department of Physics, Osaka University
Toyonaka, 560 Japan

Abstract. The photon veto detectors used in KTeV experiment were required to have high detection efficiency with high speed response. To satisfy the requirements, we used scintillation counters with blue wave length shifter fibers for their readout. This document describes the design and performance of the photon veto detectors and a possible improvement for future experiments.

INTRODUCTION

KTeV is a fixed target experiment at Fermilab dedicated to study CP violation in neutral kaon system.

To evaluate the size of CP violating parameter ϵ'/ϵ, KTeV measures the double ratio, R, defined as:

$$R \equiv \frac{\Gamma(K_L \to \pi^+\pi^-)/\Gamma(K_S \to \pi^+\pi^-)}{\Gamma(K_L \to \pi^0\pi^0)/\Gamma(K_S \to \pi^0\pi^0)} = 1 + 6Re(\epsilon'/\epsilon)$$

The goal of E832 is to measure the R with an accuracy of 6×10^{-4} combining statistical and systematic error.

To observe the four decay modes, we used a detector setup shown in Figure 1. The momenta of charged particles are measured by a spectrometer composed of four drift chambers and analyzing magnet. The position and energy of photons decayed from π^0 are measured by a CsI calorimeter.

In measurement of the R, $K_L \to \pi^0\pi^0\pi^0$ decay is a major background for $K_L \to \pi^0\pi^0$ decay since the photons from $K_L \to \pi^0\pi^0\pi^0$ decay may miss the fiducial area of the calorimeter. The photon veto detectors are used to reduce such a $K_L \to \pi^0\pi^0\pi^0$ background by detecting the escaping photons. In order to achieve the required background level of 0.1% [1], the photon veto detectors must attain a high efficiency for photons above 100 MeV [2]. KTeV also uses the photon veto detectors in a search for $K_L \to \pi^0\nu\bar{\nu}$ decay to reduce backgrounds from $K_L \to \pi^0\pi^0\pi^0$ and $K_L \to \pi^0\pi^0$.

CP450, *SciFi97: Workshop on Scintillating Fiber Detectors*
edited by A. D. Bross, R. C. Ruchti, and M. R. Wayne

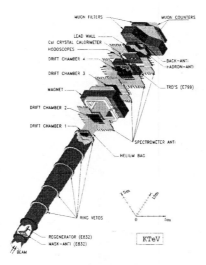

FIGURE 1. KTeV Detectors used to detect the decay products of K_L and K_S . The detector shown as Ring Veto is referred to as Ring Counter in this context, respectively.

To produce kaon beams, the Fermilab Tevatron delivers protons to our target for 1 ns bucket in every 19 ns. Thus the produced kaons enter the detector every 19 ns. The photon veto detectors are required to be efficient for every bucket.

To conclude, there are two characteristics which must be maintained in the KTeV photon veto detectors; efficiency and speed.

DETECTOR DESIGN

In order to achieve a high detection efficiency and a high speed response, the photon veto detectors were designed as sampling calorimeters consisting of leads and scintillation counters. The optical fibers are used to read out the scintillation light to photomultiplier tube(PMT). Among various photon veto detectors in KTeV, we will describe "Ring Counter(RC)" [2] in detail.

RC's have a round shape with a square hole inside, and come in two sizes. The outer radius is 1.2(1.0) m, and the size of square hole is 1.18(0.84) m for larger(smaller) RC. Viewed from upstream, each RC is divided into 16 sections azimuthally. Each section of RC consists of 24 layers of lead sheets and 2.5 mm thick plastic scintillators. For the first 16 layers, the lead sheets are 0.5 X_0(2.8mm) thick and for the rest of 8 layers, the lead sheets are 1.0 X_0 thick, which makes 16 X_0 in total. This configuration was determined from GEANT simulation [3] to attain the high detection efficiency above 100 MeV photon.

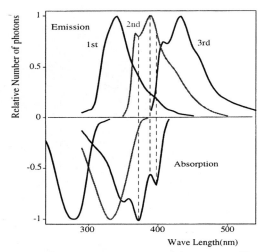

FIGURE 2. Emission and absorption spectrum of dyes in scintillation tiles and fibers

We chose Kuraray [4] SCSN-88 for scintillation tiles, and SCSF-B21 with 1 mm diameter for fibers. The tiles are wrapped with Tyvek to improve the light yield. Dupont [5] 1073D with 200 μm thickness was chosen as our Tyvek for its high light reflection. The fibers are composed of core(n=1.59) and clad(n=1.49) so that light collection is more efficient by making use of the full reflection at the boundary. There were 10 fibers put into the grooves on each scintillation tile with an interval of roughly 4 cm. Therefore, a total of 240 fibers are stacked and seen by one PMT. The fibers are bundled together into a RTV cookie, cut, and polished to be read out by a PMT. Another end of fibers is polished and aluminized in order to improve the light collection.

The SCSN-88 scintillation tile contains two dyes(1st and 2nd). The SCSF-B21 fiber contains 3rd dye as a wave length shifer(WLS). Figure 2 shows the relative emission and absorption spectrum for those dyes. As can be seen in the figure, 2nd dye has a peak emission at around 380 nm(= violet light), which is absorbed in 3rd dye. The 3rd dye emits blue light with 420 nm wave length, which is detected by PMTs' photocathodes. The use of blue fibers gives us two benifits. One benifit is that the photocathode of PMTs has a factor 2 higher quantum efficiency for the blue light than popular green light. High quantum efficiency improves not only detection efficiency itself but also the photo-statistics, thus makes the detector's resolution better. Another benifit is that we can get much faster signal than the green scintillation light. The signal for muon traversing through all 24 layer stack has a width of 20 ns at base-to-base, which is suitable for our requriment. In contrast, the width of output from green fibers is typically 40 ns at base-to-base with the same configuration.

PERFORMANCE

The light yield, efficiency, and gain stability of RC were measured.

The cosmic muon was used to measure the light yield. The stack of scintillators with fibers produced ~40 p.e. per muon passing through perpendicluar to the RC. This corresponds to 3.3 p.e. per MeV of energy deposit in the scintillation tile.

To evaluate the performance during the experiment, an inefficiency was extracted from the data of physics run with a following procedure. We looked for a photon decayed from π^0 in $K_L \to \pi^+\pi^-\pi^0$, which hit the photon veto detectors. Because of a momentum conservation, we can determine the energy and position of a missing photon by assuming it came from $K_L \to \pi^+\pi^-\pi^0$ and by making the π^0 and kaon mass as constraints. Figure 3 shows the obtained inefficiency as a function of incident photon energy. The threshold was set to an incident photon energy of 66 MeV, which was equivalent to 20 p.e., a half of the energy deposit by muon. An inefficiency expected by GEANT simulation is also shown in the same plot. Data and GEANT are in good agreement.

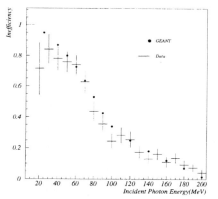

FIGURE 3. Inefficiency of RC as a function of incident photon energy for data and GEANT simulation. The threshold is set at 66 MeV(=20 p.e. equivalence).

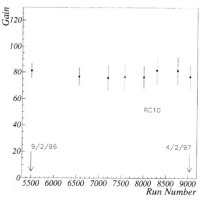

FIGURE 4. The gain of RC as a function of time. No gain adjustments were carried out during this period.

The blue fiber is known to be more susceptible to radiation damage than green fiber. However, our radiation dose level was estimated to be a few rads, and it did not affect the gain of photon veto detectors. Figure 4 shows the relative gain of RC as a function of run number where the time span corresponds to about 7 months. Even without any gain adjustments, the gain was constant within the error.

FUTURE IMPROVEMENT

The blue fibers has a shorter attenuation length(3m) than green fibers, because the emission spectrum of the WLS(3rd dye) overlapes with its own absorption spectrum as shown in Figure 2.

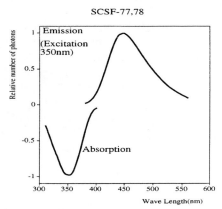

FIGURE 5. Emission and absorption spectrum of dye proposed to use in KAMI

To improve this, we are considering to use SCSN-77,78 as WLS for a future experiment. Figure 5 shows the emission and absorption spectrum of the new dye. Little overlap between the emission and absorption is expected to give a longer attenuation length, and thus a higher light yield.

SUMMARY

KTeV used blue WLS fibers to readout of the photon veto detectors made of leads and scintillation tiles. It gave twice faster pulse and twice higher quantum efficiency to photocathode of PMTs than green WLS fibers. The detection inefficiency was reproduced by GEANT simulation. No gain shift in the detectors due to radiation damage was observed.

ACKNOWLDEGEMENT

We thank all members at Fermilab and UCLA associated to design and construction of the photon veto detectors. KH acknowledges supports from JSPS fellowship.

REFERENCES

1. K. Arisaka et al., *KTeV Design Report,* FN-580(1992).
2. Y.B. Hsiung, *KTeV Internal Memo* **0121,** June 1993.
3. K. Arisaka and J. Quanckenbush, *KTeV Internal Memo* **0015,** March 1992.
4. Kuraray CO, LTD.
   Shinnipponbashi-building 3-8-2 Nipponbashi Chuo-ku, Tokyo, 103-0027 Japan
5. DuPont CO, LTD. Cooperate Headquaters.
   1007 Market Street, Wilmington, DE 19898

Light Collection from Scintillation Counters using WLS Fibers and Bars

V.Evdokimov

Institute for High Energy Physics, Protvino, Russia 142284

Abstract. Several methods of collecting light on scintillation counters using WLS fibers and WLS bars were studied. Nearly 20 prototype counters with different designs and with sizes ranging from $14 \times 11 \times 1.3 cm^3$ to $105 \times 60 \times 1.3 cm^3$ have been tested using cosmic muons and radioactive source. The efficiency of light collection on number of photoelectrons, uniformity of response, and time resolution have been measured. Test results for two new designs of light collection from scintillator based on WLS fibers around perimeter of scintillator plate and WLS fibers placed in machined on scintillator plate deep grooves are presented. Two out of the studied designs have been chosen as the basic options for the DØ muon system upgrade: light collection using two WLS bars for the forward muon scintillation counters and light collection using WLS fibers in deep grooves on scintillator for central area muon counters.

INTRODUCTION

The study of different scintillation counter designs was motivated by the DØ detector muon system upgrade. Counters for the DØ forward muon system are arranged in a projective R-φ geometry for uniform coverage in η and φ with $\Delta\eta = 0.1$, $\Delta\varphi = 4.5°$ segmentation [1]. Such a design leads to substantially different sizes of trapezoidal counters, from 9×15 cm^2 to 60×106 cm^2. Usually, different ways of collecting light from scintillator are using for small and large counters, but, it is preferable to use the same design for all counters. The same design simplifies trigger counters plane design and counter production. The forward muon trigger scintillation counters have to provide good time resolution and uniformity of response for better background rejection, to operate in the fringe magnetic field of the DØ superconducting solenoid and toroidal magnet (350 G) and have low cost and simple handling for the production almost of five thousand counters.

The requirement of good magnetic shielding substantially limits the counters design. Previously, two methods were considered: a direct light collection scheme from cut on 45° scintillator plate corner or the use of common lucid light guide. But, for good magnetic shielding, especially from the field parallel to the phototube axis, one must use a small diameter phototube and allow a small opening in the magnetic shield for light input to the phototube. Using wavelength shifting material (WLS), fibers or bars, one can concentrate the light from the scintillator to the smaller cross-section of fibers or bars and to smaller photocathode area. In practical sense, for a photocathode area

CP450, *SciFi97: Workshop on Scintillating Fiber Detectors*
edited by A. D. Bross, R. C. Ruchti, and M. R. Wayne
© 1998 The American Institute of Physics 1-56396-792-8/98/$15.00

2-3 cm² and scintillator plate side of more than 20cm² the efficiency of light collection using WLS bars is better than for direct methods of light collection. So, direct ways of light collection from scintillator are not presented in this report.

An extensive study of different methods of light collection using WLS fibers and bars for various sizes of counters was provided. Results of this work have been used as a basis for the choice of light collection from scintillator as well as to optimize the counters design for DØ muon system.

MEASUREMENTS OF COUNTER PARAMETERS

Most of measurements were made at FNAL between November 1995 and January 1996 using cosmic muons. A simple CAMAC style test stand was assembled using a VAX-3200 computer and LeCroy analogue-to-digit (ADC) and time-to-digit (TDC) modules. A telescope of 2 or 3 scintillation counters was used to trigger on cosmic ray muons. Two counters with dimensions 15×15×1cm³ were placed 50 cm apart with a test counter between them. A third counter with dimensions 5×5×1 cm³ was placed close to the test counter. For each prototype counter, the following measurements were made:

1. The efficiency of light collection on number of photoelectrons (N_{phe}) using the mean value of the pulse height (charge) for cosmic muons. A Gaussian fit (excepting Landau's tail) was performed for the pulse height spectra. The photomultiplier (PMT) was calibrated in the number of ADC channels per photoelectron using methods:

a) The mean value of single electron spectra for a light emission diode (LED);

b) From the mean value (A) and standard deviation σ using LED: $N_{phe} = (A/\sigma)^2$. This method underestimates N_{phe}, because it does not takes into account fluctuations of secondary emitted electrons from the first dynodes of the PMT .

c) Using a non-efficiency method for LED pulse height spectra:

$$1 - \varepsilon = \exp. (- N_{phe}); \qquad N_{phe} = - \ln(N_{pedestal}/N_{total}).$$

where ε is the efficiency, or the probability to have one or more photoelectrons.

The most precise method (c) is used as basic; others (a, b) are used for cross-checks.

The same photomultiplier FEU-115M [2], with better than average photocathode quantum efficiency, was used for all N_{phe} measurements. Some counters were designed for use with EMI 1″ PMT, but all data was measured using FEU-115M for better comparisons.

2. The uniformity of response for N_{phe} versus coordinate was measured using cosmic muons. For some counters, the uniformity of response was also studied by measuring PMT current using a radioactive source Ru^{106}. This method is more accurate, but it requires the uniform wrapping of scintillator and is not suitable for some designs with variable scintillator thickness or for variants using large number of WLS fibers. Additional non-uniformity caused by scintillation of thick fiber bundles on these counters is not measurable using the radioactive source.

3. The time resolution, including time shift versus coordinate. A simple constant threshold discriminator was used at a signal to threshold ratio of 5; the resolution of the trigger (0.5ns) is not subtracted.

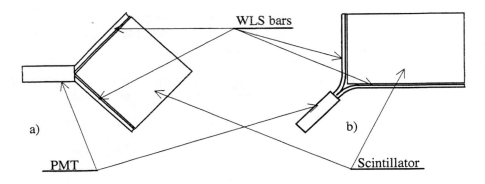

FIGURE 1. Counter design using WLS bars: a) cut at 45° bars; b) bent at 45° bars.

COUNTERS DESIGNED USING WLS BARS

Two main design variants were studied (Fig.1), (a) cut on 45° and (b) bent at 45° WLS bars along two scintillator plate sides. Some counters were made also in the 4.5° trapezoid shape.

Polymethilmethacrilate (PMMA) scintillator with dopants 1.0%PPO+0.01% POPOP (IHEP, Russia) [3] was used in variant (a), and Bicron 404A plastic is used for less thickness for the same light in variant (b). In both cases, 4.2 mm thick WLS bars are made of PMMA with dopant Kumarin 7 [4]. The decay time is 4 ns, and the absorption length is about 1 m. The opposite to the PMT ends of the WLS bars for (a) and (b) are made reflective using aluminized tape. The PMT FEU-115M with a photocathode

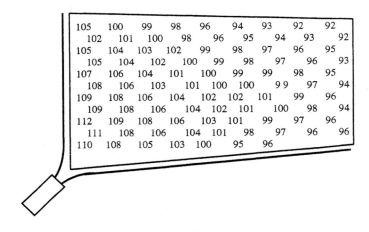

FIGURE 2. The uniformity of light response for counters using WLS bars. Measurements used the radioactive source Ru^{106} (current method). Counter 15, 465×325×12.7 mm^3, without masking. $\sigma_I/I = 4.9\%$, $I_{max}/I_{min} = 1.22$.

TABLE 1. Test results of counters using WLS bars.

Coun-ter #	Design	Sizes, mm^3 Scintillator	N_{phe} on center	Non-unifor-mity	Timing σ_t ns	Timing Δt ns
7		241×241×22 PMMA scintillator	62	±12%	1.1	2.0
13		137×137×12.7 Bicron 404A	184	±5%	0.7	0.3
14		465×325×12.7 Bicron 404A	99	±9% σ=5.5%	0.9	3.2
15	4.5°	465×325×12.7 Bicron 404A	115	±7% σ=4.9%	0.8	3.1
19	4.5°	1054×597×12.7 Bicron 404A	61	±10% σ=7.1%	1.2--1.6	7.5
21		236×163×12.7 Bicron 404A	75	±10%	0.9	1.5
23		326×236×12.7 Bicron 404A K30 WLS bars	194	±9%	0.7	1.9

diameter 20 mm, quantum efficiency 17%, and rise time 3 ns is used. The scintillator and WLS bars are wrapped in Tyvek material for better light collection, and in black paper for light tightness. A mechanically designed variant consists of aluminum container, made of two 1mm thick plates, an extruded profile around the perimeter of the counter, and a stainless steel part, welded to the soft steel magnetic shield for connection to aluminum container. Variant (a) was primarily tested. For this variant, it is difficult to provide a good magnetic shielding for the photocathode area of the PMT. For this type of PMT, it is necessary to extend the mu-metal magnetic shield for at least

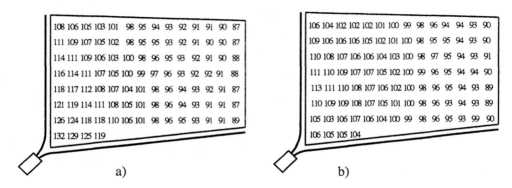

108	106	105	103	101	98	95	94	93	92	91	91	90	87
111	109	107	105	102	98	95	95	93	92	91	90	90	87
114	111	109	106	103	100	98	96	95	93	92	91	90	88
116	114	111	107	105	100	99	97	96	93	92	92	91	88
118	117	112	108	107	104	101	98	96	94	93	92	91	87
121	119	114	111	108	105	101	98	96	94	93	91	91	87
126	124	118	118	110	106	101	98	96	95	93	91	91	89
132	129	125	119										

a)

106	104	102	102	102	101	100	99	98	96	94	94	93	90
109	106	106	106	105	102	101	100	98	95	95	94	93	90
110	108	107	106	106	104	103	100	98	97	95	94	93	91
111	110	109	107	107	105	102	100	99	96	95	94	94	90
113	111	110	108	107	106	102	100	98	96	95	94	93	89
110	109	109	108	107	105	101	100	98	96	93	94	93	89
105	103	106	107	106	104	100	99	98	96	95	93	99	90
106	105	105	104										

b)

FIGURE 3. The uniformity of light response for counters using WLS bars. Measurements used the radioactive source Ru^{106}. Counter $1054 \times 597 \times 12.7$ mm^3. a) without masking. $\sigma_I/I = 9.6\%$, $I_{max}/I_{min} = 1.5$. b) with masking using paper wedges between bars and scintillator plate. $\sigma_I/I = 6.1\%$, $I_{max}/I_{min} = 1.26$; $(I_{without\ mask}/I_{with\ mask}) = 1.2$.

22 mm above the photocathode. Next, this variant must have optical contact between the cut on 45° bar ends and the photocathode due to non-perpendicular light output from the bar ends. This requirement is not acceptable for large number of counters because of its low reliability. The variant (b) is made to be free from these disadvantages.

The uniformity of light response for counters using WLS bar is shown in Fig.2 and Fig.3. Measurements of PMT current (I) are made using the radioactive source Ru^{106}. All data are normalized to a mean value. It is easy to improve the uniformity of counters by a factor of 1.5 by placing white paper wedges between the WLS bars and the scintillator plate. This masking decreases the light in the hot area in the corner of the scintillator near the PMT, but also decreases the average light output level by 20% .

Table 1 shows the overall results for counters using WLS bars. The non-uniformity is shown in terms of $\Delta = (N_{phe\ max} - N_{phe\ min})/(N_{phe\ max} + N_{phe\ min})$ for all measured points for the measurements using cosmic muons, and in terms of σ (dispersion) for the measurements using a radioactive source. The timing results are presented in terms of σ_t-dispersion for the Gaussian fit to the time distributions, and the Δt-maximum value of the timing shift versus coordinates. All tested counters demonstrated good efficiency of light collection, much more than 30 photoelectrons, which is specified as the minimally required value for the DØ muon system, and exhibit good timing.

COUNTERS USING FIBERS IN FLAT GROOVES

These counters are designed similar to the existing DØ Cosmic Cap counters [5]. WLS fibers are placed on the top side of the scintillator plate in machined 4mm wide grooves, 4 fibers in each groove, with a 4 mm space between grooves, so ½ of the

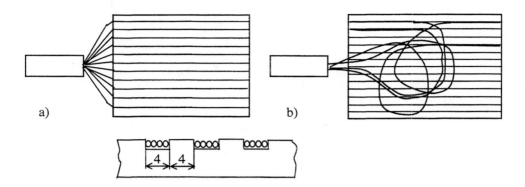

FIGURE 4. Design options for counters using flat grooves. a) one set of fibers in each groove; b) two set of fibers in each groove.

scintillator plate is covered by fibers (Fig.4). For some variants, the spacing between grooves is made 2 mm wide at the edges of scintillator for best uniformity, or two set of fibers are placed in each groove. Each set covers ½ of the groove length. The ends of fibers opposite to PMT as well as this side of the scintillator plate are made reflective using aluminized tape. Bicron 404A scintillator and 1mm diameter Bicron BCF92 WLS fibers with decay time 2.7 ns and an absorption length of 2.7 m are used.

Table 2 shows the test results for counters using flat grooves. Values are quoted as in Table 1. This design shows 36-64 photoelectrons and generally good timing. The use of so many fibers, up to 320 per counter, is not the best choice, and not only for cost and handling. Fiber bundles scintillation decrease in some cases counter characteristics such as the uniformity of response and the timing accuracy.

COUNTERS WITH FIBERS AROUND THE SCINTILLATOR

Designs of counters with fibers around the perimeter of scintillator plate are shown in Fig.5. Bicron 404A scintillator and 1mm diameter Bicron BCF92 WLS fibers are used. The main idea of this design is simple: ~1/6 of total light in scintillator plate goes to each of the 6 sides inside internal reflecting cones; the smaller is the side, the higher is the brightness of light to this side. Fibers are oriented to collect light to the small sides of plate. So, 4 of 6 sides are covered using only 12 (or less in some tests) fibers for 12.7 mm thick scintillator (~4/6 of total light goes to these fibers). Both ends of the fibers are glued in a lucid tube. This assembly is diamond-cut and placed at the PMT photocathode.

The advantages of this design, compared to the flat grooves design are as follows:
- Higher efficiency of light collection using smaller number of fibers;
- It is not necessary to cut, polish and mirror the ends of fibers opposite to the PMT;
- A smaller diameter PMT may be used;
- It is easy to place PMT on top of the scintillator, the small number of fibers makes fiber assembly more flexible.

Fibers are longer in this design, however their length is equal to the perimeter of the

TABLE 2. Test results of counters using WLS fibers in flat grooves.

Coun-ter #	Design	Sizes,mm^3 Scintillator	N_{phe} on center	Non-unifor-mity	Timing σ_t ns	Timing Δt ns
5	**320 fibers**	465×325×12.7 PMMA scintillator	36	±18%	1.1 (> 2 with bund-les)	1.2
8	**320 fibers**	465×305×12.7 Bicron 404A	40	±11%	1.1	1.2
6	**76 fibers**	229×152×12.7 Bicron 404A	55	±5%	0.7	1.1
10	**180 fibers**	470×356×12.7 Bicron 404A	60	–	1.1	2
12	**160 fibers**	457×305×12.7 Bicron 404A	54	±7%	–	–
17	**160 fibers**	914×318×12.7 Bicron 404A	64	±21%	0.8	4.7

scintillator. This design is therefore good for counters that are not very large. A mechanical design is required to protect the fibers from damage. The mechanically designed variant is similar to the counters with bars and consists of an aluminum container made of two 1mm thick plates and an extruded profile around the perimeter of the counter. The PMT is placed on top of the counter.

Test results of counters using fibers around the scintillator are shown in Table 3.

The number of photoelectrons for this design is more than that obtained using flat grooves for the same size counter but less than for design using bars. Counters using 12 fibers have 1.8 times less light yield for sizes 465×324×12.7mm^3 and 1.2 times less for sizes 1054×597×12.7 mm^3 compared to a design using WLS bars. The difference is less for larger counter because of less than for fibers absorption length of bars. Timing for this design is 10-20% worse as compared to the bars design. Fig.6 shows the light yield on the center of the counter versus the number of fibers for two different size counters. There is a saturation effect for this dependence. It is surprising

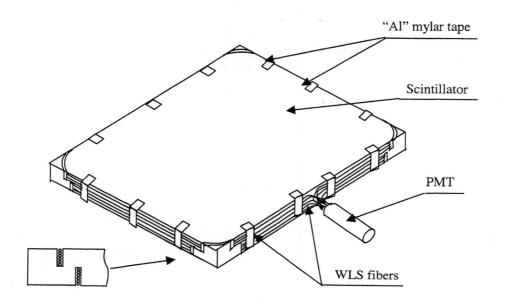

FIGURE 5. Design with fibers around the counter perimeter.

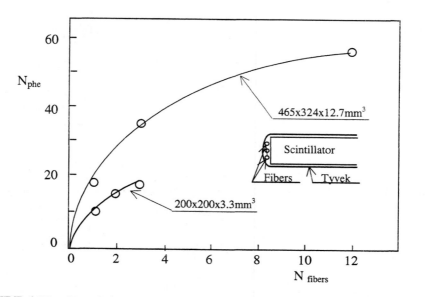

FIGURE 6. Number of photoelectrons versus number of fibers for design with fibers around the counters perimeter.

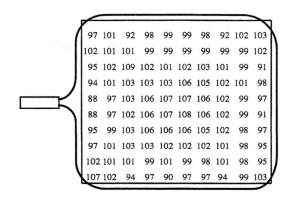

FIGURE 7. Uniformity of light response for counters using WLS fibers. Measurements used the radioactive source Ru^{106}. Counter 4, $200 \times 200 \times 3.3$ mm^3, $\sigma_I/I = 4.2\%$, $I_{max}/I_{min} = 1.21$.

TABLE 3. Test results of counters using to side oriented WLS fibers.

Coun-ter #	Design	Sizes, mm^3 Scintillator	N_{phe} on center	Non-unifor-mity	Timing σ_t ns	Δt ns
11	**12 fibers**	465×325×12.7 Bicron 404A	54	±3.5%	1.0	1.5
4	**3 fibers**	200×200 ×3.3 Bicron 404A	17	σ=4.2%	–	–
18	4.5° **12fibers**	1054×597×12.7 Bicron 404A	37	±7%	1.3-1.9	3.7
18a	**12 fibers ×2 loops= 24 fibers**	1054×597×12.7 Bicron 404A	52. (43 worst point)	±11%	1.5	3.2
20	**6 fibers ×7grooves= 42 fibers**	914×318×12.7 Bicron 404A	51	±10% σ=8%	0.8	4.5

that a counter of size $465{\times}324{\times}12.7$ mm^3 using only 3 fibers gets 34 photoelectrons and good enough timing with σ_t=1.48 ns. The uniformity of light yield for counter with sizes $200{\times}200{\times}3.3$ mm^3 is shown on Fig.7. Note that the variation of the scintillator plate thickness gave noticeable part of the measured non-uniformity for this thin counter. Uniformity for 12.7 mm thick counter 12 is 1.5 times better. It is easy to improve the light yield for counters using so few fibers by a factor of 1.8 by the use of more efficient but more expensive double-clad Kuraray Y-11 fibers [5].

COUNTERS USING WLS FIBERS IN DEEP GROOVES

Design options using fibers in deep grooves are shown in Fig.8. The main idea of this design is the same as using fibers around the scintillator plate: fibers are oriented to collect the light to the smaller sides of scintillator. Fibers for the tested prototype are placed in 1.2 mm wide 6 mm deep grooves with 50 mm spacing between grooves. Fibers are glued into the grooves using epoxy glue on 1cm of their length with 20 cm spacing between the glued points. The ends of the fibers opposite the PMT are glued into the grooves and diamond-cut flash with the scintillator plate. This end of scintillator plate (and fiber ends) are made reflective using aluminized tape. Bicron 404A scintillator and 1mm diameter Bicron BCF92 WLS fibers are used.

Test results for this prototype are shown in Table 3. Direct comparison of the number of photoelectrons versus coordinates for this prototype and for the counter of the same size but using fibers in flat grooves is shown in Fig.9. Counters with flat grooves have 20% more light on average, and 26% more in the center of counter. Both counters have the same light for the worst for light yield point on the corner opposite to PMT. Maximum to minimum ratios are ($N_{phe\ max}$ / $N_{phe\ min}$) =1.61 for flat grooves design and ($N_{phe\ max}$ / $N_{phe\ min}$) =1.16 for the deep grooves case. The uniformity is nearly 3 times better for the prototype with deep groves.

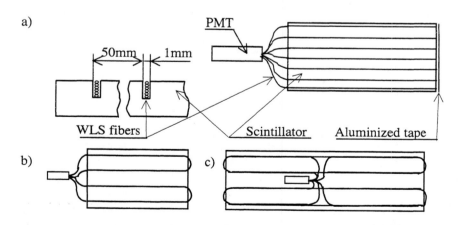

FIGURE 8. Deep grooves design. a) tested prototype; b), c) – design options.

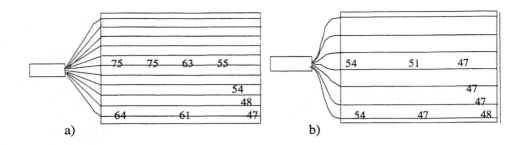

a)   b)

FIGURE 9. Number of photoelectrons versus coordinates using cosmic rays and the same size counters: a) counter 17, flat grooves, 160 fibers; b) counter 20, deep grooves, 42 fibers.

COUNTERS WITH UNPOLISHED SCINTILLATOR PLATE SIDES

The study of counters with unpolished but machined scintillator plate sides was performed in comparison to counters with polished (diamond-cut) plate sides. Namely, the top and bottom sides were polished in both cases but 4 edge sides machined only for this test. The results in tables 1-3 are mainly for the diamond-cut sides of scintillator plate. The result of this study is presented in Fig. 10. Light yield ratios are ($N_{phe\ unpolished}$ / $N_{phe\ polished}$) = 1.07±0.03 for counter 11 and ($N_{phe\ unpolished}$ / $N_{phe\ polished}$) = 1.10±0.05 for counter 14. The same ratio for small counter 13 using 2 bars is 1.03±0.03. Cheaper unpolished variants gave even better results in light yield. Uniformity is the same in both cases for counter using fibers but better for the unpolished variant of counters using bars. Tests with 2 unpolished sides of scintillator for counters using bars also were performed. The variant with all polished sides and the variant with 2 sides of plate opposite to the bars unpolished are consistent within errors. Counters using scintillator with 2 unpolished sides along WLS bars gave better results than the polished variant.

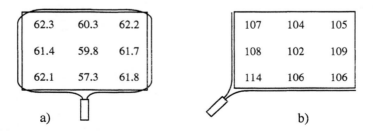

a)   b)

FIGURE 10. Number of photoelectrons versus coordinates for counters with unpolished scintillator plate sides: a) counter 11a, 12 fibers; b) counter 14a, 2 bars.

CONCLUSIONS

1. The efficiency of light collection on the number of photoelectrons, uniformity of response and time resolution have been measured for about 20 prototype counters using several methods of light collection from scintillator using WLS bars and WLS fibers.

2. New variants of light collection from scintillator were tested with good results:
   - The method based on WLS fibers around the perimeter of the scintillator plate gives $N_{phe} = 54$, $\sigma_t = 1$ ns, and the uniformity of $\pm 3.5\%$. This design is good for counters in cramped areas or in a strong magnetic field . It is easy to extend few fibers out of a high field area.
   - Light collection method using WLS fibers placed in deep grooves on the scintillator requires fewer fibers, but showed nearly the same light collection efficiency with better uniformity than the method based on flat grooves.

3. Two designs have been chosen as the basic options for the DØ muon system upgrade:
   - The method using two WLS bars bent at 45° for the forward muon scintillation counters; a few counters for forward muon system in cramped areas will be produced using the design with fibers around the perimeter of counter;
   - The method using WLS fibers in a deep grooves is used as a basic option for central muon scintillation counters.

4. Measurements using a scintillator plate with unpolished (but machined) sides was performed for some of light collection methods. The less expensive unpolished variant is not worse (even better) than the polished one in the yield of photoelectrons and uniformity for the studied methods.

ACKNOWLEDGMENTS

The author gratefully acknowledge the help of the staff at Fermilab and IHEP for prototype counters production. He would like to thank S.Denisov for support of this work, D.Denisov, A.Dyshkant, A.Galyaev, S.Gurzhiev and Yu.Gutnikov for help and for their contribution to this work.

REFERENCES

1. DØ collaboration, "The DØ Upgrade," DØ Note 2894, FERMILAB-FN-641, Fermilab, 1995.

2. S.Belikov et al., "Characteristics for the Photomultipliers FEU-115M for EMCal PHENIX", IHEP preprint 96-42, Protvino, 1996.

3. S.Belikov et al., "Characteristics of the SOFG-120 polymetylmethacrylate scintillator". Instruments and Experimental Technique 36, #3, p.386, 1993.

4. S.Belikov et al., "Physical Characteristics of the SOFZ-105 Polymethyl Methacrylate secondary Emitter". Instruments and Experimental Technique 36, #3, p.390, 1993.

5. B.S.Acharya et al., "Scintillation counters for the DØ muon upgrade", Nucl. Instr. and Meth. In Phys. Res. A 401 (1997) 45-62.

SESSION 8: PHOTOSENSORS - II

Chair: M. Petroff
Scientific Secretary: J. Marchant

Visible Light Photon Counters (VLPCs) for High Rate Tracking Medical Imaging and Particle Astrophysics

Muzaffer Atac

Fermi National Accelerator Laboratory
Batavia, IL 60510, U.S.A.
University of California at Los Angeles
Los Angeles, CA 90024, U.S.A.

Abstract. This paper is on the operation principles of the Visible Light Photon Counters (VLPCs), application to high luminosity-high multiplicity tracking for High Energy Charged Particle Physics, and application to Medical Imaging and Particle Astrophysics. The VLPCs as Solid State Photomultipliers (SSPMs) with high quantum efficiency can detect down to single photons very efficiently with excellent time resolution and high avalanche gains.

INTRODUCTION

High Energy Particle Physics experiments designed to run at high luminosity and high multiplicities require fine granularity, fast, good time resolution and good spatial resolution tracking. Scintillating fibers with the VLPC readout fulfill all these requirements. Scintillating fibers with a diameter less than 1mm can provide good multitrack resolution and high speed. The VLPCs with quantum efficiency around 80%, avalanche gain around 30,000, excellent time resolution (about 1nsec), and surface area of 1mm in diameter can provide an efficient tracking for High Energy Particle Physics (1) (2) (3). The VLPCs were developed by Rockwell International Science Center (now part of Boeing Co.) jointly with M. Atac, originally for UCLA under DOE contracts. The author has pioneered the development in 1987-88 under the contract for developing scintillating fiber tracking for High Energy Particle Physics, working together with Rockwell International Science Center, Anaheim, California. We

CP450, *SciFi97: Workshop on Scintillating Fiber Detectors*
edited by A. D. Bross, R. C. Ruchti, and M. R. Wayne
1998 The American Institute of Physics 1-56396-792-8/98/$15.00

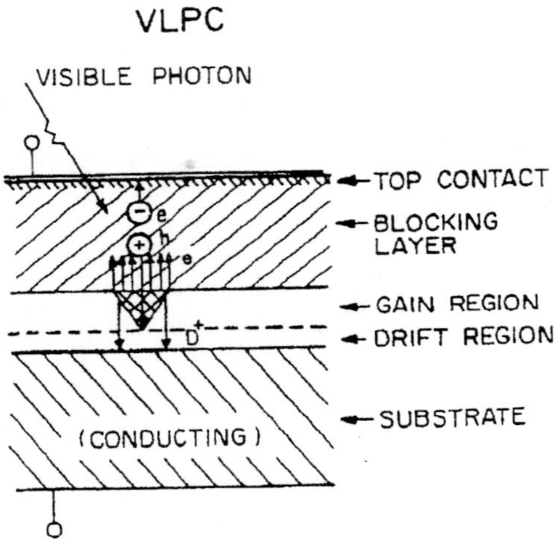

VLPC

VISIBLE PHOTON

→ TOP CONTACT
→ BLOCKING LAYER
→ GAIN REGION
← DRIFT REGION
← SUBSTRATE

(CONDUCTING)

FIGURE 1. Schematic of the operational principles of the VLPC.

called this research and development High Intensity Scintillating-fiber Tracking Experiment (HISTE)program. During the last few years the VLPCs have been purchased by E835 (4) and D0 experiments at Fermilab. Fermilab physicists and engineers have been developing large systems for D0. 100,000 pixel VLPC system is needed for D0 (5, 6). E835 has done an experiment at Fermilab using a fiber tracker with a modest number of VLPCs (about 1200 channels). COSMOS (E803) collaboration has decided to put together a scintillating fiber tracking system with 40,000 fibers and VPLCs.

In the following we will also talk about application of VLPCs to Medical Imaging and Particle Astrophysics.

Before the VLPCs became practical in using for fiber tracking, UA2 Experiment used image intensifier readout with vacuum photocathode (7). This resulted in an inefficient tracking. Most image intensifier tubes are rather sensitive to magnetic fields. Tests done with the VLPCs showed no effect up to 12kG field.

OPERATING PRINCIPLES OF THE VLPCS

The operating principles of the VLPCs are given in Reference 2, therefore we will discussed them briefly here. The VLPCs are Impurity Band Conduction (IBC) devices that are minimized in quantum efficiency in the Infrared (IR) region while maximized in quantum efficiency for the wavelengths around 550nm relative to the original device, Solid State Photo Multiplier (SSPM), which was discovered by Rockwell International

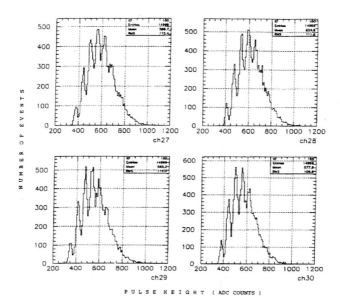

FIGURE 2. Multiple photoelectron peaks resulted when the VLPCs were illuminated with a pulsed LED. It shows uniform responses in quantum efficiency and gain from different channels obtainable under the same bias voltage.

A schematic diagram of the VLPC is shown in Figure 1. In a VLPC, a neutral donor is a substitutional ion with an electron bound to it in a hydrogen-like orbit with an ionization potential of about 0.05eV. Because of this very small energy gap, the devices need to run at cryogenic temperatures. Nominally they run at a temperature between 6 and 7K. When the concentration of impurities is sufficiently high, they form an energy band separated from the conduction band by the ionization potential. When the applied electric field is sufficiently high, about $2 \times 10^3 - 10^4$V/cm, each initial electron starts an avalanche of free electron-hole pairs within 1ns. The avalanche size could reach up to 5×10^4 when applied voltage reaches −7 volts. The avalanche may occupy about 10 micron diameter area for about few microseconds while the rest of the area is continuously available for detecting photons. The gain and the quantum efficiency (QE) of the devices taken from same wafer are very uniform at a common voltage and temperature as seen in Figure 2 (6). For this, photons from an LED illuminated the top of a cassette which housed the VLPCs.

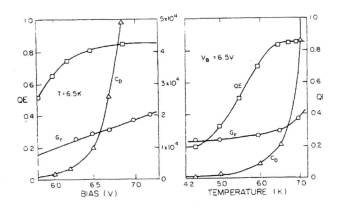

FIGURE 3. Quantum efficiency (QE) at A=560nm of an AR coated VLPC as functions of bias voltage and temperature is shown. Dark count pulse rate (C_D) and fast gain (G_F) are also indicated.

The photons then transmitted through optical fibers illuminating the VLPCs. The sensitive area of the devices was 1mm in diameter. They run in a space charge saturated avalanche mode. Due to the small gain dispersion, as seen in the figure, multi photoelectron peaks arenicely separated. Due to their capability of counting photons, we called them Visible Light Photon Counters (VLPCs).

Because the devices are impurity semiconductors and the VLPC with arsenic impurity, that has a bandgap around 50 millielectron volt, it has to be operated at cryogenic temperatures, otherwise dark pulse (thermal electron pulse) rate would be extremely high (saturating the device). Figure 3 shows the quantum efficiency, the fast avalanche gain and dark pulse rate as functions of temperature and bias voltage. Optimum operating temperature is around 6.5K and optimum bias voltage is about 6.5V. The devices were coated with 560nm antireflective (AR) coating to improve the quantum efficiency (QE) from 70% to 85%. Operating the devices at 6.5K, controlling and monitoring the temperature within 0.1K are relatively easy.

Main characteristics of the VLPCs are summarized in the following table:

Table I.

Quantum efficiency optimized for 530nm	80%
Avalanche gain	$3\text{-}5\times10^4$
Thermal electron pulse rate at 6.5K	$\sim5\times10^3$/sec.mm
Saturation pulse rate	5×10^7/sec.mm
Pulse risetime	<3nsec
Average power per pixel	1.5 microwatt
Optimum operating voltage	~6.5V
Optimum operating temperature	6-7K
Dynamic range (linear)	3000 photoelectrons
Effect by magnetic field	no effect up to 12kG

318

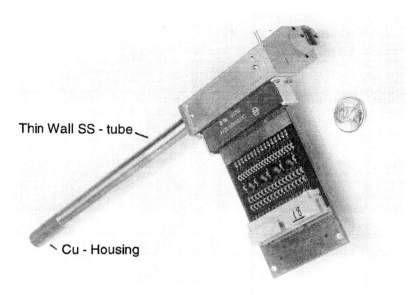

FIGURE 4. A photograph of a 32 channel VLPC cassette. The penny in the picture shows the compactness of the unit.

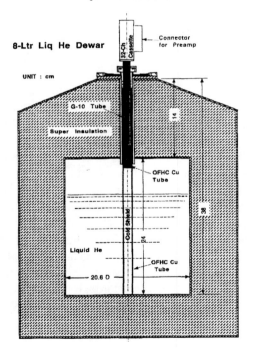

FIGURE 5. The structure of the 8-liter cryostat..

319

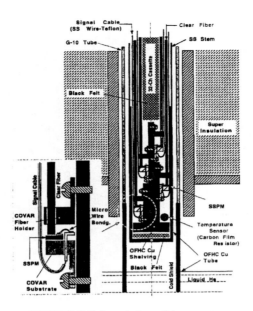

Signal Cable
(SS Wire-Teflon)

Clear Fiber

G-10 Tube

SS Stem

32-Ch Cassette

Black Felt

Super
Insulation

Signal Cable

Clear Fiber

COVAR
Fiber
Holder

Micro
Wire
Bondg.

SSPM

Temperature
Sensor
(Carbon Film
Resistor)

SSPM

OFHC Cu
Shelving

OFHC Cu
Tube

COVAR
Substrate

Black Felt

Cold Shield

Liquid He

FIGURE 6. Enlarged view of the cassette.

PROTOTYPE CRYOGENIC CASSETTE DESIGN

A 32 channel VLPC cassette was designed by the author and constructed to carry out scintillating fiber tests for determining photoelectron yield from a variety of fibers, attenuation length of the photons in the fibers, and timing and rate capabilities of the VLPCs. Figure 4 shows one of the cassettes with a penny next to it. The penny is there to show how small a cryogenic system is used. At the top of the cassette a 32 channel of a QPAO2 amplifier is shown (7). The unit is designed to pass a small amount (about 50cc) of boil off liquid Helium (He) when inserted into a liquid He dewar. Details of the cassette-dewar assembly are shown in Figure 5. With the help of the Oxygen free High Conductivity Copper (OFHC) cold shield tubing, liquid He temperature is brought to the level of the OFHC housing in which the VLPCs are kept. With this arrangement the required temperature is reached and kept fairly constant over several days, using only 2 liters of liquid He per day. This small amount of usage is due to full usage of enthalpy of boil off He going through the cassette and cooling all the components in the thin wall stainless steel 304 tubing. The details of the arrangement in the cassette are shown in Figure 6 (8).

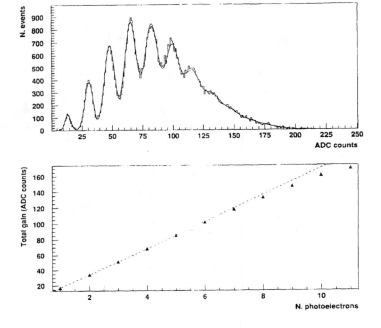

FIGURE 7. Multi-photoelectron peaks and a calibration plot. As we see in the figure, multi-photon peaks are well resolved. For this reason we call the devices "Visible Light Photon Counters" (VLPC).

SCINTILLATING FIBER TRACKING TESTS

Some scintillating fiber tracking tests were carried out using 500 micron overall diameter multiclad scintillating fibers (manufactured by Kuraray Co., Japan) and the above described cassette and cryogenic arrangement. We detected 6 photoelectrons per minimum ionizing track in the average from the middle of 280cm length of scintillating fibers (were mirrored at the end) which were coupled to 500cm length of multiclad clear optical fibers (6). An ADC counts versus number of photoelectron (pe) calibration was done before every measurement as shown in Figure 7. For this an LED was used, illuminating the optical fibers in the cassette. As seen in the figure, after the sixth photoelectron peak a small saturation appears. This is due to the amplifier saturation. The calibration was obtained by making a cut at N>2 and fitting to a convolution of Poisson and Gaussian functions:

$$f(x) = N\sum_n \frac{1}{\sigma_n \sqrt{2\pi}} exp\left[-\frac{1}{2}\left(\frac{x-n}{\sigma_n}\right)^2\right] \frac{exp(-n_{pe})}{n!} \tag{1}$$

where the free parameters were normalization factor (N) and the mean value of the Poisson distribution (n_{pe}). Sigma of each gaussian (σ_n) was fixed by determining the peaks with the LED runs for the bias voltage and temperature.

$$\sigma_n^2 = \sigma_0^2 + \sigma_c^2 \cdot n, \qquad (2)$$

where σ_c is a sigma of the n-th peak and σ_0 is a sigma value of the pedestal.

Attenuation of photons in various scintillating fibers are shown in Figure 8. The results indicate that attenuation lengths are between four and five meters and there was no appreciable change in this number for fibers of 500 microns to 1mm. Attenuation length, λ, of photons from 3HF (1500 ppm 3-Hydroxyflavone +1% P-therphenyl) scintillating fiber were measured through various diameters of multiclad clear optical fibers. The 3HF emits photons around the peak wavelength of 530nm (manufactured by Kuraray Co.). The results, as seen in the figure, indicated that the attenuation lengths of photons from the 3HF within the experimental error does not depend on fiber diameter between 0.5mm to 1mm.

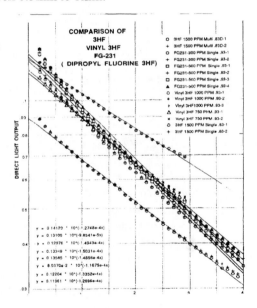

FIGURE 8. Attenuation length plots for various scintillators. There is about 80% more yield from the multi-clad fibers relative to single clad.

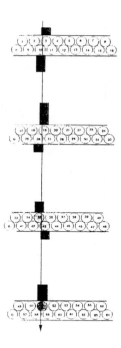

FIGURE 9. A minimizing ionizing track. The number of detected photons is indicated in the bars.

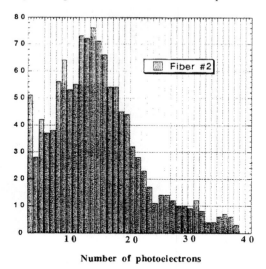

Number of photoelectrons

FIGURE 10. Pulse height distribution from VLPC fiber #2 when fiber is illuminated by minimum ionization particles (MIP).

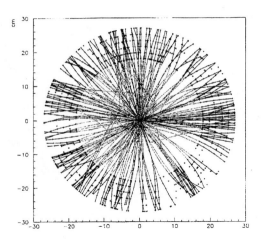

FIGURE 11A. A typical top + 6 MB event in the R-φ view. Crosses are axial hits, solid lines connect hits used, and dotted curves are extrapolation inward of the final parameters.

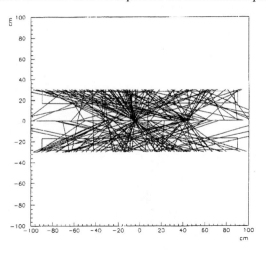

FIGURE 11B. R-Z view of a top + 6 MB event. Crosses show locations of associated stereo-hit/axial-segment points. Lines show extrapolation of fitted segment to beam line.

Some beam tests were carried out at the Meson 6 West test beam using 835 micron (scintillating core of 725 micron) multiclad 3HF fibers using author's designed cassettes (9). Four staggered doublets were used with the tests. Figure 9 shows a 15GeV hadron (most likely a pion) track with the number of photoelectrons indicated in the pulse height histograms. Figure 10 shows the number of photoelectron distribution for a fiber. Weighted average photoelectrons is more than 15. For a staggered doublet the tracking efficiency is better than 99.7%.

324

The Collider Detector Facility at Fermilab (CDF) has considered using scintillating fiber tracker and some track reconstructions were done using top quark events. Figures 11-a and 11-b show how efficiently fiber tracker could find tracks that were found by the Central Tracker (CTC). Presently Fermilab physicists and engineers are working hard on a large system design for the D0, Short Baseline Neutrino Experiment (COSMOS), and CP Violation Experiment at the Main Injector (KAMI). The number of VLPCs to be needed will be around 240,000.

USE OF VLPCS FOR MEDICAL APPLICATIONS

Single Fiber Tracker for Stereotactic Biopsy and Intraoperative Lumpectomy of Breast Cancer

Breast cancer in women of child bearing age is the second leading cause of death in the USA (10). Early detection has allowed for less extensive surgical procedures and/or decreased need for chemotherapy since a substantial majority of questionable lesions detected by mammography are benign. There is a growing interest among the health care professionals and patients in finding alternatives to surgical biopsy for diagnosing these lesions. State-of-the-art stereotactic breast biopsy is comparable in sensitivity to surgical biopsy, and the procedure is quicker, cheaper, and easier than the standard practice of preoperative, mammographically guided localization followed by surgical biopsy.

The problems mentioned above can be ameliorated by a nuclear medicine procedure using a beta detector on the end of a 0.8mm diameter scintillator and fiber optic cable (11). By positioning the detector within a few millimeters of the suspected area, small lesions, usually not detectable using gamma radiation detectors, can be identified and quantified for activity. The fiber optic cable with a small scintillating plastic fiber attached (fused) to the tip can either be inserted into a core biopsy, or can be used during ductogram to identify the duct system containing microcalcified clusters. When inserted into a surgical wand, it could be used to ensure that all residual tumor was removed during lumpectomy. This diagnostics alone is very much needed to prevent recurrence and spread of malignant tissues.

We have developed a prototype suitable probe that uses a rather small diameter biopsy needle (in the current study an 18 gauge needle with an external diameter of 1.25mm) containing a 0.83mm diameter and 3mm length 3HF (above mentioned) multiclad scintillating fiber, which is fused to the same diameter multiclad clear optical fiber of 200cm length.

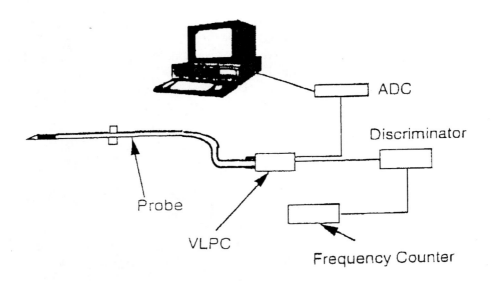

FIGURE 12. A schematic view of the biopsy needle probe together with a simple data acquisition system.

Photons emitted from the scintillating fiber by the passage of betas or positrons are transmitted through the optical fiber and are detected by a VLPC. For the set up, a cassette and the cryogenics mentioned in the tracking section above were used. The probe assembly and the rather inexpensive data acquisition system are shown in Figure 12.

The signals from the VLPC were amplified by a transimpedance amplifier (TIA), fed into a discriminator and counted by a commercial scaler. The VLPC produces an avalanche gain around 30,000 per photoelectron. We obtained less than two pulses per minute as background counts when the threshold was set above three photoelectrons. We measured experimentally that the average number of photoelectrons produced in the VLPC was more than 40, by the passage of betas through the thin scintillator. Pulse height spectrum obtained using a Bi^{207} beta source is shown in Figure 13. Only a small fraction of the 1MeV beta particle energy is left in the thin scintillator, giving rise to the pulse height spectrum.

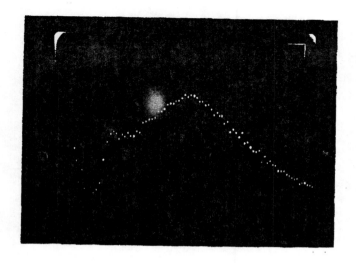

FIGURE 13. Pulse height spectrum obtained using Bi^{207} beta source. The average energy released in the scintillator fiber is about 60KeV. The peak value corresponds to 40 photoelectrons detected by the VLPC.

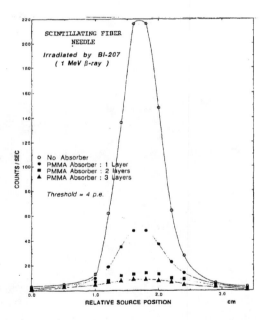

FIGURE 14. Results from the Bi^{207} source test. The curves clearly indicate that the 1MeV betas are rapidly absorbed by the 1.5mm thick Lucite sheets, and there are not many counts from gamma conversions in the scintillator although 90% of the decays from the source are gamma rays in this case.

327

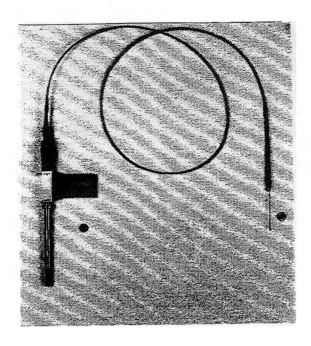

FIGURE 15. Photograph of the probe with the 2 meter long optical fiber between the biopsy needle and the VLPC unit.

FIGURE 16. Test done with a rat having an R3230 AC in the hind leg. The rat was administered 432 microcurie FDG i.v.

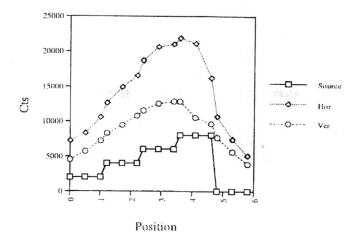

Position

FIGURE 17. Two dimensional scan, even from outside of the skin, indicates where the radionucleide concentration is.

EXPERIMENTAL RESULTS

In order to determine point spread function, we moved the probe linearly relative to 1 microcurie Bi^{207} source without and with 1.5mm thick Lucite sheets (mimicking tissue equivalent density) in between the source and the probe, and recorded the counts per second. The source diameter was approximately 4mm and it was not collimated. The results plotted, in Figure 14 show that the 1MeV betas from the source were very much attenuated after one sheet of Lucite, but we can find the source position after 4.5mm thickness. We expect that the intrinsic resolution of the probe be better than 1mm. The curves also show that the probe is sensitive to betas and not to gammas, although only 8% of the decays produce betas and the rest being the gamma activity. This feature is important due to the fact that the probe will not be sensitive to 511KeV gammas from positron annihilation when a positron source is traced. For a probe like this in a clinical condition, the cryogenic part can be cooled by liquid Helium vapor for safety. A photograph of the probe is shown in Figure 15.

As a first experiment, a preliminary test was done using a rat bearing R3230 adenocarcinoma. The experimental arrangement is shown in Figure 16. As shown in Figure 17, the biopsy needle was moved in an x,y matrix points and the counts were recorded. Even from outside of the skin the probe indicates where the radionucleide concentration is.

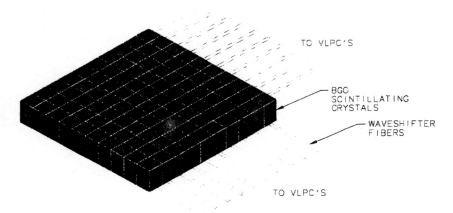

TO VLPC'S

BGO
SCINTILLATING
CRYSTALS

WAVESHIFTER
FIBERS

TO VLPC'S

FIGURE 18. The principal scheme for detecting gamma rays in a two dimensional readout. More layers can be added depending on the energy of the gammas to be detected

USE OF VLPCS FOR PARTICLE ASTROPHYSICS

A possible way of using VLPCs as photodetectors for detecting relatively low energy gammas is shown in Figure 18. In this case wavelength shifting fibers are attached to a matrix array of scintillating crystals in an x,y arrays. The crystal size can be sufficiently large to detect multi MeV gammas from outer space. Scintillating crystals like BGO can be used in this case. Rubrene doped polystyrene can be used as wavelength shifter. Optical fiber that is coupled to wavelength shifter carries the photons to the VLPCs. This idea was proposed for medical imaging by M. Petroff, but it did not work so well due to low energy gammas, 511KeV, from positron annihilation. I am convinced that it will work here due to the detection of multi MeV gammas.

REFERENCES

1. Petroff, M.D. and Atac, M. , *IEEE Trans. on Nucl. Sci.* NS-36 (1989) 163.
2. Atac, M., Park, J., Cline, D., Chrisman, D., Petroff, M. and Anderson;E., Nucl. Instr. and Meth. A314 (1992) 56.
3. Atac, M., et al.; *Nucl. Instr .and Meth.* A320 (1992) 155.
4. Mussa, R. et al., Fermilab E835 experiment, contribution to this conference.
5. Baumbaugh, B. et al., Nucl. Instr. & Meth. A 345 (1994) 271.
6. Adams, D. et al., IEEE on Nucl. Sci. Vol.42, No.4, (1995) 401,Wayne, M.R., Nucl Instr. & Meth. A387 (1997) 278.
7. Alitti, J., et al.; *Nucl. Instr. and Meth.* A279 (1989) 364.
8. Petroff, M.D., Stapelbroek, M.G., and Kleinhans, W.A., *Appl. Phys. Lett.* Vol.51 No.6 (1987) 406.
9. Atac, M., Mishina, M., Takano, T., Valls, J., Yasuokka, K. and Yoshida, T., CDF/ANAL/Tracking/Public/3569, April 11, 1996.

10. Zimmerman, T., *IEEE Trans. Nuc. Sci.* NS-37(2), (1992) 439.
11. Gubinelli, M., Tonet, O.,. Sorel, M., Atac, M., Mishina, M., and Valls; J., CDF/Pub/ Public/4154, July, 1997.
12. Atac, M., Cline, D., Pischalnikov, Y., and Rhoades, J.; Presented at the SciFi Conference at Notre Dame.
13. Titcomb, C.L., *Hawaii Medical Journal*, Vol. 49 (1990) 18.
14. Atac, M., Nalcioglu, O., and Roeck, W., unpublished report.

VLPC Characterization

A. Bross[a], S. Choi[d], G. Geurkov[c], S. Grünendahl[a], D. Lincoln[a],
R. Ruchti[b], J. Warchol[b], M. Wayne[b]

[a]Fermi National Accelerator Laboratory, [b]University of Notre Dame,
[c]University of Rochester, [d]Seoul National University

Abstract. We present the results of the study of quantum efficiency and gain of Visible
Light Photon Counters (VLPC) at high counting rates, expected in their application as
readout detectors of the scintillating fiber tracker in the DØ detector. At a projected
maximum rate of $10 \cdot 10^6$ photoelectrons/sec, the VLPC quantum efficiency only
decreases to 97% of its maximum value. Since other applications may involve higher
counting rates, we present data up to rates of $90 \cdot 10^6$ photoelectrons/sec.

INTRODUCTION

This study is part of the research and development for the scintillating fiber tracker
being built for the upgrade of the DØ Detector [1], which will be used to study proton-
antiproton collisions at a center of mass energy of 1.8 TeV at the Fermilab Tevatron.
The scintillating fiber tracker will consist of about 75000 scintillating fibers of 835 μm
diameter. The fibers will lay on eight concentric cylinders coaxial with the beam line.
The cylinder radii range from 19.5 cm to 52 cm. Photons produced in the scintillating
fibers by ionizing particles propagate through clear fiber light-guides, 7 to 11 m long,
and reach Visible Light Photon Counters (VLPC) [2]. VLPCs are arsenic doped silicon
diodes that operate at a temperature of a few Kelvin.

A photon entering a VLPC converts into an electron-hole pair. The hole drifts to a
gain layer where it starts an electron avalanche of several thousand electrons. The
quantum efficiency of the VLPC, defined as the probability that the photon incident on
the VLPC converts to a fully developed avalanche, varies from 60% to 85%, depending
on the version of VLPC. The gain of the VLPC is typically about 40,000. A distinctive
feature of the VLPC, in addition to high quantum efficiency and gain, is a low gain
dispersion of approximately 15%.

In Run II at the Fermilab Tevatron, the luminosity will reach $2 \cdot 10^{32} \text{cm}^{-2}\text{s}^{-1}$. It is
expected that the VLPC pixels used to read out the DØ fiber tracking detector will
typically see a rate of a few MHz of photoelectrons. This estimate comes from a

CP450, *SciFi97: Workshop on Scintillating Fiber Detectors*
edited by A. D. Bross, R. C. Ruchti, and M. R. Wayne
© 1998 The American Institute of Physics 1-56396-792-8/98/$15.00

detailed simulation that included the following elements: a Pythia generator; a GEANT simulation of the detector; the results of a cosmic ray test, where in the prototype of the DØ detector minimum ionizing particles generated 9 photoelectrons in the HISTE IV VLPC [3]; and the quantum efficiency of HISTE VI chips which will be used in the DØ detector in Run II. (HISTE IV and HISTE VI refer to two versions of VLPC. HISTE VI will be used in the final detector). The VLPCs reading out fibers close to the beam will have a count rate of about $10 \cdot 10^6$ photoelectrons/sec, whereas the ones reading out the outermost fibers will see about $2.5 \cdot 10^6$ photoelectrons/sec.

In this study, we examine the quantum efficiency and the gain of HISTE VI, the latest generation of VLPCs, at counting rates of up to 90 MHz. This was motivated by the possibility of a future increase of Tevatron luminosity to five times the luminosity of Run II (the so called TeV 33). In such an environment, the VLPCs would count up to $50 \cdot 10^6$ photoelectrons/sec.

EXPERIMENTAL SETUP

HISTE VI chips were produced in the form of 2 x 4 pixel arrays, with individual pixels of 1 mm diameter. The arrays were mounted on ceramic substrates providing electrical contact to each pixel via a wire-bond. The assembly of the VLPC chip and the substrate, called a hybrid, is schematically depicted in Fig. 1. The VLPC chips were tested in a cryogenic "cassette". Eight chips can be mounted in the cassette, which, for testing, is inserted into a cryostat cooled to the desired temperature by cold He gas, as shown in Fig. 1.

Photoelectrons in the VLPCs are generated by light from two orange LEDs located outside the cassette. The light from the LEDs is transported to the cassette by a liquid light guide, and then to the VLPCs by a solid, acrylic rod light-guide and a disc diffuser, both located inside the cassette. The LEDs are fired independently with an 18 ns wide voltage pulse. One, the "signal" LED, is powered with 250 Hz voltage pulses which are adjusted in such a way that about 2 photoelectrons are generated in the VLPC pixels. The other "background" LED is pulsed with a frequency of 2.5 MHz. Its voltage is set in such a way that the number of photoelectrons in one pulse multiplied by the frequency results in the desired rate of background counts. For example, 36 photoelectrons per background pulse are required in order to achieve a 90 MHz background rate.

The electrical signals from the VLPCs are brought via kapton flex circuits to QPAO2 [4] preamplifiers, and fed to LeCroy 2249 ADCs, where they are integrated for 100 ns and digitized. In normal operation, the ADC gates are centered on the pulses generated by the low rate signal LED.

Experimental Setup

VLPC Hybrid

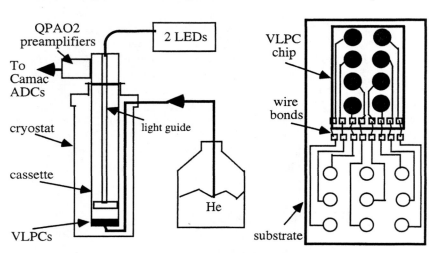

FIGURE 1. Schematic picture of the experimental setup and the VLPC hybrid.

The goal of this study is to measure how the signal pulse is affected by the high frequency background pulses occurring outside the ADC gate. Figure 2 shows schematically the time relation between the signal pulse, the background pulses and the ADC gate.

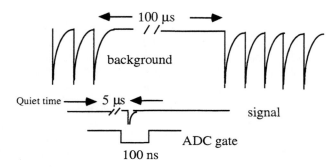

FIGURE 2. Time relation between pulses generated by background and signal LEDs, and the ADC gate.

The signal's pulse height spectrum is offset by a certain number of ADC channels because of the electronic pile-up of the background pulses. The pile-up is such that at the highest background rates, the pedestal is out of the ADCs' range. Therefore, a "quiet time" of 5 μs was set before a signal pulse (see Fig. 2), during which most of the pile-up dies off. To test the effect of the quiet time, we instrumented 16 channels with

334

linear fan-in/fan-outs equipped with a DC offset which allowed us to compensate for the pile-up level. We measured the relative quantum efficiency and gain of the VLPCs as a function of the length of quiet time, ranging from 132 ns to 10 μs. The results, taken at the background count rate of 30 MHz, are presented in Fig. 3. We do not see any significant dependence of the data on the length of the quiet time.

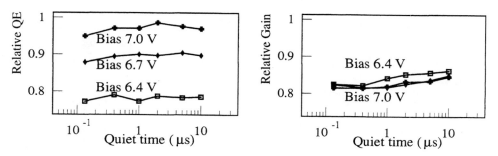

FIGURE 3. The relative quantum efficiency and gain of VLPCs as a function of the length of quiet time. The background photoelectron count rate was 30 MHz.

RESULTS

The results presented here are based on the measurements of 243 HISTE VI chips. All chips were tested at background rates of 1.3 MHz, 40 MHz, and 90 MHz, and at temperatures of 8 K, 9 K, and 10 K. A subset of 103 chips were tested at background count rates of 1.5 MHz, 10 MHz, 20 MHz, 40 MHz, and 90 MHz; the temperature was 9 K. The chips were from four different production runs, called lots. We find that chips from different lots attain full quantum efficiency at different bias voltages, therefore, where appropriate, we present the results for different lots separately.

Signal change due to the background counts

An example of a pulse height spectrum, measured while the background LED is turned off, is shown in Fig. 4a. The peaks in the spectrum correspond to 0 to 4 photons converted in the pixel (they are well separated thanks to the low gain dispersion of the VLPCs). The difference between the means of the first photoelectron peak and the pedestal is 94.9 ADC channels and is a measure of the pixel's gain. On average, 1.78 photoelectrons are generated in the pixel (we compute the number of photoelectrons by subtracting the mean of the pedestal peak from the histogram average and dividing the difference by the gain). In Fig. 4b we show the pulse height distribution from the same pixel, but with the background LED adding 90 MHz of background photoelectrons. The gain decreases to 74.1 ADC channels and on average 1.53 photoelectrons are now detected.

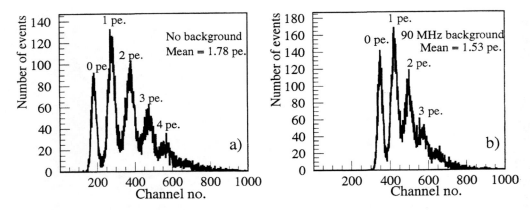

FIGURE 4. Pulse height spectra measured for a pixel operating without background (Fig. 4a) and with 90 MHz background counts (Fig. 4b).

The mean number of photoelectrons determined from the pulse height distribution, after subtraction of the dark count contribution, is proportional to the quantum efficiency of a pixel. The dark current contribution for a given pixel is determined from its pulse height spectrum measured with the signal LED turned off. We calculate the relative quantum efficiency at a given background rate in the following way: the number of photoelectrons corrected for dark current counts is divided by the mean number of photoelectrons (also corrected for dark current counts) measured without background. The background-free number is always determined at the bias voltage giving the maximum quantum efficiency for the VLPC. The gain is converted from ADC channels to charge based on calibration measurements of the QPAO2-ADC chain.

Determining the Optimal Temperature

The rate behavior of the quantum efficiency depends on the temperature of the VLPC. The higher the temperature the smaller the decrease of quantum efficiency in the presence of background counts. In Fig. 5a we show data taken at 8 K. We observe that with a background rate of 40 MHz, the quantum efficiency at a given bias is approximately equal to the efficiency without background at a bias of 0.9 V lower. The VLPC, with 40 MHz background, performs as if an internal debiasing of 0.9 V occurred in the active volume of the device. At 9 K the internal debiasing voltage is 0.6 V. (In the Series Resistor Model of the VLPC, the current generated in the avalanche region develops a voltage drop across the layers outside the avalanche region, thus dropping the effective bias in the active layers [5].) Therefore, one is forced to operate the VLPC at elevated temperatures in order to achieve the highest quantum efficiency. The data presented in this paper, unless explicitly mentioned, were taken at temperature of 9 K.

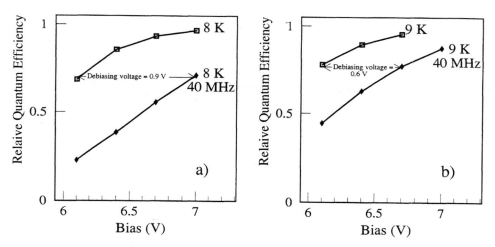

FIGURE 5. Relative quantum efficiency as a function of bias voltage at temperatures of 8 K (Fig. 5a) and 9 K (Fig. 5b). The upper line on both figures connects data taken without background counts, while the lower line connects data taken at a photoelectron background rate of 40 MHz.

Quantum Efficiency and Gain

In Fig. 6 we show the mean of the relative quantum efficiency and gain as a function of the bias voltage for 40 chips from lot 735 and 50 chips from lot 737. We notice that the chips from lot 737 require a bias 0.2 V higher than the chips from lot 735. With the background turned off, the quantum efficiency attains a maximum at a bias of 7.4 V for chips from lot 735, and at 7.6 V for chips from lot 737. At a given bias, the relative quantum efficiency decreases with increasing background level. The quantum efficiency at background rates higher than 1.5 MHz increases with the bias voltage throughout the entire bias range. For the 10 MHz background rate expected for VLPCs in the DØ detector, the quantum efficiency is 97% of the maximum efficiency if the bias is set 0.2 V higher than the optimum bias for no background operation. This value is reduced to 94% at a background rate of 20 MHz.

At 40 MHz and 90 MHz, the relative quantum efficiency decreases further. At 7.6 V bias for lot 735, and 7.8 V bias for lot 737 the efficiency drops to 84% at 40 MHz and to 70% at 90 MHz. A few chips were measured at higher bias and showed a much improved efficiency (see dashed curves in Fig. 6). We did not determine in this study what percentage of chips can operate at this highest bias because of the limited bias scan. To determine the upper limit of the bias scan we looked at the VLPC signal with an oscilloscope and increased the bias until large very wide signals appeared, signaling breakdown of the devices. We set the maximum bias to 0.2 V lower than this breakdown voltage.

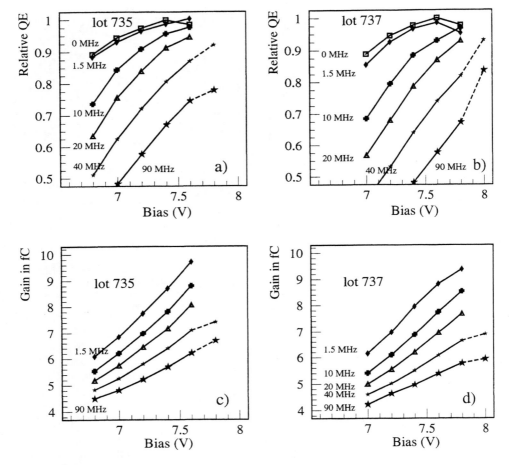

FIGURE 6. Relative quantum efficiency (Figs. 6a and 6b), and gain in fC of charge per one converted photon (Figs. 6c and 6d), are plotted as a function of bias voltage. Lines connect data taken at a constant rate of background counts; the background count rates vary from 0 MHz to 90 MHz. Data for chips from lot 735 are in Figs. 6a and 6c, for chips from lot 737 in Fig. 6b and 6d.

We observe that in presence of high background, the breakdown occurs at higher bias. In Figure 7, we show the pulse height spectrum of a pixel while the background is turned off (7a), and with 90 MHz of background counts (7b). Both spectra were taken at 8 V bias. The narrow spike, the signature of a pixel breaking down, is only seen in the spectrum with background turned-off.

At high background rates, the high quantum efficiency can be obtained only at a bias greater than at least 0.4 V above the optimum bias for no background. Such a bias can only be applied to a fraction of the chips - the majority of the chips break down. This will not pose a problem for DØ at Run II, but a VLPC more resistant to breakdown would be desirable for use in future, high-rate experiments like TeV 33.

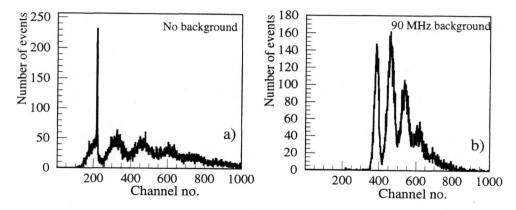

FIGURE 7. Pulse height spectrum of pixel breaking down with background turned-off (Fig. 7a). The narrow spike is the signature of breaking down. With the background of 90 MHz the pixel is not breaking down (Fig. 7b). The pixel is from lot 735 and the bias is 8.0 V.

Efficiency

The calculation of the efficiency of the scintillating fiber-VLPC detector element in the DØ application involves two quantities: the threshold that has to be set on the VLPC signal in order to cut the dark count rate to an acceptable level and the expected number of photoelectrons from a minimum ionizing particle. Simulation calculations show that a random noise level of 1% is acceptable for pattern recognition and track reconstruction. In all the calculations of efficiencies presented below, we set the threshold to keep the dark current counts below 0.35% in the experiment. This is equivalent to 0.5% in our data (the gate in the experiment will be equal to 70 ns whereas we took the data with a 100 ns gate). The expected number of photoelectrons is estimated from the cosmic ray test using the prototype of the tracker, where we observed 9 photoelectrons from minimum ionizing particles. Due to the increase in the quantum efficiency of HISTE VI compared to the quantum efficiency of HISTE IV used in that test, and due to the improved attenuation length of the recent clear light-guides, we expect 12 photoelectrons in the DØ experiment for the worst case of a track at $\eta = 0$. However, to be conservative, we continue to use 9 photoelectrons as a standard. Below we present the results for both 9 and 12 photoelectron signals.

The dark current count rate and, consequently, the threshold, was determined in runs with the signal LED turned-off. The threshold is the pulse height at which the dark count rate is equal to 0.5%. In the plots below, we express it in terms of the number of photoelectrons or in fC. The values of the threshold as a function of bias at various background rates are presented in Fig. 8. The increase of the threshold with bias is a usual feature of VLPCs. At large biases, the threshold decreases with increasing

339

background rate. This is a welcome feature, since, as we have shown in the previous section, one needs to apply higher bias at large background rates.

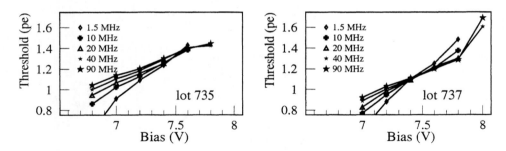

FIGURE 8. The thresholds at which the dark count rate is 50 kHz, as a function of bias voltage. Lines connect data at a constant rate of background counts.

The efficiency is the probability that the signal, assumed to have a Poisson distribution, is greater than the threshold. We present the expected inefficiency (100%-efficiency) for the signal with a mean of 9 and 12 photoelectrons. In Fig. 9 we plot with a solid line the inefficiency for a 9-photoelectron signal, of a chip from lot 737, as a function of the bias voltage for a background rate of 10 MHz (Fig. 9a) and at a background rate of 40 MHz (Fig. 9b). The dashed lines indicate the threshold values, in photoelectrons, above which the dark count rates are 0.5%. The lowest inefficiency at 10 MHz background is achieved at 7.2 V bias, and at 7.4 V bias for 40 MHz.

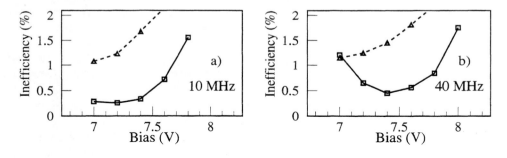

FIGURE 9. The inefficiency for a 9-photoelectron signal, of a chip from lot 737, as a function of the bias voltage, at a background rate of 10 MHz (Fig. 9a, solid line) and at the background rate of 40 MHz (Fig. 9b, solid line). The dashed lines indicate the threshold values, in photoelectrons, at which the dark count rates are 50 kHz.

The lowest inefficiency was determined for every chip, and the mean for all chips is plotted in Fig. 10a for signals of 9 and 12 photoelectrons. For the 10 MHz background rate projected in Run II for DØ detector, the expected mean inefficiency is lower than 0.5%. We also show the mean bias at which the inefficiency is lowest for a 9 photoelectron signal (Fig. 10b). These optimum bias values are then used for the measurements of relative quantum efficiency, gain and threshold as functions of background rate shown in Figs. 10c and 10d.

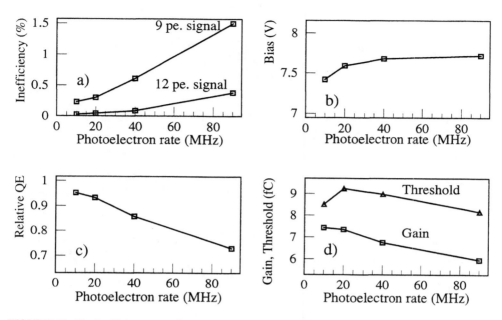

FIGURE 10. The inefficiency as a function of bias voltage for signals of 9 and 12 photoelectrons (Fig. 10a); the mean bias at which the inefficiency is lowest for 9-photoelectron signal (Fig. 10b)-called the optimum bias; the relative quantum efficiency (Fig. 10c); and gain and threshold (Fig. 10d) at the optimum bias.

Practical Considerations in Optimizing Efficiency

On the order of 75,000 channels of VLPCs will be used in the DØ fiber tracker at Run II. To optimize the overall efficiency of the detector it is useful to take into account variations in the performance of the VLPC chips.

Optimization of the maximum efficiency of the scintillating fiber-VLPC detector requires testing of chips as a function of bias voltage at the expected background rates. In Fig. 11 we show the optimum bias for a 12-photoelectron signal detected at a 40 MHz background rate for chips from four different production lots. We note that there are offsets between the lots and there is a considerable spread in the optimum bias for chips from a single lot. There is a correlation between the optimum bias and the radial position of the chip in the reactor during the epitaxial growth stage. This is illustrated in Fig. 12 for chips from lot 735.

An important feature is the spread in efficiency. We present in Fig. 13 the distribution of the inefficiency for a 9-photoelectron signal at various background rates. At background rates larger than 10 MHz, we start to see a tail of lower efficiency chips develop, but even at the highest background rates most of the chips are very efficient and would be useful for a high-rate environment such as TeV 33.

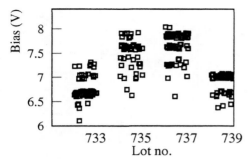

FIGURE 11. The optimum bias for chips from lots 733, 735, 737 and 739.

FIGURE 12. The correlation between optimum bias and the radial position of a chip in a reactor.

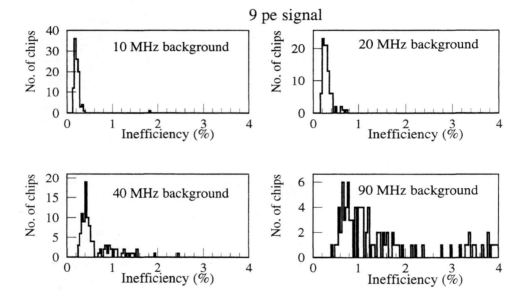

FIGURE 13. The distribution of the inefficiency for 9-photoelectron signal at various background rates.

We also find that there is a strong correlation between the inefficiency and dark current count rate. We show this in Fig. 14, where we plot the inefficiency for a 9-photoelectron signal as a function of the threshold (as in the other plots in this paper, the threshold is the pulse height at which the dark current count rate is 0.5 % in a 100 ns gate). The chips with the low threshold (low rate of dark current counts) are least efficient. The dark counts increase with bias; the inefficient chips are those that break down before the inefficiency attains a minimum. For these chips, the inefficiency as a function of bias decreases monotonically, as opposed to a chip such as shown in Fig. 9.

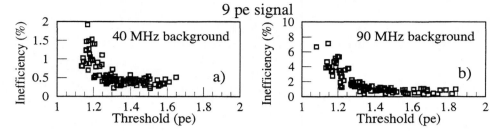

FIGURE 14. The inefficiency of 9-photoelectron signal as a function of the threshold at 40 MHz (Fig. 14a) and 90 MHz (Fig. 14b) background rates.

The susceptibility to breaking down varies from lot to lot. Among chips tested, the chips from lot 739 were more resistant to breaking down than the chips from the other lots. This results in their best performance at high background rates. This can be clearly seen in Fig. 15, where we plot the inefficiency distributions for a 12-photoelectron signal at 40 MHz background rate.

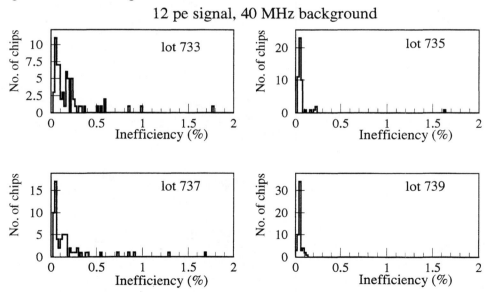

FIGURE 15. The inefficiency distributions for 12-photoelectron signal at 40 MHz background rate. The results for different lots are plotted separately.

The VLPCs that we have tested also show some slight temperature dependence. While the data presented above were taken with chips at a temperature of 9 K, we also took data at 8 K and 10 K. For each chip we found the optimum temperature at which the inefficiency is minimum. The optimum temperature was 8 K, 9K, and 10 K for 8 %, 62 % and 30 % of chips, correspondingly. Even if we choose to operate all chips at 9 K, only 5 % of the chips fail our efficiency cut (defined to be 98 % efficiency for a 12 photoelectron signal at 40 MHz background counting rate). This is a small loss in view of the simplicity of single temperature operation.

CONCLUSIONS

The performance of a sizable sample of Visible Light Photon Counters (VLPC) has been tested over a large range of background rates. Even at a background rate of 10 MHz, expected for the innermost layer of the DØ Central Fiber Tracker during peak luminosity in Run II, there will be no significant degradation in the efficiency of the tracker. To reach optimal efficiency at a 10 MHz background, the chips must be biased at about 0.2 V higher than for no background. At a background of 50 MHz, expected at TeV 33, the inefficiency would range from 0.8% down to 0.2% for signals of 9 and 12 photoelectrons, respectively. We observe variations in the performance of chips from different production lots, but these are not appreciable except at the highest background rates. While there is some variation in the optimal temperature, it is likely we will be able to operate all the VLPCs at DØ at one common temperature.

ACKNOWLEDGMENTS

We thank the support staffs at each of the participating institutions. Financial support has been provided by the U.S. Department of Energy, the U.S. National Science Foundation, Korean Ministry of Education, Korean Research Foundation and KOSEF in Korea.

REFERENCES

[1] Fermilab preprint CONF-95/177-E(1995).
[2] M. D. Petroff et al., Appl. Phys. Lett. 51 (1987) 406.
[3] B. Baumbaugh et al., IEEE Trans. on Nucl. Sci. 43 (1996) 1146.
[4] T. Zimmerman, IEEE Trans. On Nucl. Sci. 37 (1990) 439.
[5] H. Hogue, these Proceedings.

High-Rate Counting Efficiency of VLPC

Henry H. Hogue

Research and Technology Center
Boeing Electronic Systems and Missile Defense
3370 Miraloma Ave M/S HB17
Anaheim, CA 92803

Abstract. A simple model is applied to describe dependencies of Visible Light Photon Counter (VLPC) characteristics on temperature and operating voltage. Observed counting efficiency losses at high illumination, improved by operating at higher temperature, are seen to be a consequence of de-biasing within the VLPC structure. A design improvement to minimize internal de-biasing for future VLPC generations is considered.

INTRODUCTION

The Visible Light Photon Counter (VLPC) was invented[1] and developed by Boeing researchers (formerly Rockwell Science Center members) as a spin-off of silicon infrared detectors for defense satellite surveillance. Papers[2,3] in the SCIFI 93 Workshop on Scintillating Fiber Detectors (October 1993) described the operating principles of the VLPC and the characteristics and performance features that make this device almost ideal for Scintillating Fiber Tracking (SFT) applications. Continued VLPC development toward SFT requirements for the DZero Detector Upgrade at Fermilab has resulted in improved devices, called generations HISTE V and HISTE VI[4]. These devices were delivered to Fermilab for testing under conditions that simulate their operation within the DZero Upgrade SFT system. The Fermilab test results are reported elsewhere in this proceedings.[5]

One result of the system-level VLPC testing at Fermilab was a requirement to operate the VLPC at higher temperature than initially expected. Prior low-flux testing at Boeing had kept VLPC operating temperature below 7 K for low dark count rate. Fermilab researchers found that VLPC operating temperature needed to be at least 9 K to maintain high optical-pulse-detection efficiency for operation at equivalent single-photon rates >20 MHz. The dark count rate at 9 K is acceptable for the DZero SFT application.

Attempts to explain the observed VLPC performance changes with operating temperature led us to consider the significance of VLPC internal resistance and the de-biasing that it would cause under high current loads. We developed a simple model of VLPC operation that explicitly accounts for internal resistance and applied this model to the computation of rate effects. Model parameters were obtained from prior work and by fitting a new set of HISTE VI VLPC measurements for operating temperatures

CP450, SciFi97: Workshop on Scintillating Fiber Detectors
edited by A. D. Bross, R. C. Ruchti, and M. R. Wayne
1998 The American Institute of Physics 1-56396-792-8/98/$15.00

from 6 K to 11 K. This paper describes the VLPC model and the rate effects it predicts. Model results are related to the pulse-counting-efficiency measurements at Fermilab, and a VLPC design improvement to reduce rate effects for future device generations is considered.

MODEL DESCRIPTION

The basic structure of the arsenic-doped silicon (Si:As) VLPC and the characteristics of its internal electric field are shown in Fig. 1. The active VLPC structure consists of a Doped Silicon Layer, heavily donor-doped (arsenic) and lightly acceptor-doped (boron). The Doped Silicon Layer is topped with an undoped silicon layer (Blocking Layer). Donor doping is sufficient to form an impurity band, in which conduction occurs even at the deep-cryogenic VLPC operating temperature where intrinsic carriers are frozen out. When bias voltage of the indicated polarity is applied across the device, the Doped Silicon Layer functionally divides into a linear field region (Gain Region) and a constant field region (Drift Region). The electric field in the Gain Region results from the mobile charge carriers D^+ (impurity-band holes) being driven away from the positive terminal, leaving behind the immobile space charge of negatively-charged acceptor atoms. The electric field in the Drift Region, where there is no static space charge, is established by the outflow of new D^+ carriers generated by thermal ionization and impact ionization within the Gain Region. In effect, the Drift Region is an internal resistor in series with an *ideal* VLPC.

Drift Layer resistivity is dominated by thermally activated impurity conduction at the temperatures of VLPC operation. Approximately then,

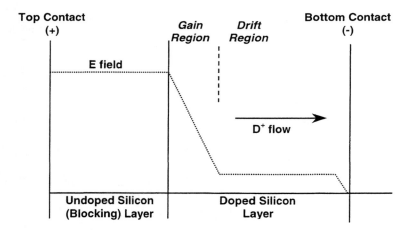

FIGURE 1. VLPC active layers and electric field profile.

346

$$\rho = \rho_0 \, e^{\,E_h/kT}, \tag{1}$$

where T is the absolute temperature, k is Boltzmann's constant, and E_h (hopping-conductivity activation energy) and ρ_0 are parameters. Parameter values

$$\rho_0 = 1000 \ \Omega\text{-cm}, \ E_h = 4 \ \text{meV} \tag{2}$$

provided a good fit to the new VLPC data and were in reasonable agreement with previous resistivity measurements and calculations for donor hole hopping between acceptor traps in impurity band conduction material.[6]

VLPC Dark Current

The VLPC dark current consists of two easily distinguished components, bias current and dark pulses. The predominant bias-current component is generated by field-assisted thermal ionization of neutral arsenic atoms in the highest-electric-field region of the VLPC Doped Layer, at its interface with the Blocking Layer. Ionization electrons have little opportunity to generate additional carriers by impact ionization of neutral arsenic atoms before moving into the undoped Blocking Layer. Consequently, the bias current is a relatively low-noise current component underlying the dark-pulse component.

Dark pulses originate from the relatively rare thermal ionization events that occur deeper in the Gain Region (or in the Drift Region). From this depth, ionization electrons are able to initiate impact-ionization avalanches as they move through the higher-electric-field part of the Gain Region. As an electron moves in the Gain Region above a critical electric field value, it obtains sufficient energy from the electric field to impact-ionize another neutral arsenic atom to generate a secondary electron. The secondary electron and the original go on to generate additional ionization events by the same process. For a typical VLPC operated at high bias, carrier doubling proceeds some relatively well-defined number of times (>17 for HISTE VI VLPC), controlled by the time required for electrons to accelerate between ionizing collisions. Local field collapse and carrier recombination may ultimately limit avalanche growth, as the density of the electron avalanche becomes sufficiently large. For a typical HISTE VI VLPC, single-photon avalanche pulses contain up to 10^5 electrons collected in less than 1 ns. Single-photon pulse distributions follow a gamma distribution (see Ref. 2), with FWHM < 40% of mean.

Photon-Initiated Pulses

Optical illumination results, directly or indirectly, in the generation of conduction band electrons deep in the VLPC, with subsequent generation of avalanche pulses. (This is, of course, the utility of the VLPC.) Each visible-light photon incident on the Blocking Layer is either reflected at the surface or absorbed and converted to an intrinsic silicon carrier (electron-hole) pair at some depth within the VLPC. For present VLPC designs, anti-reflection coating reduces reflection to <10% for green scintillation-fiber emission. For photons absorbed deep in the Gain Region or in the Drift Region, the resulting electron directly produces an avalanche.

Photons absorbed in the Blocking Layer may produce an avalanche by an indirect method. The silicon valence band hole from the initial absorption propagates into the Doped Layer of the VLPC. There it may encounter and impact-ionize a neutral arsenic atom to provide an electron for avalanche initiation. The Drift Region must be sufficiently thick for hole-initiated impact ionization to be highly probable. HISTE VI VLPC layers are optimized to provide at least 80% quantum efficiency, almost entirely by the indirect method.

DATA AND MODEL PARAMETERS

Bias current, dark count rate, photon count rate, and pulse amplitude were obtained for a HISTE VI VLPC as a function of operating temperature and applied bias voltage. Operating temperatures were varied from 6 K to 11 K in 1 K steps. The bias-voltages applied at each operating temperature were chosen to span the ~2-V operating range of the VLPC just below breakdown voltage. The model parameters for the Drift-Region resistivity (Eqn. 1 and 2) were used to correct the applied bias by subtracting the product of dark current and Drift-Region resistance. In the following discussions, we use the term *internal bias* to refer to the VLPC bias voltage corrected for the voltage drop across the Drift Region. Computation of Drift-Region resistance required use of the detector area and Drift Region thickness. The VLPC active area was a 1-mm-diameter circle, and the Drift Region thickness was estimated to be 14 μm.

Dark Current Data

Fig. 2 plots measured bias current and dark count rate versus internal bias. The curves through the data points result from fitting an exponential form Ae^{aV} to each temperature set. Both dark current components increase rapidly with temperature for a given internal bias. The dependencies on temperature and bias for the two components are different.

To aid in interpreting the dark current data, we extracted thermal activation energy vs. internal bias for each of the dark current components. The fitted curves were used to interpolate bias current and dark count rate values for internal bias voltages between 6.5 V and 8.5 V in 0.5 V steps. A functional form $Ae^{-E/kT}$, indicative of thermal activation of the primary carriers that initiate the current components, was fit to the interpolated data. E is an activation energy and A represents a bias-current or dark-count-rate coefficient. Fig. 3 plots the result of the activation energy analysis.

The different thermal behaviors of the VLPC dark current components arise from the different electric field values at the sites where these currents are initiated. Thermal ionization of dopant atoms is assisted by the electric field (Poole-Frenkel effect). The

348

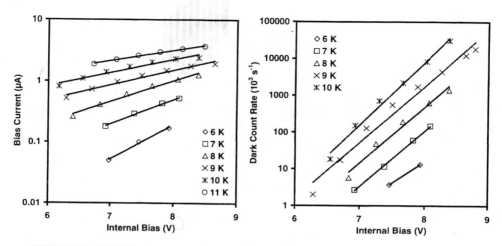

FIGURE 2. VLPC dark-current components vs. temperature and internal bias voltage.

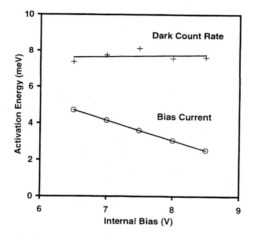

Figure 3. Thermal activation energy values extracted from VLPC dark current components.

bias current, generated at the Blocking Layer interface, has activation energy ranging from ~5 meV to ~2 meV (decreasing with bias voltage) and hugely reduced below the value (~27 meV) that would be expected if the activation site were at zero electric field.

The dark-count-rate activation energy is somewhat larger (~8 meV), although still quite reduced from the zero-field value. It is also essentially constant over the VLPC operating-voltage range. These results are consistent with an assumption that most ionization events that lead to dark counts are generated within the Gain Region, at the point where the electric field is just large enough to initiate and sustain the impact-ionization chain reaction. We refer to this field value as the *critical field*.

Optical Performance

VLPC optical performance is characterized by gain (electrons per avalanche), gain dispersion (variance of the gain), and quantum efficiency (probability that an incident photon initiates a pulse). Typical optical performance measurements for HISTE VI devices operated at low illumination, high bias voltage, and low operating temperature are ~70,000 ± 20,000 (FWHM) gain and ~80% quantum efficiency. Gain values as high as 100,000 and quantum efficiency values as high as 85% have been observed for some devices when operated near breakdown voltage.

To obtain optical-performance data for model development, we measured relative gain and quantum efficiency for the same set of applied-bias and operating-temperature values as used for the dark current measurements. A green LED was used to provide a flux of ~50,000 photons/s into the VLPC active area for measurement of pulse amplitude (proportional to gain) and pulse count rate (proportional to quantum efficiency). At the higher operating temperatures, the dark count rate exceeded the optical count rate by more than 2 orders of magnitude; therefore, many *source-on* and *source-off* counting periods had to be analyzed to obtain adequate precision on the difference (optical) signal. This procedure failed at 11 K, where dark count rate was too high for a measurable photo-generated count rate to be obtained.

We did not attempt to distinguish between dark and photo-generated pulses for the gain measurements. Since the VLPC avalanche process is the same in either case, no difference is expected. Indeed, we have not observed a difference between VLPC dark and photo-generated pulse characteristics in any other experiments.

Optical performance data (gain and quantum efficiency) vs. internal VLPC bias and operating temperature are given in Fig. 4. Once de-biasing is taken into account, the dependence of optical performance on operating temperature appears to be small. Apparently, the avalanche process is almost temperature-independent. We elected to disregard any small temperature dependence and seek phenomenological expressions (with justifiable exponential terms) to describe the composite data set. We found the following expressions for gain G and quantum efficiency η for HISTE VI devices:

$$G = G_0\, e^{aV}, \quad G_0 = 2600, \quad a = 0.4\ \text{V}^{-1} \tag{3}$$

$$\eta = \eta_0\, [1 - e^{-b(V-V^*)}], \quad \eta_0 = 0.8, \quad b = 3.2\ \text{V}^{-1}, \quad V^* = 6.3\ \text{V} \tag{4}$$

These equations are plotted as the solid lines in Fig. 4.

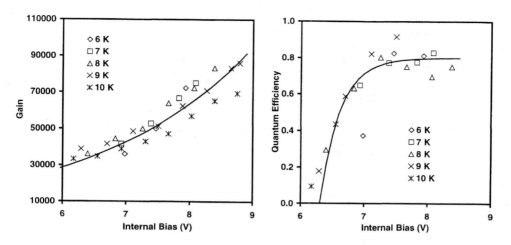

Figure 4. VLPC optical performance vs. internal bias.

RATE-EFFECTS COMPUTATIONS

Having determined the current-generation and optical-response characteristics of a model VLPC consisting of an idealized VLPC in series with an internal resistance, we applied this model to predict the effects on counting efficiency of changes in gain and quantum efficiency with illumination. Illumination R_p, expressed in photons/s, is assumed to supply additional current I_p in the VLPC Drift Region according to

$$I_p = e \, \eta \, G \, R_p, \tag{5}$$

where e is the electronic charge. R_p was varied from 10^5 photons/s to 10^8 photons/s in several steps and internal bias was adjusted to keep the external (applied) bias at a typical value of 7 V. For each adjusted internal bias value, Equations. 3 and 4 were used to calculate G and η. (This procedure neglects the effects of any avalanche pile-up within the VLPC, which would reduce quantum efficiency by a process unrelated to de-biasing and modify the current expression in Eqn. 5. We will describe avalanche pile-up and consider its possible impact on multi-photon detection efficiency in the *Results and Discussion*.)

Fig. 5 plots the calculated values of G and η vs. R_p and temperature. Each temperature-calculation set was normalized to unity at the lowest flux point, and a spline curve was used to connect the calculated values. Both optical performance parameters degrade similarly up to the maximum illumination considered here. Changes greater than 1% occur only for the two lowest temperatures, 6 K and 7 K.

351

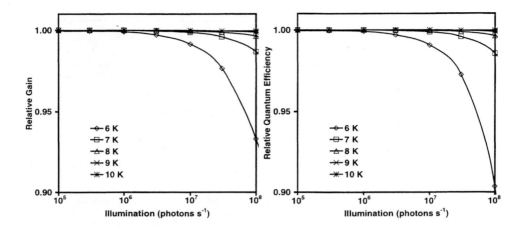

Figure 5. Calculated rate-response curves for model VLPC operated at 7-V applied bias

Counting Efficiency Considerations

In SFT applications like that of the DZero Upgrade, several photons generated from a particle crossing a scintillating fiber arrive at the associated VLPC almost simultaneously. For the photon pulses to be counted the combined photon responses must be large enough to exceed a detection threshold, set to preclude false counts from dark-count-rate pile-up in the readout electronics. In this case both the gain and the quantum efficiency of the VLPC affect the multi-photon counting efficiency. If quantum efficiency is reduced, the number of detected photons may be insufficient to provide a response above threshold. If gain is reduced, the total response from the detected photons may not be above threshold even if the maximum number of photons is detected.

In view of the importance of both gain and quantum efficiency to the multi-photon counting efficiency, we have used the product of these *quantum yield* as an indicator of VLPC optical performance for SFT applications. In Fig. 6 we have plotted the relative quantum yield variation for the model HISTE VI VLPC vs. operating temperature. Of course, VLPC quantum yield rolls off at high event rates more rapidly than does either the gain or quantum efficiency alone.

The extent to which higher operating temperature may be employed to improve the counting efficiency of VLPC-based SFT systems is limited by the dark count rate and the time resolution of the VLPC readout electronics. These factors determine the rate at which randomly occurring dark pulses will pile up in the counting electronics to form a false detection event. Since dark count rate increases exponentially with operating temperature and counting threshold must be kept large enough to exclude counting-system pile-up, a point will be reached for any VLPC-based SFT system for

352

which further increases in operating temperature degrades rather than improves counting efficiency. The Fermilab system-level measurements indicate that the optimum temperature may be reached at 9 K for HISTE VI VLPC operated with the DZero SFT electronics.

RESULTS AND DISCUSSION

Applying a VLPC model that explicitly includes internal de-biasing effects, we have estimated rate performance for HISTE VI VLPC operated under conditions simulating the DZero SFT

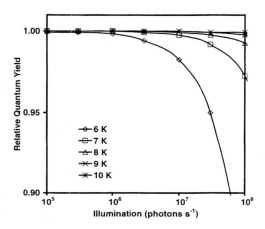

Figure 6. Calculated quantum yield performance for the VLPC model.

application. We have not attempted a detailed quantitative comparison of the magnitude of the calculated effects with the experimental data from the system-level measurements at Fermilab. To do so would require the consideration of details of the dark-count-rate pile-up threshold employed in the Fermilab testing. Nevertheless, our results are in qualitative agreement with the Fermilab measurements, and we conclude that de-biasing probably accounts for the performance variation with operating temperature seen in system-level testing.

However, we need to consider the effects of avalanche pile-up *within the VLPC*, which must become significant at sufficiently high count rates. Each avalanche results in a high density of D^+ carriers created in a very small volume of the VLPC Gain Region. This charge cluster may depress the local electric field sufficiently to prevent the occurrence of another avalanche in the same volume for the period necessary to clear the cluster out of the Gain Region. Avalanche pile-up will reduce the effective quantum efficiency at any internal bias by a term proportional to $R_p(\eta R_p + R_d)$, where R_p is the photon flux and R_d is the dark count rate. It is not apparent from inspection of the Fermilab data that avalanche pile-up within the VLPC has become significant up to the equivalent single-photon detection rate to be encountered in the DZero Upgrade SFT.

We would propose to improve future generations of VLPC by reducing the resistivity of the Drift Region. This would be expected to directly increase system count-rate capability by reducing the high-rate roll-off in VLPC performance due to de-biasing. VLPC devices with reduced Drift-Region resistivity might be operated at lower temperature, where reduced dark-count pile-up in the counting system might allow lower counting threshold and higher tolerance to VLPC gain and quantum efficiency declines with rate.

One approach to reduced Drift-Region resistivity is to increase the donor doping there. Carrier concentration would be expected to increase approximately in proportion to dopant concentration and carrier mobility would be expected to increase by a larger factor. Since both effects would reduce resistivity, an order of magnitude decrease in resistivity and a similar improvement in count-rate capability might be anticipated, provided that internal VLPC avalanche pile-up does not become limiting.

ACKNOWLEDGMENTS

The Department of Energy supported this work under Fermilab Contract B-868850. The author greatly appreciates the assistance of Dr. M. G. (Dutch) Stapelbroek and Dr. Michael D. Petroff in developing his understanding of impurity band conduction and the VLPC operational concepts upon which this effort rests. We thank the members of the DZero Fiber Tracking team at Fermilab for bringing the temperature dependence of the VLPC rate performance to the attention of the VLPC development team at Boeing and for providing helpful discussions toward understanding the nature of the effects we have attempted to describe.

REFERENCES

1. Petroff, M. D., Stapelbroek, M. G., and Bharat, R., U. S. Patent No. 4,962,304 (9 October 1990).
2. Turner, G. B., Stapelbroek, M. G., Petroff, M. D., Atkins, E. W., and Hogue, H. H., "Visible Light Photon Counters for Scintillating Fiber Applications: I. Characteristics and Performance,"in *SCIFI 93, Workshop on Scintillating Fiber Detectors*, Singapore: World Scientific Publishing Co. Pte. Ltd., 1995, pp. 613-620.
3. Stapelbroek, M. G., and Petroff, M. D., "Visible Light Photon Counters for Scintillating Fiber Applications: II. Principles of Operation," in the same work, pp. 621-629.
4. The designation used for VLPC generations HISTE is the acronym for *High-rate Scintillating-fiber Tracking Experiment*, the name of the first VLPC development program.
5. Warchol, J., "VLPC Characterization," in this proceedings.
6. Petroff, M. D., "The BIB Detector Course: Lecture 2. Physics of BIB Detectors," unpublished lecture notes.

Performance Measuremements of Histe-V VLPC Photon Detectors for E835 at FNAL

M. Ambrogiani, W. Baldini, D. Bettoni, M. Bombonati,
R. Calabrese, E. Luppi, R. Mussa and G. Stancari

Dipartimento di Fisica dell'Università, 44100 Ferrara, Italy
Istituto Nazionale di Fisica Nucleare, Sezione di Ferrara, 44100 Ferrara, Italy

Abstract. A set of 144 VLPC chips has been used by experiment E835 at FNAL for the readout of its scintillating fiber detector. During the data taking, the fibers have been exposed to rates up to 2.5×10^4 mips/sec, corresponding to 3.5×10^6 photons/sec on each VLPC pixel surface.

This paper reports about a study on rate dependence of gain and quantum efficiency throughout the run, as well as VLPC failure statistics during operation. No significant aging effects are observed. A review about the overall system performance is given.

INTRODUCTION

In the last decade, significant progress has been made to exploit the scintillating fibers in high energy physics experiments. A major contribution to this achievement comes from the photodetecting device used: the VLPCs [1] are solid state photosensitive devices, pioneered by M.Atac, produced by Rockwell International in collaboration with Fermilab, for the future upgrade of the D0 tracking system [2].

The promising results obtained from a cosmic ray test on a prototype, using Histe-IV VLPCs, done in collaboration with D0 [3], led to the approval of the construction of a scintillating fiber tracking system for the E835 collaboration. The next generation of devices, Histe-V, was expected to have reduced noise and slightly higher quantum efficiency (QE) and was conceived by the D0 experiment as an intermediate step for the development of the final version. As well as a thorough characterization of the chips, to study VLPC QE, gain and noise at different rates, the construction of a medium size detector for a real experiment was the ideal follower of the cosmic ray test. A detailed report of the performance of the detector is given elsewhere [4–6]: this paper wants to focus on the response homogeneity, short and long term stability of the photosensitive devices.

CP450, *SciFi97: Workshop on Scintillating Fiber Detectors*
edited by A. D. Bross, R. C. Ruchti, and M. R. Wayne

THE VLPC SETUP

We assembled 160 Histe-V arrays in 10 modules (referred to as *cassettes*), with 128 VLPC channels each: 9 cassettes were used, 1 was kept as a spare. The system was designed so that a large fraction of spare chips was available to compensate for channel losses during the run: out of 1152 pixels, only 860 were actually coupled to scintillating fibers.

Cassettes

CASSETTE ASSEMBLY. Each VLPC array is made of 8 round pixels, 1mm diameter, aligned in a row. The chips were indium soldered on gold plated AlN substrates. The poor quality of the metalization process will turn out to be the major source of problems for these devices. A gold wirebond brings the signal from the VLPC pixel to the substrate. To allow safe handling of the devices, the VLPC+substrate ensembles (referred to as *hybrids*) were glued on plastic (Torlon) carriers.

All hybrids were characterized [4] and sorted in order to group arrays with comparable gains and QEs within the same power supply channel. Clear fibers were then glued on the Torlon carriers, facing the VLPC pixels at a distance of approximately $100\mu m$. Direct contact was avoided in order to prevent chip damage. Ribbon cables (multi-layered copper-Kapton strips) were micro-welded on the VLPC substrates. Each cassette houses 16 such assemblies (VLPC hybrid+cable+fibers).

Controls

CRYOGENIC SYSTEM. A 12-cassette cryostat [4,7], built at Fermilab, was used to maintain the VLPCs at a temperature T=6.5K. The dewar was put at a distance of about 1 meter upstream of the interaction vertex. The cryogenic system was operated smoothly for most of the data taking period, keeping the devices at a constant temperature, within ΔT=4mK. Nonetheless, a few unexpected warm-ups occured, due to scheduled shutdowns, and to failures in the liquid helium supply system.

POWER SUPPLY. The VLPC bias voltage is provided by a 32 channel CAMAC module, custom made in Ferrara. A built-in 11-bit ADC monitors the current drawn by the chips. A maximum current of 200 μA was allowed for each channel, supplying 32 VLPC pixels. Most channels drew currents between 10 and 20 μA, i.e. 2.5 to 5 μA per VLPC chip. VLPCs were operated at 6.4 Volts until the spring 1997, when, in order to improve the signal equalization, some bias voltages were slightly adjusted. In the range between 6 and 6.5 V, I_{bias} depends exponentially on bias voltage: $\delta I/I = 2.5 \cdot \delta V$(in Volts).

ANALOG SIGNAL READOUT. The signals from the VLPCs, amplified by QPA02 preamps [8] are sent to 32 splitter-discriminator-or NIM modules: the description of the digital readout chain can be found in [6]. The analog outputs were sent to ADCs: initially, only 538 VLPC pixels (coupled to 430 inner fibers and 108 outer fibers, at wider polar angles), were read out by ADCs, mostly for calibration and diagnostic purposes. During the December 1996 shutdown, the ADC coverage was extended to almost all the fibers: the system homogeneity and stability was satisfactory enough to exploit the fibers for a dE/dx measurement. Most data were taken with a 150-nsec-long ADC gate; it was broadened to 250 nsec from March to June 1997, during the scan of the η_c resonance, taken at low luminosity. A 16% increase of the signal is measured with wider gates.

PERFORMANCE OF THE SYSTEM

The system was operated continuously for about 1 year of data taking; the total integrated luminosity amounts to 143 pb^{-1}. Experiment 835 [9] at Fermilab studies the direct formation of charmonium states from $\bar{p}p$ annihilations: most data (122 pb^{-1}) were taken with instantaneous luminosity $\mathcal{L} \simeq 2 \cdot 10^{31}$cm^{-2}s^{-1}, at $3.4 \leq \sqrt{s} \leq 3.7$ GeV, for studies on ψ' and χ states, and search of the η_c'. The rest of the data, at $\mathcal{L} \simeq 10^{31}$cm^{-2}s^{-1}, was taken in the energy range of η_c and J/ψ. The luminosity was kept constant adjusting the jet target density to compensate the decrease in the beam current.

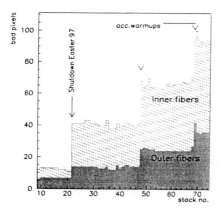 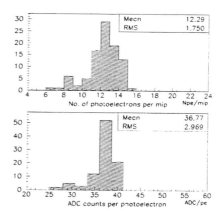

FIGURE 1. (a) Dead pixels vs stack number; (b) Detector calibration: Npe/mip (top) and ADC/pe (bottom) distributions for 1 cassette.

PIXEL LOSSES. At the end of the run, 94 out of 860 pixels no longer produced a signal, most losses occured after warmup cycles (see fig.1a), and are due to the breakage of the electrical contact between the metalized plate that supports the

VLPC and the copper-Kapton flex cable that brings the signal from the cold to the warm end of the cassette. The pixel failures did not cause significant efficiency losses, as most tracks crossed more than one fiber.

DETECTOR CALIBRATION. Various changes were done on the readout chain during the run: FERA gate delay and width, VLPC bias supply, connector cards to the FERA inputs. Samples of both hadron and electron tracks were used, to calculate the normalization constants for each channel, defined as ADC/mip ratio. The ADC/mip ratio can be decoupled in two separate contributions:

$$\frac{ADC}{mip} = \frac{ADC}{pe} \times \frac{N_{pe}}{mip}$$

The ADC/pe ratio (ADC counts per photoelectron) is a function of the VLPC and preamp gains, the signal attenuation along the cables, the gate width to the ADC. The Npe/mip ratio (no. of photoelectrons per mip) depends on the fiber light yield and transmission properties, the transparency of the optical couplings, the reflectivity of the fiber ends, and finally the VLPC QE.

The ADC/pe ratio was measured during the Easter 1997 shutdown, replacing the fiber bundles from the tracker with a pulsed LED. The ADC/pe and Npe/mip distributions are shown in fig.1b.

LONG TERM STABILITY. The integrated dose received by each fiber during the run adds up to a total of 0.8 kRad in the forward direction (0.4 at large angles). To estimate the aging effects on the whole system (i.e: on fiber light yield, on connection transparencies, as well as on VLPC intrinsic characteristics), all the contributions due to changes in the readout chain had to be deconvoluted.

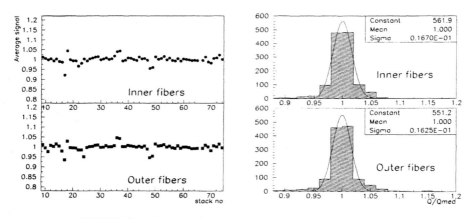

FIGURE 2. Average Signal vs time during the 1997 data taking

Fig.2a shows the time variation of the renormalized average signal, throughout the 1997 run. The average signal distribution over the whole run has a spread $\sigma_Q/Q = 1.6\%$ (fig.2b).

RATE EFFECTS. During the characterization of the devices, rate effects on the VLPCs were observed on the test stand, when all the devices were simultaneously irradiated with an LED source [10]. The final answer on all possible rate dependent effects could only come from a real running experiment. Fig.3a shows the rate of tracks per fiber (i.e: per VLPC pixel) as a function of the polar angle θ, with respect to the beam axis, in the LAB frame. The minimum bias rate per fiber is constant between 15 and 40 degrees, at both high and low CM energy.

We analyzed data from a set of runs not affected by changes in the readout chain (therefore using only 1 calibration constant), taken at both high and low $\mathcal{L}$. The average signal dependence from $\mathcal{L}$ of fibers in this angular range is shown in Fig.3b, from $\mathcal{L} = 0.5$ (upper curve) to $2.5 \times 10^{31} \mathrm{cm}^{-2}\mathrm{s}^{-1}$ (lower curve).

The excellent homogeneity along the fiber length (the average signal variation from the far to the near end - left to right in the figure - is $\approx 3.5\%$), is due to the high reflectivity of the mirrored end, as well as the high attenuation length of the scintillating fiber. Both the average signal and its slope show negligible rate dependence ($< 1\%$).

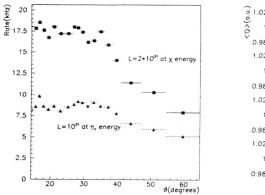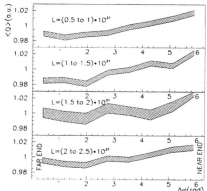

FIGURE 3. (a) Minimum bias rates (N.tracks per second per fiber) vs θ ; (b) Average signal along the fiber ($\Delta\varphi$ is the azimuthal distance from the fiber end) with increasing $\mathcal{L}$: the same normalization constant has been used for the whole data set.

CROSSTALK An estimate of the crosstalk between neighboring pixels can be given exploiting the VLPC channels that are not directly coupled to scintillating fibers. Most of these spare pixels are adjacent to channels actually coupled to fibers. Selecting events having one or both neighboring pixels hit by real tracks, we estimate an average crosstalk level about 1% (with a threshold at 80 ADC counts).

359

CONCLUSIONS

Experiment 835 at Fermilab has operated a system of 144 VLPC arrays, coupled to two layers of scintillating fibers, for approximately one year of data taking. Despite 11% of the channels were lost during cryo-cycles, the VLPCs showed impressive homogeneity and stability. Effects of radiation damage and rate dependence of the signals from the devices were investigated: the devices showed no variations at percent level.

ACKNOWLEDGEMENTS

The realization of this project would not have been possible without the invaluable help and support of the Fermilab Physics Department; the Fermilab Cryogenics Group; the D0 Tracking Group and, of course, our E835 collaborators. In particular, we want to thank Stephen Pordes (Fermilab, Physics Section), Alan Bross (Fermilab, D0 Collaboration).

A special thanks goes to the technical staff at INFN Ferrara: G.Bonora and V.Carassiti. The authors wish to thank also all the people that contributed to the assembly of the VLPC setup: L.Milano and S.Chiozzi (electronics); T.Gasteyer, P.Wheelwright (cryogenics); D.Bonsi, S.Frabetti, J.Hong (VLPC selection); S.Choi, S.Gruenendahl, J.Warchol (VLPC characterization); W.Newby, L.Ruiz, P.Ryback (cassette construction).

REFERENCES

1. M. Atac, M.D. Petroff, *IEEE Trans.on Nucl.Sci.36(1989)*,163
   M. Atac, "Scientific Applications of VLPC", in these proceedings
2. G. Gutierrez, "The D0 Scintillating Fiber Tracker", in these proceedings.
3. R. Mussa et al., *Nucl. Instr. Methods A360 (1995)*,13
   D. Bettoni et al.,*IEEE Trans. Nucl. Sci. 42(1995)*,379
4. M. Ambrogiani et al., *IEEE Trans. Nucl. Sci.* 44, 460 (1997).
5. M. Ambrogiani et al.,"The E835 scintillating fiber tracking detector", (presented by E. Luppi), *Proceedings of the 5th International Conference on Advanced Technologies and Particle Physics*, Villa Olmo (Como), Italy, October 1996.
6. M. Ambrogiani et al., "Results from the E835 scintillating fiber tracker", (presented by G. Stancari), in these proceedings.
7. T.H. Gasteyer et al., *Advances in Cryogenic Engineering 39(1994)*,619
8. T. Zimmermann et al., *IEEE Trans. Nucl. Sci NS-37 (1990)*.339
9. G. Zioulas, "First Results on Charmonium Spectroscopy from Fermilab E835", *Proceedings of the 7th International Conference on Hadron Spectroscopy*, Brookhaven National Laboratory, Upton NY, August 25-30. 1997.
10. J. Warchol et al., "VLPC characterization", in these proceedings.

Manufacturing and Testing VLPC Hybrids

L.R. Adkins, C. M. Ingram, and E. J. Anderson

Guidance, Navigation & Sensors
Boeing

Abstract. To insure that the manufacture of VLPC devices is a reliable, cost-effective technology, hybrid assembly procedures and testing methods suitable for large scale production have been developed. This technology has been developed under a contract from Fermilab as part of the D-Zero upgrade program. Each assembled hybrid consists of a VLPC chip mounted on an AlN substrate. The VLPC chip is provided with bonding pads (one connected to each pixel) which are wire bonded to gold traces on the substrate. The VLPC/AlN hybrids are mated in a vacuum sealer using solder preforms and a specially designed carbon boat. After mating, the VLPC pads are bonded to the substrate with an automatic wire bonder. Using this equipment we have achieved a thickness tolerance of $\pm$ 0.0007 inches and a production rate of 100 parts per hour. After assembly the VLPCs are tested for optical response at an operating temperature of 7K. The parts are tested in a custom designed continuous-flow dewar with a capacity 15 hybrids, and one Lake Shore DT470-SD-11 calibrated temperature sensor mounted to an AlN substrate. Our facility includes five of these dewars with an ultimate test capacity of 75 parts per day. During the course of the Dzero program we have assembled more than 4,000 VLPC hybrids and have tested more than 2,500 with a high yield.

INTRODUCTION

For VLPC devices to be an integral part of a reliable, cost-effective detector technology, assembly and test methods suitable for large-scale production must be developed. Here we report on the assembly and testing of a large number of VLPC/AlN hybrids, which have been developed for the D-Zero upgrade at Fermilab. The D-Zero upgrade will include a magnetic central tracker based on silicon microstrip and scintillating fiber tracking technologies. The Central Fiber Tracker (CFT) will have approximately 80,000 fiber optic channels, and the light from these fiber channels will be detected by VLPCs. VLPC hybrids (VLPC die mounted on AlN substrates with connector pads) will be housed in a series of 1024 channel cassettes. The error budget in the detection system places tight tolerances on the dimensions of the VLPC devices. Further, although the hybrids are designed to be replaceable, the large number required demands that each hybrid be tested and certified as operational before insertion into the cassette. Each of these issues, the production assembly with tight tolerances and testing after assembly, are treated separately below.

CP450, *SciFi97: Workshop on Scintillating Fiber Detectors*
edited by A. D. Bross, R. C. Ruchti, and M. R. Wayne
© 1998 The American Institute of Physics 1-56396-792-8/98/$15.00

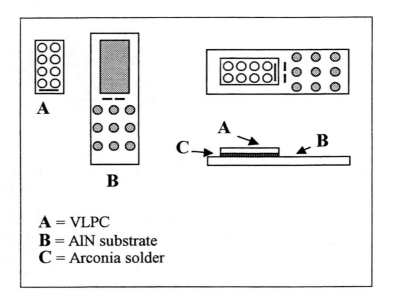

FIGURE 1. D-Zero hybrid showing the VLPC die, the AlN substrate and the solder preform

ASSEMBLY

The VLPC hybrid consists of a 0.114" x 0.197" x 0.025" VLPC die mounted on a 0.190" x 0.580" x 0.020" AlN substrate. The gold pattern on the AlN substrate has a pad with dimensions slightly larger than the die, along with a network of traces, one for each pixel. The hybrid is shown in Figure 1. The hybrid is shown above, where A is the eight pixel VLPC die, B is the AlN substrate with gold mounting pad and traces, and C is the solder preform. The VLPC has a gold electrode evaporated on its back and is attached to the substrate pad by means of the solder preform (approximately 0.002" thick). If the assembly is not done under vacuum, flux must be used to insure chip-pad adherence. In our technique, the parts are assembled in a vacuum sealer and no flux is required.

A schematic of the assembly process is shown in Figure 2. The main feature of the sealer consists of two "boats", fixtures fabricated from carbon blocks which hold the hybrids during assembly and are mounted in a vacuum chamber. The bottom boat is provided with an opening to hold a carbon substrate holder, consisting of 96 slots of the same dimension as the AlN substrates, and a carbon die holder cut to carry the VLPC die. To begin the assembly operation, the bottom boat is mounted in the vacuum chamber, and securely fastened to two electrodes. The substrate holder is then inserted into the bottom boat and filled with AlN substrates. The Die holder is placed on top of this assembly and first filled with solder preforms and then the VLPC die. Finally, the Top Boat is placed on top of the assembly. The top boat is provided with 96 holes centered on the die. These holes hold calibrated weights to provide a specific

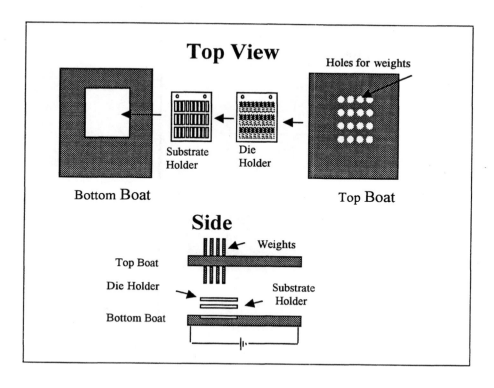

FIGURE 2. Schematic of SSI vacuum sealer

pressure to the die-preform-substrate assembly. The vacuum chamber is then pumped down and a voltage applied across the boat assembly. The carbon acts as a resistor, providing heat proportional to the voltage applied. A thermocouple monitors the temperature. Proper adhesion between die and substrate is achieved with a temperature of 210° C applied for approximately one minute. After mounting, the system is allowed to cool for thirty minutes. The parts are then removed from the holder, and the VLPC die pixels are attached to the substrate via gold leads using an automatic wire bonder. The final assembly step is pull testing the wire bonds to insure proper adhesion. The entire process for 96 hybrids takes approximately two hours. Approximately 4,000 VLPC hybrids have been assembled with the vacuum sealer approach.

The most critical dimension of the assembled hybrid is the height from the top of the VLPC die to the surface of the AlN substrate. For the D-Zero cassette, the height is required to be nominally 22 mils with a maximum height variation of less than one mil. Measurements on 85 hybrids assembled with die from six VLPC wafers gave an average height of 0.0223 inches with an extreme minimum/maximum difference of 0.00067" and a standard deviation of 0.00031. These data are shown in Figure 3. Thus, this assembly procedure is not only fast, it clearly holds the required tolerance very well.

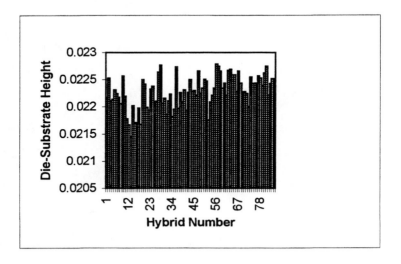

FIGURE 3. Thickness data for 85 hybrids. Height is in inches.

TESTING

After assembly, the hybrids are screen tested in a custom designed continuous-flow dewar developed for economical, high throughput VLPC device testing. A photograph of this screen test station cryostat is shown in Figure 4. Up to 16 VLPC's positioned on a hybrid carrier, can be cooled to test temperature simultaneously. The dewar is rated for testing from <5K to 350K . Under the current test configuration, one slot is dedicated to hold one calibrated Lake Shore temperature sensor and one VLPC reference detector. The remaining 15 slots hold VLPC's. The hybrids are uniformly illuminated using scintillating optical fibers. The light source is a high intensity red LED attached via BNC connector located at the LED source module.

Under test conditions the dewar is pre-cooled to 77K with liquid nitrogen (LN2) then cooled with liquid helium (LHe) to an operating temperature of 7K. Cool down rate is approximately 2 hours. PC based data acquisition and processing software written in Lab VIEW was used to operate equipment via the IEEE bus. For all functional pixels we swept applied bias voltages stepped from -5.8V to -7.6V in steps of 0.2V and recorded dark and photo-induced (source-on less source-off) count rates. The photo-induced count rate was converted to quantum efficiency (QE) by comparing it with the photo-induced count rate from the reference detector. To date, approximately 2,000 hybrids have been tested with a yield of 70% passing. In recent months, Fermilab has successfully developed and implemented a test plan to characterize VLPC's at high pulse frequencies. We are in the process of adapting our acquisition system to test in the same manner.

FIGURE 4. Screen test station cryostat

The interior of the dewar is shown in Figure 5. Here, with the outer shield cover removed, the orientation of the optical fiber inside the dewar can be clearly seen. A 1mm diameter by 610mm clear, multiclad Kuraray fiber was used under the current test setup. An optical bundle will be used in the high pulse frequency test phase. One end of the fiber is fed down through the light tube approximately 76mm and terminates flush at a diffuser lens which is mounted on the other side of the cover. The other end of the fiber is fed through an 1mm diameter opening on a flange made of G10 material. The G10 flange is mounted on the inside dewar wall of the vacuum shield. The fiber feeds through the flange and terminates against the vacuum window. The hybrid carrier with 16 slots oriented in a radial pattern is shown in Figure 6. One slot is designated for a LakeShore DT470-11 temperature sensor and VLPC reference detector and the remaining 15 slots position the VLPC hybrids. This carrier is mounted onto the dewar cold plate.

The test facility currently in operation at Boeing includes five complete screen test stations with an ultimate test capacity of 75 hybrids per day. At this stage in the D-Zero upgrade program we have tested more than 2,500 hybrids with high

FIGURE 5. Interior of dewar showing the orientation of the optical filter.

FIGURE 6. Hybrid carrier with 16 slots

yields. We are currently working with Fermilab to facilitate automatic data processing at high pulse frequencies which will further enhance the speed, reliability, and versatility of the testing program.

SUMMARY

The manufacturing effort for the D-Zero VLPC project has had two major results. (1) We have successfully applied vacuum sealer technology to the assembly of VLPC/AlN hybrids, and (2) we have designed and implemented custom screen test stations which permit the testing of parts in quantity and in a timely fashion. Using the sealer equipment we have achieved thickness tolerances of $\pm$ 0.0007 inches and a production rate of 100 parts per day, while the custom test stations now available have a daily test capacity of 75 parts. To date we have assembled more than 4,000 VLPC hybrids and have tested more than 2,500. These achievements are major steps in insuring that VLPC devices remain an integral part of a reliable, cost-effective detector technology.

SESSION 9: POSTER SESSION

Chair: M. Wayne
Scientific Secretary: K. Reynolds

New Scintillator and Waveshifter Materials

H. Zheng[1], B. Baumbaugh[1], A. Gerig[1], C. Hurlbut[2],
J. Kauffman[3], J. Marchant[1] , A. Pla-Dalmau[4], K. Reynolds[1],
R. Ruchti[1], J. Warchol[1], and M. Wayne[1],

[1]University of Notre Dame, Notre Dame, IN 46556 USA
[2]Ludlum Measurements Inc., Sweetwater, TX 79556 USA
[3]Philadelphia College of Pharmacy and Science, Philadelphia, PA 19104 USA
[4]Fermi National Accelerator Laboratory, Batavia, IL 60510 USA

Abstract. Experimental applications requiring fast timing and/or high efficiency position and energy measurements typically use scintillation materials. Scintillators utilized for triggering, tracking, and calorimetry in colliding beam detectors are vulnerable to the high radiation fields associated with such experiments. We have begun an investigation of several fluorescent dyes which might lead to fast, efficient, and radiation resistant scintillators. Preliminary results of spectral analysis and efficiency are presented.

INTRODUCTION

High luminosity colliding beam experiments in high energy physics generally push the limits of the capabilities of detector technologies. Because of the frequency of beam crossings (for example 25 ns at the Large Hadron Collider or LHC at CERN), detector response must be fast. Because of the high integrated radiation doses expected in various locations in LHC calorimetry, typically several Mrads in endcap hadron calorimetry in the CMS detector, scintillation and waveshifter materials are damaged by such exposure, with reduction in detected optical signals as a function of the dose and the time dependence of the dose.[1]

It is known that longer wavelength emission in organic scintillator can lead to improved tolerance to radiation fluence.[2,3] Such wavelengths are poorly matched to the spectral responsivity of conventional vacuum photomultiplier tubes. [3,4,5] Over the last decade, advances in photosensor technology have led to vacuum phototubes with red-extended multialkali cathodes or GaAs and GaAsP cathodes, and to solid state phototransducers such as avalanche photodiodes (APDs), visible light photon counters (VLPCs) and solid state photomultipliers (SSPMs). The advent of these devices greatly extends the effective spectral range of scintillation materials.

In this paper we present the results of initial studies of organic dyes which fluoresce in the $490 \text{ nm} \leq \lambda \leq 650 \text{ nm}$ wavelength region. These are incorporated into polystyrene either alone at low concentration (as waveshifters) or as components of multi-dye scintillators. The objective of the present study is to identify potentially useful dye materials and to characterize their spectral properties. The ultimate objective of the program is to produce fast, radiation resistant materials for a variety of experimental and technical applications. For example a replacement for the efficient and radiation resistant dye Y11/K27 is sought which has substantially faster decay time. A suitable waveshifter is sought for the yellow-green emitting 3HF.

CP450, *SciFi97: Workshop on Scintillating Fiber Detectors*
edited by A. D. Bross, R. C. Ruchti, and M. R. Wayne

Small volume polystyrene samples were polymerized in test tubes and the resulting boules were cut into several disks (typically 3-4) of 2.54 cm diameter and 1 cm thickness. The circular faces were polished smooth. A number of tests and measurements were then performed on the samples including: measurement of the optical transmission as a function of wavelength, determination of fluorescence emission and excitation spectra, measurement of fluorescence decay time, and measurement of scintillation efficiency in scintillator compositions.

OPTICAL TRANSMISSION AND FLUORESCENCE

Optical transmission through the 1 cm thickness (face-to-face) of the polystyrene-base samples was performed utilizing a Hewlett-Packard 8451A Diode Array Spectrophotometer, as well as measurement of fluorescence emission of the dyes. The fluorescence spectra were produced for a common excitation wavelength of $\lambda = 313$ nm. Figures 1-4 summarize these measurements for a variety of materials. In each case the dye concentration was 0.01% by weight in polystyrene. For reference purposes we have included samples containing the conventional dyes K27, BBQ, and DCM. The new materials under study include: M19, O240, UPER, AII, R300, DMANS, and the proton-transfer dyes BiPyD, and FFD. Of these materials, M19 and R300 are potentially interesting waveshifter materials for use with standard violet/blue emitting scintillators, and O240, UPER, AII and R300 are potential waveshifter candidates for 3HF.

FLUORESCENCE EXCITATION AND EMISSION

To assess excitation and emission maxima for the various dyes, the samples were placed in a Hitachi F-2000 Spectrophotometer for analysis. Two possible sample orientations could be studied: front-face orientation (Fig. 5) and back-face orientation (Fig. 6). In front-face orientation, surface fluorescence is observed. In back-face orientation, fluorescence in the presence of bulk optical absorption is observed. The back-face orientation is most closely related to the fluorescence measurements (displayed in Figs.1-4) and is also most closely related to actual detector applications. Hence only these measurements are described here. Table 1 presents a summary of the back-face excitation measurements for each of the samples, and peaks in the excitation spectrum which lead to emission at the indicated fluorescence maximum are tabulated.

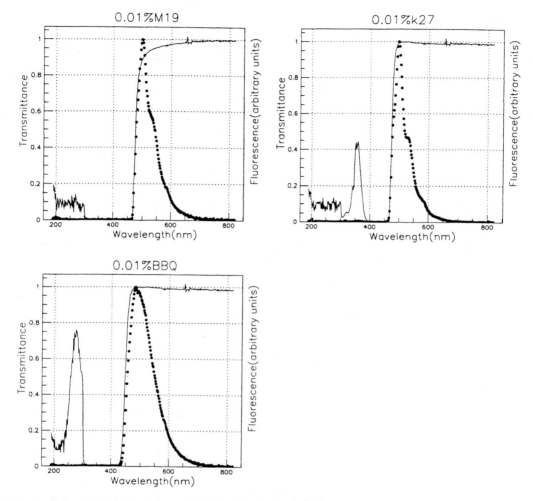

FIGURE 1. Transmission and fluorescence emission of M19, K27, and BBQ dyes (0.01% dye concentration in polystyrene). Solid line is transmittance; dotted curve is fluorescence.

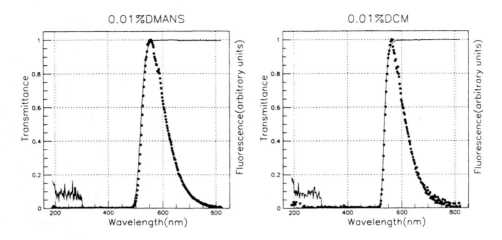

FIGURE 2. Transmission and fluorescence emission of DMANS and DCM dyes (0.01% dye concentration in polystyrene). Solid line is transmittance; dotted curve is fluorescence.

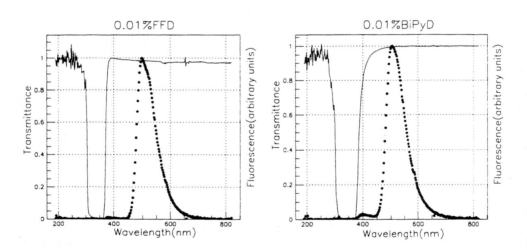

FIGURE 3. Transmission and fluorescence emission of FFD and BiPyD dyes (0.01% dye concentration in polystyrene). Solid line is transmittance; dotted curve is fluorescence. Note the shift between the emission peak and absorption edge.

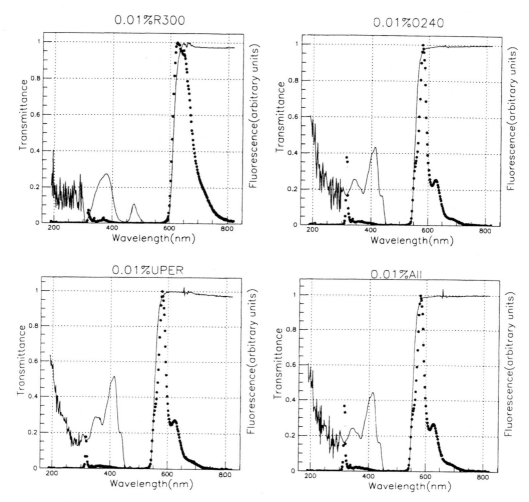

FIGURE 4. Transmission and fluorescence emission of R300, O240, UPER, and AII dyes (0.01% dye concentration in polystyrene). Solid line is transmittance; dotted curve is fluorescence.

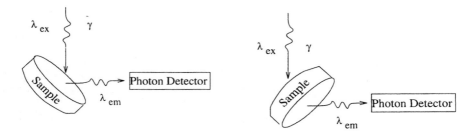

FIGURE 5. Sample orientation for front-face measurement of emission and absorption spectra.

FIGURE 6. Sample orientation for back-face measurement of emission and absorption spectra.

TABLE 1. Excitation Maxima Associated with the Emission Spectra
(All dyes are 0.01% concentration in polystyrene)

Sample	Emission Wavelength (nm)	Associated Excitation Maxima (nm) (s) denotes shoulder in emission spectrum
K27	493	338, 399, 466
BBQ	473	404, 420
M19	491	419, 423(s), 445, 460(s)
R300	618	454, 569
UPER	574	377, 470(s), 489, 525
O240	573	377, 469(s), 487, 523
AII	575	378, 470(s), 487, 530
DMANS	548	379, 470
DCM	550	425, 470
BiPyD	494	365
FFD	492	315, 356

FLUORESCENCE DECAY TIME

The samples were placed within the optical path of a Nd(YAG) laser, with excitation wavelengths of 355 nm or 532 nm depending upon the equipment setting. Fluorescence decay of all materials were measured at one or both of these wavelengths, using an optical spectrograph equipped with PMT readout. It should be noted that these wavelength values do not necessarily coincide with excitation maxima of the samples. Table 2 summarizes the decay time (τ) measurements from fits, assuming that the fluorescence decay is consistent with a single exponential. In all cases, the observed τ values were less than 10 nsec. Note that these decay time measurements are photon-induced and related to waveshifting and should not be confused with scintillation decay times. In general, scintillation decay times are expected to be a few nanoseconds slower than photo-induced decay times. In particular the dye M19, which is a potential template material to replace Y11/K27, has a value $\tau \leq 2.9$ nsec in this study. This decay time measurement is actually limited by the pulse shape of the Nd(YAG) laser itself. Additional studies with a faster excitation system will be required to determine the "true" decay time for this material.

TABLE 2. Fluorescence Decay Time Measurements

Components	Excitation Wavelength (nm)	Detection Wavelength (nm)	Lifetime (assumed exponential) (ns)
0.01%K27	355	492	7.3
0.01%M19	355	492	< 2.9
0.01%BBQ	355	480	9.0
0.01%R300	355	617	8.2
	532	617	8.0
0.01%BiPyD	355	530	4.4
0.01%FFD	355	500	5.3
0.01%O240	532	574	6.1
0.01%AII	532	574	5.8
0.01%UPER	532	574	6.2
0.01%DMANS	532	590	4.0
0.01%DCM	532	580	< 3.1

FLUORESCENCE EFFICIENCY IN SCINTILLATOR COMPOSITIONS

The above dyes were combined into polystyrene with potentially useful primary (and often secondary) dyes, to assess fluorescence efficiency of such materials within a "scintillation cocktail". Measurements were performed using a technique indicated in Fig. 7. Scintillation samples were placed in optical contact with the entrance window of a Hamamatsu R943 photomultiplier tube. This tube has a GaAs photocathode which has a flat quantum efficiency over the visible spectral range of interest. Samples were excited with a Strontium 90 beta source. Scintillation data were digitized using a LeCroy QVT analog-to-digital converter. The scintillation spectra were compared by endpoint energy measurement, since the beta decay gives a continuous electron energy distribution. For comparative purposes, the endpoint energy measurements are normalized to a conventional blue scintillator (polystyrene + PTP + TPBD). Table 3 displays the data for a variety of samples.

TABLE 3. Scintillation Efficiency Measurements. The relative efficiencies of the samples are tabulated in the right-hand column , normalized to sample 6.

Index	Components	Endpoint (0.25pc/channel)	Endpoint(i)/Endpoint(6)
1	1%Mopom	78±1	1.11±0.01
2	1%OLIGO408	89.0±0.8	1.27±0.01
3	1%DAT	49.9±0.4	0.713±0.006
4	0.01%DMANS	21.8±0.5	0.311±0.007
5	0.01%TPBD	27.4±0.6	0.391±0.008
6	1%PT,0.02%TPBD	70±1	1.00±0.01
7	1%PT,0.02%TPBD,0.01%DMANS	60±3	0.86±0.04
8	0.01%DCM	12±15	0.2±0.2
9	1%PT,0.02%TPBD,0.01%DCM	31.0±0.6	0.443±0.008
10	0.01%BiPyD	17±3	0.24±0.04
11	1%PT,0.15%BiPyD	48.6±0.8	0.69±0.01
12	1%PT,0.15%BiPyD,0.01%O240	41±2	0.58±0.03
13	1%PT,0.15%3HF,0.01%R300	54±3	0.77±0.04
14	1%PT,0.02%R300	60±3	0.86±0.04
15	1%PT,0.15%R300	63.4±0.9	0.90±0.01
16	1%BPBD,0.02%R300	73±4	1.04±0.06
17	1%BPBD,0.15%R300	67±1	0.96±0.01
18	1%PPO,0.02%R300	64±4	0.91±0.06
19	1%PPO,0.15%R300	59±4	0.84±0.06
20	1%PT,0.02%BBQ,0.01%R300	78±1	1.11±0.01
21	1%PT,0.02%K27,0.01%R300	72.4±0.9	1.03±0.01
22	1%PPO,0.02%K27,0.01%R300	79.6±0.5	1.137±0.007
23	1%OLIGO.41,0.02%R300	82±1	1.17±0.01
24	0.01%UPER	27.3±0.5	0.390±0.007
25	0.15%UPER	26.6±0.7	0.38±0.01
26	0.75%3HF,0.01%UPER	55.2±0.4	0.788±0.006
27	0.75%3HF,0.15%UPER	42.4±0.6	0.606±0.008
28	1%PT,0.15%3HF,0.08%UPER	53.6±0.9	0.76±0.01
29	0.75%BiPyD,0.02%R300	47.1±0.9	0.67±0.01
30	1%PT,0.15%BiPyD,0.01%R300	51.4±0.3	0.734±0.004
31	1%PT,0.03%BiPyD,0.01%UPER	48.2±0.6	0.688±0.008
32	0.01%FFD	32.9±0.4	0.470±0.006
33	1%PT,0.01%FFD	70.8±0.4	1.011±0.006
34	1%PT,0.15%FFD	63.1±0.6	0.901±0.008

The relative efficiency of the materials containing FFD and R300 are found to be high. In the case of R300 this is particularly true when combined with short wavelength primaries such as PT and PBPD. Since the wavelength variations between primaries and secondaries are substantial, there was concern that detected fluorescence might be dominated by unshifted emission from the primaries themselves. Followup studies on R300 were performed using an optical filter to minimize the possibility of primary light reaching the photomultiplier tube.

In Table 4 are shown a number of scintillation compositions in which R300 is the "final" dye in the fluorescence chain. Here a 550 nm filter was placed between the sample and the entrance window of the R943 PMT, to pass long wavelength light only and hence to screen out a possible fluorescence contribution from primary and/or secondary dyes in the scintillation samples.

Notable are the measured efficienies for samples 5-7 which confirm that R300 is an effective waveshifter for PT and BPBD.

FIGURE 7. Apparatus for Scintillation Efficiency Studies

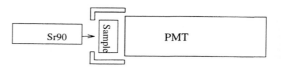

TABLE 4. Scintillation Efficiency Measurements for Various Scintillation Cocktails Incorporating R300 Dye. The relative efficiencies of the samples are tabulated in the right-hand column, normalized to sample 7. A 550 nm filter was placed between the sample and the PMT.

Index	Components	Endpoint (0.25pc/channel)	Endpoint(i)/Endpoint(7)
1	0.75%BiPyD,0.02%R300	36.6±0.4	0.600±0.006
2	1%PT,0.15%BiPyD,0.01%R300	42±2	0.69±0.03
3	1%PT,0.15%3HF,0.01%R300	46.7±0.9	0.76±0.01
4	1%PT,0.02%R300	50.8±0.3	0.833±0.005
5	1%PT,0.15%R300	59.3±0.7	0.97±0.01
6	1%BPBD,0.02%R300	60.1±0.9	0.98±0.01
7	1%BPBD,0.15%R300	61.0±0.5	1.000±0.008
8	1%PPO,0.02%R300	51±1	0.84±0.02
9	1%PPO,0.15%R300	53±1	0.87±0.02
10	1%PT,0.02%BBQ,0.01%R300	53±1	0.87±0.02
11	1%PT,0.02%K27,0.01%R300	49±1	0.80±0.02
12	1%PPO,0.027%K27,0.01%R300	52±1	0.85±0.02
13	1%OLIGO.41,0.02%R300	55±1	0.90±0.02

SUMMARY AND CONCLUSIONS

A number of potentially useful scintillator and waveshifter dyes have been studied, which could lead to new scintillation compositions of use in particle physics and in other technical applications. Of these, the most promising include: the green-emitting M19, a material with wavelength characteristics similar to Y11/K27, but with a much faster fluorescence decay; and the red-emitting R300, which appears to be a useful shifter for violet/blue scintillator as well as green-emitting scintillator.

Irradiation tests of samples containing M19 and R300 are planned, to observe the character of these materials as a function of radiation dose. Comparisons will be made with Y11/K27 which is known to be robust. M19, R300 and their derivatives warrant further study.

Finally, of the proton-transfer dyes studied, only FFD or its derivatives may offer improvement over conventional 3HF or 4-CNHBT.

ACKNOWLEDGMENTS

We thank the staff of the Notre Dame Radiation Laboratory for their assistance with the laser timing measurements. Work was supported by the U. S. Department of Energy and the U. S. National Science Foundation.

REFERENCES

1. CMS Hadron Calorimeter Technical Design Report CERN/LHCC 97-31 (1997).

2. C.L.Kim and A.D.Bross, submitted for publication in Nuclear Instruments and Methods.

3. S. Margulies, M. Chung, A. Bross, C. Kim, and A. Pla-Dalmau, SCIFI93, Workshop on Scintillating Fiber Detectors, A.D.Bross, R.C.Ruchti and M.R.Wayne, eds., World Scientific (1995), pp. 421-430.

4. For reviews of photosensors, see: H. Leutz, "Scintillating Fibres", Nuclear Instruments and Methods A 364 (1995) pp. 442-448; and SCIFI93, Workshop on Scintillating Fiber Detectors, A. D. Bross, R. C. Ruchti, and M. R. Wayne, eds, World Scientific (1995), pp. 553-650.

5. R. Ruchti, "The Use of Scintillating Fibers for Charged-Particle Tracking", Annual Review of Nuclear and Particle Science, Vol. 46, Quigg, Luth and Paul, eds. (1996), pp. 281-319.

Scintillating Fibre Hodoscopes for COSY-TOF

W. Eyrich, M. Fritsch, J. Hauffe, A. Metzger, F. Stinzing, M. Wagner
and S. Wirth

Physikalisches Institut, Universität Erlangen-Nürnberg, D-91058 Erlangen, Germany

Abstract. At the cooler synchrotron COSY (Forschungszentrum Jülich, Germany) detector arrays of scintillating fibres are used in different experiments. In this contribution we mainly report on the experience with scintillating fibre hodoscopes in the COSY Time-of-Flight (TOF) spectrometer. One of the physical topics in this experiment is the study of the associated strangeness production in the reactions $pp \rightarrow K^- \Lambda p$, $K\Sigma^- n$, $K^- \Sigma^0 p$ and $K^0 \Sigma^- p$. The apparatus consists of an inner start detector system close to the target and an outer stop detector system both mounted inside a vacuum vessel. For the inner detector system apart from microstrip components arrays of scintillating fibres are used to measure particle tracks and the decay vertices of the hyperons, thus allowing to identify and reconstruct the events of interest. Furthermore, scintillating fibre hodoscopes are installed for the online measurement of profiles and intensity of the proton beam at COSY-TOF. In both mentioned detector systems squared fibres are in operation together with multianode photomultipliers.

EXPERIMENTAL SETUP

The experimental setup for the measurement of the associated strangeness production at COSY-TOF consists of a very small liquid hydrogen target (1) with a length of 4 mm and a diameter of 6 mm, the so-called "Erlangen start detector" (2) and, in a distance of 1 m, the "Jülich quirl" (3) as stop detector. The start detector system, which in its basic version is optimized for the reaction $pp \rightarrow K^+ \Lambda p$, covers the whole phase space of the primary reaction products. As shown schematically in Fig. 1a together with the first "golden event", it consists of the "starttorte", a ring microstrip detector and a scintillating fibre hodoscope.

The "starttorte" is built up by two segmented layers of scintillator elements with a thickness of 1mm each, an outer diameter of 150 mm and a beam hole of 2 mm diameter. The two layers are twisted against each other by 15°. The silicon ring microstrip detector with a thickness of 500 µm and a beam hole with 3 mm diameter has 100 concentric rings with a pitch of 280 µm. The fibre hodoscope with a dimension of about 20x20 cm² is made up of 2x2 mm² squared fibres (Bicron BCF12, single cladding) forming two crossed planes of 96 fibres each, readout by

CP450, *SciFi97: Workshop on Scintillating Fiber Detectors*
edited by A. D. Bross, R. C. Ruchti, and M. R. Wayne
© 1998 The American Institute of Physics 1-56396-792-8/98/$15.00

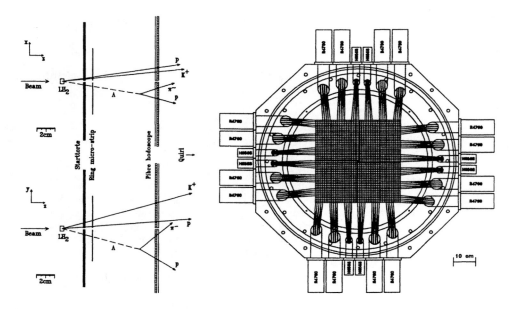

FIGURE 1a. Scheme of the start detector together with first "golden event"

FIGURE 1b. Layout of the intermediate fibre hodoscope

multichannel photomultipliers (MCPM's) of the type Hamamatsu R4760 with 16 channels each. For future measurements a second, intermediate fibre hodoscope will complete the start detector system giving additional tracking information for primary and secondary reaction products. In particular, it allows to reconstruct the tracks of K^+-particles decaying before the stop detector and extends the covered decay volume of the hyperons. The layout of the intermediate fibre hodoscope, as shown in Fig. 1b, is similar to the first hodoscope now covering an area of 38×38 cm^2 with a beam hole of 4×4 mm^2. The two crossed layers have a distance to the target of 20 cm and are made up of 192 squared fibres each (2×2 mm^2, BCF12 single cladding). The readout is permformed by MCPM's R4760 and, due to space limitations in the central region, by the new type H6568.

An additional scintillating fibre hodoscope is installed behind the TOF vessel to perform online diagnosis of the extracted COSY beam.. It consists of two crossed layers of 32 squared fibres (2×2 mm^2) each. It allows to measure the beam profile as well as the beam intensity, which reaches up to 10^7 protons/s.

Several fibre geometries have been considered to build the hodoscopes. Light output and attenuation length were measured by irradiating different samples of BCF12 fibres (single cladding) with minimum ionizing particles. As can be seen in Fig. 2, the maximum light output is achieved with the used 2×2 mm^2 squared fibres and a bundle of four 1×1 mm^2 squared fibres. The attenuation length with an effective value of

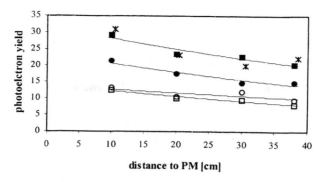

FIGURE 2. Light output and attenuation length for different fibre geometries (Bicron BCF12); squares for squared fibres, circles for circular fibres, filled symbols for 2 mm diameter, open symbols for 1 mm diameter; crosses for a.bundle of four 1x1 mm² squared fibres

about 80 cm is almost independent from the geometry. Due to the maximum fibre length of about 70 cm in the hodoscopes, they are directly coupled to the PM's instead of using clear fibres as lightguides. To optimize the light output the fibre ends have to be prepared carefully. Several steps of fine grain polish have been applied down to a grain size of 1 µm. In addition the fibres are optically coupled to the PM's using silicon grease or pads which significantly improves the light output (see Fig. 3).

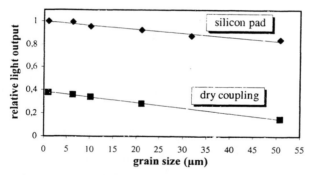

FIGURE 3. Improvement of light output by fine grain polish of the fibre end and optical coupling to the PM

DATA ANALYSIS

Because of the high granularity of all detector components, the interesting events can already be fully reconstructed with sufficient precision using only the geometrical information from the hit pattern in the various detector components. This includes especially the identification of the delayed decay of the Λ-particle, which gives a unique signature of the reaction $pp \rightarrow K^+(\Lambda \rightarrow \pi^- p) \, p$ and which is also used as the main trigger signal via the increase of the charged particle multiplicity from 2 to 4. Track parameters are obtained by interpreting the ADC-data of all detector elements. In case of the fibre hodoscopes, where it is most probable that one particle hits two neighbouring elements, a cluster analysis has to be included (see Fig. 4. for a sample ADC-spectrum).

Alltogether, the fibre hodoscopes in the TOF experiment are an important part of the detector system especially used for the reconstruction of both the reaction vertices inside the target and the hyperon decay vertices. For the primary particles traversing all detector planes a vertex resolution of about 3 mm is obtained. The tracks of the decay particles being observed only by the hodoscopes and the stop detector lead to a secondary vertex resolution of about 5 mm.

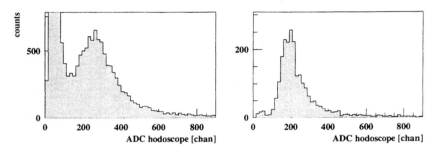

FIGURE 4. Typical raw ADC-data for a hodoscope channel (left) and ADC-spectrum for an elastic subsample after cluster analysis (right)

Further development of fibre hodoscopes is done for the upcoming COMPASS experiment at CERN to especially optimize time resolution and rate capability.

ACKNOWLEDGEMENTS

We gratefully acknowledge support from the German BMBF and the FZ Jülich.

REFERENCES

1. Jaeckle, V. et al., Nucl. Instr. and Meth. **A349**, 15 (1994)
2. Eyrich, W. et al., Physica Scripta **48**, 88 (1993);
   Metzger, A. and Wirth, S., PhD theses to be published
3. Dahmen, M. et al., Nucl. Instr. and Meth. **A348**, 97 (1994)

An effective method to read out large scintillator areas with precise timing

J. Bähr, H.-J. Grabosch, V. Kantserov[1], H.Leich,
R.Leiste, R.Nahnhauer

DESY-Zeuthen, 15738 Zeuthen, Germany

Abstract. Using scintillator tile technology several square meters of plastic scintillator are read out by only two photomultipliers with a time precision of about 1.5 nsec. Two examples are discussed to build a detector based on this technology to search for cosmic muons and neutrinos.

INTRODUCTION

The readout of plastic scintillators is done classically using adiabatic clear light guides to focus the produced photons to the photocathode of a photomultiplier (PM). Later wavelength shifting rods were applied to collect the light of larger scintillator areas. In recent years wavelength shifting fibers have been used to read out scintillator tiles of preshowers or calorimeter moduls [1,2,3].

Precise timing normally demands a large number of photons to be produced in the scintillator. Using two photomultipliers at both ends of not too large scintillator rods and a meantimer for the output signals a time precision of a few hundred psec is in reach. For large scintillator areas this method is rather expensive because many photomultipliers are needed. Using wavelength shifting fibers for the scintillator read out has the clear disadvantage that only a few percent of the produced light will be trapped to the PM photocathode. Nevertheless we will propose in the following a method to read out large detector areas with a few photomultipliers retaining a time precision of about 1 nsec.

THE MEASURING PRINCIPLE

The basic piece of a large scale detector is a small scintillator tile of about 25 x 25 x 2 cm^3 as schematically drawn in fig. 1. Two groups of wavelength shifting fibers are glued into grooves at it's surface. These fibers are connected to clear

---

[1] On leave from Moscow Physics Engineering Institute

CP450, *SciFi97: Workshop on Scintillating Fiber Detectors*
edited by A. D. Bross, R. C. Ruchti, and M. R. Wayne
© 1998 The American Institute of Physics 1-56396-792-8/98/$15.00

optical fibers of a length of several meters guiding the light produced by crossing particles to two photomultipliers. A particle hit is registered if both PM's give a coincident signal above a certain threshold and within a short time window. This demand reduces already the PM noise by orders of magnitude.

Several scintillator tiles are combined by summing up the clear read out fibers of each of the two groups (see fig. 2). A natural restriction is given only by the size of the photocatode of the used PM's in relation to the number and size of the fibers per tile. For a 2" PM with homogeneous response of the whole photocathode about 100 tiles can be combined to more than 6 m^2 detector area if four fibers of 1.5 mm diameter per tile and group are used.

RESULTS FOR SINGLE TILES

For the construction of tiles we used different types of scintillator of 1 - 2 cm thickness. The fibers tested came mainly from BICRON[2]. Finally we used the fast wavelength shifting fiber BCF 92 with a decay constant of 2.7 nsec together with 2 m long clear fibers BCF 98 both with 1.5 mm diameter.

Several fiber arrangements were tested using between 2 and 8 fibers per group. An optimum was found for four fibers per group equally distributed over the tile surface. The fibers were coupled to two Philips[3] photomultipliers XP 2020 which were read out by charge sensitive ADC's. Two additional scintillation counters on top and bottom of the tile studied are used to trigger for throughgoing cosmic particles. The efficiencies derived from measured ADC-spectra are given in the table for a 1 cm thick tile made from BICRON 408 together with those for a 2 cm thick scintillator produced at IHEP Serpukhov. In both cases values near to 99 % are found for the tile hit efficiency. The light output increases by about a factor 1.5 if double clad fibers are used.

TABLE 1. Number of photoelectrons N$_{pe}$ and efficiencies ε for two scintillator tiles read out by two groups of single clad fibers BCF 92 coupled to 2 m long clear optical fibers BCF 98

scintill.	d, cm	$N_{pe}^{(1)}$	$N_{pe}^{(2)}$	ε_1	ε_2
BC 408	1	4.7	4.7	0.991	0.991
Serpukhov	2	4.6	4.4	0.990	0.998

Keeping in mind, that the two time measurements are independent of each other we derive from their difference a time resolution of $\sim$ 1.5 nsec for the single channel signal. The timing behaviour of the scintillator signals was studied using TDC's

---

[2] BICRON, 12345 Kinsman Road, Newbury, Ohio, USA
[3] Philips Photonique, Av. Roger Roacier, B. P. 520, 19106 Brive, France

386

which provide a time resolution of 100 psec. The TDC's are read out in common stop mode. The stop signal is derived from a coincidence of the signals of the two fiber groups of a tile using very low thresholds of about 5 mV.

FIRST APPLICATIONS

A t_0–detector for the L3COSMIC-Project

Two years ago the idea was born to use this technology for a 50 m^2 scintillation detector to be installed on top of the L3 experiment at CERN[4]. This will allow to use the muon drift chambers of this experiment to measure cosmic particle momenta because a precise t_0-signal will be available. As discussed already since long [4] a rich spectrum of cosmic particle physics could be investigated in this L3COSMIC experiment. The time resolution demanded for a ± 1 % measurement of the cosmic muon momentum spectrum between 20 GeV and 1 TeV is about 1 – 2 nsec just in reach for the proposed device.

The detector will consist of 8 modules of 6 m^2 each. The first module has been constructed at DESY-Zeuthen and tested in detail with cosmic particles. The tiles have a size of 25 x 25 x 2 cm^3, therefore 96 of them are needed for a full size module. Half of the tiles were produced from Serpukhov scintillator, the other half from already used one delivered by the University of Michigan[5]. For the read out 1.5 mm double clad fibers BCF 92 and BCF 98 were taken.

The results of our measurements are given in figs. 3 and 4 for the hit efficiency and the time resolution respectively. Average values of

$$\varepsilon_{12} = 98.3 \pm 0.7 \ \% \ \text{and} \ \sigma_t = 1.4 \pm 0.1 \ \text{nsec}$$

demonstrate that we have reached the design goal.

Each 16 tiles are dense packed inside an aluminium box to cover an area of 1 m^2. Six of the boxes are assembled together to form a full size module (see fig. 5). The geometrical arrangement is done in a way that a 6 m^2 area is filled without gaps allowing to put the two times 6 fiber bundles to the two XP 2020 photomultipliers for the read out. Practically this coupling is done using a special mask arrangement Because the mask covers the whole surface of the PM's photocathode one has to make sure equal sensitivity of the PM independent of the fiber position. The high voltage divider used allows a correponding adjustment. The result is presented for one PM in fig. 6. The full size module has been tested first in Zeuthen using small (10 x 10 cm^2) scintillator paddles to allow position dependent cosmic particle triggering. We found an average hit efficiency of 99.1 $\pm$ 0.4 % and an average time resolution of 1.4 $\pm$ 0.1 nsec.

Two testruns have been performed in autumn 96 and spring 97 at CERN. A 3 m^2 and a 6 m^2 detector respectively were installed on top of the L3–magnet. Data

---

[4] see http://hpl3sn02.cern.ch/l3_cosmics/

[5] We thank L.B. Jones from the University of Michigan for providing the corresponding material.

were taken in coincidence with the barrel scintillator system of L3. Two TDC's of this system and the L3-DAQ were used to measure the arrival time of crossing particles as seen by the PM's of the t_0-detector module. With this arrangement we got a time resolution of $\sigma_t = 1.5 \pm 0.1$ nsec which confirms the laboratory results.

A Scintillator based cosmic Particle detection Yard (SPY)

The read out scheme described in section 2 has first been proposed for a cosmic neutrino detector project at earth surface. The aim was twofold. First, we wanted to detect high energetic air showers from above as well as "astrophysical" neutrinos from below the ground. Second, easy construction and installation was demanded allowing 100 % access for maintainance and exchange of components avoiding the extreme boundary conditions of present cosmic neutrino detection techniques [5,6].

The idea was to use modern scintillator technology to build a fast detector, well structured to keep the tremendous background from normal cosmic rays small. Basic elements of such a detector could be scintillator sheets made out of tiles forming a three floor tower (see fig. 7). Defining a hit as a coincidence of the two photomultipliers of one sheet as described above, a track is identified by a delayed coincidence of hits of all three planes. A sheet size of 4 x 4 m^2 would have to cope with a cosmic ray rate of about 4 kHz. A distance of 25 m between every two sheets seems to allow to do that without demanding an extreme time precision from the scintillator signal. The background rate per tower would be reduced to 6.4 Hz.

A large area detector could be build grouping many towers dense to each other. With 25 x 25 towers of the described size a total area of 10 000 m^2 is reached. The detector could measure naturally also inclined tracks. In this case the expected time delay would depend on the particle direction. Predicting hit positions in the second plane from those in the first and the last one would allow to keep the background still limited to the extension of one sheet size.

Monte Carlo calculations supported the above concept. However one has to handle a background rate of about 10^{11} per year which is difficult to simulate. Therefore we decided to build a small scale prototype detector (μ–SPY) at DESY-Zeuthen. Three planes of 1 m^2 size divided in four subsections were installed in a distance of about 10m between every two of them. The fibers of each plane were read out with eight channels of a sixteen channel Philips R4760 photomultiplier. The time resolution per plane was measured comparing hit arrival times of multihit triggers. It was found to be about 3 – 4 nsec and depends mainly on the quality of the used PM. For the data acquisition we used a OS9 based VME–system. The hit data were collected in a tandem buffer. If full, it was transfered to a SUN workstation where track reconstruction was done online. A data reduction factor of 10^4 was reached.

In 176 days we observed 4.8 x 10^9 hits. After the online filter we found 281.553 events with 324.034 tracks.The time differences between planes one–two and two–three are shown in correlation to each other in fig. 9. A clear enhancement is observed for normal cosmic rays crossing the detector from above. A very small number of hits is found in the opposite direction, as can be seen from fig. 8. Assuming the worst case, a flat background in the region of hits from neutrino interactions from below, we find a density of $\rho_{bg} = 0.009$ events/1 nsec^2 for 1 m^2 and 1 year of running. Extrapolating this number to a 10 000 m^2 detector we would get a signal to noise ratio of S/N = 0.5 for atmospheric neutrinos above 2 GeV. We have to show however, that the hit prediction for the second plane works for a detector of this size.

SUMMARY

It has been demonstrated, that small size scintillator tiles read out with two bundles of wavelength shifting fibers coupled by clear fibers of several meters length to standard photomultipliers allow to detect the crossing of minimum ionizing particles with about 99 % efficiency and a timing precision of 1.5 nsec.

A dense pack of many of these tiles gives the possibility to read out considerable large detector areas with only a few photomultipliers. Because all fibers of all tiles have the same length the properties of the detector are completely determined by the single tile behaviour independent of the position of a crossing particle.

Acknowledgement
We want to thank our colleagues K.H.Sulanke and G.Trowitzsch for their contribution to electronics and online software.

The testruns at CERN would not have been possible without the support of J.J. Blaising, P. Le Coultre and U. Uwer. Their help is gratefully acknowledged.

The SPY-project took profit from a lot of discussions with our colleagues of the Baikal-Amanda group at DESY-Zeuthen.

REFERENCES

1. Bamberger, A., et al., *NIM* **A382**, 419, (1996)
2. Blair, R., et al., FERMILAB-PUB-96-390E, (1996)
3. For present projects see also:
   Freeman, J., contribution to this workshop
   Para, A., contribution to this workshop
4. Bähr, J., et al., L3 internal note 1977, (1996)
5. Belolaptikov, I.A., et al., *Astropart. Phys.* **7**, 263, (1997)
6. Lowder, D.M., et al., Proceedings of the 17th Intern.Conf. on Neutrino Physics and Astrophysics, ed. K.Enquist, K.Huitu, J.Maalampi, Helsinki, Finland, 1996, P. 518

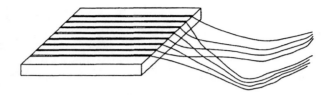

FIGURE 1. Schematic view of a scintillator tile readout by wavelength shifting fibers.

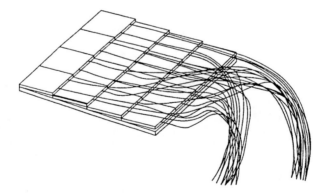

FIGURE 2. Combination of scintillator tiles to cover larger areas. Fibers from all tiles are splitted in two groups for the read out.

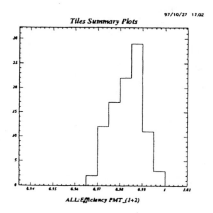

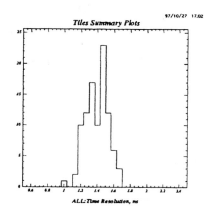

FIGURE 3. Efficiency distribution for 96 tiles of the first module of th L3COSMIC t0-detector.

FIGURE 4. Time resolution distribution for 96 tiles of the first module of the L3COSMIC t0-detector.

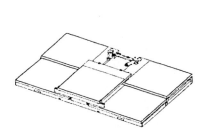

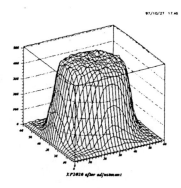

FIGURE 5. Mechanical layout of a full size 6 m² module readout with two photomultipliers.

FIGURE 6. Example for the sensitivity distribution of an XP2020 photomultiplier across its entrance window surface.

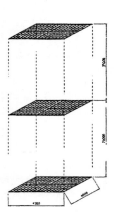

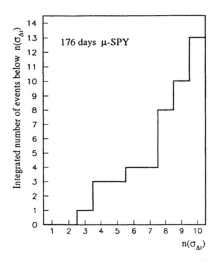

FIGURE 7. Schematic view of a three-floor SPY tower with 4x4 m² scintillator sheets made out of scintillator tiles.

FIGURE 8. Distribution of tracks (in bins of the time resolution of 3 nsec) for the radial difference to a possible neutrino signal from below the ground in the $\Delta t_{21} - \Delta t_{32}$ plane for 176 days of data taking with μ-SPY

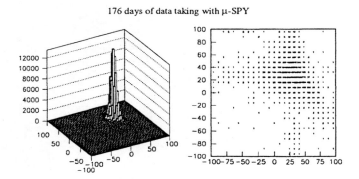

176 days of data taking with μ-SPY

Δt_{21} vs Δt_{32} for single tracks in nsec.

FIGURE 9. Distribution of tracks in the plane of time differences $\Delta t_{21} vs. \Delta t_{32}$ for cosmic particles crossing μ-SPY.

Calibration System for the Central Fiber Tracker for the DØ Upgrade

B. Baumbaugh, J. Marchant, A. Gerig, E. Popkov, K. Reynolds,
R. Ruchti, J. Warchol, M. Wayne, and H. Zheng

Department of Physics, University of Notre Dame, Notre Dame, IN 46556

Abstract. We are developing a calibration system for the Central Fiber Tracker (CFT) for the DØ Upgrade to monitor the optical integrity, channel gain, and gain stability for the 76,000 fiber channels with VLPC readout which comprise the system. Excitation is by blue Light Emitting Diodes (LEDs), with light distributed to the CFT fiber ribbons via luminous fiber panels. System design and performance will be presented.

INTRODUCTION

The DØ tracker consists of multiple layers of scintillating fiber laid on cylinders. The layers consist of axial and stereo fibers. In order to calibrate the system, a simple cost effective way to inject small amounts of light into the fibers is needed.

In the earlier tests of a 3000 channel fiber tracker, red emission LEDs were used to inject light axially into the end of the fiber[1]. As shown in Figure 1, there was a thin aluminized mylar film between the LED/waveguide fiber and the end of the scintillating fiber. This was used to increase the amount of scintillation light collected by the fiber through reflection. A small amount of light from the LED passed through the mirrored surface and into the scintillating fiber.

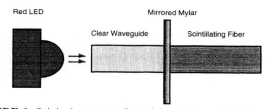

FIGURE 1. Original test setup for axial injection of light from LED.

This method was used to facilitate flexibility in calibration, to allow selected firing of the LEDs, and to allow possible track pattern simulation. This method required an LED and associated drive electronics for each fiber. In the final design this flexibility is not necessary, and, because the ends of the fiber are inaccessible and mirrored, the axial mode is not possible.

The scheme we have developed is to inject blue light from an LED through the cladded side walls of the fibers using an optical panel. The light (450nm) produced by a blue LED is poorly absorbed by the 3HF in the fibers. However, there is a small but finite absorption probability. Figure 2 shows the luminous spectrum of the 450 nm LED[2] and Figure 3 shows the transmission (absorption) and emission wave lengths of 3HF material used in the fibers. The low probability works in our favor in allowing the

CP450, SciFi97: Workshop on Scintillating Fiber Detectors
edited by A. D. Bross, R. C. Ruchti, and M. R. Wayne
© 1998 The American Institute of Physics 1-56396-792-8/98/$15.00

light to traverse multiple layers when shifting the 450nm to 530nm in the fibers. The low light level single photoelectron signal at the VLPC is just what we desire for calibration purposes.

This low level signal allows us to test several parameters of the system, which include channel on/off (fiber/VLPC defects), gain, gain stability, and rate dependence. By placing these light sources at both ends of the fibers we can assess the fiber performance throughout the life of the experiment.

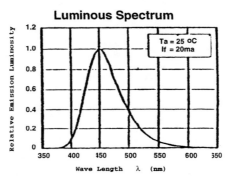

FIGURE 2. Relative emission spectra of the Nichia NLPB520S blue LED[2].

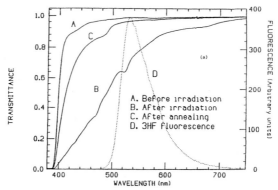

FIGURE 3. Transmittance and fluorescence emission of 3HF.

CONCEPT

The concept is to use light panels that are approximately one inch wide and several inches long to disperse the light from a single LED over many fibers. These panels have good light diffusion and uniformity characteristics. The panels are laid over a section of fiber and a single LED located in at the end of the light guide is pulsed locally (see Figure 4). A pulse driver is located at each LED with the pulse fed from a coax cable and a DC voltage is provided to control LED intensity. The panels are flexible enough to conform to the curvature of the cylinder, and will be held in place by a "strap" currently under design.

The light panels will be located at both ends of the scintillating fibers to facilitate testing of fiber integrity. This will also allow rate dependence studies, as one end can be pulsed as a noise source while the other end accepts trigger pulses to test signal integrity simulating high rate environments.

The system will consist of approximately 300 light panels and LED pulse driver systems. The voltage to the LED drivers will be controllable to allow LED intensity variations to compensate for minor panel differences, and to test varying light levels. These 300 LED/Drivers will act to calibrate approximately 76,000 channels of scintillating fibers.

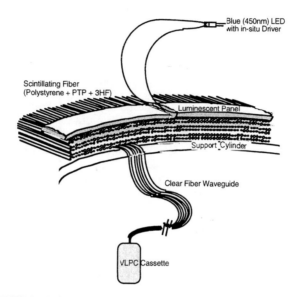

FIGURE 4. Light panels positioned over fiber ribbons arranged on cylinder.

FIBER EXCITATION METHOD

We have developed a novel alternative to axial injection and the technique is illustrated schematically in Fig. 5. When blue light is injected through the wall of a PTP+3HF scintillating fiber a small fraction of the blue light is absorbed by the 3HF which acts as a waveshifter in this case and results in scintillation light produced in the fiber. Most of the blue light passes through the fiber, and is available to excite the layer below. In this way a blue light source on the outside of each tracker cylinder can be used to excite several layers of scintillating fibers beneath. This technique has been studied using blue LEDs at Notre Dame with 6 layers of scintillating fibers. These tests provided excellent results.

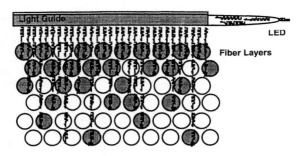

FIGURE 5. Schematic of New injection method.

For the calibration system to be useful it is necessary to get light into all the fibers on each cylinder, so a method of distributing the blue LED light with reasonable uniformity is necessary. This will be done with the use of flat optical panels, which are small (typically 10 cm long by 3 cm wide) ribbons of 100 μm diameter fibers specially treated so that light entering the ends of the fibers leaks out the fiber walls. When the end of the panel is illuminated, a very uniform light source is produced over its entire area. Samples from two commercial vendors of these panels were tested in summer 1997, and up to 6 layers of scintillating fibers were excited simultaneously by a single optical flat panel placed above them.

The length and width of the flat optical panels has not yet been finalized, but it is clear that with this technique a single LED can calibrate up to several hundred fibers. This reduces the complexity of both the mechanical structure and the drive electronics of the calibration system. The flat panels will be housed in a flexible support structure which is banded around the scintillating fiber cylinder, as illustrated in Fig. 6. The panels will be tested and mounted into these structures at Notre Dame, and the completed assemblies will be mounted onto the fiber detector at Fermilab. The drive electronics for the LEDs, which will be situated outside the detector volume, will be designed at Notre Dame.

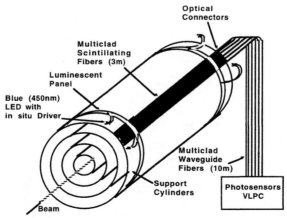

FIGURE 6. DØ CFT calibration system.

TESTS

In order to verify that the concept would indeed work we went through several steps. First we tested several LED types to determine which available wavelength would work best and if enough light would pass through the layers to facilitate calibrating all layers with a single LED. We then tested the use of light panels in a simulated fiber super-layer with waveguides and VLPCs. Finally we tested two manufacturer's light panels to quantify the uniformity of the light emission across the face of the panel.

Testing the Concept

Figure 7 shows the original test setup for the concept that the blue light from the LED would excite the fiber and traverse through the upper layers enough to allow testing of all layers beneath.

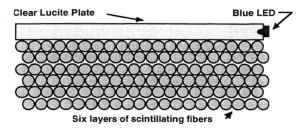

FIGURE 7. Original test setup.

The test layers consisted of six individual layers of 96 fibers each. Since the VLPC system we used has four eight element VLPC arrays we could only instrument 32 channels. The design allowed us to cover a horizontal width of 96 fibers by instrumenting every 12th fiber and using these 8 fibers for one 8 channel VLPC array. By using these groups of 8 fibers corresponding to a layer, we could look at any four layers simultaneously to determine the amount of light reaching the top layer, bottom layer, and any two layers in between. This set up allowed us to determine that indeed light reached all layers, although as expected the lower layer received less than the upper.

In the initial test we used a clear Lucite block with mounting holes drilled to accept the LEDs. The edges of the block were coated with Bicron BC-600[3] reflective coating. The upper surface was covered by a layer of white Tyvek to maximize the light transmitted to the fibers. The LEDs were placed in a series of mounting holes to see how well the light was distributed to all fibers.

We tested LEDs with emission wavelengths of 430 nm, 450 nm and 470nm. The 430nm, and 450nm devices were GaN devices capable of fast pulsing (<20ns) which is necessary for use with VLPCs if gain and light level are to be measured. The 470 nm devices available at the time were of Silicon Carbide, and were too slow to really accurately determine the individual number of photons collected.

The LEDs used were T-1 3/4 style from Nichia[2]. The one chosen was a part number NLPB520S with an emission wavelength of 450nm and a 45 degree emission angle. Other styles and emission angles as well as wavelength variations are available.

This test proved that the concept did indeed work and that the LED best suited for our use was the 450nm device. This wavelength has the best match to convert a small amount of the light and still allow most to be transmitted to the layers below.

Light Panel Tests

We tested several light panels. Most were from Poly-Optical[4], but a Lumitex[5] panel was also tested. Each type has it's own characteristics.

This test was identical to the test shown in Figure 7, but with the clear Lucite plate replaced with a light panel. This test proved that the light panels would work to provide sufficient illumination over a large section of the super-layer, but we would need to use more than one level of intensity in order to effectively test all fibers. This is not a major issue, and one that was not unexpected as the inner layers will see less of the total light due to absorption by the outer layers. Once we have a more complete prototype of the superlayer we will test to determine the exact amount of light absorbed by each layer, and how best to compensate for that.

Figure 8 Shows the results from one VLPC channel using the test setup. As can be seen, individual peaks can be used to determine gain and pedestal.

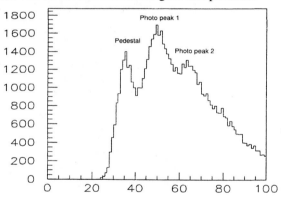

FIGURE 8. One channel of VLPC data using the test setup.

Uniformity   Tests

We also used a CCD camera and image capture hardware to measure the uniformity of light emitted across the face of the light panel. Figure 9 illustrates the setup we used for this process.

The light panel was held in place onto an aluminum plate that had a groove for vacuum milled into it. The vacuum held the panel flat (they tend to have a slight curve in them) and stationary. The imaging hardware and software in the computer allowed us to evaluate the characteristics of the light panels.

For these tests we used a new white light LED from Nichia, part number NSPW520S[2]. This wavelength worked best for use with the CCD camera due to the wavelength sensitivity of the CCD and the white LED intensity.

These tests showed that the light panels tested were relatively uniform in intensity from end to end and from side to side. Many panels have "hot spots" where the light appears as a point source much brighter than the surrounding area. Figure 10 shows the relative optical panel intensity averaged by column (top to bottom) for the Poly-Optical Products light panel and Figure 11 shows the same information averaged by rows (side to side). Figure 12 shows the uniformity for the Lumitex optical panel averaged by column and Figure 13 the same data averaged by rows.

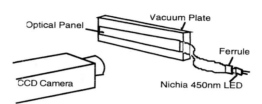

FIGURE 9. CCD Imager test setup.

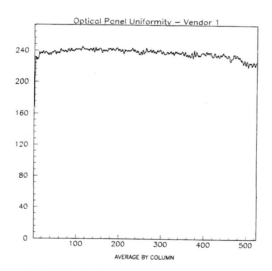

FIGURE 10. Poly-Optical Column averages.

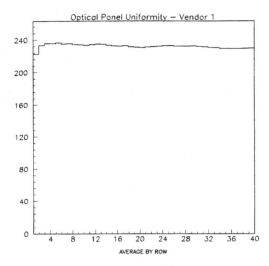

FIGURE 11. Poly-Optical Row averages.

The Lumitex panel demonstrated better uniformity as far as "hot spots", as well as from side to side and end to end. The end to end characteristic for both manufactures was acceptable for our purposes, and the side to side variations will be "averaged" out by the fibers below, due to the panel orientation at 90 degrees to the scintillating fibers.

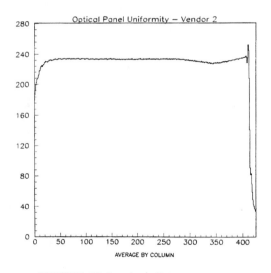

FIGURE 12. Lumitex Column averages.

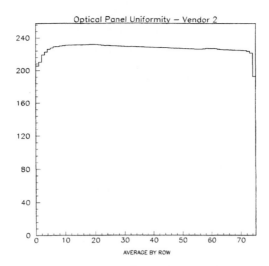

FIGURE 13. Lumitex Row averages.

FUTURE PLANS

We are currently looking into the optimal panel design to best cover the area at both ends of the DØ Central Fiber detector and to minimize cost. We are hoping to make the system with a built in redundancy (such as using two LEDs per panel), since access to the detector will be difficult and rare. We may also use quartz fibers to channel the light to the panels from outside the detector. This has the advantage of keeping the driver electronics and LEDs where they are more accessible, and would keep the drive electronics out of the detector volume to reduce the chance of creating electronic interference for the detector electronics.

We are also designing a "band" system to hold the light panels in place around the cylinder. This will probably be made from a flexible plastic with an over-center clamp to allow simple assembly and disassembly.

The same concept of using light panels to cover a large area is also being evaluated for use in a test system for fiber ribbons that will be used in waveshifter tiles for the CMS HCAL. These "pigtails" as they are referred to are a series of waveshifting fibers coming from a connector that are bonded to clear fibers. They are first assembled and then tested before they are glued into the mega tile structures. In order to test these assembled "pigtails" they are laid out in channels and a blue LED is scanned over the length of the waveguide/waveshifter to check that the shifter is functioning properly and that the waveguide and bond are acceptable.

REFERENCES

1. B. Baumbaugh et al., IEEE NSS Trans. Nucl. Sci **42,** 383 (1994).

2. Nichia America Corporation, 1000 New Holland Ave., Lancaster, PA, 17601

3. Bicron Corp., 12345 Kinsman Road, Newbury, OH   44065

4. Poly Optical, Inc., 17475 Gillette Avenue, Irvine, California 92614-5633

5. Lumitex, Inc., 8443 Dow Circle, Strongsville, OH, 44136

SESSION 10: CALORIMETRY - I

Chair: J. Freeman
Scientific Secretary: E. Popkov

Test Beam Performance of the CDF Plug Upgrade Hadron Calorimeter

Pawel de Barbaro

Department of Physics and Astronomy
University of Rochester
Rochester, NY 14627
e-mail: barbaro@fnal.gov

for the CDF Plug Upgrade Group[1]

Abstract. We report on the performance of the CDF End Plug Hadron Calorimeter in a test beam. The sampling calorimeter is constructed using 2 inch iron absorber plates and scintillator planes with wavelength shifting fibers for readout. The linearity and energy resolution of the calorimeter response to pions, and the transverse uniformity of the response to muons and pions are presented. The parameter e/h, representing the ratio of the electromagnetic to hadronic response, is extracted from the data.

INTRODUCTION

In the next Tevatron collider run (Run II), the time interval between particle bunches will be as short as 132 nanoseconds. The response of the existing (Run I) CDF electromagnetic and hadronic sampling gas calorimeters in the plug region is too slow (≈ 400 nsec) to accommodate this bunch crossing. The detectors are being replaced with new calorimeters using scintillator tiles with wavelength shifting (WLS) fiber readout technology [1–3]. The upgraded Plug Calorimeter consists of an electromagnetic section (EM) with a Shower Maximum Detector and a hadronic section (HAD). The compact design of the new calorimeter and its increased pseudorapidity coverage also allows for increased muon coverage in the upgraded CDF II detector. In this communication we report on the performance of the Plug Upgrade Hadron Calorimeter in a test beam. The performance of the CDF Plug Upgrade EM calorimeter and the wire source calibration procedure are presented elsewhere [4,5].

---

[1] Members of the following CDF institutions participate in the Plug Upgrade Project: Bologna U., Brandeis U., Fermilab, KEK, MSU, Purdue U., Rochester U., Rockefeller U., Texas Tech U., Tsukuba U., UCLA, Udine U., Waseda U. and Wisconsin U.

CP450, *SciFi97: Workshop on Scintillating Fiber Detectors*
edited by A. D. Bross, R. C. Ruchti, and M. R. Wayne
© 1998 The American Institute of Physics 1-56396-792-8/98/$15.00

HADRON CALORIMETER DESIGN

The angular (θ) coverage of the CDF End Plug Upgrade Calorimeters is from 38° to 3°, corresponding to the pseudorapidity range $1.1 < |\eta| < 3.5$. The upgraded hadronic compartment adds two plates to the 20 existing two inch iron absorber plates. Total thickness of the upgraded hadron calorimeter is approximately 7.1 λ_{INT}. An additional 1 λ_{INT} is provided by the EM compartment.

Light from scintillator tiles (6 mm thick Kuraray SCSN-38 with typical transverse size of 20 cm$\times$20 cm) is collected by embedded WLS fibers (Kuraray Y-11 0.83 mm in diameter). Outside the tiles, the WLS fibers are spliced to clear optical fibers. Light is further transported to decoder boxes located outside of the detector via mass terminated connectors and clear optical fiber cables. The decoder boxes re-group the fibers from all layers into towers to be read out by photomultiplier tubes. The East and West Hadron Plugs consist of a total of 864 readout channels.

The performance requirements of the Hadron calorimeter call for a stochastic term in the relative energy resolution function of less than 80% and a constant term less than 5%. The non-linearity in the energy response of the calorimeter to pions in momentum range 10-200 GeV/c is required to be less than 10%. To guarantee good muon identification, the light yield of each tile is required [6] to be above 2 photoelectrons per minimum ionizing particle.

In order to meet the above performance requirements, a strict quality control program was implemented during the production stage of the Hadron calorimeter optical system. The absolute light yield of each batch of scintillator and the light yield and transmission of the WLS and clear fibers were tested. The WLS fibers were spliced to clear fibers glued into optical connectors. These fiber/connector assemblies ("pigtails") were tested with a UV lamp scanner. The distribution of the relative light yield of over 18,000 individual fibers had an RMS of 3.2%. The light yield of all the tiles after the insertion of WLS fibers into the scintillator was measured using a Cs^{137} γ source. The distribution of the relative light yield of tile/fiber assemblies had an RMS of 6.1%.

TEST BEAM SETUP

A 60° sector "carbon copy" of the Plug Upgrade Hadron calorimeter modules installed in the CDF hall has been assembled for test beam studies at the MT beamline at FNAL. Data were recorded using electron (5-120 GeV/c), pion (5-220 GeV/c) and muon (10-220 GeV/c) beams. Particle momenta were determined using single wire drift chambers with precision of $\approx$ 0.2%. A 6 X_o lead converter instrumented with a scintillation counter was used to reject large positron contamination in hadron tunes, especially at low momenta. For part of the run, a Cerenkov counter was installed to measure both the proton contamination in the hadron tunes and the difference in the calorimeter's response to protons and pions.

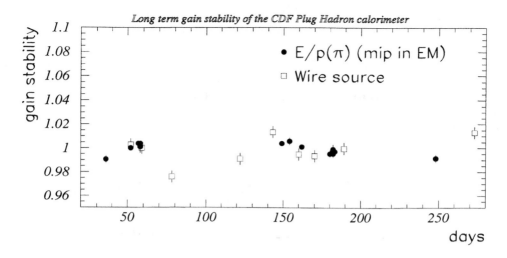

FIGURE 1. Long term stability of the gain of the CDF Plug Upgrade Hadron calorimeter.

The test beam provides the absolute energy calibration of the EM and Hadron calorimeters. A system incorporating removable radioactive γ source which is attached to a wire ("wire source") and can be inserted into tubes mounted on the scintillators is used to illuminate each tile in the calorimeter. The wire source calibration provides a mechanism to transfer the absolute energy scale of the calorimeters from the test beam module to the modules in the CDF hall.

The wire source is also used to equalize the gain (relative calibration) of all calorimeter towers prior to the data taking. The accuracy of the "wire source" calibration is checked by comparing the measured response of individual calorimeter towers to particles (electrons in case of the EM and pions in case of the Hadron calorimeters). The measurement indicates that wire source calibration is accurate within 2%. In addition, the wire source provides an independent tool to monitor the gain stability of individual towers in the Hadron calorimeter. During the test beam study (lasting over 250 days), the stability of the absolute gain of the calorimeter (as determined by the wire source) was constant to within 2%. This is consistent with the measurements based on the energy/momentum ratio, E/p, using test beam pions, as shown on Figure 1.

The CDF Plug Upgrade electronics will use 120 nsec integration time, while the Rabbit based electronics used at the test beam had 2.2 μsec integration time. A gate length study indicated an increase in the Hadron calorimeter response to pions by approximately 6%, when signals integrated for 120 nsec and 1 μsec are compared. Therefore the absolute calibration constant of the Hadron calorimeter (short gate) needs to be corrected by 6%, with respect to the calibration constant of the Hadron calorimeter measured in the Test Beam module (long gate).

PERFORMANCE OF THE HADRON CALORIMETER

Figure 2 shows preliminary results on the linearity of the energy response (E/p), and relative energy resolution (σ_E/E) of the Hardon calorimeter for incident pions. The absolute energy scale of the EM compartment is set using 50 GeV/c electrons. The absolute energy scale of the Hadron calorimeter is established using a subset of 50 GeV/c pion sample (pions which interact in the Hadron calorimeter only). The square symbols indicate that response of the calorimeter to pions which interact in the Hadron calorimeter only. Triangle symbols indicate the response of the combined EM+Hadron calorimeters to pions interacting in either in the EM or in the hadronic compartment. As shown in the figure, the response of the calorimeter to pions which interact in the Hadron calorimeter only is $\approx 3\%$ higher than the response to pions which interact in either the EM or hadronic compartment. The non-linearity of the response of the calorimeter to pions in momentum range 10-220 GeV/c is less than 10%, meeting the CDF Plug Upgrade requirement criteria. The relative energy resolution of pions which only interact in the Hadron calorimeter can be described by the function $\sigma_E/E=(74\pm1.0)\%/\sqrt{E} \oplus (3.8\pm0.3)\%$. Full pion sample (i.e. pions which interact in either the EM or Hadron calorimeters) have a slightly better energy resolution, $\sigma_E/E=(68\pm0.7)\%\sqrt{E} \oplus (4.1\pm0.2)\%$. We attribute this improvement to the fine sampling (4.5 mm Pb) of the EM calorimeter, relative to that of the Hadron calorimeter.

Note however, that linearity of response and good energy resolution for pions for the combined EM+Hadron calorimeters can not be achieved if the ratio of the response to electromagnetic and hadronic components of the pion showers in the EM calorimeter, e/h(EM) is very large. As in the case of the lead tungstate crystal ECAL calorimeter for the CMS [7] or the liquid argon ECAL calorimeter for the ATLAS [8], large e/h of the EM compartments significantly degrades the response of such combined EM+Hadron calorimetric systems to jets.

Comparison of the response to protons and pions

The test beam studies were conducted with positive hadron beam tunes. Protons constitute a significant fraction of the beam, especially at high energies. A correction for the proton contamination must be made in order to extract the response of the calorimeter to pions. During a part of the run, a Cerenkov counter was used to tag (offline) pions and protons. The response of the CDF End Plug Upgrade Calorimeter to protons and pions was measured [9] for incident momentum tunes of 5.6 GeV/c and 13.3 GeV/c.

The difference in the response is primarily due to two effects. The first effect is caused by the difference in the available energy for deposition in the calorimeter. For protons the available energy is the kinetic energy, and for pions it is the total energy; i.e. $\sqrt{P^2 + m_p^2} - m_p$ for protons versus $\sqrt{P^2 + m_\pi^2}$ for pions. The second effect originates from the different fraction of π^0 mesons produced in proton versus

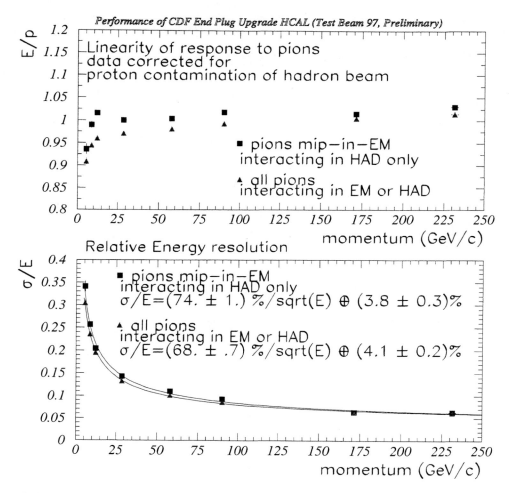

FIGURE 2. Linearity and relative energy resolution of the CDF Plug Upgrade calorimeter to pions (1997 Test Beam, Preliminary results).

pion induced showers. For a non-compensating calorimeter ($e/h \neq 1$), this difference leads to higher response for pion showers. A third effect resulting from the different interaction lengths of pions and protons is negligible. The measured ratio of proton/pion response, $\langle E_\pi \rangle / \langle E_p \rangle$ was 1.227±0.006 at 5.6 GeV/c and 1.104±0.006 at 13.3 GeV/c. The results are consistent with the expectation from the available energy, the fraction of π^0, and interaction length effects.

The e/h parameter for the CDF Hadron Calorimeter

The observed non-linearity of the response of the Hadron calorimeter to pions can be explained by the difference in the response of the calorimeter to the electromagnetic and hadronic components of the showers, ($e/h \neq 1$) in combination with the fact that the average fraction of π^0's in pion induced showers increases as a function of pion momentum. By measuring the ratio of the response of the Hadron calorimeter to pions and electrons as a function of pion momentum, e/h parameter of the calorimeter can be extracted. The measured response of the Hadron calorimeter (using hadrons which do not interact in the EM compartment) is first corrected for proton contamination of the hadron beam. It is also corrected for the longitudinal shower leakage, since the total depth of the Hadron calorimeter is only 7.1 λ_{INT}. The response of the Hadron calorimeter to electrons was measured in dedicated runs for which the electron beam was pointed directly [2] into the Hadron calorimeter. The response of the Hadron calorimeter to electrons is linear to within 2%, and approximately 20% higher than the response to pions.

Figure 3 shows the pion/electron response ratio of the Hadron calorimeter, after the proton contamination and longitudinal shower leakage corrections have been applied. The extracted value of e/h depends on the functional form assumed for the fraction of π^0 as a function particle momentum. There are two available parameterizations, one by Wigmans [10] which uses $ln(E)$ form, the other by Groom et al. [11] which uses exponential parameterization. The value of the extracted e/h (using data points between 10 and 230 GeV varies from 1.34 ± 0.01 (Wigmans) to 1.42 ± 0.015 (Groom). Both parameterizations are consistent with the measured pion/electron response ratio at 5 and 8 GeV.

Transverse Uniformity of Response

A fine scan of the response of the Hadron calorimeter with pion and muon beams has been used to measure the transverse uniformity. Figure 4 shows two-dimensional (in ϕ and θ) response maps of the Hadron calorimeter using 150 GeV/c muons (upper left plot), and 50 GeV/c pions (upper right plot). The scanned area fully covers a single calorimeter tower. The four crossing lines on the map indicate

---

[2] The EM module covered only 45^0 in ϕ of the 60^0 hadronic module.

410

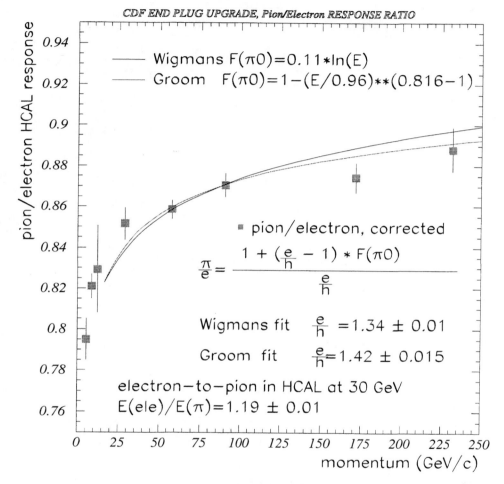

Wigmans F(π0)=0.11∗ln(E)
Groom F(π0)=1−(E/0.96)∗∗(0.816−1)

■ pion/electron, corrected

$$\frac{\pi}{e} = \frac{1 + (\frac{e}{h} - 1) * F(\pi 0)}{\frac{e}{h}}$$

Wigmans fit $\frac{e}{h}$ =1.34 ± 0.01

Groom fit $\frac{e}{h}$=1.42 ± 0.015

electron−to−pion in HCAL at 30 GeV
E(ele)/E(π)=1.19 ± 0.01

FIGURE 3. Pion/electron response ratio of the CDF Hadron calorimeter. The hadron data has been corrected for the proton contamination of the beam, and for longitudinal shower leakage. The data is used to extract the *e/h* parameter for the Hadron calorimeter.

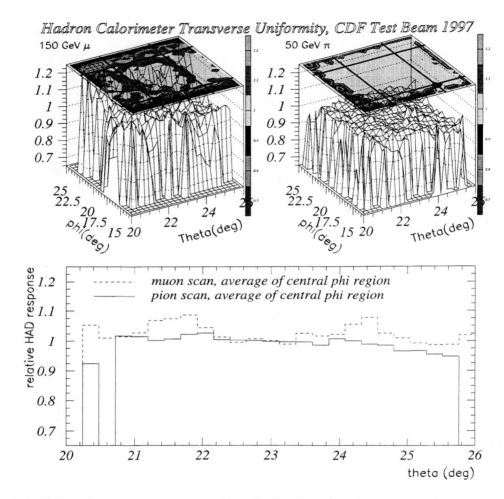

FIGURE 4. Transverse response map of a single CDF Plug Upgrade Hadron calorimeter tower using 150 GeV/c muons (upper left plot) and 50 GeV/c pions (upper right plot). The lower plot shows the data averaged over the central ϕ bins, as a function of the θ. The muons data (broken line) indicates an increased response of the tile at $\theta=21.5^0$ and $\theta=24.5^0$, corresponding to the location of the tile boundaries. The pion data (solid line) indicates uniform (within 2%) response of the Hadron calorimeter to pions. The drop in response at $\theta=25^0$ is caused by transverse leakage (out of the θ edge of the detector).

the boundary lines between neighboring towers. The lower plot shows the data averaged over the central ϕ bins, as a function of the θ. The muon data (broken line) indicates an increased response of the tiles at $\theta=21.5^0$ and $\theta=24.5^0$, corresponding to the location of the tile boundaries. The pion data (solid line) indicates a uniform (within 2%) response of the Hadron calorimeter to pions. The drop in response at $\theta=25^0$ is caused by transverse leakage out the sides, near the θ edge of the detector.

The plots indicate that the response to muons is uniform within RMS of 5% and response to pion showers is uniform to within $\pm$ 2%. The 8% relative light increase at the tile boundaries observed in the muon scans is consistent with earlier measurements done using a 2 MeV/c electron beam (from a beam line using a β radioactive source) for R&D studies of the optical system.

CONCLUSIONS

The CDF Plug Upgrade sampling calorimeter is based on tile/fiber technology. It has been constructed under a set of strict QC procedures. The calorimeter modularity (megatiles, connectors, optical cables, decoder boxes) greatly simplifies the final assembly at the CDF collision hall. Results from the recently completed Test Beam studies indicate that the detector performance satisfies the design requirements. For the Hadron calorimeter, the linearity of response to pions in the momentum range 5-250 GeV is $\leq$ 12%. The relative energy resolution of the combined EM+Hadron calorimetric system to single pions can be parameterized as $\sigma(E)/E = 74\%/\sqrt{E} \oplus 4\%$. The transverse uniformity of the response to incident pions is constant to within 2%. The installation of the calorimeter is on schedule and we expect to be ready for data-taking in the CDF hall at the end of 1999.

REFERENCES

1. *The CDF II Detector Technical Design Report*, FERMILAB-Pub-96/390-E.
2. *Scintillator Tile-Fiber Calorimeters for High Energy Physics, The CDF End Plug Upgrade, Selected Articles.*, ed. by P. de Barbaro and A. Bodek, UR-1309, Oct. 1994.
3. H. Budd et al., *Fiber R and D for CMS HCAL*, these Proceedings, UR-1513.
4. Y. Fukui et al., *The CDF Plug Upgrade EM Calorimeter*, these Proceedings.
5. V. Barnes et al., *Wire source calibration of the CDF Plug Calorimeter*, these Proceedings.
6. A. Bodek and P. Auchincloss, NIM, A**357**, 292 (1995)
7. P. de Barbaro et al., *Performance of a Prototype CMS Hadron Barrel Calorimeter in a Test Beam*, Proceedings of the CALOR97, Tucson, AZ, UR-1512, Nov 1997.
8. Z. Ajaltouni et al., NIM, A**387**, 333-351(1997); see for example: Fig 2, page 339.
9. J. B. Liu et al., *Testbeam Results for the CDF End Plug Hadron Calorimeter*, CALOR97 Proceedings, U. of Rochester preprint UR-1509, Dec 1997.
10. R. Wigmans, NIM, A**265**, 273-290(1988).
11. T.A. Gabriel, D.E.Groom, NIM, A **338**, 336-347(1994).

413

Test Beam Performance of CDF Plug Upgrade EM Calorimeter

Yasuo Fukui

*Fermilab/KEK, P.O.Box 500 Batavia
IL60510, e-mail: fukui@fnal.gov*

for the CDF Plug Upgrade Group[1]

Abstract. CDF Plug Upgrade(tile-fiber) EM Calorimeter performed resolution of $15\%/\sqrt{E} \oplus 0.7\%$ with non-linearity less than 1% in a energy range of 5-180 GeV at Fermilab Test Beam. Transverse uniformity of inside-tower-response of the EM Calorimeter was 2.2% with 56 GeV positron, which was reduced to 1.0% with response map correction. We observed 300 photo electron/GeV in the EM Calorimeter. Ratios of EM Calorimeter response to positron beam to that to $^{137}C_S$ Source was stable within 1% in the period of 8 months.

INTRODUCTION

In order to obtain better calorimeter data in the plug and forward/backward region ($| \eta | \geq 1$) with the shorter bunch spacing of 132 ns (than the current bunch spacing of 3500 ns) in the Fermilab Run II collider runs, the CDF group upgraded the gas sampling Plug and Forward Calorimeters with the Upgrade Plug(tile fiber) Calorimeter [1]. We calibrated a test beam module of the CDF Upgrade Plug Calorimeter at the Fermilab MTest beam line from December 1996 till September 1997.

TEST BEAM SETUP

The test beam module consists of 45 degree(ϕ) section of EM Calorimeter and 60 degree(ϕ) section of Hadron Calorimeter. Figure 1 shows a cross section of the calorimeter. The calorimeter has depth segmentation of PPS(Plug PreShower), PES(Plug EM Shower max), PEM(Plug EM), and PHA(Plug HAdron).

---

[1] Members of the following CDF institutions participate in the Plug Upgrade Project: Bologna U., Brandeis U., Fermilab, KEK, MSU, Purdue U., Rochester U., Rockefeller U., Texas Tech U., Tsukuba U., UCLA, Udine U., Waseda U. and Wisconsin U.

CP450, *SciFi97: Workshop on Scintillating Fiber Detectors*
edited by A. D. Bross, R. C. Ruchti, and M. R. Wayne
© 1998 The American Institute of Physics 1-56396-792-8/98/$15.00

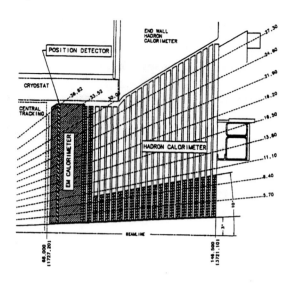

FIGURE 1. Side view of the Plug Calorimeter (cross section)

Table 1 shows parameters of the calorimeter including material and photon collection system.

PPS is a scintillator layer with the same $\eta - \phi$ segmentation of PEM with a structural iron plate as an radiator. PES(Plug EM Shower max) is made of two layers of arrays of thin scintillator bars with 22.5 degree crossing angle of bars in two layers placed at the depth of EM shower maximum. EM Calorimeter covers $1.1 \leq | \eta | \leq 3.5$, and Hadron Calorimeter covers $1.3 \leq | \eta | \leq 3.5$. ϕ segmentation($\Delta\phi$) is 7.5 degree at $| \eta | \leq 2.1$ and 15 degree at $| \eta | > 2.1$ in both EM and Hadron Calorimeter. Momentum tagging system of the MTest beam line gives $\Delta p/p$ of 0.2 %. FWHM beam size is 2.5 cm in horizontal and 1.3 cm in

TABLE 1. EM Calorimeter parameters

Segmentation	$\sim 8 \times 8 \ cm^2$
Total Channels	960
Thickness	$21 \ X_0, 1 \ \lambda_0$
Density	$0.36 \ \rho_{Pb}$
Samples	22 + Preshower
Active layer	4 mm SCSN38 (EM), 10 mm BC408 (PreShower)
Passive layer	4.5 mm Pb
Photo tube/Gain	R4125/2.5×10^4(EM), R5900-M16/1×10^5(PreShower)
Light Yield (pe/MIP/tile)	~ 5
Non-linearity	≤ 1 %
Resolution	$15\%/\sqrt{E} \oplus 0.7\%$

415

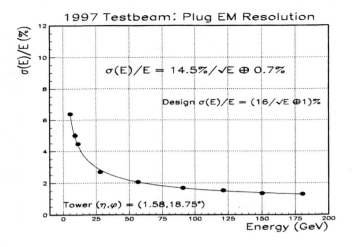

FIGURE 2. EM Energy Resolution as a function of e^+ beam energy

vertical.

RESPONSE TO ELECTRONS

Energy Resolution

Figure 2 shows Energy Resolution ($\sigma(E)/E$) of (PPS and PEM) as a function of beam energy with e^+ beam at the center of a reference tower. (PPS and PEM) energy was summed up in a $3(\eta) \times 3(\phi)$ tower window around a reference tower with beam. We obtained $(14.5 \pm 0.2\%)/\sqrt{E} \oplus (0.7 \pm 0.1\%)$ in the reference tower, and $(14.6 \pm 0.2\%)/\sqrt{E} \oplus (0.8 \pm 0.1\%)$, and $(15.8 \pm 0.2\%)/\sqrt{E} \oplus (0.4 \pm 0.2\%)$ in two other towers. Measured stochastic terms and constant terms are smaller than the *Design Values* of $16\%/\sqrt{E} \oplus 1\%$.

Linearity

Figure 3 shows non-linearity $[\equiv (EM(e)/p - 1)]$ of response of (PEM and PPS) and non-linearity of PEM alone as a function of e^+ beam energy at the center of a reference tower of the test beam module. By using PreShower response, we can reduce the non-linearity significantly at beam energy of 56 GeV or below. Energy scales were calculated to be 1.46 pC/GeV on (PPS and PEM) response to e^+ beam at 56 GeV in the reference tower in a reference run.

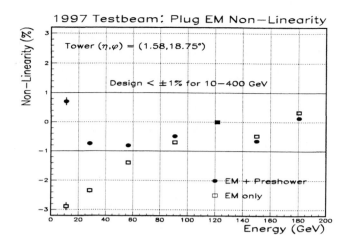

FIGURE 3. Non-Linearity as a function of e^+ beam energy

PreShower channel weight was adjusted in order to achieve our *Design Goal* of obtaining less than or equal to 1 % of non-linearity with a beam energy range from 10 GeV to 400 GeV. Non-linearity of (PPS and PEM) is normalized to e^+ beam at 120 GeV in the figure.

Response Uniformity

Inside four towers, we measured uniformity of the PEM response to e^+ beam at 56 GeV by moving the Upgrade Plug Calorimeter with respect to the beam in small steps in η and ϕ with overlapped beam area. Incident beam position was reconstructed by using the ShowerMax detector(PES). Figure 4 shows normalized response of PEM as a function of normalized η and ϕ inside and on tower boundaries of four towers, and an average of those four responses in normalized η and ϕ. In the figure, tower boundaries correspond to η or ϕ at $\pm$ 0.5. ϕ direction was reversed in two towers so that we obtain the same configuration of WLS(Wave Length Shifter) fiber routing inside a tower as that of the reference tower.

Normalization was done at the center of the tower. Size of the each bin in $\eta \times \phi$ was around 1 cm $\times$ 1 cm where EM(e)/p was fitted to a Gaussian with around 100 momentum reconstructed e^+ beam events to obtain an average PEM response in a bin. The average of the normalized response of PEM in normalized η and ϕ in four tower was used as a response map of the PEM. EM response is highest around boundaries where WLS fibers exit from tiles, and lowest at four corners of tiles as

417

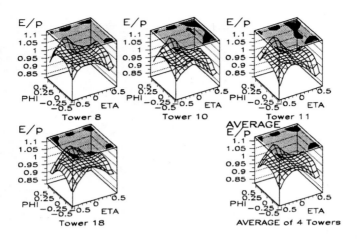

FIGURE 4. Normalized EM response to e^+ beam inside 4 towers and the average EM response in normalized η and ϕ

shown in the figure. We can obtain transverse uniformity of PEM response of 1.0 % inside a reference tower with e^+ beam at 56 GeV with the response map correction, where the transverse uniformity of PEM response was 2.2 % without response map correction. *Design Value* of inside a tower response of EM was less than 2.5 %.

Response Stability

Figure 5 shows the response of EM Calorimeter (E/p) to 56 GeV e^+ beam as a function of time in the day number of 1997, as well as estimated EM response to beam based on the response of EM Calorimeter to $^{137}C_S$ Source. Data points covers almost 8 months in 1997. RMS of ratios of EM response to 56 GeV e^+ beam to EM response to $^{137}C_S$ Source is 0.4 % in the period when both EM response to 56 GeV e^+ beam and EM response to $^{137}C_S$ Source increased around 4 %. The Ratio of the calorimeter response to beam to that to wire source allows us to bring the energy scale(pico Coulomb/GeV) into B0 collision hall where the Plug Upgrade Calorimeter will be used in the Run II.

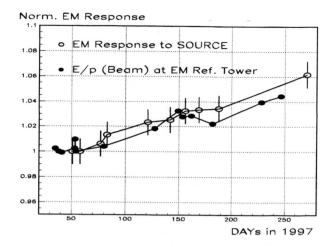

FIGURE 5. Normalized EM response to e^+ Beam and Source as a function of DAY number in 1997

CONCLUSION

The response of the PreShower and the EM Calorimeter ($\Delta\varphi = 45\,\mathrm{deg}$) to e^+ beam are better than or equal to the *Design Values*.

We obtained Energy Resolution of $14.5\%/\sqrt{E} \oplus 0.7\%$, where *Design Value* was $16.0\%/\sqrt{E} \oplus 1.0\%$.

On response linearity, the *Design Goal* of (Non-linearity $\leq 1\,\%$ in the energy range of 10 - 180 GeV) was achieved.

RMS/meam of transverse response uniformity inside a tower was measured to be 2.2 % without response map correction, and $\approx 1\%$ with response map correction, where the *Design Goal* was $\leq 2.5\,\%$.

We observed 5 photo electrons/MIP (150 GeV μ beam), where greater than 3 was expected as the *Design Value*.

RMS/mean of the ratio of the EM Calorimeter response to e^+ beam to the response to $^{137}C_S$ Source was 0.4 % in a period of 8 months, which shows the accuracy of the energy scale calibration of the EM calorimeter with the beam.

REFERENCES

1. The CDF II Collaboration, *The CDF II Detector Technical Design Report*, FERMILAB-Pub-96/390-E.

419

Calibration of the CDF Tile-Fiber Endplug Calorimeters Using Moving Radioactive Sources

Virgil Barnes[1], Alvin Laasanen, Arnold Pompos and Matthew Wilson

Physics Department, Purdue University, West Lafayette, Indiana 47907

Abstract. The use of moving radioactive gamma sources to assess, calibrate and monitor scintillating tile calorimeters is discussed, and the techniques and equipment are described. The capabilities of the technique are illustrated using Cs^{137} sources with the CDF Endplug Upgrade EM and Hadron calorimeters at testbeams and at a cosmic ray test stand. Source measurements of all the tiles in testbeam modules which are exact replicas of the calorimeters, predict the relative responses of EM towers to 50 GeV positrons and muons, and of Hadron towers to 50 GeV pions, with RMS accuracies of 1.3%, 1.8% and 2.0%, respectively. Source measurements will be used in lieu of testbeam measurements for the initial calibration of all towers in the final calorimeters. Source measurements of single tiles are reproducible to 0.4% and will be used to monitor gain changes of the photomultiplier tubes.

Introduction

The ability to measure the response of every scintillating tile in a calorimeter to a moving radioactive source permits in principle a variety of functions including:

- Quality checking of the tile layers and wavelength shifting readout fibers and optical cables at production and after assembly into towers, including discovery of optical miscablings

- Monitoring of changes in the response profiles within towers due to radiation damage, age, or scintillator brightening in magnetic fields

- Initial calibration transferred from a small testbeam module to all towers, and equalization of tower responses by adjustment of phototube gains

---

[1] Presenter, for members of the CDF Endplug Calorimeter Upgrade Group: Bologna U., Brandeis U., Fermilab, KEK, MSU, Purdue U., Rochester U., Rockefeller U., Texas Tech U., Tsukuba U., UCLA, Udine U., Waseda U. and Wisconsin U.

CP450, *SciFi97: Workshop on Scintillating Fiber Detectors*
edited by A. D. Bross, R. C. Ruchti, and M. R. Wayne
© 1998 The American Institute of Physics 1-56396-792-8/98/$15.00

- Monitoring of changes in phototube gains

A wire source driver system capable of delivering a pointlike gamma source near every tile in a calorimeter has been adopted for the CDF endplug upgrade calorimeter, and is planned for the CMS detector at LHC [1], among others.

For the CDF Plug Upgrade Calorimeters, all of the above capabilities will be used, and the source system will also be used for the Plug's PreShower and ShowerMax position detectors.

Description of the CDF Plug Upgrade calorimeters is left to the two other CDF articles in these procedings [2], and can also be found in the CDF II Technical Design Report (TDR) [3].

The Calibration System

The source calibration system is described in the CDF II TDR [3] and in earlier conference proceedings [4,5]. Briefly, it is a 10-inch storage reel which can both push and pull up to 11 m of "wire" containing a pointlike radioactive source into any one of 380 different channels. DC electric motors power the reel and indexing functions, via one cable which can be over 100 meters long, from a box which has manual control switches and also a remote control interface to either CAMAC or to a PC. Wire position and index channel number are read out by reversible counters, which are then read out either by CAMAC or PC.

The Cs^{137} "wire" sources are located at the tips of very thin flexible 6 or 9 m long stainless steel hypodermic tubes (OD=0.71 mm) and have active lengths of about 5 mm.

Each quadrant of each plug has a driver capable of inserting a radioactive gamma source past every scintillating tile in selected layers when the detector is closed. Other roving drivers can access the remaining layers via holes in the "skin" when the detector is open. Thin 0.050-inch OD metal tubes are embedded in the plastic cover sheet of every scintillator megatile package, and are coupled to the source drivers via 1/8-inch acetal tubing. The permanently coupled layers also use steel extender tubes running under the skins to the rear of the plugs.

Additionally, a collimated gamma source scanner is used at production to permanently establish the response ratio of every tile to the collimated source and an uncollimated wire gamma source. The collimated source generally provides an almost absolute excitation of each tile, without the systematic response variations due to tile size and source tube placement inherent in the uncollimated wire source ($\sim$15% variation between large towers and small towers, 1 - 3% variation front to back within a tower, and $\sim 10\%/mm$ variation with separation when the source tube is embedded in plastic and is within 1 mm of the scintillating tile). However, the collimated cone of gamma radiation was large enough to partly miss the tiles of the innermost EM towers, which therefore require one extra systematic correction based on testbeam measurements.

421

If the source tubes do not move relative to the tiles, the source system provides an almost absolute calibration of most towers (up to one overall calibration constant for the EM calorimeter and one for the Hadron calorimeter, which must come from the testbeams). The calibration constant of a tower is the convolution of the individual tile responses (corrected to the collimated source response), with the average electron or hadron shower profile. For muons, an unweighted sum over tiles is used, corrected for the muon path length in the tiles.

The initial source calibration is useful to set phototube gains and to permit clean trigger thresholds at the start of collisions.

Gain monitoring via light injection to the phototubes is used to maintain the calibrations of all towers, quickly and as frequently as desired. The CDF light injection system has a Nitrogen laser and a light fiber distribution system with excellent equalization to all towers, approximately 7% RMS. Monitoring of the light injection by multiple pin diodes is generally good, but may need occasional verification and correction using the source system. We plan to do source insertion as infrequently as possible since it is a relatively slow process and since it involves mechanical wear on the system.

Eventually, in situ calibration can be done using physics processes such as electronic decays of Z bosons or possibly of J/ψ. Collecting the statistics for in-situ physics calibrations may involve time spans which are long compared to possible gain drifts. The light injection system and/or the source calibration system are essential to monitor such drifts. Since the constant term in the EM calorimeter resolution is $\sim 1\%$, the stability must be monitored to better than 0.5%

Measurements of Source Responses

The collimated source measurements are done with the lead collimator stationary and close to the center of each tile. All source response measurements are DC current measurements.

The wire source current measurements are done with the wire moving at roughly constant speed, typically 10 cm/s. Operational amplifier signal conditioners with typically 1 MegOhm precision resistors feed either a PC-based DAS 1801HC multiplexing DVM (150 Hz sampling rate) or are built into the DC current channels of the CDF RABBIT system (20 Hz sampling rate). A PC source scan past a single tower in the EM calorimeter is seen in Figure 1, with eleven 150 Hz readings averaged per display point.

The two peaks seen are from the insertion and retraction of the wire source into the source tube in one megatile. The single highest data point in the peak is taken to be the tile response (more complicated peak finders often work less well!) The reproducibility of the source measurements is excellent: in Figure 2 the distribution of fractional differences in peak currents (extension vs. retraction) for 1157 EM calorimeter tiles has a RMS of 0.4%. For further accuracy, the average of the two peaks is used each time a tile is measured.

Source wire intercalibrations using one megatile (20 towers) typically show a tile-to-tile RMS of 0.5%, implying an overall source intercalibration accuracy of $\sim 0.1\%$ The 30.2 year half-life of Cs^{137} was taken into account by adjusting all of the data to January 1,1997.

A permanent data base of Collimated/Wire source response ratios is established at production testing time. After the calorimeter is assembled and coupled via optical cables to an optical-fiber layer-to-tower translation box at the phototubes, a complete set of tower "fingerprints" is taken with the wire source visiting every tile of every tower. At this stage, the source, which travels past well-defined sequences of towers, can detect miscablings and mistakes in the translation box.

While the gains of some phototubes are observed to drift by several percent over a period of months, the fingerprint patterns within a tower are very stable. In the testbeam module in 1997, this is demonstrated in Figure 3 which shows a tile-by-tile comparison of March/January results for the 60 EM towers in the testbeam calorimeter. After removing phototube gain shifts between January and March, there remain 21 degrees of freedom corresponding to the 22 tiles per tower. The tower fingerprint patterns have fitted $\sigma/Mean = 0.35\%$ and RMS = 0.66% for tiles within a tower. The comparable distribution for 72 towers in the Hadron calorimeter testbeam module (not shown) has $\sigma/Mean = 0.26\%$ and RMS = 0.43%.

Thus, a "minisourcing" at a later time into one or two layers effectively recovers the complete fingerprint of the tower. Minisourcing is the method by which tower gains (or responses) are monitored. The typical round-trip time for one wire insertion is 3 to 5 minutes, and there are twelve source tubes per quadrant in one calorimeter layer in the final detector.

The observed fingerprint patterns must be corrected for the spillage of gamma radiation into adjacent scintillator layers within the tower. We have measured the "sharing fractions" to unsourced layers by optically coupling only one scintillator layer to the phototube and sourcing many layers. In the EM calorimeter, the spillage adds typically 20% extra signal from each neighboring tile and 27 to 35% extra, overall, from each side of the sourced tile. Thus the front and back tiles of a uniformly made tower will appear to be weak by a factor $\sim 1/1.3$, On the other hand, tile-to-tile non-uniformities will appear to be a factor $\sim 1/1.6$ less than they actually are. The sharing effect is unfolded by solving 22 simultaneous linear equations using the measured sharing fraction matrix.

In the Hadron calorimeter, the sharing to one nearest neighbor tile is $\sim 4\%$ and is very small to the next neighbors, so we neglect the effect.

Figure 4 has two Lego plots of one 15-degree sector of the EM testbeam calorimeter, before and after unfolding the sharing. In both cases the Collimated/Wire ratio has been applied to all tiles. Unfolding reduces the "sag" in measured response at the front and rear of the towers, and increases the fluctuations from tile to tile, all as expected.

A quantitative measure of the uniformity of construction of the EM towers is given by the distribution of unfolded fingerprints (corrected to Collimated responses). To remove the effects of the 7% spread in phototube gains, the mean

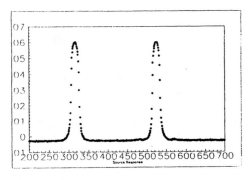 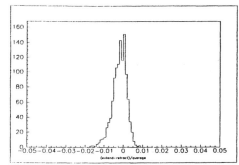

FIGURE 1. Typical response in μA of an EM tile to the moving Cs^{137} source.

FIGURE 2. Fractional difference between the tile response to the extension and retraction of the Cs^{137} source

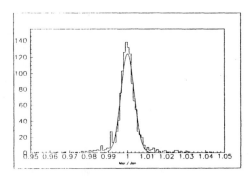

FIGURE 3. Ratio of tile response from March fingerprint compared to January fingerprint for the EM calorimeter. The mean response of each tower has been renormalized to 1.0.

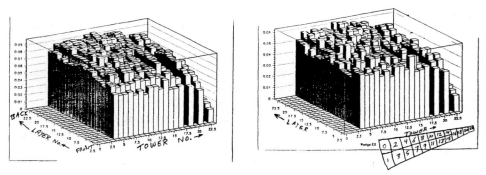

FIGURE 4. Lego plots of the tile response to the wire source in μA (transformed to collimated response) for EM Wedge 0, before and after unfolding the radiation sharing. The tower numbers are shown on the x-axis (and in the inset), and the layers on the y-axis.

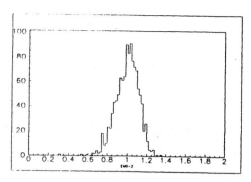

FIGURE 5. Distribution of unfolded tile responses (transformed to collimated responses) for the EM testbeam calorimeter. The mean tile response of each tower has been renormalized to 1.0.

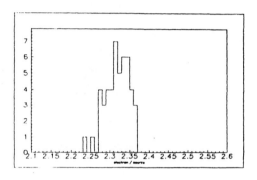

FIGURE 6. The ratio of the measured 50 GeV positron response to the source-predicted response for the EM testbeam calorimeter (arbitrary units).

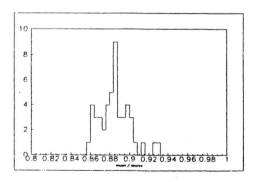

FIGURE 7. The ratio of the measured muon response to the source-predicted response for the EM testbeam calorimeter (arbitrary units).

response for each tower is renormalized to 1.0. For one quadrant of the final detector east EM Plug at the Cosmic Ray Test Stand, the RMS spread of tiles within a tower is 9.5%. For the 60-degree EM testbeam calorimeter (see Figure 5) the RMS spread is 11.5%. The goal for the final detector is an RMS less than 10% to keep the constant term in the energy resolution below 1%.

Comparisons of Source Predictions and Testbeam Responses

Source predictions for the EM towers are weighted sums of the individual tile responses (corrected to the collimated source response), where the weights are given by an experimentally measured 25 GeV electron shower profile [6]. The predictions are compared tower by tower via ratios with the 50 GeV testbeam responses, in Figure 6. The innermost and outermost towers of the calorimeter are omitted from the comparison because of collimated source spillage from small or truncated tiles, respectively. The RMS/Mean of this distribution is 1.3%.

For the Hadron calorimeter, the weights are given by a paramaterization from the Particle Data Tables. The 50 GeV pion responses are determined only for the 18 central towers with 5x5 tower energy sums, to contain the lateral shower size; and a minimum-ionizing signal in the EM compartment was required. The RMS/Mean of the beam/source comparison is 2.0%.

For muons in the EM calorimeter, an unweighted sum over tiles is used, corrected for the muon path length in the tiles. The beam/source comparison is shown in Figure 7 and the RMS/Mean of the distribution is 1.8%

Conclusions

The source calibration system appears to meet its many goals for quality checking, accurate initial calibration, and better than 0.5% gain monitoring at the new CDF endplug calorimeters. The uniformity of construction of the EM calorimeter towers is measured to be adequate for the desired 1% constant term in the energy resolution.

Acknowledgements

We wish to thank Qifeng Shen for his valuable early work on this project. We also wish to thank the Fermilab engineers and technicians who have made many contributions to the success of this effort. This work was supported in part by the U. S. Department of Energy.

426

REFERENCES

1. "CMS, The Hadron Calorimeter Technical Design Report", CERN/LHCC 97-31, CMS TDR 2, 20 June 1997.
2. P. de Barbaro, these procedings; Y. Fukui, these procedings.
3. "The CDF II Detector Technical Design Report", FERMILAB-Pub-96/390-E November 1996.
4. Virgil E. Barnes, Alvin T. Laasanen and James Ross, "Proceedings of the Second International Conference on Calorimetry in High Energy Physics", ed. Antonio Eriditato, World Scientific, (1992) 195.
5. Virgil E. Barnes, "Proceedings of the Fifth International Conference on Calorimetry in High Energy Physics", ed. Howard A. Gordon and Doris Rueger, World Scientific, (1994) 338.
6. The shower profile was measured by Masa Mishina et al. using the previous CDF gas calorimeter plug.

The Use of WLS Fibers in a Hadronic Calorimeter for the HyperCP Experiment

C. Durandet [a], M. Crisler [b], E. C. Dukes [a], T. Holmstrom [a],
M. Huang [a], K. S. Nelson [a], D. Rajaram [a], N. Saleh [a], and
Y. Tzamouranis [a]

[a] *University of Virginia, Charlottesville, Virginia 22901, U.S.A.*
[b] *Fermi National Accelerator Laboratory, Batavia, Illinois 60510, U.S.A.*

Abstract. Preliminary results are presented on the operational aspects of an iron-scintillator sampling hadronic calorimeter used in the HyperCP experiment at Fermilab during the 1996-1997 fixed target run. The calorimeter used wavelength shifter fibers for light collection from scintillator sheets. Details of how the 2 m × 2 mm fibers were polished, sputtered, and used for the readout are discussed. The average reflectivity of the sputtered fibers was 0.85±0.05, and the average attenuation lengths were 3.48±0.34 m. The calorimeter was designed to trigger on the proton (anti-proton) from Λ ($\overline{\Lambda}$) decays, suppressing triggers from secondary interactions and background muons.

I INTRODUCTION

The primary goal of HyperCP is to look for direct CP violation in non-leptonic decays of Ξ and Λ hyperons. CP violation should manifest itself as a difference in the angular distributions of the daughter baryons between parent hyperon and anti-hyperon. We expect to achieve a sensitivity of 2×10^{-4} in the difference of the slopes of the Λ ($\overline{\Lambda}$) decay angular distributions. To achieve this sensitivity, an enormous sample of 75 billion triggers has been accumulated with a high-rate trigger and data acquisition system. A key component of the trigger was a hadronic calorimeter used to trigger on the large energy of the proton (anti-proton) from Λ ($\overline{\Lambda}$) decays.

II E871 SPECTROMETER

Figure 1 shows the elevation and plan views of the HyperCP spectrometer. A charged hyperon secondary beam of 150 GeV/c momentum with a 19.5 mrad angle to the horizontal was defined by a target assembly followed by a collimator with a 4.88 μsr solid angle acceptance embedded in a 6 m long 1.67-T dipole magnet.

CP450, SciFi97: Workshop on Scintillating Fiber Detectors
edited by A. D. Bross, R. C. Ruchti, and M. R. Wayne
© 1998 The American Institute of Physics 1-56396-792-8/98/$15.00

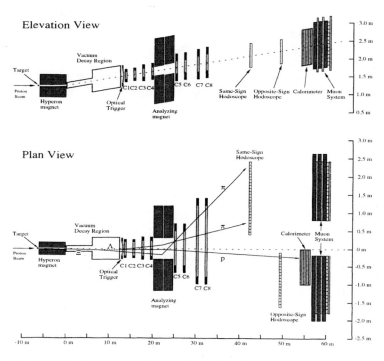

FIGURE 1. Elevation and plan views of the HyperCP spectrometer.

Immediately downstream of the decay region were four high-rate proportional wire chambers, followed by an analyzing magnet composed of two BM109 dipoles having a combined p_t kick of 1.43 GeV/c, followed by another four high-rate proportional wire chambers. The analysis magnet had sufficient strength to ensure that the protons and pions from the Ξ, Λ, and K decays were always well separated from each other as well as from the charged beam in the downstream portion of the spectrometer. This allowed a simple, yet selective, trigger to be formed by requiring the coincidence of charged particles in the hodoscopes at the rear of the spectrometer and on either side of the channeled beam. A hadronic calorimeter on the proton side was used to make the trigger "blind" to muons and to reduce the trigger rate from interactions of the channeled beam in the spectrometer. A simple muon system at the rear of the spectrometer allowed access to rare and forbidden decay physics.

III HADRONIC CALORIMETER

Due to the high flux expected for HyperCP, the hadronic calorimeter was designed to be used at the trigger level. In order for this to work, the calorimeter had to be fast since fluxes greater than 300 kHz were encountered. The calorimeter had to have a large dynamic range since it had to detect muons (for calibration purposes), and it had to contain hadronic showers in order to avoid trigger biases. Good position resolution was not necessary.

The calorimeter contained 64 layers of 2.41 cm Fe and 0.5 cm PS scintillator. Its total length was 238.92 cm, and its transverse size was 99.0×98.0 cm^2. It was split into 8 cells, four longitudinal and two lateral, each cell containing 256 fibers. The division into 8 cells was designed for simplicity of readout. The total interaction length was $9.6\lambda_I$, and the corresponding radiation length was $88.5X_o$. It had a sampling fraction of 3.54%. A side view of the calorimeter is shown in Fig. 2 where the particles are incident perpendicular to the absorber plates.

IV SCINTILLATORS AND FIBERS

The calorimeter incorporated a combination of scintillator and fiber, where the scintillator was the active medium. The scintillator, chosen for its speed, light output, and cost was Kuraray SCSN-38 [1]. It has a polystyrene base, butyl-PBD first fluor and PBD second fluor, with an emission frequency centered at 428 nm. Bicron BCF-92 [2] (G2) wavelength shifting fibers embedded in keyhole shaped grooves were used to carry light to acrylic light guide mixers. The fibers were 2 mm in diameter and 2 m long. This particular fiber is very bright and has a short decay time of τ=2.7 ns. It absorbs light at 405 nm and emits at 492 nm. The use of fibers rather than light guides allowed for a more hermetic calorimeter, they minimized the needed photocathode area, and they greatly simplified the mechanical design. There was a total of 32 fibers in each plane of scintillator with a spacing between grooves of 3 cm.

The fibers were read out at one end with the other end polished and sputtered with aluminum. The polishing process involved installing the fibers in a 1.8 m $\times$ 1.9 cm PVC tube that could hold up to 75 fibers. The ends to be polished were inserted in a ceramic sleeved mold to provide rigidity, and kapton tape was used to seal the bottom. Water was added to the mold, up to a depth of 7.6 cm. In steps, never exceeding a total depth of 2.5 cm, the mold was slowly immersed in liquid nitrogen, remaining there for 15 minutes. This ensured that the individual fibers were frozen in place as a complete unit for polishing.

After the fibers were frozen, the mold was inserted into a polishing machine [3] which used a 2-cutting head process. The first head of four carbon steel blades provided the rough cut, making a series of three passes, each pass cutting ~1/16". The second head of two diamond blades provided the fine cut, making two passes, the first pass taking a 1 mil step and the other taking a 1/2 mil step. Polished

430

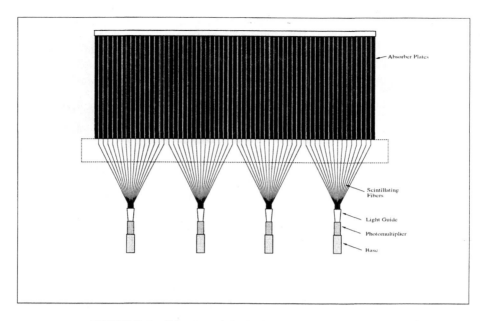

FIGURE 2. Side view of the hadronic calorimeter.

fiber ends were inspected individually for defects using a magnifying glass. Rejects, accounting for 2%, were polished again.

The fibers were aluminized in a 10-foot tubular vessel utilizing a sputtering magnetron gun driven by a DC power supply at 300 Watts. The average sputtering time was 20 minutes. For good reflectivity, 99.999% pure aluminum targets were situated a distance of 15.2 cm away from the substrate. The sputtering process was started after achieving a 10^{-6} Torr vacuum and then backfilling with "sputtering grade" Argon gas to 2 milliTorr pressure. The coating was approximately 250 nm thick and was monitored using an oscillating quartz crystal sensor device. Mylar and kapton tape were used as packaging to protect fibers from being coated insitu and in protecting the handling of fibers in installation. Ultimately, groups of 500 fibers were coated at a time. UV-curing epoxy was put on the sputtered ends for protection.

The sputtered fibers were then tested for reflectivity. Since this was a destructive procedure, only 8-10 fibers were tested from each batch. The fibers were measured twice, once with the sputtered end intact, and the other with the end cut off. A setup consisting of a UV lamp, a photodiode, a picoampmeter, and a PC was used for this measurement. The fibers were tested by exciting them with the UV lamp source and measuring their output current. The lamp source was moved in 10 cm increments along the length of the sputtered fiber, and the subsequent output currents were measured. The sputtered end of the fiber was cut off at a 45° angle about 2 mm from the end and painted black. The light output was measured again. Taking the ratio of the light output at each source position for the two measurements yielded average reflectivities of 0.85±0.05 and attenuation lengths (long component) of 3.48±0.34 m. This corresponded to a difference in light output of 6% from top to bottom of the calorimeter, well within specifications.

V CALORIMETER READOUT

Each cell of 16×16=256 fibers was potted in opaque epoxy which was attached to an acrylic light guide that mixed the light before reaching the green extended R329 Hamamatsu [4] PMT's. A high-rate transistor base was designed to provide good gain stability with rate, with about a 300 μA maximum stable operating rate [5]. The signals from the calorimeter were sent to the fast electronics room via RG-8 cabling, where they were split. The summed output of the signals was discriminated to provide an energy trigger.

The other portion of the signals from the calorimeter was sent to ADC's. These were 14-bit flash ADC's, with 1 μs conversion times, designed and fabricated to digitize the calorimeter signals for the data acquisition system.

The calibration of the calorimeter was performed by using particles of well measured energy impacting it, such as the protons from Λ decays. Cell-to-cell gains were adjusted during the run using muons which produced 1.6 photoelectrons per plane of scintillator. A light pulser system using a Nitrogen laser and PIN diodes was also used to monitor the time dependence of the calibration.

VI CALORIMETER TRIGGER PERFORMANCE

Figure 3 shows how the calorimeter performed. The minimum proton momentum from a Λ decay was around 70 GeV/c. The trigger threshold was set to a conservative value of about 50 GeV. The left plot is the trigger acceptance as a function of energy for charged particles entering the calorimeter. Typical efficiencies were greater than 99%. The right plot is the uncalibrated sum of the ADC's for a momentum cut between 100 GeV/c and 120 GeV/c. The preliminary (uncalibrated) estimate of the calorimeter resolution is $\sigma/E = 80\%/\sqrt{E} \oplus 2.5\%$.

The calorimeter trigger proved to be a vital component of the main triggers, reducing the trigger rate for Ξ's and K's by about a factor of six. The fast response of the wavelength shifting fibers allowed us to use the calorimeter as a trigger in a very useful way.

VII ACKNOWLEDGEMENTS

We thank the Fermilab staff for their contributions to this experiment and for machining the keyhole shaped grooves in the scintillators and in the fiber polishing

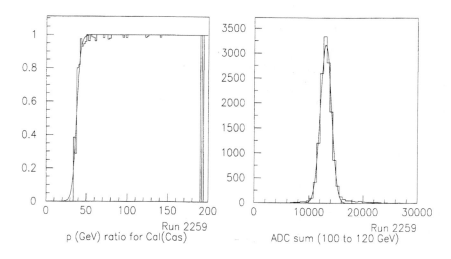

FIGURE 3. Calorimeter energy and resolution.

and sputtering processes. Thanks especially to Phyllis Deering for the machining of the keyhole shaped grooves in the scintillators, and thanks to Eileen Hahn for the sputtering of the fibers. This work was supported in part by the U.S. Department of Energy and the National Science Foundation.

REFERENCES

1. Kuraray International Corp., 30th Floor, PanAM Building, 200 Part Avenue, NY 10166.
2. Bicron, 12345 Kinsman Rd., Newbury, OH 44065.
3. Designed and built at Fermilab by Carl Lindenmeyer et al.
4. Hamamatsu Corp., 360 Foothill Rd., Bridgewater, NJ 08807-0910.
5. D. Rajaram, "Design and Performance of a High-Rate Photomultiplier Base," Master's Thesis, (1996) pg. 51.

SESSION 11: CALORIMETRY - II

Chair: V. Hagopian
Scientific Secretary: T. Lin

The KLOE fiber
electromagnetic calorimeter

Marco Incagli

Istituto Nazionale di Fisica Nucleare - Pisa (Italy)
For the KLOE EMC group[1]

Abstract. The construction and equipment of the KLOE electromagnetic calorimeter has ended in March 1997. In parallel to the construction, all modules have been tested at the Cosmic Ray Test Stand (*CRTS*) facility, in Frascati National Laboratories (Rome). The construction technique, based on scintillating fibers alternated to very thin (0.5 mm) grooved lead planes, is described and the main results both from the *CRTS* and from a preliminary Test Beam with low energy electrons and muons are reported in this note.

CP PHYSICS AT KLOE: THE ROLE OF THE EMC

The main goal of the KLOE experiment at the DAΦNE ϕ–factory is to study CP violation in the Kaon system, in particular by measuring $\Re(\epsilon'/\epsilon)$ with an accuracy of $\mathcal{O}(10^{-4})$. The *golden channel* in which $\Re(\epsilon'/\epsilon)$ will be measured is the CP violating decay $\phi \to K_s K_l \to (\pi\pi)(\pi\pi)$, using the standard technique of the *Double Ratio*. The electromagnetic calorimeter (*emc*) will have to identify

---

[1] The KLOE emc group is composed by:
M. Antonelli[c], G. Barbiellini[j], S. Bertolucci[c], C. Bini[f], C. Bloise[c], P. Branchini[h], R. Caloi[f], G. Cabibbo[g], P. Campana[c], F. Cervelli[e], P. Ciambrone[c], C. Colantuono[f], G. De Zorzi[f], G. Di Cosimo[f], A. Di Domenico[f], O. Erriquez[a], S. Di Falco[e], A. Farilla[a], A. Ferrari[f], P. Franzini[f,d], M. L. Gao[b,c], P. Gauzzi[f], S. Giovannella[c], E. Graziani[h], S. W. Han[c,b], H. G. Han[b,c], M. Incagli[e], W. Kim[i], G. Lanfranchi[f], J. Lee-Franzini[c,i], T. Lomtadze[e], S. Miscetti[c], F. Murtas[c], A. Passeri[h], F. Scuri[j], E. Spiriti[h], L. Tortora[h], G. Venanzoni[e], Y.G. Xie[c], P.P. Zhao[c,b]. [a] Dipartimento di Fisica dell'Università e Sezione INFN, Bari ; [b] Institute of High Energy Physics of Academica Sinica, Beijing, China ; [c] Laboratori Nazionali di Frascati dell'INFN, Frascati ; [d] Physics Department, Columbia University, New York ; [e] Dipartimento di Fisica dell'Università e Sezione INFN, Pisa ; [f] Dipartimento di Fisica dell'Università e Sezione INFN, Roma I ; [g] Dipartimento di Fisica dell'Università e Sezione INFN, Roma II ; [h] Istituto Superiore di Sanità e Sezione INFN, ISS, Roma ; [i] Physics Department, State University of New York at Stony Brook ; [j] Dipartimento di Fisica dell'Università e Sezione INFN, Trieste/Udine.

CP450, *SciFi97: Workshop on Scintillating Fiber Detectors*
edited by A. D. Bross, R. C. Ruchti, and M. R. Wayne
© 1998 The American Institute of Physics 1-56396-792-8/98/$15.00

and reconstruct the neutral decays $K_{L,S} \to \pi^0 \pi^0$. In the ϕ two body decay the neutral Kaons are emitted at low speed ($\beta\gamma \simeq 0.22$) and will have a mean flight path of $\simeq 3.5\,m$; as a consequence the photons emitted from the π^0 decay will be *soft*. Therefore the emc has to be efficient in the energy range $20 - 300\,MeV$. Moreover the emc has to *measure* the K_L decay point (*neutral vertex*) with a precision of $\simeq 1\,cm$, in order to identify the Fiducial Volume relevant for the *Double Ratio*[2]. Knowing the K_L flight direction, from the charged decay of the K_S, the γ arrival point on the calorimeter and its flight time, it is possible to identify the *neutral vertex* with a simple algorythm described, for example, in [1]. This implies that the *emc* has to have excellent time and position resolution. Note also that, because of the long K_L flight path, the photons will impact on the calorimeter face from all directions, making a projective geometry useless.

The main background to the *golden channel* is the CP conserving decay $K_L \to \pi^0 \pi^0 \pi^0$ in which two of the final photons are missed. This channel is $\simeq 200$ times the *golden channel*, so, in order to reach the design accuracy on $\Re(\epsilon'/\epsilon)$, it has to be suppressed at the level of 10^{-4}. In order to reach this rejection factor the calorimeter has to be as hermetic as possible.

The *emc* requirements can be summarized as follows:

- full efficiency in the range 20–300 MeV;

- excellent time resolution ($\simeq 70$ps $/\sqrt{E(GeV)}$);

- determination of the γ conversion point with an accuracy of $\simeq$ 1cm $/\sqrt{E(GeV)}$;

- good energy resolution ($\simeq 5\% /\sqrt{E(GeV)}$) to apply kinematical constraints;

- hermeticity to reject $K_L \to \pi^0 \pi^0 \pi^0$;

- fast response for trigger purposes, mainly to reduce the Bhabha rate.

All of the above has to be accomplished by a calorimeter having dimensions of $\sim 4m \times 4m$!

THE KLOE CALORIMETER

Given the above requirements, the choice has been made of using a fine sampling lead–scintillating fibers calorimeter with $1\,mm$ fibers and $0.5\,mm$ thick lead foils, kept together by optical glue (Bycron BC600). The (volume) ratio *fiber : lead : glue* is $48 : 42 : 10\%$, allowing for a very high samplng fraction of 13%. The calorimeter density is $\rho = 5\,gr/cm^3$, and its equivalent interaction length is $X_0 = 1.5\,cm$.

---

[2] The same situation occurs also in the K_S decay, but the short lifetime of this particle ($\simeq 0.6$cm) makes the identification of its Fiducial Volume less problematic.

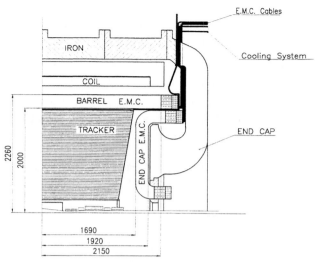

FIGURE 1. Section view of KLOE detector.

The KLOE *emc* consists of a central part, the *barrel*, and two *end-caps* (see fig.1). The *barrel* covers the angular region $43^o < \theta < 90^o$ and it is organized in 24 modules of trapezoidal cross section $4.3\,m$ long and $23\,cm$ thick (corresponding to $15X_0$), approximating a cylindrical shell of $4\,m$ inner diameter. Fibers run parallel to the beam. Each *end-cap* consists of 32 modules of different lengths which run parallel to the vertical axis (y-axis). They also are $23\,cm$ thick and are bent outward at the two ends, becoming parallel to the *barrel*. The resulting C shape provides an hermetical coverage of the calorimeter.

It is important to note that the fibers run *perpendicular*, or at large angles, with respect to incident particles, differently from standard *Spaghetti Calorimeters* that point toward the Interaction Point.

The fibers are read, at the two ends, by photomultipliers coupled to the calorimeter through light guides glued on the calorimeter sides. The time difference at the two ends is proportional to the coordinate along the fiber. The readout granularity is 4.4×4.4 cm^2, for a total number of 4800 photomultipliers.

Several commercial plastic fibers have been tested at CERN PS in summer 1994 [2]. All of the tested fibers, emitting in the blue region, have shown a fast time response ($\simeq 2\,ns$ time rise) and a light yield of $3.5 - 4.5\,pe/mm$ at $50\,cm$ from PM.

After these tests the choice has been made of using Kuraray SCSF-81 fibers in the first half of the calorimeter (closer to the Interaction Point) and Pol.Hi.Tech. 0046 fibers in the second half.

A subsample of all the $\simeq 2.5$ million fibers used in the calorimeter has been tested to check for quality standards. The test setup consists of a β source (^{90}Sr), which can be positioned in reference points along the fiber, and a PM optically air-coupled. The intensity of the current is registered for each reference point and then fitted with the sum of two exponential functional forms. The longest attenuation length λ and its corresponding current I are reconstructed with a reproducibility of 2% and 4%, respectively. This check has been perfomed on two fibers per batch ($\simeq 1\%$ of the total fibers).

Values obtained for the Pol.Hi.Tech. fibers have a wider spread in the $\lambda - I$ plane (see fig. 2) and show average worst characteristics with respect to the Kuraray ones (except for price). In order to guarantee a uniform response among calorimeter modules of different lengths, we select fibers which give the same amount of current when illuminated at their farthest length. Therefore the worst fibers have been assigned to the shortest *end-cap* modules, with the condition that these fibers have to lie above the solid line, shown in figure 2, corresponding to $\simeq 25$ photoelectrons at 1 meter from the PM.

MODULES CONSTRUCTION

Mechanical assembly

The mechanical assembly of the 24 *barrel* modules has been carried out at the *Pol.Hi.Tech.* factory, while the 64 *end-cap* modules have been built in parallel at the INFN laboratories of Pisa, Rome and Frascati.

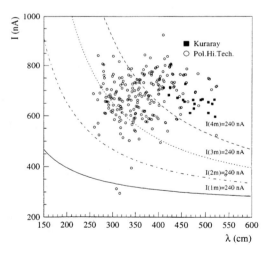

FIGURE 2. Distribution of tested fibers in the $\lambda - I$ plane. Isoquality lines are shown that correspond to the functional form $\lambda I = const$. The solid line draws the limit between accepted and rejected fiber batches.

First of all the lead foils, of $0.5\,mm$ thickness, are machined to the proper grooved shape by our *lead-o-matic*. Tolerances on foil construction are kept tight in order to facilitate the assembly: grooves do not deviate more than $0.5\,mm$ from a straight line, while the pitch is precise to the level of $20 - 30\mu m$. The assembly starts by glueing the first lead foil on the $3\,cm$ thick ($2.5\,cm$ for the *end-cap*) aluminium supporting plate. After the first layer is set, optical epoxy glue (Bicron BC600) is distributed on the plane, a layer of fibers is positioned in the grooves and glue is spread again before positioning the next lead foil. This procedure is repeated until a stack of 10-20 planes is formed. Before the glue starts curing, a uniform pressure of $\simeq 1\,ton$ is applied for 2 hours on the calorimeter surface, to allow for a more uniform glue distribution. This procedure is repeated until a thickness of $23\,cm$, corresponding to $\simeq 200$ lead-fiber planes, is reached.

The only relevant difference between *barrel* and *end-cap* construction is that, for end-cap modules, the two ends are mounted on removable surface and they are bent before the pressure is applied.

Modules equipment

After construction, modules are milled to their final dimensions, wrapped in aluminium tape $0.1\,mm$ thick, for light tightness, and then the read-out system is mounted on the two ends: light guides are glued and PM's holders are mounted in an aluminium box.

The light guides consist of a mixing part and a Winston cone concentrator. This system allow for a high efficiency ($\epsilon \simeq 80\,\%$), even with an area reduction factor of 4, and for a uniform photon distribution on the photocathode.

The calorimeter is placed inside the superconducting coil providing a magnetic field of $0.6\,T$. In the PM area the residual magnetic field is up to $2\,KGauss$, with an angle with respect to PM axis up to $25°$. Because of this, fine mesh Hamamatsu R5946 photomultipliers, specially designed for KLOE, are used. They have been tested with a solenoid and they show a response indipendent from the magnetic field, in the KLOE working region.

TEST OF MODULE 0

At the beginnig of 1994 the first full size *barrel* module has been built, equipped with the final light guides, PM's and front-end electronics. It was taken to the *Paul Scherrer Institute* (Zurich, CH) in July and tested with electron, muon and pion beams in the energy range $100 - 450\,MeV$.

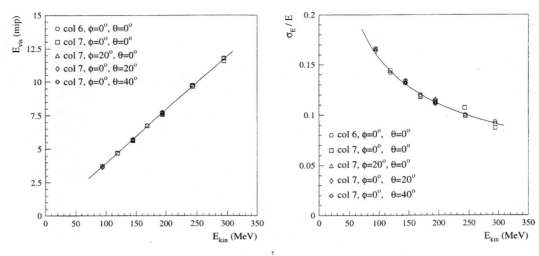

FIGURE 3. Energy response (a) and energy resolution (b) for electrons.

Energy response and resolution

The module was calibrated using muons of $450\,MeV$ energy that, in KLOE calorimeter, behave as *Minimum Ionizing Particles* (*MIP*s).

The energy response to electrons is linear and indipendent from the incidence angle (see fig. 3.a). The slope of this distribution indicates that a *MIP* release, in a given cell, the same energy as a $32\,MeV$ photon, corresponding to a sampling fraction of 13% and to a light yeld of $\simeq 1600\,pe/GeV$, for Kuraray fibers, and of $\simeq 1075\,pe/GeV$, for Pol.Hi.Tech. fibers (integrating on all the cells). The energy resolution is $\sigma_e/E = 5\%/\sqrt{E(GeV)}$ and it is fully dominated by sampling fluctuations (fig. 3.b).

This excellent resolution is mainly due to the very fine sampling and to the correspondingly high sampling fraction. Also the usage of scintillating fibers, instead of scintillator slabs, allows for the construction of very homogeneus devices with good time performances.

Time resolution

The time resolution was studied using the energy weighted sum of the sum of the TDC readouts at the two ends (A and B) of a given cell:

$$T = \frac{\Sigma_i (T_A + T_B)_i (E_A + E_B)_i}{\Sigma_i (E_A + E_B)_i} \tag{1}$$

442

A simple gaussian fit nicely reproduces the time distribution. After quadratically subtracting the trigger jitter ($\simeq 180\,ps$), the time resolution can be parametrized as $\sigma_T = 72ps/\sqrt{E}$, showing its dependence on the number of photoelectrons N_{pe}.

THE COSMIC RAYS TEST STAND

In parallel with the mass construction, each module has been tested at the *Cosmic Rays Test Stand* facility (*CRTS*), built in summer 1995 in Frascati Laboratories. It consists of a telescope made by twelve $60\times7\times3\,cm^3$ NE110 scintillator slabs and 4 planes of $310\times64\,cm^2$ Limited Streamer Tubes (LST) with longitudinal and transverse coordinate read out by means of 1cm pitch aluminium strips. A barrel module could be inserted between the two halves of such a telescope (see fig.4). Between the barrel module and the bottom part of the telescope, a layer of 5cm thick lead was inserted to harden the spectrum of the cosmic rays triggering the apparatus. On top of this structure one or two *end-cap* modules can be placed and tested at the same time.

The trigger requires a top-bottom coincidence of scintillators or LSTs, for a total rate of $\simeq 20\,Hz$. Each module was tested for at least one week (five full days) in a *Standard Calibration Run* (*SCR*), corresponding to $\simeq 3 \times 10^6$ events, over the whole module, and $\simeq 3 \times 10^4$ events in each cell.

During data taking ADC and TDC spectra have been collected for each read-out element. The ADC spectra are used to obtain the *MIP* peak at half the module length ($z = 0$), the attenuation length (λ) and the number of photo-electrons (N_{pe}); from the TDC spectra the light velocity in the fibers, the time

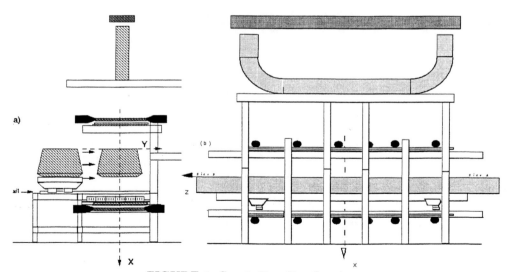

FIGURE 4. Cosmic Rays Test Stand.

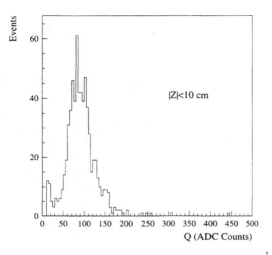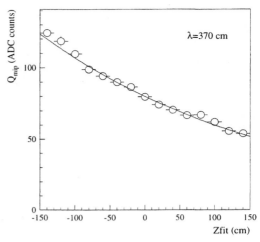

FIGURE 5. MIP peak in a cell in a slice of 0±10 cm. In (b) the peaks, taken for a set of slices in z, are fitted with an exponantial function to get λ.

resolution and the position resolution along the fibers are derived.

Attenuation length and Number of photoelectrons

The ADC spectrum at $z = 0 \pm 10\,cm$ of a *MIP* is shown in fig.5.a. Plotting the peak of this distribution, for different slices of z, the attenuation length λ is obtained (see fig.5.b). In a SCR, λ can be measured to an accuracy of 2% ($\simeq 6\,cm$).

N_{pe} can be obtained by plotting the ratio r_{ADC} of the ADC readout at the two ends of a given channel. In this ratio, taken at $z = 0$, the Landau fluctuations cancel out, and one gets: $\sigma_{r_{ADC}}/ < r_{ADC} >= 1/\sqrt{N_{pe}}$.

Figure 6 summarizes the values of λ (a and b) and of N_{pe} (c and d) for all barrel and end-cap modules. The large variation of the λ values in end-cap modules is due to the assignement of the lowest quality fibers to the shortest modules. This allows for a uniform distribution in N_{pe}: on average 30-35 *pe* are collected for a *MIP* at $z = 0$.

Figure 6 shows also that the quality of the barrel modules is higher with respect to Module 0, in terms of light yield and attenuation length. This is due to the improved quality of the fibers and to the special care devoted in handling the scintillating fibers, as they resulted to be damaged by the ultraviolet component of the natural and artificial light.

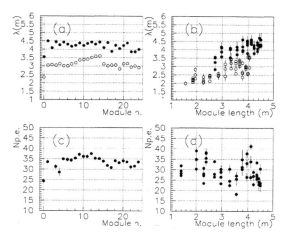

FIGURE 6. Attenuation length and number of photoelectrons for barrel (left) and end-cap (right) modules.

Time resolution

Since the LST trigger has a large intrinsic time jitter, the time resolution can only be evaluated by looking at time differences, in which the trigger jitter cancels out. The parameter that has been studied is the difference between the mean times of two superimposed cells crossed by the same track:

$$\Delta T_i = \frac{1}{2}(T_i^A + T_i^B) - \frac{1}{2}(T_{i+1}^A + T_{i+1}^B) \tag{2}$$

Assuming that the cells have the same time response, the width of this difference, divided by $\sqrt{2}$, corresponds the time resolution σ_t.

Taking into account that the energy release of a *MIP* in one element corresponds to the energy released by a $32\,MeV$ photon, a slight improvement in the energy resolution (dominated by sampling fluctuations) and a large improvement in the time resolution (dominated by photoelectron statistics) is observed with respect to Module 0. The expected resolutions are: $\sigma_E/E = 4.7\%/\sqrt{E}$ and $\sigma_t = 52ps/\sqrt{E}$.

Z-coordinate resolution

The coordinate along the fiber is measured by the left-right time difference: $Z = 1/2(T_A - T_B - \Delta T_0)v_f$, where v_f is the effective velocity of light in the fibers.

In a *SCR* v_f can be determined to an accuracy of 0.5% (see fig.7.a) and the constant ΔT_0 with a statistical uncertainty of $\simeq 10\,ps$. The average value for

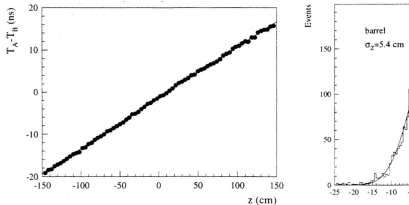

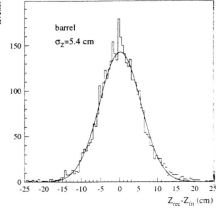

FIGURE 7. Effective speed of light in fibers (a) and z resolution.

v_f is $16.9\,cm/ns$, and the Z-resolution for *MIPs* is $\sigma_z = 5.4\,cm$ (see fig.7.b), corresponding to $\sigma_z = 0.9\,cm/\sqrt{E}$.

End-Cap tests

A particular care has been taken in studying the behaviour of the curved part of the end-cap modules. Two different configurations have been analized by selecting cosmics of appropriate directions:

- Central events: the particle hits the end-cap module in the horizontal section; no differences from barrel behaviour are expected;

- Side events: the ray hits the cell in the curved or in the vertical sections (see fig.4); a larger amount of energy per cell is expected.

The energy linearity as a function of the track length inside the cell has been studied both for single-hit events and for all end-cap events. By selecting cosmic rays having incidence angle $\theta < 45^o$ (with respect to normal incidence) it is possible to study the energy deposited for a track length l in the interval $4.5 - 7\,cm$ (see fig.8.a). The energy response is linear and this linearity is well verified also for tracks passing through the curved part (fig.8.b) up to a track length of $\simeq 20\,cm$. Above this value a saturation effects is observed. This effect corresponds to the saturation of the preamplifiers mounted on the PM base, saturation that occurs at 2 Volts, corresponding to $\simeq 1300$ ADC counts. However this pulse height corresponds to an energy release of $\simeq 150\,MeV$ in a single cell, and it is very rare that *real events* in KLOE detector will leave more than that energy.

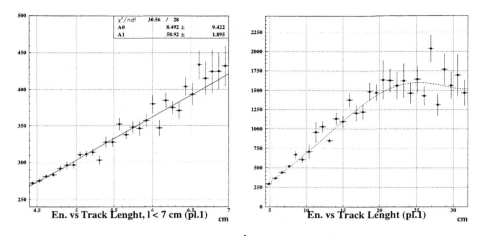

FIGURE 8. Energy linearity as a function of track length inside the module for central (left) and for all (right) end-cap events.

CONCLUSIONS

The CP violation physics program of KLOE experiment at DAΦNE requires excellent e.m. shower detection performances. A hermetic, fine sampling calorimeter has been constructed and extensively tested first at PSI beam (Zurich, Switzerland) and then at the Cosmic Ray Test Stand facility in Frascati. A time resolution $\sigma_t = 55ps/\sqrt{E(GeV)}$, a space resolution along fibers of $\sigma_z = 0.9cm/\sqrt{E(GeV)}$ and an energy resolution of $\sigma_E/E = 4.7\%/\sqrt{E(GeV)}$ have been observed. The first calorimeter element has been mounted on the iron yoke on october 16, 1997. The installation will be completed by the end of December 1997.

REFERENCES

1. The KLOE collaboration, *The KLOE detector: technical proposal*, LNF-93/002 (1993).
2. Antenelli, A. et al., *Nucl. Instr. and Meth.*, **A370**, 367 (1996).

Hadron Calorimetry using scintillator Tiles and WLS fibers: the Tilecal/ATLAS

Tilecal/ATLAS collaboration

presented by M. Varanda
LIP and Univ. of Lisbon, Lisbon, Portugal

Abstract. Scintillator Tiles read out by WLS optical fibers are the active medium of the Tilecal, the barrel hadron calorimeter of the ATLAS detector. The scintillator tiles are oriented perpendicularly to the beam axis and light is detected by photomultiplier tubes.

Several optical components have been tested in the laboratory and in calorimeter prototypes. The first real scale prototype of a barrel module was built and tested by the first time in 1996. The response to beams of muons and pions allows to study the performance of the calorimeter, namely its uniformity. Preliminary results are presented.

INTRODUCTION

The Tilecal, the barrel part of the hadron calorimeter of the ATLAS detector, is a sampling calorimeter using 3mm thick scintillator tiles read out by 1mm diameter WaveLength Shifting (WLS) fibers as active medium and iron as absorber [1]. The barrel part is subdivided in 64 modules, each one made of stacks of repeating elements, with 18 mm thickness (periods). Each period is a stack of four layers: the first and third layers are formed by large trapezoidal steel plates (masters), the second and fourth layers are formed by small trapezoidal steel plates alternating with scintillator tiles (figure 1). An innovative feature of the design is the orientation of the tiles which lie in the r-ϕ plane, perpendicularly to the beam axis [2].

The first real scale module (the barrel module 0) was tested by the first time in 1996. The barrel module 0 has a front face area of $560\times22cm^2$ spanning $\frac{2\pi}{64}$ in the azimuthal angle with a radial depth of 7.6λ at η=0. Three depth segmentations are defined with thicknesses equal to 1.5λ for the first sampling, 4.2λ for the second sampling and 1.9λ for the third one at η=0. The barrel module 0 has a projective tower read out with a granularity in $\Delta\eta \times \Delta\phi$ equal to 0.1$\times$0.1 for the first two samplings and 0.2$\times$0.1 for the third sampling. There are 11 tile sizes in the radial depth. The transversal segmentation within the calorimeter is achieved by grouping

CP450, *SciFi97: Workshop on Scintillating Fiber Detectors*
edited by A. D. Bross, R. C. Ruchti, and M. R. Wayne
© 1998 The American Institute of Physics 1-56396-792-8/98/$15.00

fibers reading out different tiles in one photomultiplier. Two photomultiplier tubes (PMT) are required to read out each cell: one to the right side and the other to the left side of the tiles. The PMT's are housed in the drawer inside the girder structure located in the outer face of the modules.

In this paper the optical components of the calorimeter are described and the results on the light yield uniformity and response of the barrel module 0 to pions and muons are presented. Other aspects on the performance of the calorimeter are described in ref [3].

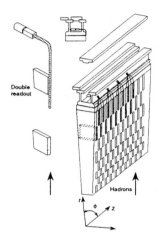

FIGURE 1. The periodic structure of the Tilecal.

THE OPTICAL COMPONENTS

The scintillator tiles for the Tilecal are produced by injection molding technique, using polystyrene as scintillator matrix base, and PTP and POPOP as dopants. The ionizing particles crossing the tiles induce the production of light in the base material of the tiles, with wavelength in the UV range which subsequently is converted to visible light by the scintillator dyes. The wavelength shifting process is assured by the combination of 1.5% PTP and 0.05% POPOP dopants. The light generated by ionizing particles is internally reflected between the tile surfaces and transmitted to the WLS fibers. The uniformity of the light output of a tile depends on the local light yield uniformity and on the tile transparency. The optical parameters are sensitive to the base material quality, to the injection conditions, to the homogeneity of the dopants distribution in the tile and to the tile thickness.

Central scans with a ^{90}Sr source in the tiles show non-uniformities exceeding the design requirement of 5% near the read out fibers (figure 2). The tiles are wrapped with Tyvek, a diffuser material with an opacity of 97% and a reflectivity of 95%.

To reduce the non-uniformities, a mask was applied on the tyvek wrapper to absorb part of the light which would otherwise be reflected back into the tile [4]. The ideal mask should cover a complex surface constituted by regions where the light output is more than 5% above the tile average. For the barrel module 0 only a trapezoidal black strip was used and it was decided that no masking was required on the 5 smallest tiles, because their uniformity was acceptable [5].

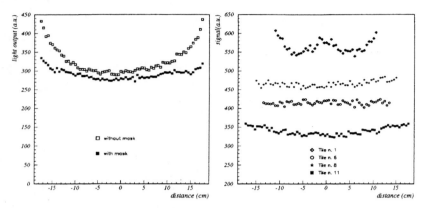

FIGURE 2. Central scan for some tiles of the barrel module 0. On the left, a scan for tile size #11 with and without mask; on the right, only the three largest tiles are masked.

The WLS fibers collect the scintillation light from each side of the tiles and transmit it to a PMT. Each fiber collects light from one or two tiles and this is accomplished by the fluor in the fiber which absorbs the blue light from scintillator and re-emits it to longer wavelengths. The light propagated through the fiber is read out by a PMT in one of the fiber end. The other fiber end is aluminized in order to increase the light collection in the PMT. The light collection should be efficient, fast, with low attenuation along the fiber length and the fibers should be radiation hard.

Sample fibers from three producers (Bicron, Kuraray and Pol.Hi.Tech) have been (and are still being) studied in laboratory in a comparative way. These fibers are characterized in what concerns to light output, attenuation length, radiation hardness and mechanical flexibility. By their quality, the double clad Kuraray fibers Y11(200) doped with 600ppm UV Absorber were chosen to equip the module 0.

Many factors contribute to fiber-to-fiber response fluctuations. These include fiber diameter, fiber light yield, attenuation length and aluminization quality.

The light yield measured (I), when fibers are excited at different distances (x) from the PMT (figure 3), can be parameterized by a sum of two exponentials with reflection (R) at the aluminized end, given by the following equation [6]:

$$I = I_{o,s}e^{\frac{-x}{L_{short}}} + I_{o,l}e^{\frac{-x}{L_{long}}} + R\{I_{o,s}e^{\frac{-(2L-x)}{L_{short}}} + I_{o,l}e^{\frac{-(2L-x)}{L_{long}}}\} \quad (1)$$

The flexibility of the WLS fibers for the Tilecal is essential because the routing of

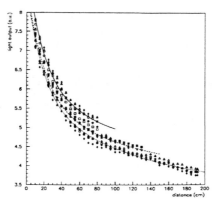

FIGURE 3. Light output produced by 1m, 1.5m and 2m long aluminized fibers. The fibers are excited with light produced by a Tilecal scintillator.

the fibers to the PMTs requires that they are bent with curvature radii larger or equal to ~5cm causing some mechanical stress. The bending can affect the optical properties of the fibers, since it introduces mechanical damage to the fiber surface, namely small micro-cracks in the cladding [1].

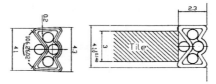

FIGURE 4. Boxing of the profiles into the grooves of the module.

To allow the optical coupling between the fiber and the tile, profiles made of high impact white polystyrene doped with TiO_2 to provide a good diffuser reflectivity have been designed and prototypes have been constructed (figure 4). The elasticity of the plastic material allows an easy insertion of the profiles into the grooves of the calorimeter. The profiles box into the grooves of the module by a simple pressure action. The dimensions of the profiles should cope with the grooves dimensions and tolerances. Each profile gives support to four fibers, but only one of the fibers is in optical contact with the tile, collecting the light that is produced in the tile. The profiles are in development phase.

THE LIGHT YIELD UNIFORMITY

The light output of a Tilecal cell depends on many parameters among which the most important are the tile and the fiber response. To get the cell geometry

451

of the module 0, tiles of several sizes and fibers of several lengths are grouped in the same photomultiplier, leading to non-uniformities within cells and different light outputs for different cells. Light budget calculations (the product of the tile response by the fiber response) were done to design the fiber routing configuration. The minimization of the non-uniformities within a cell and the maximization of the total light output are the most important conditions used to design the fiber routing. The uniformity response of the light yield of the barrel module 0 was studied using two sources of information:

- the light yield produced by the ^{137}Cs source passing through the tiles of the module;

- the light yield produced in the cells by 180 GeV muons impinging the barrel module 0 at $\theta = 90°$;

The ^{137}Cs data was taken passing the source through all the tiles during the calibration of the barrel module 0. In the following analysis the integral of the calibration curves normalized to the width of the cell was used. This quantity is proportional to the amount of light produced by the source and collected in the PMT, and to the gain of the PMT [1,7].

To estimate the energy loss by muons in each cell for each tile size, a fit with a Moyal function was done to the total energy deposited in each cell and the most probable value (MOP) was taken from the fit (fig. 5, left). The signal produced by muons in each cell was normalized to the size of the cell (18mm×number of tiles in the cell).

The light yield studied with the ^{137}Cs source shows non-uniformities with a rms value of the order of 8% while the non-uniformities on the light yield measured with muons have a rms value of the order of 10% (preliminary results). A study of the correlation between the Cesium and the muon data shows a correlation with a rms value of the order of 7%.

RESPONSE OF THE BARREL MODULE ZERO TO MUONS AND PIONS

The Tilecal can identify muons with energies above 2 GeV [8]. These muons lose in average about 2 GeV in the whole calorimeter. The energy loss spectrum approximately follows a Landau distribution with high energy tails due to energetic δ rays as well as radiative losses. Figure 5 shows the most probable value of the energy losses which increase with the energy of the beam due to the increasing probability of radiative processes. The most probable value varies from ∼2.4GeV to ∼2.8GeV for incident muon energies between 10 and 180GeV, which represents a variation of ∼ 15% over the whole range studied.

The η uniformity of the barrel module 0 response to 180 GeV muons and 100 GeV pions has been studied as a function of the pseudo-rapidity η and impact

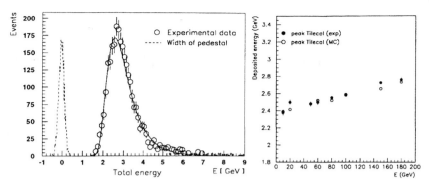

FIGURE 5. Detection of muons in the Tilecal: Left: Muon energy loss spectrum for 100GeV muons at $\eta=0.55$, a fit with a Moyal function is superimposed; Right: Most probable value in module 0 as a function of the muon beam energy.

point in the front face of the module [9]. For polar angles close to zero there is a modulation on the signal produced by 180 GeV muons which is a consequence of the structural periodicity of the module and the correspondent oscillation in the scintillator sampling fraction. Muons that crossed the module traversing almost only iron produce small signals (figure 6) while muons that cross mainly scintillator produce large signals. This is very clear for η near zero, where the amplitude of the modulation represents about 60% of the mean energy loss by muons in the module. The modulation has periodicity of $\sim$9mm, which is one half period of the barrel module 0, and the respective period increases with η.

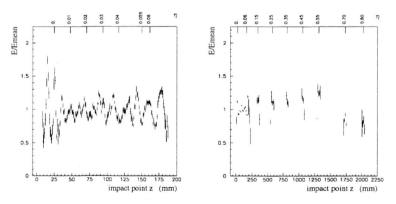

FIGURE 6. η response of the barrel module 0 to 180 GeV muons (test beam data). (a) small η values; (b) whole η range.

For 100 GeV pions a modulation on the signal for small η values can be seen (figure 7). This effect is expected to be strongly suppressed in the ATLAS environment because the electromagnetic compartment is in front of the Tilecal. The pion shower will start in the electromagnetic calorimeter and than the response will be less sensitive to the variations in the sampling fraction.

The non-uniformities measured over the whole η range of the barrel module 0 present a rms value of the order of 2%. The decrease in the signal for $\eta <0.09$ and $\eta >0.85$ is due to the lateral and longitudinal leakage. Note that only one half of the barrel module 0 is instrumented with optics and electronics, and for small η values the shower enters the non-instrumented part of the module.

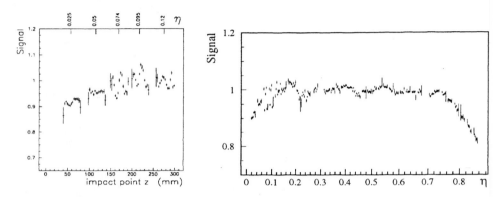

FIGURE 7. η response of the barrel module 0 to 100 GeV pions (test beam data- preliminary results). Left: small η values; right: whole η range.

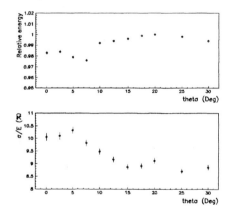

FIGURE 8. Top: Mean energy of 50 GeV pions in module 0 as a function of θ. Bottom: The corresponding relative resolution $\frac{\sigma}{E}$. The values are normalized to the value at 20°.

Figure 8 shows the variation of the energy deposited by 50 GeV pions in the module 0 normalized to the 20° point. The response is uniform except in the range from 0° to 8° where the response of the calorimeter is 2% smaller, because in this range the pions showers enter the non-instrumented part of the module. The bottom figure shows the corresponding resolution which is about 9% in the range above 15° and slightly worst bellow 8° ($\sim 10\%$), due to the large variations in the sampling fraction for small η values and to the lateral leakage [1].

CONCLUSIONS

The Tilecal barrel module 0 was instrumented with optics and electronics and was tested during 1996. Preliminary results for the response to muons and pions are in agreement with the expectations. The most probable value of the muon energy loss in module 0 increases with the energy of the beam showing an increase of about $\sim 15\%$ for muon energies between 10 and 180GeV. The η response of the barrel module 0 to muons shows a modulation for small η values that is a consequence of the structural periodicity of the module. The response of the barrel module 0 to pions is uniform over almost the whole θ range, except in the zone where the shower enter the non-instrumented part of the module. The resolution of the barrel module 0 for 50GeV pions is about 9%.

ACKNOWLEDGEMENTS

I would like to thank to LIP and to JNICT the financial support.

REFERENCES

1. ATLAS Tile Calorimeter Technical Design Report (TDR), CERN/LHC/96-42, December 1996
2. O. Gildemeister, F. Nessi-Tedaldi, M. Nessi, Proceedings of the Second International Conference on Calorimetry in High Energy Physics, Capri, 1991
3. A. Gomes, "Optics of the Tilecal Calorimeter: Components and Performance", in Proceedings of the VII Int. Conf. on Calorimetry in HEP, Tucson, USA, November 1997
4. B. Di Girolamo, E. Mazzoni, Measurements of scintillating tiles, ATLAS Internal note Tilecal-no-65, 1995
5. J. Silva, graduation thesis, F. C. Univ. Lisbon, 1996
6. M. David, A. Gomes, A. Maio, A. Henriques, Y. Protopopov, D. Jankowski, R. Stanek, Comparative Measurements of WLS Fibers, ATLAS Internal note Tilecal-no-34
7. G. Blanchot et al., Cell Intercalibration and Response Uniformity Studies Using a Movable ^{137}Cs Source in the Tilecal 1994 Prototype, ATLAS Internal note Tilecal-no-44, 1994
8. Z. Ajaltouni et al., Response of the ATLAS Tile Calorimeter to muons, Nucl. Inst. and Meth., A338(1997)64-78
9. M. Varanda, A. Maio, A. Henriques, Uniformity of the barrel Module 0 response to muons and pions - test beam data and simulation, ATLAS Internal note Tilecal (in preparation)

Forward and Central Preshower Detectors for the D0 Upgrade

Ken Del Signore for the D0 collaboration

University of Michigan, Dept. of Physics, Ann Arbor, Michigan 48109

Abstract. Within the upgraded D0 detector at Fermi National Accelerator Laboratory, forward and central preshower detectors will be used for fast level 1 triggering of electrons. These detectors consist of approximately 25000 channels of extruded scintillator strips with embedded wave length shifter fiber readout. Readout is via clear fiber lightguide to Visible Light Photon Counters. An overview of each system will be presented. Results of prototype detectors to cosmic rays will be presented. Scintillator/fiber manufacture and assembly will be discussed.

INTRODUCTION

The upgraded D0 detector [1] will be used in the high luminosity RunII and Tev33 eras at Fermi National Accelerator Laboratory, tentatively scheduled for 2000-2000, 2003-2005, respectively. The D0 Central (CPS) and Forward (FPS) Preshower Detectors [2,3] are designed for use in the Level 1 trigger to tag events which contain high p_T electrons. The detectors will also be used offline for tagging soft electrons from b jet decays and as the first layer of the calorimeter to help restore electromagnetic energy resolution otherwise degraded by the solenoid.

The CPS and FPS detectors are made from fine grained scintillator strips, of triangular cross section, with wavelength shifter fiber (WLS) readout. Pb radiators are used to enhance electron showering. Light collected in the WLS fibers is piped via clear fiber light guides ($\sim$ 10m) to the D0 Visible Light Photon Counter (VLPC) system. The VLPC system provides $\sim$ 80% QE and a gain of about 10^4.

Successive strips are rotated 180 degrees such that the triangular cross sections nest into one another. This provides a vernier effect to electron showers and also minimum ionizing particles (mip). A prototype detector, including ten meter clear light guides and readout by HISTE-IV (60% QE) VLPCs, was installed in the D0 fiber tracker cosmic ray test stand [4]. The light yield was measured to be $\sim$ 4.5 photons per millimeter of scintillator traversed. The position resolution to mip tracks using two adjacent strips in a vernier mode was 580um.

DETECTOR GEOMETRY

The layout of the CPS and FPS detectors can be seen in Figures 1, 3.

The CPS consists of three concentric cylinders, of diameter 28.5" and length 103",situated between the new D0 solenoid and the calorimeter. Each cylinder consists of a nested layer of triangular strips sandwiched between 1/32" stainless steel skins. The detector covers the region $| \eta | \leq 1.2$. The strips within each cylinder are situated in axial and u-v stereo views, the stereo angle is $\sim$ 22.5 degrees.

CP450, *SciFi97: Workshop on Scintillating Fiber Detectors*
edited by A. D. Bross, R. C. Ruchti, and M. R. Wayne
© 1998 The American Institute of Physics 1-56396-792-8/98/$15.00

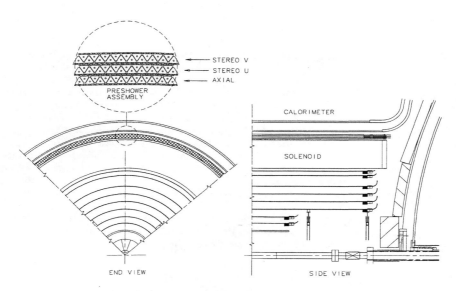

FIGURE 1. End and side views of the Central Preshower Detector. The cylindrical detector is positioned in the 5.5cm gap between the solenoid and central calorimeter.

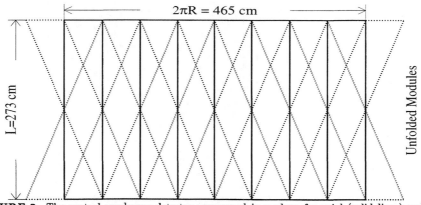

FIGURE 2. The central preshower detector unwrapped in a plane for axial (solid lines) and stereo (dotted lines) octants. Note that each octant crosses three octants in the other two views and that the stereo octant edges are precisely aligned with the axial octant edges. The arrangement yields a stereo angle of $\sim 22.5°$.

A cylindrical Pb radiator ($1\ X_0$) is mounted between the solenoid and CPS to enhance electron showering. The total radiation length (solenoid + Pb) for normally incident electrons is $2\ X_0$. A Level 1 electron trigger is formed by the spatial match of a hi p_T track in the D0 Fiber tracker to a shower in an axial strip.

To facilitate detector assembly, each cylindrical layer is formed from eight octant modules. The stereo modules rotate through 90 degrees along the z-axis of the solenoid, so that they attach to every other axial module. Figure 2 shows a rolled out view of the axial and stereo modules.

Each module contains 160 scintillator strips. The WLS fibers within each strip are split in the middle of the detector and readout from each end. The total number of readout channels is 7680.

The FPS covers the region $1.5 \leq |\ \eta\ | \leq 2.5$. In the region $1.65 \leq |\ \eta\ | \leq 2.5$ there are four layers in a u-v u-v arrangement, with a $2X_0$ Pb absorber between the second and third layers. In the region $1.5 \leq |\ \eta\ | \leq 1.65$ only the third and fourth layers are present. The stereo angle is 22.5 degrees. The detector is spherical to conform to the shape of the D0 Endcap Calorimeters.

The 2π spherical surface is made from sixteen ϕ wedges, each subtending 22.5 degrees, Fig. 4. One WLS fiber reads out each strip. The total channel count is 14,368. The choice of sixteen ϕ wedges is a compromise between the total channel count versus decreasing the occupancy of each strip.

A Level 1 electron trigger is formed by requiring a mip signal in the front two layers be spatially matched to a shower signal in the back layers. Photons are identified by a shower in the back layers with no signal in the front. Pions will typically leave mip signals in the front and rear.

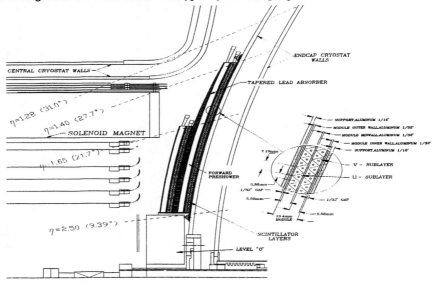

FIGURE 3. Detail of the Forward Preshower Detector. The four layer detector is spherical to match the D0 endcap calorimeter.

SCINTILLATOR STRIP MANUFACTURE AND PREPARATION

The scintillator strips used in the CPS and FPS detectors were manufactured by an extrusion process pioneered by FNAL and D0 collaborators. Pellets of polystyrene are impregnated with the primary and secondary fluors. The concentrations used were 1% PT and 0.015% DPS. These pellets are then extruded into strips at RDN Corporation [5]. Relatively precise, $\sim 25\mu$m,

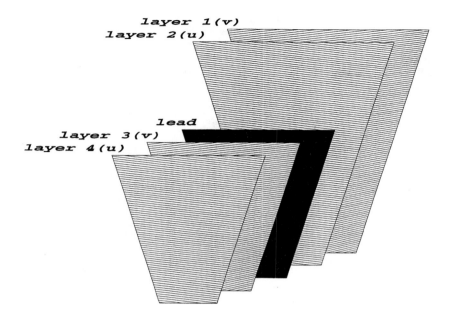

layer 1(v)
layer 2(u)

lead
layer 3(v)
layer 4(u)

FIGURE 4. Schematic picture of one azimuthal wedge. The lines represent the scintillator strips

tolerances on the strips can be held. A small obstruction in the extrusion flow produces the hole running the length of the strip. Figure 5 shows the strip cross section used for the CPS. The strips were then machine wrapped with aluminized mylar at FNAL.

Following wrapping, the strips are heat formed to the desired shape. Due to both the extrusion and wrapping procedures, the strips tend to be bowed. The heat forming has the additional benefit of relieving stresses within the plastic. The CPS and FPS detectors require unique strip shapes. The CPS axial layer required only straight strips, while the stereo layers require four unique helical shapes. The FPS requires two unique spherical shapes.

The heat forming is accomplished by constraining the strips to the desired shape with appropriate fixtures, then slowly heating to a temperature of $\sim$ 180-185 degrees, followed by a slow cool down. A slow heating cycle allows for uniform temperature to be achieved throughout the strip. Care must be taken with the fixture that is holding the strips to the desired shape as any defects will be impregnated in the strips. Expansion of the plastic is also of concern.

MODULE ASSEMBLY

The CPS modules were built on three special assembly tables. Each table has a cylindrical surface of appropriate radius and arclength. Modules are built by first epoxing (3M DP-190) a bottom layer of strips onto a stainless steel skin. A polystyrene jig, with registration grooves

459

milled in, is set over the strips and registered to precise pins situated on the table surface. This assembly is then vacuum bagged. The uniform pressure applied by the vacuum bag seats each strip into the jig. The strip to strip registration achieved in this manor is ~50μm. After an overnight cure, a top layer of strips is set in place using the bottom layer for registration.

The mylar wrapping at the ends of the strips tended to be non-uniform. This necessitated gluing the strips on with ~4" of excess length. The strips were cut to the proper length with a diamond disk in a dremel.

Following this, each hole was cleaned and inspected for obstructions. Fiber bundles, in groups of sixteen, were then inserted. Care had to be taken during insertion as the fibers were prone to damage. After insertion a visual inspection of the polished connector, for non-uniform brightness of individual fibers, proved to be a robust test for damaged fibers.

The FPS modules are constructed in a slightly different manor, owing to the differences in geometry [3].

A spherical assembly table is made on a large lathe. Onto the surface are set jigs that constrain the edges of the first and last strips and the ends of every fifth strip of the bottom layer. The top and bottom layer are glued at same time between two .01" g-10 substrates. A vacuum bag compresses the assembly during curing. With the first, last, and every fifth strip constrained to the proper position, and given the natural symmetry of the top and bottom layers, the strips align themselves during compression. Prototypes built in this manor have a strip to strip registration of .25μm.

The final trapezoidal shape is then cutout using a CNC water jet cutter. This technique employs a thin jet of water deployed from a small nozzle at ~ 50,000 psi. Outer dimensions can

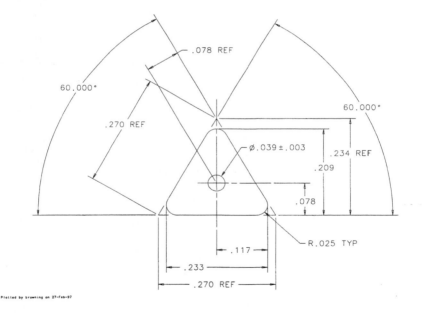

FIGURE 5. Cross sectional dimensions of the equilateral triangular strip with a hole in the center for housing a WLS fiber

460

be held to ~ .015". This procedure sometimes leaves a small amount of sand (added to the water as an abrasive) in the WLS fiber holes, which necessitates cleaning by an air gun.

Full details of CPS and FPS detector assembly, mounting hardware, etc, can be found in [2,3].

FIBER BUNDLE/CONNECTOR PREPARATION

The WLS fibers used for both the FPS and CPS are Kurrary Y-11, 250ppm, S-type [6]. The non readout end of the fibers are ice-polished in batches on approximately 400. The polished end is then silvered by Al sputtering. A small bead of Elmers glue was applied to the silvering as a protectant. Care had to be taken with the fibers as the silvering was still somewhat fragile. Approximately 70% of the light incident on the silvering is reflected back into the fiber [4].

The connectors used to transition the WLS fiber to clear fiber were developed by D0 collaborators at the University of Illinois at Chicago (UIC) [7]. The connectors are made by an injection molding process that allows accurate reproduction in large quantities. Figure 6 shows a schematic of the FPS 16 channel connector. The connectors, made of black ABS plastic, can be made to tolerances of ~ .001", and offer typical light transmission of 95% ± 1%.

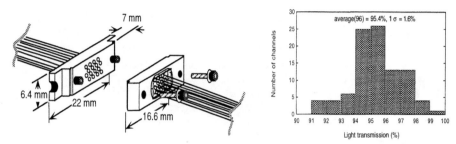

FIGURE 6. Schematic of FPS 16 channel WLS-clear fiber connector and a typical measurement of transmission efficiency.

REFERENCES

1. *The D0 Upgrade*, FERMILAB-Pub-96/357-E.
2. *Design Report of the Central Preshower Detector for the D0 Upgrade*, D0-Note 3014.
3. *Technical Design Report of the Forward Preshower Detector for the D0 Upgrade*, D0-Note 3445.
4. M. Adams, et al., *A New Detector Technique Using Triangular Scintillating Strips to Achieve Precise Position Measurements for Minimum Ionizing Particles*, FERMILAB-Pub-95/285-E, September 1995, to be submitted to Nucl. Instrum, and Meth. A.
5. RDN Corporation, 160 Covington Drive, Bloomingdale, IL 60108, (630) 595-4876
6. Kurraray Intl., 200 Park Ave., New York, NY 10166 (212)986-2230
7. S. Margulies and M. Chung, *Development of a multichannel fiber-to-fiber connector for the D0 upgrade tracker*, Photoelectronic Detectors, Cameras, and Systems, eds. C.B. Johnson and E.J. Fenyves, Proc. SPIE 2551, pp. 10-16, 1995. M. Chung and S. Margulies , *Development of a multichannel fiber-to-fiber connector for the D0 upgrade tracker*, 1995 IEEE Nuclear Science Symposium, Vol 43, No. 3, June 1996

The CMS Central Hadron Calorimeter

Jim Freeman

Fermi National Accelerator Laboratory, Batavia, Illinois 60510
Representing the CMS Hadron Calorimeter Group

Abstract. The CMS central hadron calorimeter is a brass absorber/ scintillator sampling structure. We describe details of the mechanical and optical structure. We also discuss calibration techniques, and finally the anticipated construction schedule.

OVERVIEW

The CMS detector is a general purpose experiment that will operate at the Large Hadron Collider at CERN. The heart of the CMS detector is a large volume 4T superconducting solenoid. The solenoid is 13 meters long with an inner radius of 3 meters. Inside the solenoid are tracking detectors, followed by an electromagnetic calorimeter (ECAL) made of lead tungstate crystals and finally the hadron calorimeter (HCAL). Figure 1 shows a quarter section of the CMS detector design.

A major function of the HCAL is the measurement of missing transverse energy. For this measurement, Gaussian resolution is not as important as elimination of low energy tails in the response function. With this in mind, the CMS HCAL design strives to eliminate dead material that causes energy loss, and to maximize the calorimeter thickness. The location of the HCAL inside the magnetic field requires that the calorimeter be non-magnetic. The absorber chosen is "cartridge brass", 70% brass and 30% zinc.

The central HCAL covers the η range of $-3< \eta <3$ and $0 < \phi < 2\pi$. The central HCAL is physically composed of 2 regions, the barrel ($|\eta|<1$), and the endcap, which extends to $|\eta| = 3$. The very forward region of CMS, $3 < |\eta| < 5$, is covered by a quartz fiber calorimeter, and is discussed elsewhere in these proceedings. [1] The tower granularity is chosen to be $\delta(\eta)$ times $\delta(\phi) = 0.87 \times 0.87$. The segmentation is commensurate with the granularity of ECAL, and is also sufficient for jet reconstruction. [2] Properties of the HCAL are shown in Table 1.

The active medium of the calorimeter is scintillator plastic, read out with wavelength shifting (WLS) optical fibers. This technique uses scintillator tiles to sample the shower. Wave-shifting fiber imbedded in the tile traps the scintillator light and clear fiber spliced to the WLS fiber carries the light to photo-transducers. The technique has been refined and applied to the CDF endcap upgrade calorimeter at

CP450, *SciFi97: Workshop on Scintillating Fiber Detectors*
edited by A. D. Bross, R. C. Ruchti, and M. R. Wayne

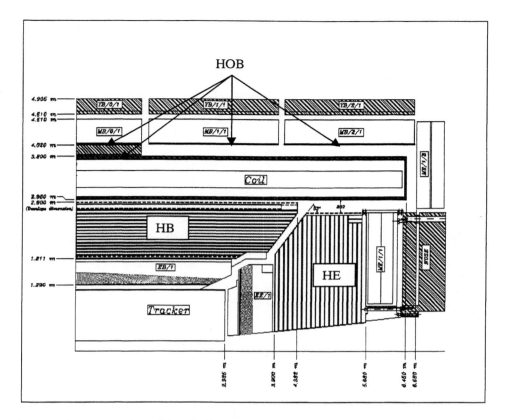

Figure 1. Quarter section of the CMS central detector.

Fermilab. [3,4,5] With this technique, the samples are thin. They require only 0.9 cm of thickness between absorber plates, so a high-density calorimeter is maintained.

The choice of scintillator for readout provides a calorimeter that is fast, stable and reliable, and radiation-resistant. More than 90% of the scintillation light will be collected within a 50 ns time window. The radiation dose for the scintillators at the high η end of the endcap calorimeter is large, 2 to 4 Mrads for the life of the experiment. This dose will damage the scintillators. We use a strategy of multiple longitudinal samples of the calorimeter in this region. By recalibrating the relative weights of the multiple samples on a yearly basis, we will be able to minimize the effect of radiation damage on the calorimeter resolution constant term. [6]

Figure 2 shows the mechanical structure of a wedge of the central HCAL barrel. The barrel HCAL is made of $\delta(\phi) = 20°$ wedges. Each wedge extends from $\eta = 0$ to the high η boundary, $\eta \cong 1$. A wedge is about 4.5 meters long, and installed in CMS will extend from an inner radius of 1.8 meters to 2.9 meters. Each wedge weighs

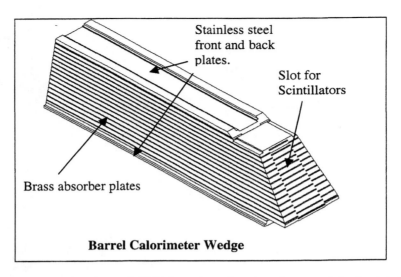

Figure 2. Isometric view of HCAL wedge showing front and back stainless steel plates, interior brass absorber plates, and slots for scintillator trays.

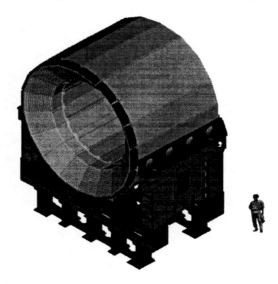

Figure 3. A half barrel of the hadron calorimeter resting on its installation cradle.

about 30 metric tons. A wedge is composed of 6 cm thick inner and outer stainless steel plates (for mechanical strength), with 6 cm thick brass absorber samples inside.

The entire internal structure is bolted together using about 2000 bolts. A set of 18 wedges are then bolted together to form a half-barrel ring, weighing about 500 metric tons. There are two such rings, one of which is shown in Figure 3.

The absorber structure is designed with alternating "staggered" slots for the scintillators. Thus the (4 tower wide in phi) wedge starts with a slot to accommodate scintillators for the middle 2 towers, then 1/2 of an absorber sample later in depth, have separate slots for the outer 2 towers. The staggering of the absorber plates provides a rigid mechanical structure, with no projective dead regions. Each slot runs the full length (in η) of the wedge. Long thin scintillator "tile trays" will be inserted from the high η end, and optical cables will carry the scintillation light to the photodetectors.

Tight mechanical tolerances on the plates (and lack of distortion because of bolting) allows the design to have $\leq$ 2 mm of air gap between adjacent wedges. In addition, only a 9.5 mm high slot is needed to accommodate the 7.5 mm thick scintillator packages.

The construction of the endcap hadron calorimeter is logically similar to the barrel. Each endcap is a monolithic structure, bolted together from "pizza slice" shaped plates of brass. Again the plates are staggered to provide alternating slots for the scintillator tile trays. The endcap HCAL has 8 cm thick brass sampling.

Figure 4 shows the total number of interaction lengths of material provided by the ECAL (1.0 λ) and the HCAL. We see that at $\eta=0$, the combined calorimeter is somewhat thin, $\approx 7.0\ \lambda$. Therefore our design places an additional scintillator sample outside the solenoid, the Outer HCAL Barrel (HOB). Figure 4 also shows the resulting total number of interaction lengths in the calorimeter. The placement of HOB is shown in Figure 1.

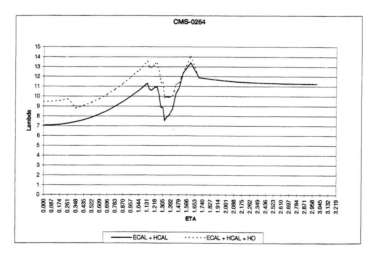

Figure 4. Interaction lengths vs eta for the ECAL +HCAL, and for the ECAL + HCAL +Outer Calorimeter.

OPTICAL SYSTEM

The CMS HCAL optical system contains of approximately 70,000 individual scintillator tiles. Since the optical systems for the barrel and endcap HCAL are very similar, we will concentrate on the description of the barrel. The optical system is illustrated in Figure 5. Scintillation light from the scintillator tiles is captured in an embedded WLS fiber. The WLS fiber is mirrored on one end with an aluminum sputtered coating. The other end of the fiber is thermally welded to a clear fiber that carries the light to the high eta edge of the tile tray, terminating in an optical connector. An optical cable attached to the connector carries the light to the readout box at the outer radius of the HCAL wedge. Finally, the light from all tiles in a tower readout segment are ganged together and illuminate a pixel of a multi-pixel hybrid photodiode (HPD), the optical transducer used in the central HCAL.

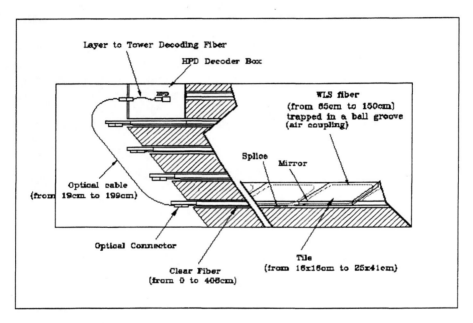

Figure 5. The layout of the optical readout of the HCAL.

The scintillators are organized into tile trays, as shown in Figure 6 for the barrel calorimeter. The trays are either 1 or 2 tiles wide in ϕ to match the slots shown in Figure 2, and the full length of the barrel (16 tiles) in η.

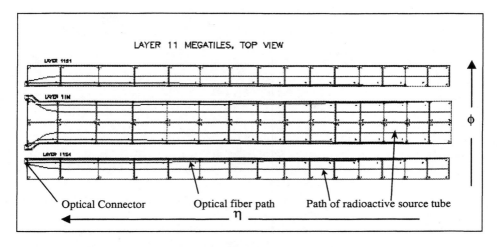

Figure 6. Scintillator tile trays to be inserted in the high-eta end of the wedge absorber.

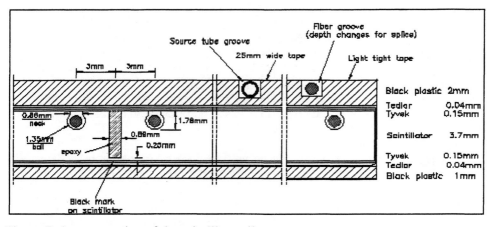

Figure 7. A cross section of the scintillator tile tray.

    The cross section of the tile tray is shown in Figure 7. The 3.7 mm thick scintillator tiles are covered with white reflective Tyvek plastic, wrapped in opaque Tedlar black film, and then sandwiched between top and bottom cover plates. The entire package is connected together by small through-bolts. The thin bottom cover plate provides mechanical protection. The 2 mm thick top plate protects the tiles and supplies a path for the fibers from the tiles to travel to the high $|\eta|$ end of the tile tray. There, the fibers are terminated into multi-fiber optical connectors. Optical cables carry the light onward to photodetector decoder boxes where the light from each tile is organized into readout towers. Our baseline choices for optical materials are Kuraray SCSN81 for the scintillator, Kuraray multi-clad Y-11 (K-27 fluor) for the wave-shifting fiber, and Kuraray multi-clad clear fiber. With these materials, typical light-yields are about 2 photoelectrons per minimum-ionizing particle per scintillator layer.

CALIBRATION

Quality control and calibration schemes are built into the optical system from the beginning. After assembly of the tile trays, a collimated Cs^{137} source is used to test the tiles. By measuring the induced radioactive source current after a photodetector, the collimated source measurement establishes the absolute response of the tiles. At the same time a moving wire source is used for cross calibration. Figure 6 shows the 2 mm top cover plate carrying "source tubes". The source tubes are stainless steel tubes that terminate into source-tube connectors that allow them to be to "plumbed" to tubes from a moving wire source system.

In the moving wire source system, an approximate point source at the end of a wire moves through the source tubes to excite the scintillators. The source position relative to the tile is well controlled by the permanently attached source tube. Because of the fixed geometry of the source tube relative to the tile, the ratio of response of collimated to wire source is stable. The moving wire source (Cs^{137}) is basically an isotropic source. The scintillator response changes by approximately 1% per 0.1 mm of separation between the scintillator and the source. Thus meaningful measurements with the wire source require that the stability of the placement of the wire source relative to scintillator must be controlled to order 0.1 mm. This is achieved by permanent attachment of the source tubes on the tile trays.

After the tile-trays have been installed into the absorber, the wire source is again used to test for system stability. Since the geometry of the wire source tube relative to the scintillator is unchanged, the measurement of wire-source response allows for reference back to the original QC test using the collimated source. A small number of the source tubes will be accessible during operation of CMS. These tubes will be periodically re-tested to verify stability.

A laser flasher system will also be used to directly excite the photodetectors. This system will be used on a periodic basis to track the gain of the photodetector/amplifier/digital readout system.

The stability of ratio of wire source measurement to collimated source measurement (i.e. to the true tile absolute response) provides a convenient way to carry test beam calibrations to the actual CMS detector. A subset of the barrel wedges will be extensively studied in test beam. Their calorimeter towers will have their pion response measured with test beams. In addition the wire-source system will measure the "fingerprint" of each tower's optical longitudinal response. An effective "pion-weighted source response" will be calculated by convoluting the measured longitudinal optical profile with an average pion longitudinal profile. Then the ratio of actual test beam response to pion-weighted source response will be used to carry test beam calibration to wedges in CMS that were not exposed to beams. From past experience on CDF, an initial absolute calibration of 2 - 3% is expected for wedges that are not calibrated by the test beams. [7] This is small compared to the constant term of the HCAL resolution.

The initial absolute calibration can be improved by using in-situ physics calibrations while CMS is operating. [8] One likely signal is $\bar{t}t$ production, where the top quarks then decay into $W+b$. One W is required to decay into jets, while the other is required to decay into lepton + neutrino to provide a trigger. For events that have 2 tagged b-jet's, CDF has shown that it can readily reconstruct the W boson that decays into jets. They measure a rms/mean of about 9 GeV / 81 GeV for 8 reconstructed top events in 100 pb^{-1} of data.

A similar analysis has been performed using a simulation of the CMS detector. There, even in the presence of the ~30 minimum-bias events anticipated at the ultimate luminosity of 10^{34}, the W into 2-jet decay can be reliably reconstructed. The results expected for one month of LHC running at 10^{33} is shown in Figure 8. The LHC, $\bar{t}t$ production will supply a ready source of di-jet events that reconstruct into a fixed mass and will help with calibration as well as understanding systematics of jet clustering. Other methods such as jet-jet and jet-gamma energy balance are also possible.

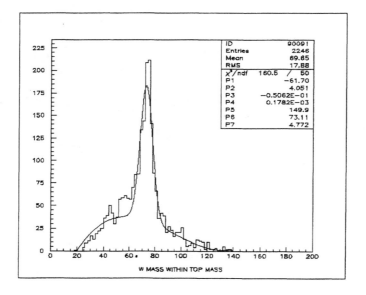

Figure 8. Reconstructed W decays into 2 quarks for double-tagged $\bar{t}t$ events

PHOTODETECTORS

The barrel photodetectors are positioned at the high $|\eta|$ / large radius of the barrel HCAL, inside the 4T magnetic field. The corresponding location for photodetectors for the endcap calorimeter is at the large radius, large z corner. Because of the placement of the photodetectors, conventional photomultiplier tubes are unusable. Instead a recently developed transducer, the hybrid photodiode, HPD, has been adopted. The HPD is a proximity-focused device consisting of a vacuum envelop, a conventional photocathode, and a reverse-biased silicon diode. A high accelerating voltage (of order 10kV) is supplied between the photocathode and the silicon diode. Photoelectrons emitted by the photocathode gain kinetic energy falling through the electric field. This kinetic energy is converted into electron-hole pairs when the photoelectron impacts the diode. The generated electric pulse is read off the diode, amplified, and sent to digitizing electronics. Typical gains for the HPD are from 1000 to 2000. Figure 9 shows the internal design of a HPD.

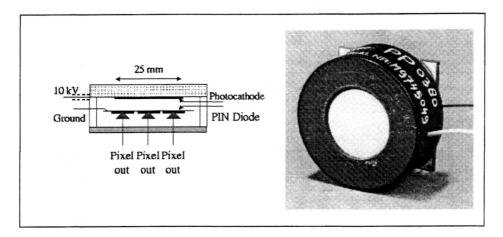

Figure 9. HPD internal structure and photograph of DEP multi-pixel HPD.

Because the low gain of the HPD requires the use of an associated amplifier, they are effectively noisier than conventional photomultiplier tubes. The estimated noise is about 1 photoelectron. A minimum ionizing signal in the calorimeter is about 10 photoelectrons, so a MIP can be clearly seen with an HPD. Figure 10 shows a test beam measurement of the muon signal through 8 layers of scintillator using an HPD for readout. [9]

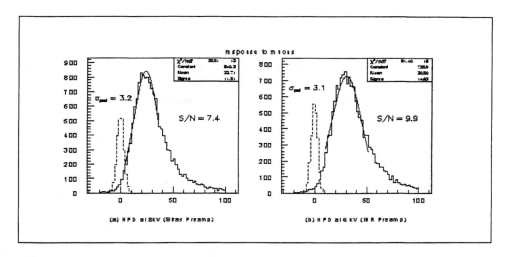

Figure 10. Test beam response for minimum ionizing tracks through the HCAL. The HPD signal is amplified and then digitized.

Because of our reliance on radioactive sources for quality control and calibration, the HPD must be able to accurately measure DC currents (supplied by the radioactive sources). The HPD has a typical leakage current of about 1 - 5 nA, with a jitter in the current of only order 10 pA. The source-induced currents are of order 5 nA, comparable to the total leakage current. Under these conditions, the HPD's have been shown to allow source current measurements of accuracy 1%. [9]

Status

The HCAL barrel design has been completed. A full size barrel wedge is being fabricated in the factory and should be complete in September 1998. Another full size prototype is planned for spring 1999. The fabrication of the 36 production wedges will start in late 1999, and should be complete in late 2001. The completed wedges will be mated with the optical packages that will be built concurrently. This integration process will take place at CERN in 2000 and 2001. The wedges will be joined into half barrels in 2002, and installed into CMS in 2003.

The endcap calorimeters will be assembled in the CMS assembly hall during 2002, and installed into CMS in 2003. Electronics, trigger, and calibration systems will be integrated in 2004, in advance of detector operation in 2005.

Table 1 summarizes the properties of the CMS Hadron calorimeters.

HCAL Summary Table

	HB	HOB	HE	HF
η coverage η-φ segmentation	0 - 1.3 ~0.087 x 0.087	0 - 1.1 ~0.087 x 0.087	1.3 - 3 ~0.087 x 0.087	3 - 5 ~0.175 x 0.175
Min. thickness HCAL + ECAL + ECAL + HO	 5.8 λ 6.8 λ 10.5 λ	 - 	 10.3 λ 11.4 λ 11.4 λ	 9λ total
Structure	18 wedges per half-barrel 26 tonnes/wedge brass absorber 940 tonnes total	1 to 2 layers of scintillator in the barrel muon iron	monolithic endcaps 290 tonnes/end brass absorber	2 halves/end opening horizontally 98 tonnes/end brass absorber
Depth Segments (in HCAL)	H1(0.1 λ), H2	1	H1(0.1 λ), H2, η<2.5 H2,H3, H4, 2.5<η<3	EM (15 X₀) HAD (7.5λ) Tail (2λ)

Sampling thickness
  Absorber
  Scintillator
  WLS Fiber
	
	5 cm, 6 cm brass
	4 mm SCSN81
	0.94mm Y11
		
		~20 cm iron
		10 mm BC408
		0.94mm Y11
			
			8 cm brass
			4 mm SCSN81
			0.94mm Y11
				Fe matrix
				0.3 mm quartz fibers on 2.5mm square grid

sampling fraction	7.5%	-	5.5%	0.85%
No. channels	5184	2160	3774	1920
No. tiles	40,000	2,500	26,000	2 x 10⁶ fibers
No. samples	17	1,2	20	-
Scintillator Area	2950m²	600 m²	1300 m²	3850 km fiber
PE/Sample/MIP	2pe / tile/MIP	10pe / tile/MIP	2 pe / tile/MIP	-
PE/GeV	20 pe/GeV	20pe/GeV	15pe/GeV	0.25 pe/GeV
Signal/noise for MIP	50 pe/1.5 pe (H2)	10 pe/1.5 pe	60pe/1.5 pe (H2)	0.5 pe /0.1 pe
Nominal resolution HCAL + ECAL	 90% ⊕ 4% 112% ⊕ 4%	 - 	 1.17 ⊕ 6.6% 	 200 ⊕ 3%
e/h (effective)	1.3	-	1.3	2
Photodetector gain	HPD 2000	HPD 2000	HPD 2000	PMT 5 x 10⁴
Max rad dose 5x10⁵ pb⁻¹ (=10 yr. operation)	0.01 Mrad	~0	3 Mrad	100 Mrad

Dynamic range
  (pe)
  (GeV)
	
	1pe - 70000 pe
	0 - 3.5 TeV
		
		1pe - 20000pe
		0 - 1 TeV
			
			1pe - 50000pe
			0 - 3.5 TeV
				
				1 pe - 1500 pe
				0 - 7 TeV

BIBLIOGRAPHY

1. D. Winn, et al., CMS Quartz Fiber Calorimeter, these proceedings.

2. Compact Muon Detector Technical Proposal, CERN/LHCC 94-38

3. G. Foster, J. Freeman and R. Hagstrom, Nucl. Phys. B, **A23**(1991) 93

4. J. Freeman, et al. The CDF Upgrade Calorimeter, Proc. 2nd Int. Conf on Calorimetry in HEP, Capri, Italy, 1991

5. P. de Barbaro, et al. Nucl. Instrum. Methods, **A315** (1992) 317

6. The Hadron Calorimeter Project Technical Design Report, CERN/LHCC 97-31, p25.

7. V. Barnes, CDF Endplug Calibration, these proceedings.

8. J. Freeman and W. Wu, In situ Calibration of the CMS HCAL Detector, FNAL-TM-1984

9. P. Cushman, et al., CMS HPD Development, these proceedings.

SESSION 12: MEDICAL

Chair: M. Wayne
Scientific Secretary: L. Coney

Overview of Nuclear Medical Imaging Instrumentation and Techniques*

William W. Moses

Lawrence Berkeley National Laboratory, University of California, Berkeley, CA 94720 USA

Abstract. Nuclear medical imaging is a well established method for obtaining information on the status of certain organs within the human body or in animals. This paper presents an overview of two commonly used methods, namely SPECT (single photon emission computed tomography) and PET (positron emission tomography), as well as the emerging method of intra-operative probes with imaging capability. The discussion concentrates on the instrumentation requirements for these systems and on the potential for incorporating scintillating, wavelength-shifting, and fiber optic light guides into them.

INTRODUCTION

Nuclear medical imaging is a generic term that covers many imaging techniques, with the common theme being that ionizing radiation originating from within the body is detected and imaged in order to determine something about the physiology or anatomy of the subject (1-5). The ionizing radiation is often in the form of a radioisotope that is incorporated into a biologically active compound (*i.e.* a drug) that is introduced into the body (in trace quantities) either by injection or inhalation. This compound then accumulates in the patient and the pattern of its subsequent radioactive emissions is used to estimate the distribution of the radioisotope and hence of the tracer compound.

Since the image that is produced is of the distribution of a drug within the body, nuclear medical imaging is capable of targeting where certain metabolic processes occur and measuring the rate at which these processes take place. Thus, it is able to determine whether the biochemical function of an organ is impaired, while many other forms of medical imaging (such as x-ray, ultrasound, or magnetic resonance techniques) are usually confined to determining the physical structure of the organ. It is therefore most frequently used in organs and diseases where biological function is of primary importance and information on physical structure is either irrelevant or ambiguous. Examples are neurological diseases (such as Alzheimer's disease) where physical affects are only observable on a microscopic level, heart disease (where the relative vigor of the tissue is of primary importance), or oncology (cancer), where the metabolism rate gives

* This work was supported in part by the U.S. Department of Energy under Contract No. DE-AC03-76SF00098, and in part by Public Health Service Grants No. R01-CA48002, R01-CA67911, and P01-HL25840 from the National Institutes of Health.

CP450, *SciFi97: Workshop on Scintillating Fiber Detectors*
edited by A. D. Bross, R. C. Ruchti, and M. R. Wayne
1998 The American Institute of Physics 1-56396-792-8/98/$15.00

valuable information on whether tissue is cancerous and how it responds to treatment.

While the instruments used to detect and image this radiation are often similar to those used for high energy or nuclear physics experiments, there are many important differences between the requirements for medical imaging and the requirements for detecting subatomic particle interactions. This paper describes two common nuclear medical imaging techniques, namely SPECT and PET, with emphasis on the requirements of the detection system and the reason for those requirements. Some discussion of the potential for use of various forms of optical fibers (scintillating, wavelength-shifting, and purely transmitting light guides) in these applications is given, concluding with a description of a new imaging system (an intra-operative probe) that utilizes optical fiber technology.

GENERAL CONSIDERATIONS

Imaging emissions from a radioactively labeled drug necessarily involves selecting a radioisotope, and the choice of the radioisotope used for a given nuclear medical technique involves a tradeoff of several factors. The radioisotopes most commonly used are gamma ray emitters (or positron emitters, whose net result is a pair of annihilation photons), as gamma rays can penetrate the body and be imaged with external detectors more easily than other forms of ionizing radiation. In general, high energy gammas penetrate the body easily (and so have high detection efficiency), but low energy emissions are convenient to detect in small detector volumes and are easy to shield. Most medically used isotopes have monochromatic emissions between 60 keV and 511 keV. Radioisotopes with long half lives are convenient for drug manufacture and distribution, as both the synthesis of complex biochemicals and the delivery from the place of manufacture to the patient can be time consuming. Radioisotopes with short half lives usually result in a lower radiation dose to the patient, as practical considerations limit imaging times to less than two hours and emissions that occur after the imaging session is complete contribute to patient dose but not to image quality. Most medically used isotopes have half lives between 1 minute and several days. Finally, chemistry of the radioisotope is important, as it must be incorporated into complex, biologically active compounds. Isotopes of hydrogen, carbon, oxygen, and nitrogen are most desirable, as most biochemicals contain at least one of these elements, but many other elements have been incorporated into biologically active compounds.

The detection system should be very efficient, as safety concerns cause regulatory agencies to limit the total radiation dose that can be administered to a patient. Thus, reduced statistical noise cannot be obtained by increasing the activity (*i.e.* the signal source), and so must be obtained by maximizing detection efficiency. At the gamma ray energies used for nuclear medical imaging (60–511 keV), the attenuation length in tissue is 5–10 cm. This is similar to the distance that gamma rays often need to travel to exit the body, so a significant fraction of the internally emitted gamma rays interact within the patient. Due to the low effective atomic number of tissue, most of these interactions result in Compton scatter, and most of the scattered photons continue to undergo Compton scattering until they eventually leave the body. While the net radiation flux impinging upon the detector usually contains 10%–50% unscattered gammas (which can be used to form an accurate image), the majority are Compton photons that form a background. The detection system must then be capable of reducing this Compton background, usually by measuring the energy of each detected photon with 8%–20% full-width at half-maximum (fwhm) resolution. Finally, a image cannot be

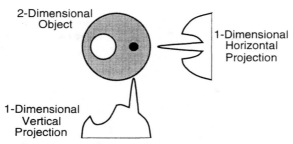

FIGURE 1. Computed Tomography. The 1-dimensional horizontal and vertical projections (the line integral of the density along parallel horizontal and vertical lines) are shown adjacent to the 2-dimensional object that they were taken of. Computed tomography is the process of reconstructing the 2-dimensional object given its 1-dimensional projections from all angles.

formed unless the direction of photon travel is known, so the detector system must be capable of determining this direction.

Both SPECT and PET make use of the mathematical technique of computed tomography to form images (4, 6, 7). Briefly stated, computed tomography is a process whereby a 2-dimensional image of an object is formed using multiple 1-dimensional projection images of that object. This is demonstrated in Figure 1. The 2-dimensional object in question is a large circle of uniform density in which a medium circle of lower density and a small circle of higher density are embedded. The one dimensional projection of the object in the vertical direction is shown below the object. This projection is what would result if a planar x-ray of the object were taken, and represents the line integral of the 2-dimensional object's density along parallel, vertical lines that cross the object. The hemisphere corresponding to the large circle is clearly seen, modified by a dip due to the low density region and a spike due to the high density region. To the right of the object is its one dimensional projection in the horizontal direction. Again the hemisphere with the associated dip and spike are seen, but the positions of the dip and spike have changed because of the change in viewing angle. What the details are beyond the scope of this paper, taking one dimensional projections at all angles around an object (not just the two shown in Figure 1) provides enough information to reproduce a 2-dimensional image of the object.

SPECT

SPECT (Single Photon Emission Computed Tomography) (7, 8) employs radioisotopes that are single gamma emitters. The most commonly used isotopes are ^{99m}Tc (141 keV), ^{201}Tl (135 keV and 167 keV), and ^{123}I (159 keV), although other isotopes with energies ranging from 80 keV to 511 keV are occasionally employed. Although SPECT radioisotopes tend to be challenging to incorporate into biologically active compounds, they have been included in drugs that show metabolic activity (such as ^{99m}Tc-Sestamibi, which concentrates in mitochondria), blood flow (such as ^{201}Tl for the heart and ^{99m}Tc-HMPAO for the brain), presence of certain forms of cancer (such as ^{67}Ga-Citrate, ^{123}I-MIBG, and ^{99m}Tc-MDP), liver function (such as ^{99m}Tc-sulphur colloid), and a variety of other imaging agents.

A typical SPECT camera is shown schematically in Figure 2. The emitted gamma rays are detected by two-dimensional position sensitive detectors. The direction of the

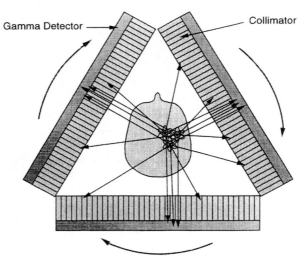

FIGURE 2. SPECT Camera Schematic. A collimator blocks gamma rays that are not traveling normal to its surface, and those gammas that pass through the collimator are detected with a planar position sensitive detector. The assembly rotates around the patient to provide the many projections necessary for computed tomography.

gamma rays is determined by a collimator placed between the detector array and the patient — photons that are not traveling in the desired direction are absorbed by the collimator. The collimator and detector combination (and their associated electronics and mechanical support) form what is known as a gamma camera "head". While multiple heads increase the detection efficiency of a SPECT camera (and the cost), there is little efficiency gain above three heads, so most SPECT cameras have between one and three heads. Each head measures a planar projection of the activity in the patient, and by simultaneously rotating the heads, the requisite set of projections to perform computed tomography is obtained.

The most commonly used gamma detector for SPECT is the Anger camera, which is named for its inventor (9). A schematic of an Anger camera is shown in Figure 3. A

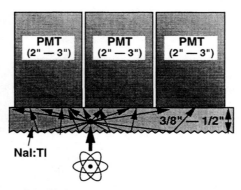

FIGURE 3. Anger Camera. Scintillation light from gamma ray interactions is detected by multiple photomultiplier tubes. The interaction position is determined by the ratio of the analog signals, and the energy by the analog sum of the signals.

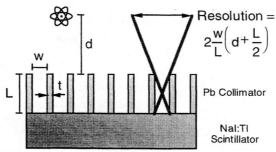

$$\text{Resolution} = 2\frac{w}{L}\left(d+\frac{L}{2}\right)$$

Pb Collimator

NaI:Tl
Scintillator

FIGURE 4. Collimator Cross Section. The spatial resolution and efficiency of SPECT are determined by the collimator.

plate of NaI:Tl scintillator crystal is optically coupled to many photomultiplier tubes. When a gamma ray interacts in the crystal, the resulting scintillation photons are emitted isotropically and are detected by several of the photomultiplier tubes. The position of the gamma ray interaction is then determined by the analog ratio of the photomultiplier tube output signals, and the gamma ray energy is determined by the sum of these signals.

The thickness of the NaI:Tl crystal determines the efficiency of the camera. As its attenuation length for 140 keV gamma rays is 4 mm, the 9–12 mm thick plates commonly employed in SPECT systems provide nearly complete absorption. The scintillator surface not coupled to the photomultiplier tubes strongly affects the light distribution observed by the photomultiplier tubes and is often prepared with a proprietary surface treatment to optimize both position and energy resolution. Special care is needed near the edges of the scintillator crystal, as reflections from the side of the crystal cause the dependence of the measured light distribution on the interaction position to be much weaker near the sides than near the middle. A typical Anger camera has 9% fwhm energy resolution and 3.5 mm fwhm position resolution for 140 keV gammas.

While the Anger camera is an important component of a SPECT imager, the performance of a SPECT camera is almost entirely determined by the collimator (10). Figure 4 shows a schematic drawing of a parallel hole collimator. The collimator is usually made from cast lead with hexagonal cross section holes arranged in a honeycomb fashion. Each hole can be thought of as having a diameter w and a length L, separated from its neighbors by a lead septum of thickness t (typically 0.2 mm). The spatial resolution is determined by the collimator geometry and is given by

$$\text{Resolution} = 2\frac{w}{L}\left(d+\frac{L}{2}\right), \tag{1}$$

where d is the distance from the collimator surface to the object being imaged. From Equation 1 we can see that the resolution scales linearly with the aspect ratio w/L of the collimator, and as L is typically 1–3 cm and d is typically tens of centimeters, increases roughly linearly with the distance d from the collimator. While arbitrarily high spatial resolution can be achieved with a collimator by reducing w/L, more modest spatial resolution is usually obtained because this aspect ratio greatly affects the efficiency. The fraction of gammas that are transmitted through the collimator is

$$\text{Efficiency} = \left(\frac{w}{2L}\right)^2, \tag{2}$$

so while the resolution improves linearly with w/L, the efficiency decreases quadratically. For a typical "all-purpose" collimator, the spatial resolution is 6.2 mm fwhm at a distance of 5 cm and the efficiency is 0.023%. From this we see that the spatial resolution of a SPECT image is not determined by the resolution of the Anger camera (which is known as the intrinsic resolution), but by the collimator.

Based on current SPECT cameras, the gamma ray detector requirements for SPECT are, in order of decreasing importance, (1) >85% detection efficiency (to minimize statistical fluctuations), (2) <15 keV fwhm energy resolution (to reject Compton scatter events), (3) <3.5 mm fwhm intrinsic position resolution (to avoid degrading the collimator resolution), (4) <$15/cm^2 parts cost (SPECT cameras are widely available commercially), and (5) <2000 µs cm^2 dead time product (single event dead time multiplied by detector area affected — low collimator efficiency implies low counting rates).

PET

PET (Positron Emission Tomography) (11) employs radioisotopes that are positron emitters. The most commonly used isotopes are ^{18}F, ^{11}C, ^{15}O, and ^{13}N, although other isotopes are occasionally employed. A great number of biologically active compounds have been synthesized with these isotopes, they have been included in drugs that show metabolic activity (such as ^{18}F-FDG, a sugar analog), blood flow (such as ^{15}O-water), a variety of neurotransmitters (such as ^{18}F-Dopamine), and a large variety of other imaging agents (12).

A typical PET camera, shown schematically in Figure 5, consists of a planar ring of small photon detectors, with each photon detector placed in time coincidence with *each* of the individual photon detectors on the other side of the ring. When a pair of photon detectors simultaneously detect 511 keV photons, a positron annihilated somewhere on the line connecting the two detectors. The method of using time coincidence between two detectors (rather than a collimator and one detector) to restrict events to a line is known as electronic collimation, and is much more efficient than the mechanical collimation used in SPECT. The set of all lines connecting detectors (known as chords) makes the requisite set of projections to perform computed tomography for a single plane. Multiple detector rings are stacked on top of each other to obtain images from multiple slices, and thus a three-dimensional image of the patient. Planes of tungsten

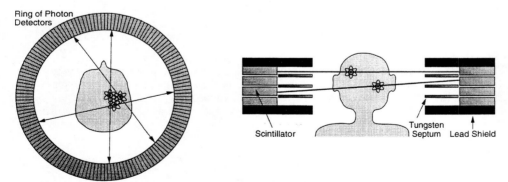

FIGURE 5. PET Camera Schematic. Positron annihilations yield back to back 511 keV photons, which are individually detected in a ring of photon detectors, shown on the left. Pairs are identified by time coincidence. Multiple rings are stacked up, as shown on the right, to create a 3-dimensional image.

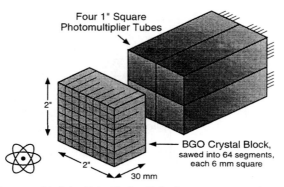

Four 1" Square
Photomultiplier Tubes

2"

2"

30 mm

BGO Crystal Block,
sawed into 64 segments,
each 6 mm square

FIGURE 6. PET Detector Module. Scintillation light from gamma ray interactions is detected by multiple photomultiplier tubes. The interaction position is determined by the ratio of the analog signals, and the energy by the analog sum of the signals.

septa placed between detector planes are often used to shield the detectors from Compton scattered photons emanating from other parts of the body, and images taken in this geometry are often known as "2-D PET" images. Coincidences between nearly adjacent "cross-plane" rings are usually added to the closest "direct plane" to increase detection efficiency. If the septa are removed, the efficiency is greatly increased (as coincidences from widely separated planes can be accepted), but the backgrounds also increase significantly. However, the signal to noise ratio improves in some situations, and this mode of operation is known as "3-D PET."

The most commonly used PET detector module is known as a block detector, and a schematic is shown in Figure 3 (13). A block of BGO scintillator crystal is partially sawn through to make a group of quasi-independent crystals that are optically coupled to four photomultiplier tubes. When a gamma ray interacts in the crystal, the resulting scintillation photons are emitted isotropically but the saw cuts limit (but does not entirely prevent) their lateral dispersion as they travel toward the photomultiplier tubes. The position of the gamma ray interaction is then determined by the analog ratio of the photomultiplier tube output signals, and the gamma ray energy is determined and a timing pulse generated by the sum of these signals.

The thickness of the BGO crystal determines the efficiency of the camera. As its attenuation length for 511 keV gamma rays is 1.2 cm, the 30 mm thick BGO crystals commonly employed in PET systems provide nearly complete absorption. The coincidence timing efficiency is limited by the decay time of BGO — the photoelectron (p.e.) rate immediately after a 511 keV photon interaction is approximately 0.5 p.e./ns. A typical PET detector module has 20% fwhm energy resolution, 2 ns fwhm timing resolution and 5 mm fwhm position resolution for 511 keV gammas.

Based on current PET cameras, the gamma ray detector requirements for PET are, in order of decreasing importance, (1) >85% detection efficiency (to minimize statistical fluctuations), (2) <5 mm fwhm position resolution (to obtain good spatial resolution), (3) <$100/cm^2 parts cost (PET cameras are widely available commercially), (4) <1 μs cm^2 dead time product (single event dead time times detector area affected — high counting rates are often encountered), (5) <2 ns timing resolution (to identify coincident pairs), and (6) <100 keV fwhm energy resolution (to reject Compton scatter events) (14).

OPTICAL FIBER USE IN NUCLEAR MEDICAL IMAGING

The potential for use of optical fibers (including scintillating fibers, wavelength-shifting fibers, and fiber optic light guides) in nuclear medical imaging equipment is somewhat limited by the low energies of the radiation that must be detected and imaged. Most geometries that utilize optical fibers end up delivering less than 10% of the light impinging on one end of the fiber to the opposite end of the fiber, even when using high quality, double clad fibers. This reduction in efficiency would often degrade the signal to the point where the statistical noise becomes too large to measure the gamma ray energy with sufficient resolution.

As an example, let us consider SPECT, in which 140 keV gammas must be imaged with ≤10% fwhm energy resolution. If statistical noise was the only effect that contributed to energy resolution degradation, this would imply that ≥550 quanta must be detected. If NaI:Tl scintillator is used (one of the most luminous scintillators known), its 38,000 photon/MeV conversion efficiency implies that ~5,300 scintillation photons will be produced by a 140 keV interaction. If a photomultiplier tube with 20% quantum efficiency is used to detect these photons, approximately 1000 photoelectrons will be detected *provided* that the coupling between the scintillator and photomultiplier tube is 100% efficient. Obviously, the <10% coupling efficiency afforded by optical fibers will yield too small a signal to meet the energy resolution requirement.

Optical & Wavelength-Shifting Fiber Use in PET

However, there are several proposals for nuclear medical imaging equipment that incorporate optical fibers. The first is a PET detector that uses quartz fibers to couple individual LSO scintillator crystals with 2 mm square cross section to a position sensitive photomultiplier tube (15). These crystals must be packed together into a circular, gap-free array around the patient (in this case, a small animal), and the significant dead area at the perimeter of the position sensitive photomultiplier tube does not allow them to be directly coupled to the crystals. Therefore, 2 mm diameter, 24 cm long, double clad quartz fibers are used to couple the scintillator crystals to the photosensors. This arrangement, shown in Figure 7, moves the photomultiplier tubes to a larger radius to allow the scintillators to be packed without gaps.

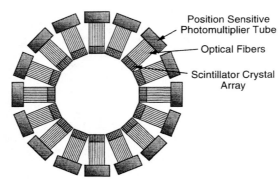

FIGURE 7. PET Camera Utilizing Optical Fibers. Scintillator crystal arrays are coupled to position sensitive photomultiplier tubes with arrays of optical fibers. This allows the scintillators to be close packed even though the photomultiplier tubes have significant dead area.

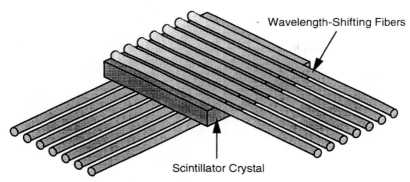

Wavelength-Shifting Fibers

Scintillator Crystal

FIGURE 8. PET Camera Utilizing Wavelength-Shifting Fibers. Scintillation emissions from a thin plate of scintillator crystal are absorbed by orthogonal arrays of wavelength-shifting fibers. These fibers re-emit photons of a lower energy, with the interaction position determined by coupling the fiber arrays to position dependent photomultiplier tubes that identify the fibers that emit photons. Several layers are stacked on top of each other, and additional photosensors are required to measure energy and timing.

Other proposed PET detector modules utilize wavelength-shifting fibers coupled to thin plates of LSO scintillator crystal (16, 17). Arrays of fibers are placed in orthogonal orientations on either side of the plate, and since the plate is only a few millimeters thick, the spread of scintillation light in the crystal is small and the majority of the signal is collected in a single wavelength-shifting fiber. The fiber absorbs this primary scintillation light and re-emits lower energy photons, some of which are transported down the length of the fiber. A position-sensitive photodetector is coupled to each array of fibers and determines the position of interaction based on which fibers signals are observed in. Multiple scintillator / fiber layers are stacked on top of each other in order to achieve the requisite gamma interaction fraction. Additional photosensors are employed to measure the total energy of the gamma and produce a timing signal, as the signal transmitted by the fibers is only tens of electrons, which is insufficient for sufficiently accurate timing or energy measurement.

Optical & Scintillating Fiber Use in Intra-Operative Probes

Finally, a combination of scintillating and transparent optical fibers have been proposed for use with an intra-operative probe system (18). When surgically removing cancerous tumors, it is desirable to remove all cancer cells associated with the tumor. Often extra, non-cancerous tissue around the tumor is removed in order to ensure this. However, removal of additional tissue can have serious side-effects in some organs, such as the brain. One method for maximizing tumor / normal tissue removal is to inject the patient (before surgery) with a positron emitting drug that accumulates in tumors (such as ^{18}F-FDG). After the bulk of the tumor has been removed, a positron-sensitive probe is inserted into the surgical opening and placed next to the suspect tissue to determine whether it has high radiotracer concentration (*i.e.* is cancerous) and so should also be removed. A positron-sensitive (rather than gamma-sensitive) imager is used because there is little room in the incision, so the large detector volumes necessary for efficient 511 keV photon detection and for shielding the detector from the tremendous flux of annihilation photons from the rest of the body are not practical. Because of the short range of positrons in tissue (~0.5 mm), a detector placed in contact with the tissue

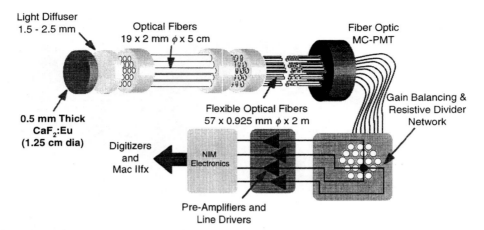

FIGURE 9. Intra-Operative Probe. A thin CaF_2:Eu scintillator crystal detects positrons from tissue that has been injected with a tumor-avid, positron-emitting drug. The resulting scintillation light is coupled using optical fibers to a position-sensitive photomultiplier tube, and Anger logic used to determine the interaction position. A bundle of BGO scintillator fibers placed between the light diffuser and the transparent optical fibers can be used to reject 511 keV photon interactions in the CaF_2:Eu.

form a "contact" image the underlying positron distribution — computed tomography is not used.

The schematic of such a detector is shown in Figure 9. The scintillator chosen is a thin plate of CaF_2:Eu because its relatively low density and atomic number make it relatively insensitive to gamma rays, while its high light output (19,000 photons/MeV) provides a relatively high signal. An array of optical fibers carries the scintillation light to a position sensitive photomultiplier tube, and the position of the interaction determined by Anger logic (*i.e.* the analog ratio of signals observed in the position sensitive photomultiplier tube). While this system is effective, there is still significant background due to annihilation photon interaction in the CaF_2:Eu. In order to reduce this background, the first centimeters of optical fiber are replaced with optical fibers made of BGO scintillator. This scintillator is highly efficient at detecting annihilation photons, and the difference in scintillation decay times (300 ns for BGO, 1 μs for CaF_2:Eu) allows the use of a "phoswich" technique, in which a single photomultiplier tube can separate signals in the two scintillator crystals based on their decay time. When the signal from the CaF_2:Eu (assumed to be from a positron interaction) is required to be in time coincidence with ~511 keV energy deposit in the BGO scintillator (assumed to be from an annihilation photon from the same positron interaction), the image contrast improves significantly.

CONCLUSION

Nuclear techniques have been used for over 50 years to image biochemical processes in living organisms. Most involve imaging photons with energies of 100–511 keV, as these photons can be produced by radioisotope decay, penetrate the body with reasonable probability, and be detected with high efficiency and good spatial resolution. Detectors for these applications have seen many years of development and refinement, and generally involved high detection efficiency (to maximize image quality for a given patient dose), good energy resolution (to minimize background from Compton scatter in the patient), low cost (as there is a competitive commercial market for these devices), and moderate (3–5 mm) spatial resolution. While optical fibers (scintillating fibers, wavelength-shifting fibers, and fiber optic light guides) are rarely used in nuclear medical imaging equipment due to their relatively low (<10%) scintillation light collection efficiency, they are used in some detector designs where other considerations (typically packaging concerns) outweigh the desire for high light collection efficiency.

ACKNOWLEDGMENTS

I would like to acknowledge Dr. Stephen Derenzo, Dr. Thomas Budinger, and Dr. Ronald Huesman for many interesting conversations, and Dr. Martin Tornai for his insights into intra-operative probes. I would like to thank Dr. Randall Ruchti and Dr. Mitchell Wayne for the opportunity to consider the impact of optical fibers on nuclear medical imaging. This work was supported in part by the Director, Office of Energy Research, Office of Biological and Environmental Research, Medical Applications and Biophysical Research Division of the U.S. Department of Energy under contract No. DE-AC03-76SF00098 and in part by the National Institutes of Health, National Cancer Institute under grants No. R01-CA48002 and R01-CA67911, and National Institutes of Health, National Heart, Lung, and Blood Institute under grant No. P01-HL25840.

REFERENCES

1. Sandler, M. P., Coleman, R. E., Wackers, F. J. T., et al., *Diagnostic Nuclear Medicine*, Baltimore, MD: Williams & Wilkins, 1996, 1549 pp.
2. Hendee, W. R. and Ritenour, R., *Medical Imaging Physics*, St. Louis, MO: Mosby Year Book, 1992, 781 pp.
3. Krestel, E., *Imaging Systems for Medical Diagnosis*, Berlin: Siemens Aktiengesellschaft, 1990, 636 pp.
4. Macovski, A., *Medical Imaging Systems*, Englewood Cliffs, NJ: Prentice Hall, 1983, 256 pp.
5. Webb, S., *The Physics of Medical Imaging*, Bristol: Institute of Physics Publishing, 1993, 633 pp.
6. Cormack, A. M. *J. Appl. Phys.* 34, 2722–2727 (1963).
7. Kuhl, D. E. and Edwards, R. Q. *Radiology* 80, 653–662 (1963).
8. Budinger, T. F., *Single Photon Emission Computed Tomography*, In <u>Diagnostic Nuclear Medicine</u>, (Edited by Sandler, M. P., Coleman, R. E., Wackers, F. J. T., Patton, J. A., Gottschalk, A. and Hoffer, P. B.), Baltimore, MD: Williams & Wilkins, 1996, pp. 121–138.
9. Anger, H. O. *Rev. Sci. Instr.* 29, 27–33 (1958).
10. Tsui, B. M. W., Gunter, D. L., Beck, R. N., et al., *Physics of Collimator Design*, In <u>Diagnostic Nuclear Medicine</u>, (Edited by Sandler, M. P., Coleman, R. E., Wackers, F. J. T., Patton, J. A., Gottschalk, A. and Hoffer, P. B.), Baltimore, MD: Williams & Wilkins, 1996, pp. 67–79.
11. Cherry, S. R. and Phelps, M. E., *Positron Emission Tomography: Methods and Instrumentation*, In <u>Diagnostic Nuclear Medicine</u>, (Edited by Sandler, M. P., Coleman, R. E., Wackers, F. J. T., Patton, J. A., Gottschalk, A. and Hoffer, P. B.), Baltimore, MD: Williams & Wilkins, 1996, pp. 139–159.
12. *J. Nucl. Med.* 32, 561–748 (1991).

13. Cherry, S. R., Tornai, M. P., Levin, C. S., et al. *IEEE Trans. Nucl. Sci.* NS-42, 1064–1068 (1995).

14. Moses, W. W., Derenzo, S. E. and Budinger, T. F. *Nucl. Instr. Meth.* A-353, 189–194 (1994).

15. Cherry, S. R., Shao, Y., Silverman, R. W., et al. *IEEE Trans. Nucl. Sci.* 44, 1161–1166 (1997).

16. Worstell, W., Johnson, O. and Zawarzin, V., "Development of a high-resolution PET detector using LSO and wavelength-shifting fibers," in *Proceedings of The 1995 IEEE Nuclear Science Symposium and Medical Imaging Conference* (Edited by Moonier, P. A.), San Francisco, pp. 1756-1760, 1995.

17. Williams, M. B., Sealock, R. M., Majewski, S., et al. *IEEE Trans. Nucl. Sci.* 45, 195–205 (1998).

18. Levin, C. S., Tornai, M. P., MacDonald, L. R., et al. *IEEE Trans. Nucl. Sci.* 44, 1120–1126 (1997).

Performance Study of a Prototype PET Scanner Using CsI(Na) Scintillator with Wavelength-Shifting Fiber Readout

V. Zavarzin , O. Johnson, H. Kudrolli and W. Worstell

Boston University Physics Department and Center for Photonics
590 Commonwealth Avenue, Boston, MA 02215

Abstract. We have designed and built a high resolution dual-head PET scanner using wavelength-shifting optical fibers coupled with thin plates of CsI(Na). Each detector head has the sensitive area of 114 x 114 mm^2 and is being read out by 8 Hamamatsu multi-anode PMTs R5900-L16 using charge division method. Additional 4 single-anode 60 mm square PMTs per head provide trigger and total energy measurement. Being initially developed for Positron Emission Mammography the design of the system is also of general interest for gamma-imaging. The principles of operation are discussed and results of some measurements are presented.

INTRODUCTION

Perpendicular ribbons of wavelength-shifting fibers (WLSF) may be used to determine the position of gamma-ray interactions within thin scintillator crystals, with the obtainable spatial resolution comparable to the crystal thickness [1-2]. By building a multilayer stack of alternating thin, flat crystal layers and perpendicular fiber ribbons, one can measure gamma-ray interaction energies and positions in 3 dimensions with good accuracy and efficiency. Depth-of-interaction sensitivity is needed to preserve the spatial resolution of PET detectors when accepting events whose lines-of-response intersect detector elements at significantly oblique angles. This is particularly significant in the design of devices for Positron Emission Mammography (PEM) where two parallel plate detector elements above and below a breast may be used to detect coincident gammas [3-4]. For a device whose field of view is comparable to the detector separation, events with significant parallax may be acquired and used to resolve features at different depths-of-field between the two plates. A schematic illustration of our device geometry is given in Fig.1. For flat polished crystals, total internal reflection guides scintillation light emitted at shallow angles with respect to the crystal surface toward the crystal edges, while light emitted nearly normal to the surface exits the crystal above and below the gamma ray interaction point. This exiting light position may be sensed locally by WLS fibers, which absorb the primary scintillation photons and isotropically re-emit secondary photons at longer wavelengths.

Perpendicular fiber ribbons (X-fibers and Y-fibers) on opposite sides of a thin, polished crystal layer may thus be used to measure gamma-ray interaction positions in two dimensions; further fibers at the edges of crystal layers (Z-fibers) may be used to

CP450, *SciFi97: Workshop on Scintillating Fiber Detectors*
edited by A. D. Bross, R. C. Ruchti, and M. R. Wayne

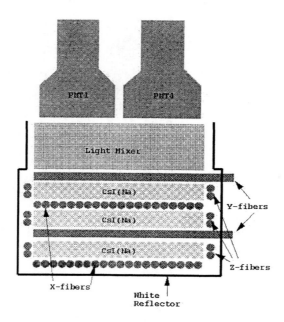

FIGURE 1. Schematic Illustration of detector geometry. Three thin layers of CsI(Na) scintillator alternate with perpendicular fiber ribbons. Edge fibers are used to determine layer-of-interaction, while Anger PMTs view the laminated fiber/crystal block through a light mixer; the block walls are painted with diffuse reflector.

sense scintillation light which was totally internally reflected within a layer, providing information on depth-of-interaction within a multilayer stack. Alternatively, one may read out fiber ribbons from different depths within a multilayer stack with different photosensors, but this is more costly than the edge-fiber decoding method.

Within a WLS fiber, a small fraction of the secondary photons emitted at angles near that of the fiber axis are piped to fiber readout photosensors, while most emerge transversely from the fibers and traverse both scintillator and fibers within a multilayer stack transparently. This unpiped light can be collected with a standard Anger array on the module surface to provide energy and trigger information as well as approximate interaction coordinates. These approximate coordinates can then be used to de-multiplex the readout of several fiber ribbons which may then be read out with the same position-sensitive multianode photomultiplier. Multiplexed fiber readout is important for limiting both system component costs and electronics channel counts. The above method decouples the event energy, timing, and triggering functions, which are performed by the Anger system, from the precise event position determination in 3 dimensions, which is performed by the fiber readout system. This is important since the light yield from the WLSFs, although easily sufficient for precise coordinate determination and high efficiency, is insufficient for precise event energy and timing determination.

The performance achievable with WLSF readout of scintillator crystals depends critically upon the choice of scintillator material, the quality of crystal surface treatment and related optical coupling parameters, and finally upon the device geometry and in particular the crystal layer thickness. Our initial measurements and device characterization, reported here, were performed with a device with just 3 crystal layers; this resulted from our limiting the number of layers and the amount of multiplexing in the readout so as to study in detail the effects of multiplexing in depth.

DETECTOR CONSTRUCTION

Measurements were carried out with two detector modules operating in coincidence. Each module consists of 3 CsI(Na) scintillator plates, each 3mm thick and having a $114 \times 114mm^2$ lateral extent. These crystals were polished and then coated with a high-index (n=1.65) epoxy. Crystal layers were alternated with perpendicular WLSF ribbons (Bicron BCF91A, 1mm diameter multiclad), with each ribbon 112mm in width and approximately 30cm in length. Acrylic frames with 4mm thickness were used to assemble a multilayer stack of scintillator crystals alternating with fiber ribbons, and to hold depth-encoding fibers in position at two crystal edges. The inside edge of each acrylic frame (nearest the crystal) was painted with a diffuse white reflector consisting of titanium oxide powder in an epoxy base.

After assembly, each multilayer stack was cast in clear epoxy (Stycast 1267). When the epoxy cured, two module edges opposite the fiber readout were milled to remove most of the acrylic frames. The fiber ends on the milled edges were then polished and optically coupled to plane mirrors (aluminized mylar). The base of each module was optically coupled to a diffuse white reflector, as was the outside of each acrylic light mixer.

Each module was coupled with optical grease to an Anger-type array of 4 60mm square PMTs (Phillips XP3697/PA). Preamplifiers were mounted on each photomultiplier's base to boost gains. Detector modules, these Anger PMTs, and fiber readout PMTs were mechanically mounted in a support structure with signal, high-voltage, and low-voltage feedthroughs.

Each 112mm fiber ribbon was split into 8 bands each containing 14 fibers. The 16 bands of X-fibers and the 16 bands of Y-fibers were grouped into 8 sets of 4-fold multiplexed bundles, with a pair of bands within a layer (lateral multiplexing) bundled with the corresponding pair from another layer (depth multiplexing). Each bundled set of 4 bands was held in a custom Delrin fiber vise, polished, and coupled with optical grease to a Hamamatsu R5900-L16 multianode photomultiplier. The 14mm bundle width ran across the discrete anodes of the multianode PMT, each of which corresponds to a 1mm x 16mm photocathode region. The 16 anodes of each multianode were then read out with 1D charge division: the sum of the charge division signals A1+A2 indicated the light yield from each bundle, while the normalized coordinate (A2-A1)/(A2+A1) measured the average lateral position of light across the bundle [5].

DATA ACQUISITION ELECTRONICS

The signals from each Anger tube were split, with one copy used for triggering and the other delayed for ADC measurement. The 4 trigger signals from each detector module were analog summed, shaped, and discriminated; coincidences between discriminated analog sums from the two modules provided triggers. Additional copies of each discriminated analog sum were sent to TDCs (LeCroy 2228A) for timing measurements. Copies of the Anger signals, as well as all charge division signals from the fiber readout PMTs, were delayed by 200 ns and digitized within 1 μs gates by LeCroy 2249W ADCs. A total of 2 TDC channels and 47 ADC channels were read out through CAMAC for each event.

DATA PROCESSING

All fiber readout PMTs were calibrated in a separate run to give the number of photoelectrons collected from each fiber bundle for each event. For each event, the X- and Y- bundles with the most light were identified, and the Anger and depth information used to demultiplex and identify which of 4 bands within a bundle produced the signal. The Anger system spatial resolution was sufficient such that the demultiplexing was quite unambiguous. The charge division information from the selected X- and Y- fiber readout PMTs were then converted to coordinates across each fiber ribbon and combined with the position of each fiber ribbon within the module to give global X-, Y- and Z- coordinates.

Analytic single-pass 3D reconstruction is impossible for the limited-angle geometry of two parallel detector plates, since the geometry doesn't allow using Fourie-methods. Nonetheless, 3D information is present in the parallax of oblique rays, and can be exploited through iterative reconstruction algorithms. We have used an Inverse Monte Carlo algorithm [6] based on ML/EM but operating in image space.

EXPERIMENTAL RESULTS

Detector system performance was measured in response to a line source; separate measurements tested performance of the Anger energy/trigger/timing subsystem and of the fiber coordinate measurement subsystem. Fig.2 below shows the energy resolution (i.e. the ADC sum for the 4 Anger PMTs within a module, after calibration) and the Anger system light yield in photoelectrons. The energy resolution obtained at trigger level for those events which generated a trigger was better than 20%, including nonuniformities in detector response. It should be noted that information on event positions both laterally and in depth is available on an event-by-event basis, so that the obtainable offline energy resolution should be even better. The low-energy tail of our trigger senstivity reflects the limitations of the very simple shaping circuit used. The timing performance for two detectors in coincidence is also shown in Fig.2, demonstrating 25ns FWHM time resolution. The fiber coordinate measurement achieved nearly 3mm FWHM spatial resolution in response to a line source (Fig.3).

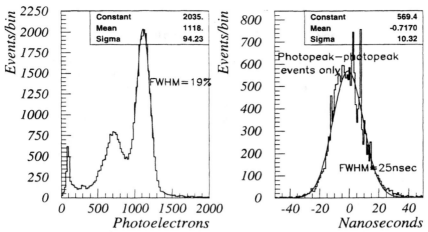

FIGURE 2. Collected light distribution (on the left) at trigger level for the analog sum of 4 Anger PMTs within a detector module. The photopeak is clearly visible and easily selected by a simple threshold trigger circuit. Time difference (on the right) between signals from discriminated anode sums for two coincident detector modules.

Measurements with a point source though showed better spatial resolution, approaching 2mm FWHM when restricted to one small section of one module coincident with another small section on the opposing module. We attribute this to variations in crystal surface finish and to some readout fiber system positioning and calibration errors. The average fiber light yield for the fiber band with most photoelectrons in a photopeak event was 10 photoelectrons, as also shown in Fig.3.

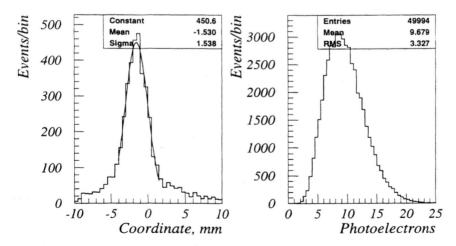

FIGURE 3. Measured line spread function (on the left) in response to a line source spanning nearly the width of the field of view. Measured light yield distribution (on the right), for one end of one fiber bundle as read out through one multianode photomultiplier.

493

DISCUSSION

Separation of functions between energy/trigger/timing (with the Anger array) and interaction coordinate measurements (with the WLSF readout) allows easier optimization. High spatial resolution results from small single crystal layer thickness, while high efficiency results from a large total crystal thickness. The initial studies reported here indicate the capability of 3-layer modules, and we are continuing this work on modules containing more crystal layers. Other scintillator material properties which affect the detector performance are its density and its photofraction. A Compton scatter where the secondary gamma is converted elsewhere within the detector will produce a signal which fakes the photopeak. Positioning errors due to detector Compton scattering are lessened when using a scintillator with a higher photofraction, and a denser scintillator can be used to decrease the number of layers and/or to decrease layer thickness while maintaining detector efficiency. All these reasons, in addition to its greater speed relative to CsI(Na), make LSO an excellent candidate for future development.

CONCLUSIONS

We have constructed two multilayer CsI(Na) PET detector modules with WLSF readout, and characterized it's performance. Wavelength-shifting fiber readout has intrinsic depth-of-interaction sensitivity, which is important when scintillators are brought into close proximity to an object being imaged. Finally, device performance seems to be sufficient for application to Positron Emission Mammography.

ACKNOWLEDGMENTS

We would like to thank Dr. Anatoly Putmakov of BINP (Novosibirsk, Russia) for providing us a custom-made fast PC-CAMAC interface. We would also like to thank Dr. Jack Correia of the Massachusetts General Hospital for his gracious assistance in supplying us with F-18 for our studies.

REFERENCES

1. W. Worstell, S. Doulas, O. Johnson, and C.-J. Lin, Scintillator Crystal Readout with Wavelength-Shifting Optical Fibers, *Proc. IEEE 1994 Med. Imag. Conf.*, 1869-1873.
2. W. Worstell, O. Johnson, V. Zavarzin, Development of a High-Resolution PET Detector using LSO and Wavelength-Shifting Fibers, *Proc. IEEE 1995 Med. Imag. Conf.*, 1756-1760.
3. Thompson CJ, Murthy K, Weinberg IN, Mako F Feasibility Study for Positron Emission Mammography, *Med. Phys.* **21(4)**, 529-537 (1994).
4. Thompson CJ, Murthy K, Clancy RL, Robar JL, Bergman A, Lisbona R, Loutfi A, Gagnon JH, Weinberg IN, Mako R, Imaging Performance of PEM-1: A High Resolution System for Positron Emission Mammography, *Proc. IEEE 1995 Med. Imag. Conf.,*1074-1078.
5. D. Grigoriev, O. Johnson, W. Worstell and V. Zavarzin, , *IEEE Trans. Nucl. Sci,***44(3),** 990-993 (1997).
6. W. Worstell, H. Kudrolli, V. Zavarzin, Monte Carlo-Based Implementation of the ML-EM Algorithm for 3-D PET Reconstruction, presented at the IEEE conf. on Medical Imaging, Anaheim CA, November 3-9, 1996

SESSION 13: PHOTOSENSORS - III

Chair: J. Elias
Scientific Secretary: L. Coney

TEST OF HPD

James Freeman, Daniel Green and Anatoly Ronzhin

Fermi National Accelerator Laboratory
Batavia, Illinois 60510

Abstract. The hybrid photodiode (HPD) is the baseline transducer for the CMS Hadron Calorimeter. The results of a study of the HPD vacuum, timing and increase of the response in magnetic field parallel to the HPD electric field are presented below.

We have published the results of the DEP HPD test in the sources (1,2,3). The HPD has many advantages in comparison with other photodetectors. The main advantage among them is the following:
1. Single photoelectron detection at room temperature.
2. Very uniform PC sensitivity.
3. Excellent timing properties.
4. Linear gain dependence (stability).
5. Low sensitivity to magnetic field.
6. Small value of crosstalk for multipixel device.
7. Perfect linearity in a huge dynamic range.

We have to underline the problems which should be solved before the final application of the HPD in the CMS. We tested a few experimental samples of the DEP HPD. For the experimental samples we discovered some effects which will be discussed below.

The tails in the amplitude spectra of the experimental 25 pixel HPD (Fig. 1) were observed (Figs. 2a,b,c). The low intensity tail was also observed in the Single Photoelectron Distribution (SPD, Fig. 3).

It is difficult to observe these effects by measuring the several nicely separated photoelectrons (with average number of the photoelectrons, say 6-9, Fig. 4, (4)) because:
1. The low intensity part of the SPD that shows the effect is hidden by the main part of the spectra.
2. It is difficult to distinguish between contributions to low energy and high energy part of the spectrum.
3. The tail is disappeared when photoelectron statistics increases.

The best way to see the effect clear is to use the SPD (Fig. 3). The data has to be presented not in linear but in log scale because the effect is small. The SPD is the most sensitive tool to observe the effect because:
1. The shoulder can be perfectly observed.

CP450, *SciFi97: Workshop on Scintillating Fiber Detectors*
edited by A. D. Bross, R. C. Ruchti, and M. R. Wayne
© 1998 The American Institute of Physics 1-56396-792-8/98/$15.00

2. There are no questions about low or high energy tail.
3. The statistics under the peak can be simply normalized to the statistics under the single photoelectron distribution.

25 PIXELS HPD DEP

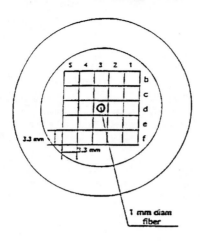

25 pixels HPD DEP

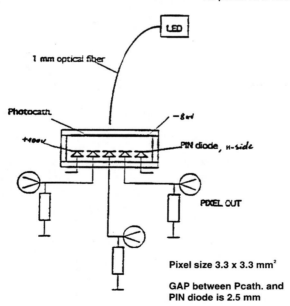

Pixel size 3.3 x 3.3 mm²

GAP between Pcath. and
PIN diode is 2.5 mm

FIGURE 1. Front and schematic views of 25 pixel HPD DEP. Pixel size is 3.3 mm x 3.3 mm.

The ratio is related with quantity, which will be discussed below. Let us consider the Fig. 3. The mean value of the tail of the spectra is around 15 photoelectrons.

The only reasonable source to produce such a number of secondaries is ion feedback (Fig. 5) or in other words the poor HPD vacuum. The single photoelectron sometimes ionizes the residual gas when accelerating between the photocathode and silicon. The potential of the ionization is in the order of 10 eV. The initial photoelectron produces additional positive ions that will accelerate in the electric field and can knock out several additional electrons from the photocathode depending on the energy of the ion. The transit time for the single photoelectron signal is less than 1 ns, but for the tail signals it is around 10 ns which is also consistent with the proposed model.

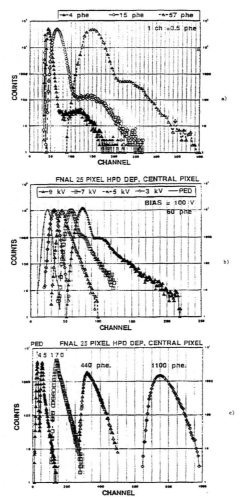

FIGURE 2. a, b and c. The amplitude spectra of the central pixel of the 25 pixel HPD under different light illumination.

499

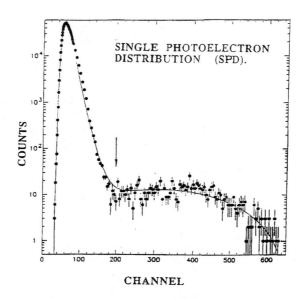

FIGURE 3. Single photelectron spectrum.

The Monte-Carlo simulations are in very good agreement with the experimental data. What are the consequences of the poor HPD vacuum?

1. The level of the tail in the amplitude spectra is dependent on the number of the initial photoelectrons. The shape of the spectrum changes when the number of the photoelectrons changes. What is more dangerous - the spectrum is not under control if vacuum changes during the course of the experiment.
2. The energy scale is not linear when the tail exists.
3. The photocathode effects can appear under ion feedback and nobody can guarantee the stability of the HPD in the case.

The effect of poor vacuum was observed only for 25 pixel HPD with the experimental anode pixel outlet made of glass with kovar pins. It was not observed with outlet made of ceramics. The next generation of the HPD for the CMS will be made with ceramic outlet.

Nevertheless we have some worry about the vacuum so our request to DEP is to introduce the getter in the final device to keep high vacuum. The SPD will be used as the tool to measure the HPD vacuum.

We plan to work with HPD of the T-type. The type of charge carriers, which are holes, determines the timing properties of the HPD. The output signal duration is around 30 ns for 140 V bias (Fig. 6). For the CMS application it should be 2 times smaller. It is known that a 10 keV electron penetrates into the silicon by only 1 μm. At the same time the thickness of the used silicon is the order of 300 μm. To improve the HPD timing thinner silicon can be used. The initial plan of DEP to improve the timing is to change the parameters of the silicon itself. But thinner silicon remains as another opportunity.

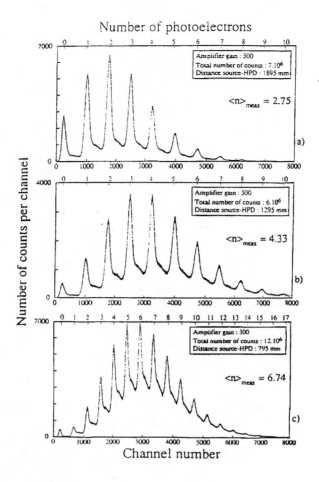

FIGURE 4. HPD amplitude spectra taken from Reference 4.

Some new effects of the HPD behavior in the longitudinal magnetic field were observed. We tested 7-pixel HPD in 5 Tesla magnet. A several percent increase of the HPD signal increase when the magnetic field is parallel to the HPD electric field was noticed (Fig. 7).

The study of the effect was continued at Fermilab. The single channel HPD with 25 mm diameter photocathode (23 mm diameter of the silicon sensitive area) was tested (Fig. 8). The photocathode (PC) was illuminated by fiber connected with light emitting diode (LED) or by LED itself. The light spot on the PC was 1 mm for fiber or 3 mm when using the LED. The PC was illuminated in its geometrical center. The warm magnet with a pole tip diameter 150 mm and distance between poles of 100 mm was applied. The HPD placed in the magnet geometrical center. The nonuniformity of the field was less than 1% in the HPD area.

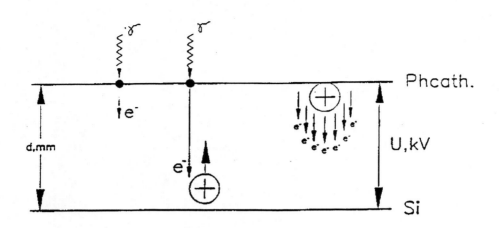

FIGURE 5. The illustration of the ion feedback model.

The small increase of the signal (at a level 1-2%) was measured (both for the fiber and the LED, Figs. 9, 10) when misalignment between electric and magnetic fields was less than 0.3 degree. The signal increases when the angle increases. An oscillation of the signal was observed for non-zero angles. The angular dependence of the signal is shown on Fig. 11.

The LED behavior in a magnetic field is not well investigated but our preliminary data show that we did not see any big signal difference for both LED and fiber for the magnetic field up to 1 Tesla. **The increase of the signal can be due to the PC or the silicon. To check it the HPD photocurrent measurement were done.**

The HPD was used like the photodiode with ohmic contact faced to the PC grounded through the Keithley ammeter. The measurements were done with the magnetic field parallel to the HPD electric field (Fig. 12). The small decrease of the photocurrent (which are limited by our accuracy 0.3%) with the value of the field can be neglected in future consideration. **We can conclude now the HPD signal increase is mostly due to the silicon.**

One can consider processes in the silicon that can increase the output signal with the longitudinal magnetic field. One of the processes is backscattered electrons in the silicon. For the 13 keV photoelectrons initial energy the fraction of the backscattered electrons is 18%. The fraction deposits in average 0.4 of their 13 keV energy within the sensitive diode volume (4). The electrons lose more energy in the silicon dead layer than the photoelectrons when depositing in output signal because of incident angles (Fig. 13).

The backscattered electrons compress along the Z-axis by the longitudinal magnetic field in accordance with its transverse momenta (Fig. 14).

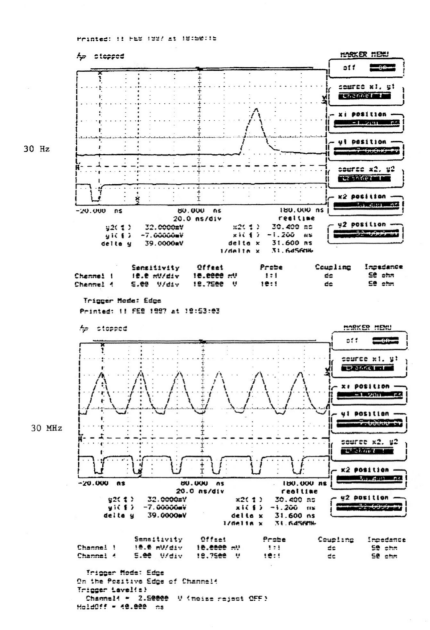

FIGURE 6. The oscilloscope traces of the HPD signals at 30 Hz and 30 MHz frequency.

The HPD has 23 mm diameter of the silicon sensitive area, 5.6 mm gap between PC and silicon. The maximum distance across silicon is 11.2 mm for the backscattered electrons (Fig. 13). The maximum transverse momentum of the electrons is 60 keV. When applying the 1 Tesla longitudinal field the radius of the rotation around Z-axis (B-field direction) will be 0.2 mm for the electrons (Fig. 14).

7 pixel HPD DEP, central pixel

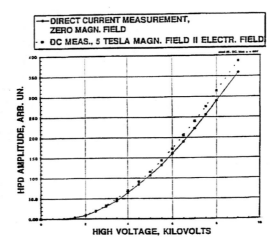

FIGURE 7. The dependence of the HPD's signal on photocathode high voltage with 5 Tesla longitudinal magnetic field and without it.

One can assume some leaving backscattered electrons from the silicon sensitive area without the magnetic field. The magnetic field returns them back to the area. The effect of the increase of the signal is of the order of 2%, which is consistent with estimation based on the data (5). We note the effect of the increase of the signal for longitudinal magnetic field is not due to the difference in energy losses of the backscattered electrons in the silicon dead layer (100 nm of the ohmic contact). The incident angle of the electrons on the silicon surface will be the same with and without the longitudinal magnetic field (6).

As a result the increase of the signal at zero angle can be explained due to the additionally collected backscattered electrons on the silicon sensitive area which are lost without the field. A little bit more signal increase for the small angles probably relates with redistribution of the energy losses of the backscattered electrons in the silicon dead layer because of the angles of the incidence changes when electric and magnetic fields are not parallel. Oscillations of the signal are caused by the photoelectron rotation due to the magnetic field.

Good tool to check the magnetic field effects is the HPD gain curve. Without the field the curve shows the threshold related with the thickness of the silicon dead layer, and the slope of the curve presents energy to produce electron hole pairs. Magnetic field changes the photoelectron trajectories so they can lose more energy in the silicon dead layer. Magnetic field also affect on the trajectories of the backscattered electrons. Both these effects manifest itself as change of the threshold and the slope of the gain curve.

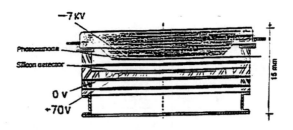

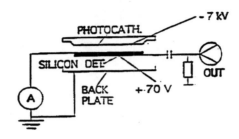

FIGURE 8. Schematic view of the single channel HPD.

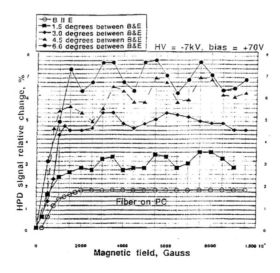

FIGURE 9. The HPD signal relative change in magnetic field in dependence on the value of the field for zero angle and small angles between electric and magnetic fields. The PC illuminated by 1 mm diameter fiber at PC geometrical center.

505

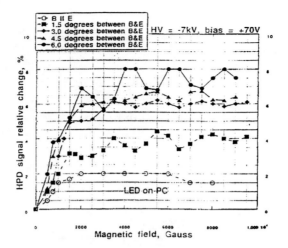

FIGURE 10. The HPD signal relative change in magnetic field in dependence on the value of the field for zero angle and small angles between electric and magnetic fields. The PC illuminated by 3 mm diameter light spot of the LED at PC geometrical center.

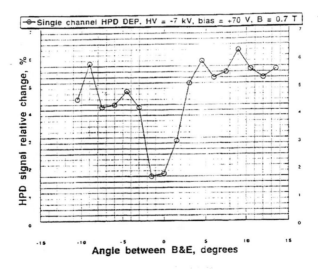

FIGURE 11. The HPD signal relative change in dependence on the angle between the HPD electric field and magnetic field. The value of the magnetic field is 0.7 Tesla.

506

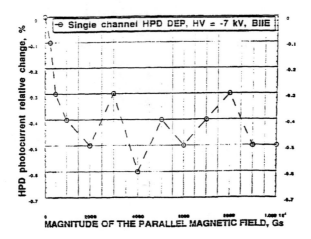

FIGURE 12. The HPD photocurrent relative change in magnetic field parallel to the HPD electric field in dependence on the value of the magnetic field.

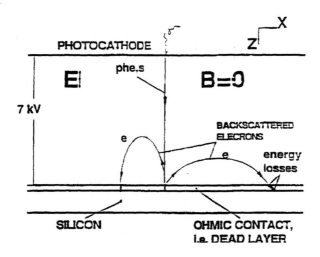

FIGURE 13. The illustration showing the tracks of the backscattered electrons without magnetic field.

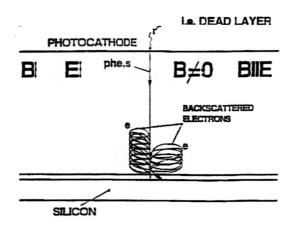

FIGURE 14. The illustration of the compression of the backscattered electrons by the magnetic field parallel to the HPD electric field. The backscattered electrons returning back into the silicon sensitive area cause the increase in the HPD output signal by the magnetic field.

REFERENCES

1. Baumbaugh, A., Binkley, M., Elias, J., et al., "Test of DEP Hybrid Photodiode," Fermilab TM-2013, 1997.
2. Freeman, J., Green, D., Ronzhin, A., et al., "Test of the DEP Hybrid Photodiode in 5 Tesla Magnet," Fermilab TM-2027, 1997.
3. Freeman, J., Green, D., Ronzhin, A., "Effect of HPD Signal Increase in a Magnetic Field Parallel to the HPD Electric Field," Fermilab TM-2039, 1998.
4. Ambrosio, C.D., Gys, T., Leutz, H., et al., *NIM*, **A345**, 1994, pp. 279-283.
5. Darlington, E.H., *J. Phys. D: Appl. Phys.* **Vol. 8**, 1975, pp. 85-93.
6. Green, D., "Backscattering in Hybrid Photodetector Devices," Fermilab FN-662, 1997

A Single-Photon Multichannel Detector: the Megapixel EBCCD

P. Annis[3], A. Bay[7], D. Bonekämper[8], S. Buontempo[9], C. Currat[7],
R. v. Dantzig[1], A.V. Ekimov[10], A. Ereditato[9], J.P. Fabre[5],
D. Frekers[8], A. Frenkel[11], F. Galeazzi[11], F. Garufi[9], J. Goldberg[6],
S.V. Golovkin[10], V.N. Govorun[10], K. Hoepfner[6a], K. Holtz[8],
J. Konijn[1], E.N. Kozarenko[4], I.E. Kreslo[4b], B. Liberti[11],
M. Litmaath[5], G. Martellotti[11], A.M. Medvedkov[10], P. Migliozzi[9],
C. Mommaert[3c], J. Panman[5], G. Penso[11], Yu.P. Petukhov[4],
D. Rondeshagen[8], V.E. Tyukov[4], V.G. Vasil'chenko[10], P. Vilain[3],
J.L. Visschers[1], G. Wilquet[3], K. Winter[2], T. Wolff[8d],
H. Wong[5e], and H.J. Wörtche[8]

presented by I.E. Kreslo

[1] *NIKHEF, Amsterdam, The Netherlands;* [2] *Humboldt-Universität, Berlin, Germany;*
[3] *IIHE (ULB-VUB), Bruxelles, Belgium;* [4] *JINR, Dubna, Russia;*
[5] *CERN, Genève, Switzerland;* [6] *Technion, Haifa, Israel;*
[7] *Université de Lausanne, Lausanne, Switzerland;*
[8] *Westfälische Wilhelms-Universität, Münster, Germany;*
[9] *Università "Federico II" and INFN, Napoli, Italy;* [10] *IHEP, Protvino, Russia;*
[11] *Università di Roma "La Sapienza" and INFN, Roma, Italy.*

[a] now at DESY, Hamburg, Germany.

[b] now at INFN, Roma, Italy.

[c] now at CERN, Genève, Switzerland.

[d] member of the Graduiertenkolleg of elementary particle physics at Humboldt-Universität, Berlin.

[e] now at Institute of Physics, Academia Sinica, Taiwan.

CP450, *SciFi97: Workshop on Scintillating Fiber Detectors*
edited by A. D. Bross, R. C. Ruchti, and M. R. Wayne
© 1998 The American Institute of Physics 1-56396-792-8/98/$15.00

Abstract. The CERN[1] RD46 Collaboration is developing a new high resolution tracking detector based on glass capillaries filled with liquid scintillator. In this framework we have built a read-out chain based on a newly developed device: the Megapixel Electron-Bombarded CCD (EBCCD). This device is a hybrid image-intensifier tube, with a 1024×1024 pixels CCD chip in place of the phosphor screen. The tube has a 40 mm diameter photocathode, is gateable and zoomable from 0.62 to 1.3. The EBCCD tube is able to detect single photoelectrons, with a signal to noise ratio better than 10. This Megapixel multichannel photon counting device is attractive for many applications in high-energy physics, astrophysics, biomedical diagnostics, and very low-light imaging.

INTRODUCTION

In present experimental particle physics, scintillating fibres are more and more frequently used in tracking detectors of various configurations. One possible option for high-resolution tracking has been investigated by the RD46 Collaboration [1], which developed the technique of liquid-core fibres, i.e. glass capillaries filled with liquid scintillator. In this framework, a new optoelectronic readout system has been developed, based on a Megapixel Electron-Bombarded-CCD (EBCCD) tube supplied by Geosphaera[2].

In the first section of this paper we briefly describe the last detector prototype, comprising the capillary bundle, an Image Intensifier and the EBCCD tube. In the following sections we report the characteristics of the Megapixel EBCCD device and the results of the laboratory tests performed to investigate its capability to detect single photoelectrons.

THE CAPILLARY DETECTOR

High resolution tracking detectors based on capillaries, have been successfully tested by the RD46 Collaboration, during 1994-1997, in the wide-band neutrino beam at the CERN Super Proton Synchrotron [2]. The last prototype, equipped with an EBCCD tube, is shown in Fig. 1. The active part of the detector is a coherent capillary bundle, made by Schott[3], comprising more than 10^6 glass capillaries filled with a 1-methylnaphtalene based liquid scintillator. The individual capillary has a diameter of about 30 μm. The bundle is about 1 m long and is placed longitudinally to the beam, as an active target for neutrino interactions. The image from the bundle is read out by an optoelectronic chain, consisting of a first generation image intensifier with high quantum efficiency and demagnification of 0.6 followed by a Megapixel EBCCD tube. The phosphor decay time of the

---

[1] European Organization for Nuclear Research, Genève, Switzerland.

[2] Geosphaera Research Centre, Moscow, Russian Federation.

[3] Schott Fiber Optics Inc., Southbridge, MA, USA.

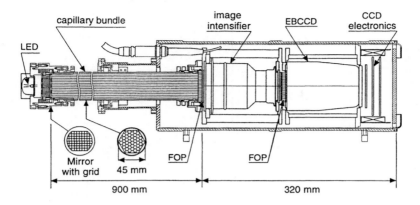

Figure 1: Schematic section of the capillary detector.

first image intensifier acts as an optical memory, which allows to wait for a trigger decision which takes ~ 1 μs. When the trigger conditions are satisfied, the gate of the EBCCD tube is opened for a fixed time interval and the image is acquired. In Fig. 2 we present the image of a neutrino interaction, as recorded by the set-up.

THE EBCCD TUBE

The design of the Megapixel EBCCD chip and the zoom tube was described in detail in [3]. Here we repeat only briefly the main points. The EBCCD chip contains 1024×1024 sensitive pixels, 13.1×13.1 μm^2 each, which correspond to a total active area of 13.4×13.4 mm^2. The chip is thinned down to ~ 8 μm and backside electron-bombarded. It is mounted (Fig. 3) into a metal-ceramic housing and installed in place of the output phosphor in a first generation electrostatically focusing zoom tube. The fibre-optic input window has a multialkali photocathode with a useful diameter of 40 mm. The quantum efficiency of the photocathode reaches its maximum value of 14 % at 500 nm. A potential of -15 kV is applied to the photocathode. The EBCCD tube can be gated by applying pulses of 1.6 kV to the focusing electrode. The device has been tested for 2000 hours at the illumination of 2×10^{-2} lux, and no variations of the tube sensitivity and dark current have been detected. The magnification of the tube can be varied from 0.62 to 1.3. The geometric distortion of the tube is less than 3%. With an illumination of 2×10^{-4} lux and a magnification of 1.3 a spatial resolution of the order of 50 lp/mm (at 15% MTF) was measured.

The chip has two adjacent active zones of 512×1024 pixels each and two output registers, one for each zone. A fast, low noise, compact electronic circuit based on Altera MAX 7000[4] programmable gate arrays was designed for the control and

---

[4]Altera Corporation, San Josè, California 95134, USA.

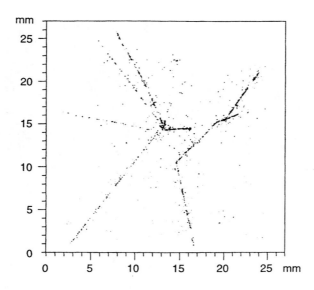

Figure 2: Image of a neutrino interaction in the capillary bundle.

readout of the EBCCD. Low noise amplifiers and an analog signal processor (correlated double sampling) for noise suppression were used. The readout of the 10^6 pixels is performed at a clock frequency of 10 MHz through one of the two output registers. Alternatively, the two zones of the active area can be shifted in opposite directions and read out in parallel, at the same clock frequency, through the two output registers. The total readout time is 104 ms in the first case and 52 ms in the second case.

To reduce the background level given by the dark current of the EBCCD and the noise associated to it, the EBCCD chip is operated at a temperature around 0 °C using a Peltier cooler in contact with the metal-ceramic housing of the chip. This reduces the dark current by a factor ~ 10 and the total noise by a factor ~ 2.

SINGLE PHOTOELECTRON SENSITIVITY

Three different measurements have been made on the EBCCD tube: a) calibration of its readout chain, b) dark current measurement, and c) single photoelectron response. In all of them the tube and its electronics have been placed in a dark box. To test the single photoelectron response of the system, only a few thousand photoelectrons, randomly distributed on the EBCCD area, were detected per frame. This was obtained by illuminating the fibre optic input window of the tube with a low intensity diffused light generated by a LED.

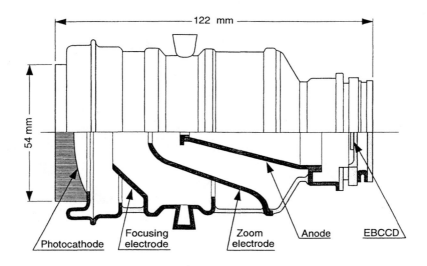

Figure 3: Schematic vew of the EBCCD tube.

The EBCCD tube is read through one of the two output register. After 104 ms the reading is finished and the entire cycle is repeated many times in one run of measurements. The same readout conditions are used for the dark current and calibration measurements. The data acquisition system is based on a commercially available framegrabber Matrox[5] Pulsar connected to the PCI bus of an IBM PC. Analog signals from the EBCCD are digitized by the 8 bit ADC of the Pulsar module.

Before each run of measurements, eight "dark frames" are acquired with no photocathode illumination and with the high voltage of the tube turned off to prevent the collection of electrons due to thermal emission from the photocathode. These dark frames are then averaged pixel by pixel and recorded in the PC memory as a "background frame" which will represent, for a given run, the nominal pedestal position for each pixel. This background frame is subtracted pixel by pixel from each frame acquired and the resulting images are stored for subsequent analysis.

The calibration constant, i.e. the ratio between the number of ADC counts and the number of electrons in a pixel potential well, has been determined by measuring the fluctuation of the output signal. We found ~ 1 μV per electron at the output node of the CCD, corresponding to ~ 0.07 ADC channel per electron. From this value we deduce that the average number of background electrons per pixel (dark current) is ~ 13800. The total fluctuation of the output signal is ~ 180 (r.m.s.) equivalent electrons, part of which (~ 130) being due to the noise of the readout electronics [4].

When a photoelectron hits the silicon substrate, it produces a cloud of electrons which are captured by one or more potential wells, forming a cluster of adjacent

---

[5]Matrox Graphics Inc., Dorval H9P 2T4, Quebec, Canada.

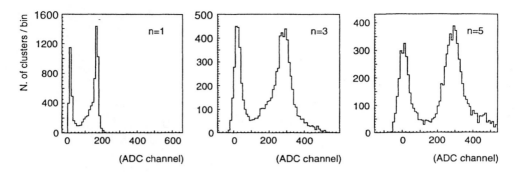

Figure 4: Distribution of the cluster charge for n=1, 3 and 5 pixel clusters. The single-photoelectron peak is well separated from the background.

pixels. In order to detect such a cluster we first search for local maxima (LM) in the charge distribution on the CCD. Then we define an n-pixel cluster as a group of n pixels adjacent to the selected LM, with its total charge being the sum of the n pixels charges. In Fig.4 we present the distribution of cluster charge for different n. From this figure one deduces that 3-pixel clusters already contain $\sim$ 80% of all charge generated by one photoelectron. The total gain of the tube is estimated to be $\sim$ 4000 electrons per photoelectron. The signal-to-noise ratio (S/N) is defined as the ratio between the average cluster charge and the total noise (180 equivalent electrons per pixel). For 3- and 5-pixels clusters we observe (Fig. 4) S/N=14 and 10, respectively.

In Fig. 5a the average cluster shape is shown. It is obtained by adding 5×5 pixels centered on the LM. The cluster has a FWHM of 13 μm in the x direction and is wider (26 μm FWHM) and slightly non-symmetric in the y direction. Very likely this effect is due to the response of the amplifier system at high clock frequency and not to the spatial distribution of the electron cloud in the silicon chip. To confirm this we plot on the $x - y$ plane (Fig. 5b) the position of the centroid of the cluster charge distribution relative to the centre of the LM pixel. This distribution appears to be very narrow and fully contained in that pixel. This fact indicates that electron clouds produced by photoelectrons have a size significantly smaller than a pixel and that the cluster shape (Fig. 5a) is essentially due to the readout system. This is also confirmed by the fact that S/N for n=1 pixel cluster is extremely good.

The single-photoelectron response has also been studied by selecting clusters of adjacent pixels above a given threshold. In this case the clusters have variable size and can be due to one or more coalescent photoelectrons. The cluster charge distribution obtained with this method is shown in Fig. 6, where single, double and even triple photoelectron peaks can be observed. The distance between the peaks is equal to $\sim$ 320 ADC counts, in good agreement with the result of the first method.

514

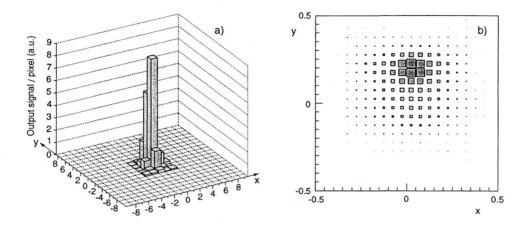

Figure 5: a) Average cluster shape. b) Distribution of the coordinate of the centroid of the clusters inside the pixel corresponding to the local maximum. In a) and b) the units on the x and y axes are equivalent to a pixel size (13 μm).

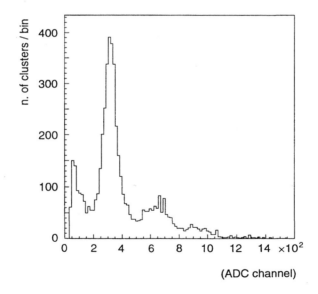

(ADC channel)

Figure 6: Distribution of the total charge contained in the clusters, as defined in the text. The single-photoelectron peak is well separated from the background. The peaks due to two and three photoelectrons can also be seen.

CONCLUSIONS

We have investigated the response to single photoelectrons of a Megapixel EBCCD zoom tube together with its readout electronics. We found a tube gain of about 4000 electrons per photoelectron at a voltage of 15 kV and a noise of 180 electrons per pixel at 10 MHz pixel readout frequency. A signal to noise ratio greater than 10 was obtained. With an appropriate threshold on the signal pulse height, the noise can almost be eliminated with a loss of the order of few percent in single-photoelectron counting. The single-electron cluster size ranges between 13 μm and 26 μm (FWHM) on the CCD plane, depending on the considered direction on that plane. Other EBCCD tubes based on this Megapixel EBCCD chip are under study. Among them we can cite a magnetic focusing EBCCD used as the readout system of an optoelectronic delay tube for high rate experiments in high energy physics [5], and a large EBCCD tube with a 80 mm diameter photocathode and demagnification of 5 for biomedical applications.

REFERENCES

1. R. Van Dantzig et al., RD46 Proposal, CERN/LHCC 95–7, P60/LDRB (1995).

2. S. Buontempo et al., Nucl. Instr. and Meth. **A 360** (1995) 7.
   P. Annis et al., Nucl. Instr. and Meth. **A 367** (1995) 377.
   P. Annis et al., Nucl. Instr. and Meth. **A 386** (1997) 72.
   P. Annis et al., Nucl. Phys. B (Proc. Supp.) **54** (1997) 86.
3. S. V. Golovkin et al., Proc. SPIE **2551** (1995) 118.
4. S. Buontempo et al., to be submitted to Nucl. Instr. and Meth.
5. A.G. Berkovski et al., Nucl. Instr. and Meth. **A 380** (1996) 537.

SESSION 14: ASTROPHYSICS - I

Chair: M. Atac
Scientific Secretary: E. Ivanov

Scintillating Fibers and Their Use in the Cosmic Ray Isotope Spectrometer (CRIS) on the Advanced Composition Explorer (ACE)

W.R.Binns[1], E.R. Christian[3], W.R. Cook[2], A.C. Cummings[2], B.L. Dougherty[4], P.F. Dowkontt[1], J.E. Epstein[1], P.L. Hink[1], B. Kecman[2], J. Klarmann[1], R.A. Leske[2], M. Lijowski[1], R.A. Mewaldt[2], M.A. Olevitch[1], T.T. von Rosenvinge[3], E.C. Stone[2], M.R. Thayer[2], and M.E. Wiedenbeck[4]

1.Washington University, Campus Box 1105, 1 Brookings Drive, St. Louis, MO 63130
2. California Institute of Technology, Mail Code 220-47, Pasadena, CA 91125
3. Goddard Space Flight Center, Code 661, Greenbelt, MD 20771
4. Jet Propulsion Laboratory, Pasadena, CA 91109

Abstract. The Cosmic Ray Isotope Spectrometer (CRIS) experiment was launched aboard the NASA Advanced Composition Explorer satellite on August 25, 1997. The experimental objective of CRIS is to measure the isotopic composition of galactic cosmic ray nuclei for elements with charge $3 \leq Z \leq 28$ over the energy range ~50-500 MeV/nuc. The instrument consists of a scintillating fiber hodoscope to determine particle trajectory, and four stacks of silicon wafers for multiple dE/dx and E_{tot} measurements. This instrument is the first to use scintillating fibers in space. The CRIS instrument has a large geometrical factor of ~250cm²sr. The spatial resolution obtained by the fiber hodoscope is ~100μm. The mass resolution achieved is ~0.12 amu for Carbon and 0.30 amu for the heaviest isotopes measured. Mass histograms of selected isotopes are presented.

INTRODUCTION

The CRIS experiment was launched aboard the NASA Advanced Composition Explorer satellite (1) on August 25, 1997 and is now orbiting about the L1 libration point ~10^6 miles from Earth along the Earth-Sun line. The experimental objective of CRIS is to measure the isotopic composition of galactic cosmic ray nuclei for elements with charge $3 \leq Z \leq 28$ over the energy range ~50-500 MeV/nuc. The CRIS geometrical factor is ~250cm²sr which is more than 20 times larger than previous isotope instruments. This large geometrical factor was made possible in part by the use of a scintillating fiber hodoscope which has considerably larger acceptance geometry than solid state hodoscopes used in previous instruments. This enables us to obtain isotopic abundance measurements with considerably higher statistical precision than has been previously achieved. For some isotopes which are already well measured, the measurements will have reduced statistical uncertainties. However, there are a number of very rare isotopes in this charge range which have not been well measured by previous experiments. For these isotopes, the CRIS measurements will be the first measurements obtained with good statistical precision. In two years of data collection under solar minimum conditions CRIS should collect ~5 x 10^6 stopping heavy nuclei

CP450, SciFi97: Workshop on Scintillating Fiber Detectors
edited by A. D. Bross, R. C. Ruchti, and M. R. Wayne
© 1998 The American Institute of Physics 1-56396-792-8/98/$15.00

(Z≥3). The particles which are detected are predominately cosmic ray nuclei which represent a direct sample of high energy matter originating outside our solar system in the Milky Way galaxy.

There are a variety of issues can be addressed using these measurements. These include:

- Nucleosynthesis of galactic cosmic ray source material.
- Cosmic ray life-times and mean gas density in the galaxy.
- Cosmic ray acceleration time-scales and reacceleration.
- Establish the pattern of isotopic differences between the galactic cosmic rays and solar system matter.
- Compare isotopic patterns in galactic and solar material to test models of galactic evolution.

THE CRIS EXPERIMENT

The CRIS experiment (2) consists of a scintillating optical fiber trajectory (SOFT) detector and four stacks of Lithium drifted Silicon detectors. Each of the four stacks consists of fifteen 3mm thick silicon wafers for multiple dE/dx and E_{tot} measurements (3). Figure 1 shows cross-sectional views of the CRIS instrument. The SOFT system consists of a hodoscope comprised of three x,y scintillating fiber planes (6 fiber layers) and a trigger detector composed of a single fiber plane (2 fiber layers). The hodoscope and trigger fibers are coupled to an image intensifier which is then coupled to a CCD for hodoscope readout, and to photo-diodes to obtain trigger pulses. The SOFT detector system has two fully redundant readouts, only one of which is operative at any given time because of limited power and bit rate on ACE.

The SOFT Detector

The scintillating fibers consist of a polystyrene core doped with scintillation dyes (BPBD and DPOPOP; emission peak 430nm) and an acrylic cladding (2,4) and are fabricated by Washington University. The fibers have a 200μm square cross-section including a 10μm cladding wall thickness. The cladding of the hodoscope fibers is coated with a black extra-mural absorber (EMA) to prevent cross-talk between fibers. The fibers are bonded together with an elastomeric adhesive (Uralane 5753). Each of the three hodoscope planes is composed of two layers of orthogonally crossed fibers which are bonded to a 25μm thick Kapton substrate. The spacings between the hodoscope planes are 3.9 cm and 3.3 cm as shown in the figure below. The center fiber plane is unequally spaced from the outer planes to enable us to resolve ambiguities in trajectory determination which can occur for low-Z nuclei due to electron "hopping" in the microchannelplate image intensifier. The 26 cm wide hodoscope output for a single fiber layer is split into 11 "tabs" with width ~2.4cm of contiguous fibers. These tabs are stacked and bonded together to form rectangular outputs for each layer with each rectangle having dimensions ~0.3 cm x ~2.4 cm. The two sets of six fiber layer outputs (H1x, H1y, H2x, H2y, H3x, H3y) are each routed to one of the cameras. The six outputs are stacked and held together in a rectangular block. The block is then cut and polished. The output is then coupled to the image intensifier (Figure 2). The trigger fiber plane is essentially identical to the hodoscope planes with the exception that the fibers are not coated with EMA so that we can obtain the maximum light output. The

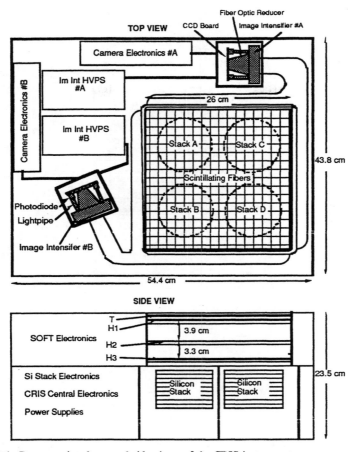

FIGURE 1. Cross-sectional top and side views of the CRIS instrument.

trigger fiber outputs are formatted in a similar way to the hodoscope fibers as shown in Figure 2. Spacers are placed between the hodoscope and trigger fiber outputs to separate the readout of hodoscope and trigger fibers. Four extra tabs of fibers (not shown) were attached to LEDs and were used for functional testing and alignment checks of the system.

Figure 3 shows a cross-sectional view of the image intensifier assembly. The image intensifiers (5) are 40mm diameter, dual microchannel plate (MCP), gateable devices (Photek Model # MCP-340S) with fiber optic windows on the input and output. The rear MCP is double thickness (0.8 mm) so that the gain of the tube is equivalent to that for a three-MCP intensifier. These tubes have blue peaked, S-20 photocathodes with a sensitivity of ~50-60mA/W at 450nm (roughly 15% quantum efficiency). The intensifiers exhibit about 20 to 30 dark counts/cm²/s at room temperature and a much smaller number at temperatures of ≤0°C (the image intensifier and CCD are passively cooled in space to about -15° C). The output phosphor is P-20

521

phosphor and a thin film aluminum anode is deposited over the phosphor. The which has a 1/e decay time of about 50 μs, with a low level tail extending for about 1 ms. The

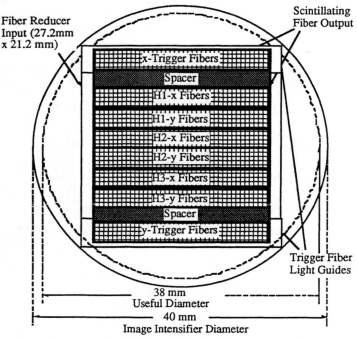

FIGURE 2. The hatched areas show the fiber formatting onto the photocathode of the image intensifier. Additionally, the reducer used to image the intensifier output onto the CCD and the trigger fiber light guides which couple the trigger fibers to the photodiodes are shown.

output window is coated on the external face with a thin, optically transparent metallic layer which is grounded to eliminate corona from the anode. The tubes are each powered by a high voltage power supply (Model # 2404) made by K&M Electronics (6). The image intensifier was ruggedized so that it could survive the ACE launch vibration and shock levels.

The voltages used to power the intensifier are shown in Figure 3. The cathode voltage with respect to the MCP-in is -200V when it is gated on and +40V when it is gated off. The gating time is ~1μs. The gain of the image intensifier is controlled during flight by adjusting the voltage across the MCPs. The maximum photon gain that can be achieved is ~ $2x10^6$ but we are operating at a gain of typically $6x10^5$. The image intensifer output which corresponds to the area covered by the hodoscope fibers on the input is coupled onto the large end of a 34.5mm to 11mm (diagonal dimension; 3 to 4 aspect ratio) fiber-optic reducer using Dow-Corning 93-500 adhesive. The reducer output is coupled to a fiber optic window installed on the CCD. The CCD that we are using is a Thomson TH-7866 (244 x 550 pixel array) with individual pixel size 16μm (width) x 27μm (height). A single fiber projects onto an area equivalent to 4 pixels wide by 2.4 pixels high, after the image is reduced by the fiber optic reducer. The output area of the image intensifier which corresponds to the two trigger fiber inputs are coupled to Hamamatsu S-3590-01 photodiodes using acrylic light guides. For further details see Reference 2.

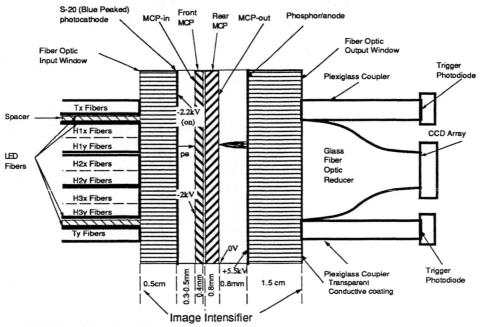

FIGURE 3. Side view of image intensified CCD system.

Method of Obtaining Real-Space Coordinates from SOFT

The first step in obtaining real space coordinates is to develop a pixel to fiber map. The left panel of Figure 4 shows a superposition in CCD pixel space of many oxygen events obtained from a MSU NSCL test (7,8). The beam was scanned over the active area so that particles traversed every fiber. Centroids were calculated for each pixel cluster and plotted. The right panel shows a magnified view of a small part of the fiber array. We see that most centroids can be clearly identified with an individual fiber. Cuts were taken through these data in x and y separately along each fiber row and column and the data were histogrammed. The data were then fit with a gaussian fitting routine to obtain the center position of each fiber. From these center positions we defined cells about each position as shown in the right panel. A centroid from an event falling in a given cell is identified with that particular fiber.

The second step is to develop a fiber-space to real-space map. This is obtained using a fiducial hole plate made of lead which has 1 mm diameter holes spaced in a 1.27 cm square grid. The hole plate was placed in an oxygen beam at MSU. Figure 5 shows data from the hole plate fiducial run at MSU. The projection of the holes onto the fiber array is clearly seen. Histograms in x and y were made of each hole and the center of the distribution was taken to be the center of the hole. The application of these two maps enables us to transform from pixel space to real space.

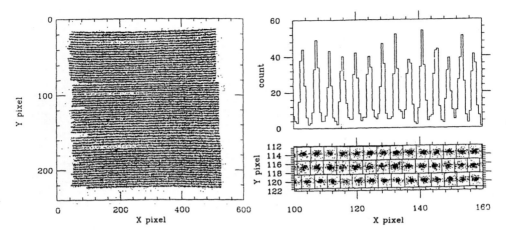

FIGURE 4. Pixel to fiber map. The left panel is a cross plot of pixel cluster centroids in CCD pixel space. This figure is a superposition of many events obtained by scanning an oxygen beam over the full 26cm x 26cm area of the fiber hodoscope. Each row of dots corresponds to a single fiber tab of width ~2.4 cm. The right panel shows an expanded view of a small region in pixel space. It is seen that the individual fibers are clearly resolved as shown in the crossplot and histogram.

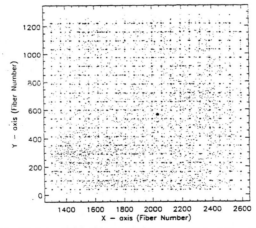

FIGURE 5. An absolute calibration of SOFT was obtained by placing a lead mask with regularly spaced holes in the beam and irradiating the hodoscope with oxygen nuclei. The regularly spaced dark spots are the superposition of events which passed through the holes.

SOFT In-Flight Performance

The spatial resolution that is required for SOFT to contribute less than 0.1amu to the mass resolution for iron nuclei at 45° is ~130μm. Figure 6 shows the spatial resolution obtained for iron nuclei in-flight using the method of residuals. We see that the resolution obtained meets this requirement.

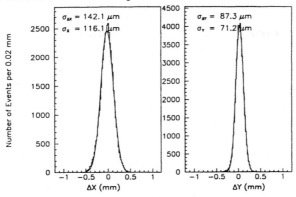

FIGURE 6. Spatial resolution obtained using the method of residuals for x and y coordinates.

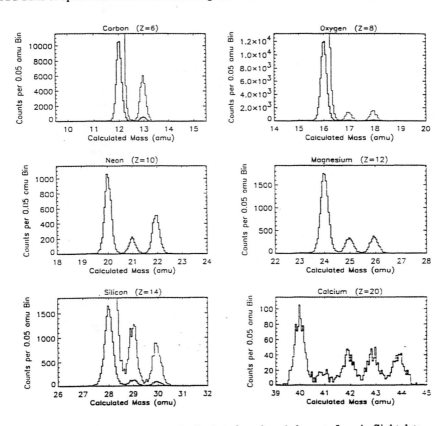

FIGURE 7. Histograms of isotope distributions for selected elements from in-flight data.

In Figure 7 we plot mass histograms for a sample of the CRIS data for selected elements. We see that the individual isotopes are cleanly resolved and have excellent statistics. The mass resolution which we obtain ranges from 0.12amu for carbon to 0.30amu for iron nuclei. In subsets of the data that are restricted to smaller incidence angles (e.g., $<45°$) the iron mass resolution improves to $\sim$0.25 as expected. This excellent mass resolution, combined with the large collecting power of CRIS, is expected to provide significantly improved knowledge of the isotopic abundances of galactic cosmic rays and should enable us to test models of cosmic ray origin, acceleration, and propagation.

ACKNOWLEDGEMENTS

This work was supported by NASA Contract NAS5-32626 and Grant #NAG-6912.

REFERENCES

1. Stone, E.C., Frandsen, A.M., Mewaldt, R.A., Christian, E.R., Margolies, D., Ormes, J.F., Snow, R., , "The Advanced Composition Explorer", Space Science Reviews (To be published). Also see Stone, E.C., Burlaga, L.F., Cummings, A.C., Feldman, W.C., Frain, W.E., Geiss, J., Gloeckler, G., Gold, R.E., Hovestadt, D., Krimigis, S.M., Mason, G.M., McComas, D., Mewaldt, R.A., Simpson, J.A., von Rosenvinge, T.T., Wiedenbeck, M.E., "The Advanced Composition Explorer", *Particle Astrophysics*, AIP Conference Proceedings Vol. 203, pp.48-57, 1989.
2. Stone, E.C., Cohen, C.M.S, Cook, W.R., Cummings, A.C., Dougherty, B.L., Grumm, R.L., Milliken, B.D., Radocinski, R.G. , Wiedenbeck, M.E., Christian, E.R., Shuman, S., Trexel, H., von Rosenvinge, T.T., Binns, W.R., Crary, D.J., Dowkontt, Pl, Epstein, J., Hink, P.L., Klarmann, J., Lijowski, M., and Olevitch, M.A., "The Cosmic Ray Isotope Spectrometer for the Advanced Composition Explorer", Space Science Reviews (To be published).
3. Allbritton, G.A., Anderson, H., Barnes, A., Christian, E.R., Cummings, A.C., Dougherty, B.L., Jensen, L., Lee, J., Leske, R.A., Madden, M.P., Mewaldt, R., Milliken,, R., Nahory, B.W.,O'Donnell, R., Schmidt, P., Sears, B.R., von Rosenvinge, T.T., Walton, J.T., Wiedenbeck, M.E., and Wong, Y.K., "Large diameter lithium compensated silicon detectors for the NASA Advanced Composition Explorer (ACE) mission", IEEE Trans. Nucl. Sci., Vol. 43, pp. 1505, June 1996.
4. Davis, A.J., Hink, P.L., Binns, W.R., Epstein, J.W., Connell, J.J., Israel, M.H., Klarmann, J., Vylet, V., Kaplan, D.H., Reucroft, S., "Scintillating Optical Fiber Trajectory Detectors", *Nuclear Instruments and Methods*, Vol. A276, pp.347-356, 1989.
5. Photek Ltd., 26 Castleham Road, St. Leonards-on-Sea, East Sussex TN38 9NS, United Kingdom.
6. K&M Electronics, Inc., 11 Interstate Drive, West Springfield, MA 01089.
7. Hink, P.L., Beatty, J.J., Binns, W.R., Klarmann, J., "The ACE-CRIS Scintillating Fiber Hodoscope: A Prototype Calibration at the MSU NSCL", *24th ICRC Proc.*(Rome), Vol. 3, pp. 653-656, 1995.
8. Hink, P.L., Binns, W.R., Klarmann, J., and Olevitch, M.A., "A Calibration at the NSCL", in *Gamma-Ray and Cosmic-Ray Detectors, Techniques, and Missions*, Ramsey, B.A. and Parnell, T.A. (editors), Proc. SPIE 2806, Denver, Society of Photo-Optical Instrumentation Engineers, pp. 199-208.

The Use of Optical Fibers in the Trans Iron Galactic Element Recorder (TIGER)

S.H. Sposato[1], L.M. Barbier[3], W.R. Binns[1], E.R. Christian[3],
G.A. de Nolfo[2], P.F. Dowkontt[1], J.W. Epstein[1], P.L. Hink[1], M.H. Israel[1],
J. Klarmann[1], D.J. Lawrence[1*], R.A. Mewaldt[2], J.W. Mitchell[3],
S.M. Shindler[2], R.E. Streitmatter[3], C.J. Waddington[4]

[1] *Washington University, Box 1105, One Brookings Drive, St. Louis, MO 63130*
[2] *California Institute of Technology, Mail Code 220-47, Pasadena, CA 91125*
[3] *Goddard Spaceflight Center, Code 661, Greenbelt, MD 20771*
[4] University of Minnesota, Minneapolis, MN 55455

* *Currently at Los Alamos National Laboratory, Mail Stop D466, Los Alamos, NM 87545*

Abstract. TIGER, the Trans-Iron Galactic Element Recorder, is a cosmic-ray balloon borne experiment that utilizes a scintillating Fiber Hodoscope/Time of Flight (TOF) counter. It was flown aboard a high altitude balloon on September 24, 1997. The objective of this experiment is to measure the elemental abundances of all nuclei within the charge range: $26 \leq Z \leq 40$. This initial balloon flight will test the detector concept, which will be used in future balloon and space experiments. The instrument and the fiber detector are described.

INTRODUCTION

The Trans-Iron Galactic Element Recorder, TIGER, is a balloon-borne cosmic-ray experiment designed with both scientific measurement goals and engineering goals. Its science goal is to measure the elemental abundances of ultra-heavy cosmic rays with $Z \gtrsim 26$, over an energy range of 300 MeV/n to 8 GeV/n with 0.2e charge resolution. One of the key measurements in this interval is the abundance ratio Co/Ni. This ratio is important because it can put limits on the time between the nucleosynthesis and acceleration of cosmic rays. Previous measurements differ by a factor of two. (See Englemann *et al.* 1990, Leske *et al.* 1992, and Webber *et al.* 1990). The measurement TIGER makes of the Co/Ni ratio can help resolve this discrepancy.

TIGER's engineering goal is to prove that ultra-heavy Galactic Cosmic Rays (GCR) can be measured with excellent charge resolution using a full scale, large area detector. The version of TIGER described here is a prototype for future long duration balloon flights and space experiments.

TIGER was flown from Fort Sumner, NM on September 24, 1997 using a 40 million-ft^3 balloon. It achieved float altitudes between 115,000 ft and 133,000 ft for 23.5 hours.

THE TIGER DETECTORS

The TIGER instrument contains eight different detectors. (For more information

CP450, SciFi97: Workshop on Scintillating Fiber Detectors
edited by A. D. Bross, R. C. Ruchti, and M. R. Wayne

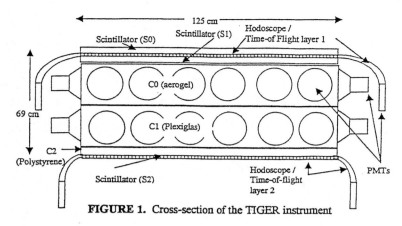

FIGURE 1. Cross-section of the TIGER instrument

see Lawrence 1996.) Figure 1 shows a cross-section of the instrument. It contains three scintillation counters with wave-length-shifter-bar readout. There are also three Cherenkov detectors: 1 Plexiglas radiator in a light box, 1 aerogel radiator in a light box, and 1 wavelength-shifter-bar readout Cherenkov detector with a Plexiglas radiator. The scintillation counters and the Cherenkov counters together provide charge and velocity measurements of the cosmic rays, provided that pathlength through the instrument is known. (See Figure 2.) In order to determine charge with the desired resolution, the position of the particle in each plane must be known to within a few mm so that pathlength corrections can be made. TIGER has two detectors to accomplish this: a Coarse Hodoscope, which contains two planes of fibers and gives position to within 11cm, and a Fine Hodoscope, which also contains two planes of fibers and gives position to within a few mm. Each fiber plane gives a measurement of X and Y position. The Fiber Hodoscope also makes a TOF measurement that can be used along with the scintillation counters to give a redundant charge measurement (Figure 2.)

The Fiber Hodoscopes

The Fine Fiber Hodoscope is made of two planes of 1.5 mm square cross section scintillating optical fibers, which were fabricated at Washington University. Each plane has two layers of fibers, one each in the X and Y directions. The hodoscope size is 1.1 m square. In order to measure charge accurately, the path length of the cosmic rays through the instrument must be known to within a few millimeters. The smallest spatial resolution TIGER can measure is the fiber tab. Each tab contains 6 fibers, making it 9mm wide. The tabs are then grouped into fiber segments, with 16 tabs per segment. Each segment is 14.4 cm wide. There are 128 fiber tabs in both the X-direction and the Y-direction.

TIGER has a unique fiber-coding scheme that allows the 128 tabs to be read-out with 8 Photomultiplier Tubes (PMTs). Figure 3 shows a schematic of the coding scheme. Depending on where the particle goes through the plane of fibers, ideally a

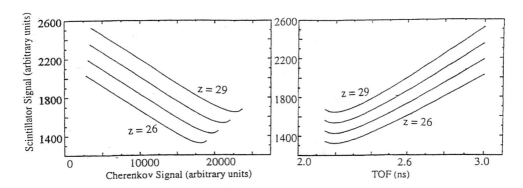

FIGURE 2. Calculated Cherenkov signal versus scintillator signal and calculated TOF versus scintillator signal respectively.

certain combination of 4 PMTs will fire. These four signals identify the tab that was hit, but not the segment. Since there are 8 segments in each coordinate, the particle could have 8 different positions. As a result, the Fine Hodoscope has an 8-fold degeneracy if used by itself.

The Coarse Hodoscope breaks this degeneracy. It contains fiber paddles made of 1.5 mm^2 cross-section fibers with one PMT at each end. The Coarse Hodoscope paddles lie right on top of the Fine Hodoscope segments. There are 10 paddles per layer. The combination of 1 Coarse Hodoscope signal and 2 Fine Hodoscope signals uniquely determines the position of the particle in one coordinate. The Fine Hodoscope identifies the tab and the Coarse Hodoscope identifies the segment. Thus, the fiber hodoscopes on TIGER can identify position to within a few mm by reading out 128 tabs using only 10 Coarse Hodoscope PMTs and 8 Fine Hodoscope PMTs. This is important for long duration balloon flight and satellite experiments since it reduces power and bit rate requirements.

The TOF Detector

The fibers in the Fine Hodoscope are also used to make a Time of Flight measurement. The TOF measurement can be used in conjunction with the scintillation counters to give a redundant charge measurement. A cosmic ray nucleus goes through the instrument and ideally 8 of the 32 Phillip's XP2020 PMTs will fire. The last signal to come into the logic board forms the Time-to-Digital Converter (TDC) common start. The signals from the 32 PMTs form the 32 individual stops. The signals from the PMTs are also pulse-height analyzed using charge integrating Analog to Digital Converters (ADCs) so that timing walk corrections can be applied.

The TOF value can be written as:

$$TDC_i = STOPi - START .$$ (1)

And if :

$$TDC_0 = (TDC_8 + TDC_7 + TDC_6 + TDC_5) - (TDC_4 + TDC_3 + TDC_2 + TDC_1),$$ (2)

529

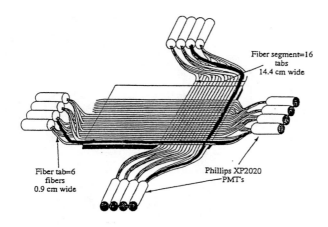

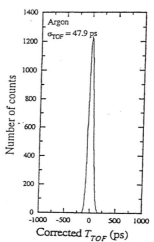

FIGURE 3. Schematic diagram of one plane of the Fine Hodoscope. Only a portion of the fibers Shown. All the PMTs are show.

FIGURE 4. A Gaussian has been fit to the histogram of the TOF values for AR. The resulting sigma is shown on the plot.

where $TDC_{5\text{-}8}$ are the TDC values for the PMTs that fired on the bottom plane, and $TDC_{1\text{-}4}$ are the TDC values for the PMTS that fired on the top plane, then the particle time-of-flight- is:

$$T_{TOF} = \frac{1}{4}(TDC_0 - T_f - T_S - T_W - T_D).$$ (3)

TDC_0 is the measured TDC difference. T_f is a timing offset due to fibers in different tabs being of different lengths. This can be corrected by mapping out the segments using in-flight particles, or by just measuring the lengths of the fibers. T_s is the time it takes the signal to travel through the PMTs and cables to the discriminators. This can be measured using an electronic pulse. T_W is a timing walk offset cause by signals of different heights triggering the discriminators at different times. Pulse-height analysis of the signals from the PMTs corrects this offset. Lastly, T_D is the time it takes the signals to travel through the delay lines to the TDCs. Measuring the length of the delay lines determines this constant.

TESTING THE TOF DETECTOR

In order to make charge measurements within the desired resolution. the TOF detector on TIGER needs a 35 ps timing resolution. Some factors that affect this resolution are light dispersion, PMT rise time, electronic noise and light attenuation. The TIGER TOF detector has many advantages that improve the timing resolution. First, it is made of fibers, which have very little dispersion. Second, TIGER uses XP2020 PMTs, which are specifically designed for timing measurements, to reduce

PMT rise time, transit time and transit time jitter. Lastly, electronic noise in the TOF system is reduced by using electronics also specifically designed for timing measurements.

At the time the TIGER TOF detector was designed, no fiber detector had been able to achieve 35 ps timing resolution. To prove that it could be done, a prototype fiber detector made of 2 ribbons of 1.5mm and 2mm round fibers was taken to the Michigan State University (MSU) Cyclotron for calibration in June of 1993. The midpoints of the ribbons were exposed to 155 MeV/n Ar, O, and C nuclei. The Ar run is the most important measurement because the dE/dx of Ar is similar to that for higher energy cosmic ray Fe. 50 ps resolution was achieved with Ar. (See FIGURE 4.) To study attenuation effects in the fibers, the midpoints of the ribbons were moved perpendicular to the beam. 50 ps resolution was again achieved with Ar.

The TIGER instrument has two planes of fibers and therefore makes two independent timing measurements. Because of this, TIGER should be able to get a factor of $\sqrt{2}$ better than 50 ps and therefore should theoretically be able to achieve a timing resolution of 35 ps.

FUTURE MODELS OF TIGER

The model of TIGER described here is an engineering prototype for long duration balloon experiments and for future space experiments. In building and testing the TIGER Fine Hodoscope/TOF detector, much was learned to improve the design and construction of future fiber detectors. For example, making all the fibers in a detector the same length simplifies the position measurement and the TOF measurement.
A fiber hodoscope/TOF system modeled after the one on TIGER is planned for use on future space experiments. The fiber TOF system will be used only to distinguish up from down moving particles, so it will only need a timing resolution of 300 ps.

In future long duration balloon experiments, a fiber hodoscope similar to the one on TIGER will be used as well. This one will be "fully coded", with 28 PMTs, or 14 per side, reading out 196 fiber tabs. The signal from each PMT will be pulse height analyzed. There will be no TOF system included in future balloon experiment for several reasons. First, balloon experiments do not need to distinguish up from down moving particles. Second, a TOF measurement used for charge identification would not have good enough charge resolution. And lastly, and most importantly, long duration balloon experiments require light-weight, low power detector systems. TOF systems usually add unwanted power and weight.

REFERENCES

Englemann, J.J., Ferrando, P., Soutoul, A.., Goret, P., Juliusson, E., Koch-Miramond, L., Lund, M., Masse, P., Peters, B., Petrou, N., Rasmussen, I.L., *Astronomy and Astrophysics*, **233**, 96 (1990).
Lawrence, D.J., *Ph. D. Thesis*, Washington University in St. Louis, MO. 1996.
Leske, A.L., Milliken, B., and Wiedenbeck, M.E., *The Astrophysical Journal*, **390**, L99 (1992).
Webber, W.R. and Gupta, M., *The Astrophysical Journal*, **348**, 608,(1990).

SESSION 15: APPLICATIONS - II

Chair: D. Lincoln
Scientific Secretary: H. Zheng

The Calibration and Monitoring System for the PHENIX Lead-Scintillator Electromagnetic Calorimeter[1]

G.David, E.Kistenev, S.Stoll, S.White, C.Woody

Brookhaven National Laboratory, Upton, New York 11973

A.Bazilevsky, S.Belikov, S.Chernichenkov, A.Denisov, Y. Gilitzky, V.Kochetkov, Y.Melnikov, V.Onuchin, A.Semenov, V.Shelikhov, A.Soldatov

Institute for High Energy Physics, Protvino, Russia

Abstract. A system for calibrating the PHENIX lead-scintillator electromagnetic calorimeter modules with cosmic rays and monitoring the stability during operation is described. The system is based on a UV laser which delivers light to each module through a network of optical fibers and splitters and is monitored at various points with silicon and vacuum photodiodes. Results are given from a prototype system which used a nitrogen laser to set the initial phototube gains and to establish the energy calibration of calorimeter modules and monitor their stability. A description of the final system to be used in PHENIX, based on a high power YAG laser, is also given.

I. INTRODUCTION

The lead-scintillator electromagnetic calorimeter for the PHENIX experiment (1) is a shashlik-type detector consisting of 15552 individual towers which covers an area of approximately 48 m^2. The calorimeter will be used to measure electron and photon production in relativistic heavy ion collisions at RHIC, and will be an integral part of the particle identification and trigger system for PHENIX. The calorimeter will also be used to measure high pT photon production and other electromagnetic processes in high energy polarized proton collisions as part of the spin physics program at RHIC.

The calorimeter has an nominal energy resolution of 8%/$\sqrt{E}$(GeV) and a timing resolution of <100 ps for electromagnetic showers(2). A precision calibration and monitoring system has been developed to achieve an absolute energy calibration of less than 5% for day-one operation at RHIC, and to maintain an overall long-term gain stability of less than 1%. The design of the system will be described, and results will be presented from various tests and module calibrations using cosmic rays.

---

[1] This work was supported under D.O.E. contract DE-AC02-76ch00016

CP450, *SciFi97: Workshop on Scintillating Fiber Detectors*
edited by A. D. Bross, R. C. Ruchti, and M. R. Wayne
1998 The American Institute of Physics 1-56396-792-8/98/$15.00

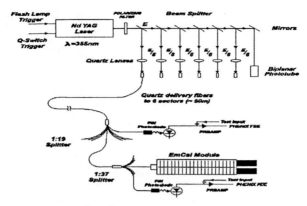

FIGURE 1. Laser calibration, monitoring, and light distribution system.

II. SYSTEM DESIGN

The heart of the calibration and monitoring system is a UV laser which supplies light to the calorimeter through a network of optical splitters and fibers. The calorimeter is composed of six sectors, each consisting of a framed, 3 x 6 array of 18 supermodules. Each 67 cm x 67 cm supermodule is a 6 x 6 package of individual modules, each containing four readout towers and four phototubes. Injected laser light excites the plastic scintillator of each module via a "leaky fiber" which "leaks" light with a distribution in depth that simulates an electromagnetic shower. The intensity of light from the laser is monitored at the initial beam splitter with a biplanar photodiode, and at each intermediate splitter and supermodule with PIN diodes. The overall layout of the system is shown in Fig. 1.

A. Laser and Primary Beam Splitter

The UV laser used as the light source for the initial precalibration of the calorimeter modules with cosmic rays was a Laser Science VSL-337ND pulsed nitrogen laser (λ=337 nm). For the final system, a Continuum Surelite II-10 (2 W, λ=355 nm) YAG laser will be used. The laser beam intensity is controlled by a set of fixed attenuators and a remotely controlled variable attenuator. Light from the laser is initially split into six equal intensity beams by a set of partially reflecting mirrors. The beam fraction extracted by each mirror passes through a quartz lens and is focused to a point just in front of a quartz fiber. These six 50 m fibers will be 600 mm silica-core/silica-cladding high OH⁻ fibers (3M Specialty Fiber FG-600-UAT) which have numerical aperture of 0.16 and an attenuation length of 60 dB/km at 355 nm. In addition, they are designed for high power applications (up to 5 GW/cm^2) in order to handle the extremely high instantaneous power of the focused laser beam. The spot size should be small (< 0.8 times the fiber diameter) to prevent light from entering the cladding and damaging the fiber. Also, the injection angle (θ) must not be too large or too small compared to the numerical aperture of the fiber (typically 0.3 NA < θ < 0.8 NA).

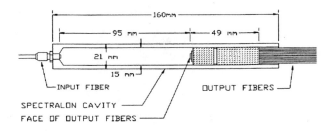

FIGURE 2. First level optical splitter (1:21)

B. Optical Splitters

Two levels of optical splitters are used to distribute the light to each of the individual calorimeter modules. The "first level" splitter, shown in Fig. 2, is designed such that the input fiber projects light onto a bundle of 21 output fibers. The bundle of 5 m x 1 mm dia. plastic fibers is mounted inside a cavity made of UV reflecting Spectralon(3). The reflective cavity collects and diffuses the light leaving the input fiber, improving the efficiency and uniformity of the output fibers. The uniformity is strongly dependent on the cavity depth for short distances, but is limited by other factors, such as end preparation and polish of the fibers, for larger distances. For PHENIX, a cavity spacing of approximately 70 mm will be used, which gives an individual fiber efficiency of 2.8×10^{-3}, a uniformity with a sigma/mean of 5.3%, and a max./min. ratio of 1.25.

Eighteen outputs of the first level splitter are connected to second level splitters inside each of the supermodules (one is connected to a photodiode for monitoring and two are spares). The 1:21, second level 1:38 splitters, and other connecting fibers are the same Hoeschst-Celanese EN-52 jacketed plastic fiber. The design of the second splitter is similar to the first except that the input fiber is included within the bundle of output fibers and projects light onto the back surface of the Spectralon cavity. As shown in Fig. 4, this allows for a more compact design in which the splitter acts as a simple integrating cavity with good output fiber uniformity. Of the bundle of 38 0.6 m long output fibers, 36 go to the calorimeter modules within each supermodule, one goes to a reference PIN diode inside the supermodule, and one goes to an external PIN diode outside the supermodule. The performance of the second level splitter is similar to the first level splitter. The individual output fiber efficiency is $\sim 2.2 \times 10^{-3}$, and the uniformity has a sigma/mean ratio of 7.3% and a max./min. ratio of 1.32.

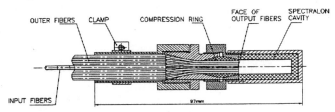

FIGURE 3. Second level optical splitter (1:38).

C. Calorimeter Modules

The output fibers of the second level splitter are connected to a "leaky fiber" which is inserted into a space along the central axis of each module and is designed to let light "leak" out in such a way as to simulate an electromagnetic shower penetrating into the module. This is accomplished by mechanically scribing a spiral scratch along a 38 cm x 2 mm dia. plastic fiber. Light escapes from the fiber and excites the plastic scintillator tiles in the 4 surrounding calorimeter stacks. The scribe pattern was tuned to give a depth profile which resembled an electromagnetic shower with an energy of 1 GeV. The pattern was determined experimentally by measuring the light output as a function of position along the fiber. Figure 4 shows a cutaway view of a calorimeter module showing the location of the leaky fiber along with the lead-scintillator stacks and readout wavelength shifting fibers.

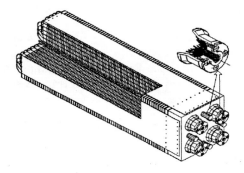

FIGURE 4. Cut-away interior view of calorimeter module showing stack of scintillator and lead plates, wavelength shifting fiber readout, and leaky fiber inserted in central hole.

The efficiency for converting UV light injected into the module to visible light at the output bundle of wavelength shifting fibers was ~ 7.6 x 10^{-4}. This efficiency includes the leaky fiber efficiency, the conversion efficiency from UV light to blue scintillation light inside the scintillator, and the conversion efficiency of the scintillation light into green light in the wavelength shifting fibers. The overall efficiency to convert the laser light to photoelectrons in a module is of order 4.2 x 10^{-12}. Given that the calorimeter has an intrinsic light output of ~ 1500 photoelectrons per GeV, ~ 0.2 mJ per pulse from the YAG laser is required to deliver 1 GeV of equivalent energy to each module. Linearity studies and calibration of the calorimeter for high energy polarized proton running at RHIC can require up to 80 GeV per module, or 16 mJ per pulse from the laser.

III. MONITORING PHOTODIODES

Since the energy of the laser varies from pulse to pulse with an rms. variation of ~ 4%, and the average energy can change over periods of hours by more than 10%, it is necessary to measure the pulse-by-pulse light output of the laser with a stable reference device. The biplanar phototube (Hamamatsu R1328U-02) is a highly linear vacuum photodiode with a high current photocathode. As shown in Figure 1, it is positioned at the end of the set of six primary beam splitter mirrors and serves as the primary reference monitor of the beam intensity.

PIN photodiodes (Hamamatsu S-1223-01) are used to monitor the light intensity after each level of splitting. One diode is used to measure the output of each of the 1:21 splitters for each sector. One output of each 1:38 splitter goes to an internal PIN photodiode located underneath the front cover of the supermodule and is used as the reference for that supermodule. A second 1:38 splitter output is connected to an external photodiode located on the interior wall of the sector. This diode is an accessible secondary reference for the supermodule.

The PIN diodes are read out using a high speed voltage amplifier (Elantec 2075). Each diode readout circuit also contains a stable current-mirror source which is switched by an external ECL trigger to deliver a calibration pulse to the voltage amplifier. This calibration pulse is used to monitor the stability of the amplifier and to conveniently test the rest of the readout chain.

IV. ENERGY CALIBRATION AND GAIN MONITORING

The laser calibration and gain monitoring system has been used extensively to study several prototype calorimeter modules in beam tests at the Brookhaven AGS, and is presently being used to establish the initial phototube gain settings for calorimeter modules in preparation for operation at RHIC. Every supermodule is equipped with the same set of 144 phototubes to measure each tower's light output using cosmic ray muons. This is done at a fixed gain setting for each tube, and the same setting is used for each supermodule. At the same time, a spectrum of laser events is collected for each tower and for the internal reference photodiode. The ratio of the signal from the module towers to the signal from the photodiode is independent of the intensity of the light from the laser and serves as the reference to re-establish the same phototube response (given by the product of gain times quantum efficiency) for the final set of phototubes which are eventually installed in the supermodule. The final tubes are selected from a set of pre-measured tubes and are installed in the supermodule in three groups of 48 tubes with similar gains. This grouping is necessary because only three high voltage settings are available for each supermodule, so all 48 tubes within a given group operate at the same high voltage. Therefore, when the final tubes are installed, there is a large dispersion in the actual response of each tower to the original muon calibration. When the response is renormalized to the original reference response using the laser, as shown in Fig. 5a., the dispersion is reduced to only 2.3%. This residual dispersion is due mainly to the variation in the quantum efficiencies of the phototubes, but is more than a factor of two better than the design goal of 5% in predetermining the initial energy calibration of the calorimeter for RHIC operation.

To determine the ability of the calibration system to monitor and correct for phototube gain variations, the change in muon response over time of a supermodule was measured. The module was calibrated using cosmic ray muons with the standard set of phototubes used for precalibrating all supermodules. The tubes were left on at their nominal operating voltage. Periodic measurements were then made of the muon calibration and compared to the initial values. Figure 5b. shows, for all phototubes in the supermodule, the deviation from initial calibration over a period of one week. The figure includes uncorrected data and the same data renormalized using the laser calibration system. After renormalization, the gain drift variation is reduced to 0.9%.

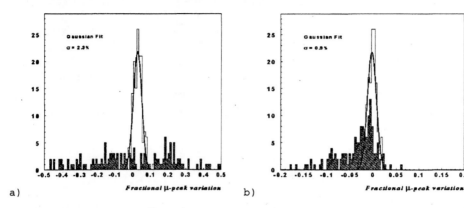

a)   b)

FIGURE 5. Deviation from initial muon calibration of all channels of one supermodule; a) after installation of final phototubes, b) with no change in phototubes or phototube high voltage after one week. Shaded area is without renormalization using laser calibration and non-shaded histogram is with renormalization.

V. CONCLUSIONS

A precision calibration and monitoring system has been designed for the PHENIX lead-scintillator electromagnetic calorimeter. It is being used with cosmic ray muons to predetermine the initial energy calibration of the calorimeter to better than 5% and to monitor gain drifts to ~ 1%. The system has been used to precalibrate more than half of the 108 modules which will be installed at RHIC. It has performed more than a factor of two better than the design goal and has demonstrated the ability to monitor gain drifts in the phototubes to the required precision. It is therefore expected that the final system will meet the necessary performance requirements for the PHENIX calorimeter. The final system should be available for commissioning and testing at least one year before the initial colliding beams at RHIC.

REFERENCES

1 PHENIX Conceptual Design Report, BNL 48922, January 29, 1993
2 G.David et.al.,"Performance of the PHENIX EM Calorimeter",
      IEEE Trans. Nucl. Sci. NS-43 (1996) 1491-1495.
3 *Spectralon* is a product of Labsphere, P.O.Box 70, North Sutton, MA 03260.

A FAST DIGITIZATION SYSTEM FOR THE g-2 SCINTILLATING TILE HODOSCOPE

J. KINDEM

University of Minnesota, Minneapolis, MN 55455

Abstract. A fast digitization system for a high-efficiency scintillating tile hodoscope has been developed for the muon g-2 experiment. Light pulses from the tiles are amplified by a Philips 64-channel photomultiplier tube. Electrical signals are both discriminated and stored locally in custom NIM modules as four 16-bit words. Digital information is then pumped via shielded "phone" cables to a remote custom VME module which serves as a data buffer and computer interface. Testbeam performance of PSD and digitization system is discussed.

INTRODUCTION

The goal of the Brookhaven muon g-2 experiment (E821) [1] is to measure the anomalous magnetic moment of the muon to 0.35 ppm. The anomalous magnetic moment of the muon is measured by capturing muons in a precision magnetic field and measuring the difference between the muon spin precession frequency and the muon momentum precession frequency. The number of forward-going, high-energy decay electrons (> 1.6 GeV) striking a set of calorimeters spaced evenly around the inner perimeter of the storage ring has the functional form:

$$N(t) = N_o e^{-t/\tau}[1 - A cos(\omega_a t + \phi)], \tag{1}$$

where τ is the muon lifetime and ω_a is the g-2 precession frequency.

Position Sensitive Detectors (PSD) are finely-segmented tile-fiber hodoscopes installed on the front face of the calorimeters. Their purpose is to reject events where two or more decay electrons strike the calorimeter within the same 5 ns time bucket. Two low-energy electrons, which would otherwise be below threshold, could resemble one higher-energy electron and introduce a systematic bias in the frequency measurement between early and late times, since the event rate is exponentially decaying with the muon lifetime. In addition, vertical position information provides a limit on the electric dipole moment (edm) of the muon; a change of several millimeters in the mean of the vertical distribution would correspond to an edm of 10^{-20}e-cm.

CP450, *SciFi97: Workshop on Scintillating Fiber Detectors*
edited by A. D. Bross, R. C. Ruchti, and M. R. Wayne
© 1998 The American Institute of Physics 1-56396-792-8/98/$15.00

DESCRIPTION OF PSD FIBER HODOSCOPE

The elements of the PSD are plastic scintillator tiles, 7 mm x 8 mm in area and 13 cm or 22.5 cm long (for x and y elements respectively) with embedded 1mm diameter green wavelength shifting fiber [2]. Each scintillating tile starts off as two halves, each 7mm x 4mm in cross section, which are grooved lengthwise using a ball-end mill. A wavelength shifting fiber is glued into the groove and the other half glued on top using optical epoxy (EPO-TEK 310). After assembly, the far end of the tile is polished and coated with 8000 $\mathring{A}$ of aluminum. The remaining sides of the tile are painted with white reflective paint (NE 560) which prevents crosstalk between tiles and aids in light collection. A stub of green fiber protrudes from the non-aluminized end of the tile.

To assemble the PSD, the fiber stub of each tile is glued into counter-sunk holes in a bar of black Delrin[1]. The fiber/Delrin/epoxy surface is flycut to provide an optically flat surface. A second Delrin bar with identical hole spacing is machined to hold a set of matching diameter 1.5 m long clear fibers which are read out by a Philips 64-channel photomultiplier tube XP1723/D1 with a gated base. Transmission loss at the tile-lightguide interface was measured to be approximately 25%.

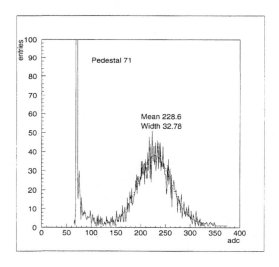

FIGURE 1. Integrating ADC distribution for typical tile exposed to triggered betas from a Ru-106 source

Before assembling the detector, tests were performed on sample tiles using a Ruthenium source. A trigger fiber behind the tile ensured that only electrons which made it through both scintillators were included in the sample. Tiles were read out

---

[1] DuPont Engineering Plastics, 1007 N. Market St., Wilmington, DE 19898

by a high resolution, photon counting tube with a bialkali photocathode having a nominal quantum efficiency of 25% in the blue and 12% in the green (Hamamatsu R1332). The phototube was calibrated in the lab by measuring the response to one and two photoelectrons. A plot of the charge integrating ADC distribution for a typical tile exposed to the source is shown in figure 1. The R1332 was attached directly to the fiber stub. The signal mean corresponds to 38 photoelectrons, as determined by a calibrated photon counting PMT.

DIGITIZATION ELECTRONICS

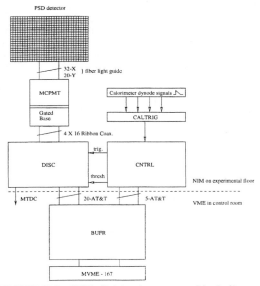

FIGURE 2. PSD digitization system block diagram

The overall readout scheme is presented in figure 2 with subsystem detail in figures 3-5. In the g-2 experiment, the PSD operates as a triggered device in order to eliminate hits from the low-energy background. The trigger is initiated by a pulse above threshold in the calorimeter. When a trigger is received, each of the four custom NIM discriminator modules (DISC) processes the PMT signals through LeCroy MVL407 comparators. If an input signal exceeds the threshold, it fires the corresponding one-shot (with an adjustable width from 20-40 ns). The active one shots (10 ns minimum overlap) are recorded in the 16-bit hit pattern of a 2K FIFO memory. Stored hit patterns are asynchronously clocked to the VME module at 1 MHz. The trigger rate is limited to 50 MHz due to the trigger one-shot, and the maximum write speed of the FIFO is 66 MHz. MVL407s were used because they are fast, low-crosstalk devices that do well with small signals.

The calorimeter trigger is formed by the CALTRIG module. It sums the four dynode outputs of the calorimeter and level discriminates, producing an ECL signal which goes to the CNTRL module through shielded twisted-pair cable and thence to the DISC modules. The trigger discriminator level is set by the computer via the CNTRL module which provides the relevant adjustable voltage level.

The CNTRL module also fans out the CALTRIG trigger pulse into single-ended ECL for input into the four DISC units. Separate programmable threshold control is provided for each DISC module (one threshold per 16 channels). A threshold is set using 7-bit on-board digital to analog converters (DAC's) with a range of 0 to 160mV. Digital read-back is provided by directly converting the threshold lines with 6-bit analog to digital converters.

The custom NIM modules are located in a bin on the experimental floor and communicate with a custom VME module (BUFR) in the control room via shielded 4-pair cables with "phone" connectors[2] and AT&T 41LP and 41LR line drivers and receivers. The BUFR module acts as a buffer and interface module for the computer and NIM modules. The four 16-bit hit patterns are buffered into 2K-deep FIFO's and read-out across the VME backplane using a MVME 167. Since there is no recorded time stamp, a duplicate signal from the calorimeter trigger is recorded by the calorimeter electronics. Each trigger is assigned to a PSD hit pattern and matched in time to the corresponding hit in the calorimeter offline.

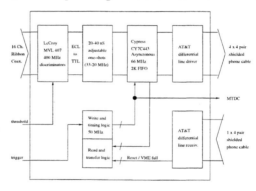

FIGURE 3. DISC: Custom NIM discriminators block diagram

TEST BEAM PERFORMANCE OF PSD WITH DIGITIZATION SYSTEM

PSD's were extensively tested in May of 1995 and 1996 in the B-2 testbeam at Brookhaven, where each channel was read out through CAMAC charge integrating

---

[2] category 5 cable with RJ-46 connectors

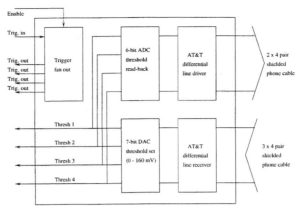

FIGURE 4. CNTRL: Custom NIM trigger and threshold block diagram

VME Readout and Threshold control

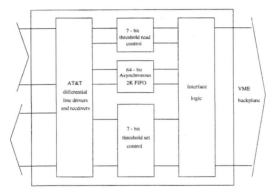

FIGURE 5. BUFR: VME module block diagram

ADC's. Since readout in the experiment was done with the digitization system through level discriminators, it was important to determine where the thresholds should be set in order to maximize efficiency and minimize multiple-hit counting (i.e. false multiple hits due to crosstalk). We also wanted to quantify the performance with level discrimination compared to charge integration techniques. Figure 6 shows the efficiency and multiple-hit rejection for the digitization system (top) and ADCs (bottom). The y-axis represents the efficiency. The x-axis represents the threshold setting in units of DAC counts (1.2mV per count) for the digitization scheme and in units of percentage of the mean for the ADC scheme. Both methods led to almost identical results for both efficiency and multiple-hit rejection (use the crossing of the the the curves to set the relative scale of two plots). The efficiency depends on photomultiplier voltage and threshold settings. When the threshold setting is optimized, the efficiency for a single tile is over 91% at 1300V and 98%

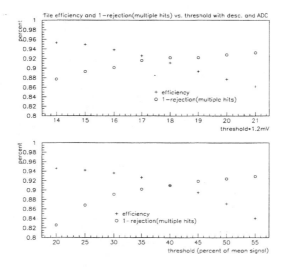

FIGURE 6. Efficiency for readout through digitization system (top) and charge integrating ADCs (bottom)

at 1450V (only ADCs were used at 1450V). Efficiency cannot ever reach 100% due to the gaps between PSD tiles (approximately 0.1mm). If the inefficiency of the gap is folded out, a single x-tile is greater than 99% efficient.

CONCLUSION

The PSD and digitization system provided position information and multiple-hit rejection for the Brookhaven muon g-2 experiment through geometric segmentation. The PSD itself is efficient, and the electronics provides fast digitization of the analog signal and easy implementation in a VME-based acquisition system. The digitization system does not have a noticeable impact on the efficiency of the PSD.

REFERENCES

1. E821 collaboration, Design Report BNL AGS E821, March 1995.
2. P. Cushman, S. Giron, J. Kindem, D. Maxam, D. Miller, C. Timmermans, A HIGH EFFICIENCY TILE-FIBER HODOSCOPE READ OUT BY MULTICHANNEL PHOTOTUBES, Nucl. Inst. and Meth. A 378 (1996) 116-130

Applications of Boron-Loaded Scintillating Fibers as NDA Tools for Nuclear Safeguards

Douglas R. Mayo*, Norbert Ensslin*, Ronald F. Grazioso†, A. Sharif Heger†, David J. Mercer*, Michael C. Miller*, Phyllis A. Russo*, and Martin R. Sweet*

*Los Alamos National Laboratory[1]
Los Alamos, New Mexico 87545
† University of New Mexico
Albuquerque, New Mexico 87131

Abstract. Nuclear safeguards and nonproliferation rely on nondestructive analytical tools for prompt and noninvasive detection, verification, and quantitative analysis of nuclear materials in demanding environments. A new tool based on the detection of correlated neutrons in narrow time windows is being investigated to fill the niche created by the current limitations of the existing methods based on polyethylene moderated ^{3}He gas proportional tubes. Commercially produced Boron-loaded (^{10}B) plastic scintillating fibers are one such technology under consideration. The fibers can be configured in a system to have high efficiency, short neutron die-away, pulse height sensitivity, and mechanical flexibility. Various configurations of the fibers with high density polyethylene have been considered which calculationally result in high efficiency detectors with short die-away times. A discussion of the design considerations and calculations of the detector efficiency, die-away time, and simulated pulse height spectra along with preliminary test results are presented.

INTRODUCTION

Neutron coincidence/multiplicity counting (NCC/NMC) is a nondestructive assay (NDA) approach applicable to the bulk assay of plutonium [1]. The samples in question are residues, aged compounds, and other impure forms. The NCC method addresses safeguards needs by quantifying plutonium stored in thousands of cans and drums. NCC systems have the advantages of economy and safety, and the benefits of ruggedness and reliability.

---

[1] This work supported by the U.S. Department of Energy Office of Safeguards and Security under contract W-7405-ENG-36 .

CP450, *SciFi97: Workshop on Scintillating Fiber Detectors*
edited by A. D. Bross, R. C. Ruchti, and M. R. Wayne
© 1998 The American Institute of Physics 1-56396-792-8/98/$15.00

The detection of multiple neutrons that are correlated in time from their point of origin, i.e., the fission of an actinide nucleus, is a powerful tool in nuclear safeguards and nonproliferation. Neutrons, unlike gamma rays, penetrate dense shielding or large quantities of nuclear materials. The requirement of neutron coincidences is effective in eliminating background events. Passive NCC/NMC of plutonium uses the count rate of correlated neutrons from spontaneous fission to determine the mass of even-A isotopes, with ^{240}Pu usually the major even isotope present. Active NCC/NMC of uranium by interrogation with uncorrelated (i.e., not fission) neutrons at energies below threshold for the fission of ^{238}U induces fission in ^{235}U to produce a count rate of correlated neutrons that is a measure of the mass of ^{235}U.

The current detector technology typically thermalizes neutrons in a polyethylene moderator and captures the neutrons in a ^{3}He gas tube detector. The detector and moderator are macroscopically distinct and separate, resulting in a wide (in time) coincidence gate for counting correlated neutrons because of the long (50-100 μs die-away) time between thermalization and capture. Although these detector systems provide discrimination against gamma rays, high neutron detection efficiency, and reliable, stable, long-term performance in variable environments, numerous field applications, including many *in-situ* measurements, are precluded by the long die-away time.

Because of the relatively long neutron die-away times in polyethylene-moderated thermal (^{3}He) neutron counters, NCC assay sensitivity is limited by high accidental coincidence rates from α,n (uncorrelated) neutron yields. The uncorrelated neutron yields can greatly exceed the yields of (correlated) fission neutrons produced by these impure materials. Sensitivity is governed by the neutron detection efficiency and the total neutron count rate, which determines the accidental coincidence rate. The sensitivity of the NMC assay decreases as the fraction of uncorrelated neutrons in the total neutron signal rises. For materials with neutron yields dominated by uncorrelated neutrons, the NMC assay precision and its sensitivity (in g of ^{240}Pu or ^{235}U) are lessened by the greater excess of uncorrelated neutrons.

An approach to overcome the uncorrelated neutron limitation in thermal counters is to homogenize the moderator and detector materials to shorten die-away time, thus reducing the accidentals fraction. For this approach to be successful, it must be done without sacrificing the high neutron detection efficiency and essential gamma-ray discrimination capability provided by (^{3}He) counters. A boron-loaded plastic scintillator that combines detector (^{10}B) and moderator (plastic) at a molecular level for order-of-magnitude reductions in neutron die-away time has been developed and investigated at Los Alamos National Laboratory (LANL) [2]. When used in monolithic forms such as rods or blocks, high neutron detection efficiency is readily achieved, neutron die-away time is reduced by more than a factor of ten, and benefits of neutron spectroscopy are also realized. To eliminate signals from gamma rays, capture events are tagged by simultaneous detection of capture gamma rays with an inorganic scintillator bonded to the plastic. A prototype counter that uses this approach is being tested at LANL [2].

Boron-loaded plastic can be extruded in the form of fibers. Reducing the diameter of a cylinder of plastic from rod to fiber dimensions results in a reduction in the amplitudes of pulses that arise from gamma-ray interactions, possibly enabling threshold gamma-ray discrimination. The reduction in diameter of a fiber limits the track length of electrons passing through a single fiber, and consequently the signal measured is less. From a practical standpoint, the fiber's characteristics lend themselves to a unique combination of capabilities: short coincidence resolving times, pulse-height discrimination, logical sorting, three dimensions of position sensitivity, and flexibility of configuration. Additional capabilities could include detection of neutrons with high efficiency and discrimination against gamma rays, both of which are essential criteria that are presently achieved with existing neutron-capture detectors.

Gamma-ray discrimination is further enhanced by the logical sorting of fiber readouts to eliminate multi-fiber gamma-ray events within a layer. The enhancements provided by the new detector will extend the measurement sensitivity, characterize the neutron energy spectrum, enable the imaging of the actinide materials, and support multiple applications by mechanical reconfiguration.

DETECTOR SIMULATION AND OPTIMIZATION

A well counter for a can that is 18-cm diameter x 25-cm tall was simulated to check the feasibility of the proposed detector. A crude design is shown in Figure 1. The alternating layers of scintillating fiber and polyethylene are achieved by wrapping a stack of 3-m long boron-loaded fiber ribbons interleaved with thin flexible polyethylene sheets. The photon energy deposition in fibers and neutron capture probabilities in the ^{10}B were calculated using the MCNP code [3].

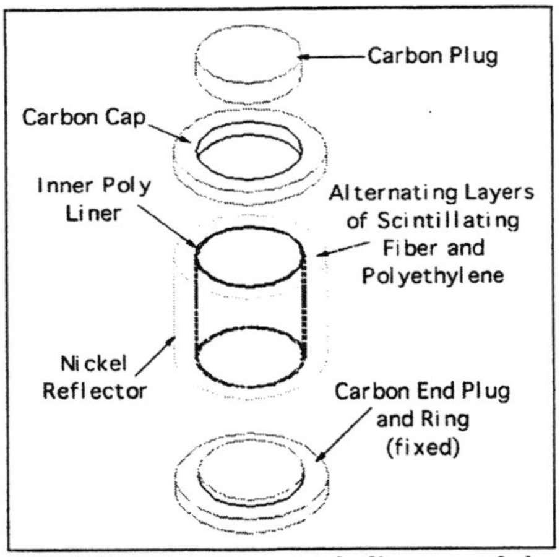

Additional constraints were placed on the design with an attempt to maintain a cost of the overall system, comparable to that of a ^{3}He NCC system. A detector made solely of ribbon layers would be costly. As such, thin sheets of polyethylene between ribbon layers were considered as a means to increase efficiency by moderating neutrons and decrease cost by lessening the amount of fiber needed. Configurations were tested in simulations to maximize the efficiency (ε) while minimizing the die-away time (τ), as shown in Figure 2, to

FIGURE 1.: Conceptual diagram of detector for 18-cm diam. x 25- cm tall cans.

549

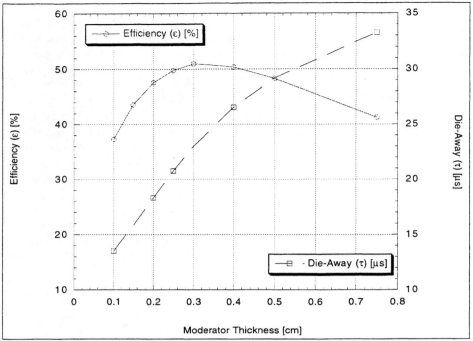

FIGURE 2. Efficiency (ε) and die-away (τ) for a 32 ribbon detector (2 stacks of 16) as a function of polyethylene sheet thickness between ribbon layers. The simulations were performed for 0.5-mm fibers.

produce an optimal detector that meets the design criteria.

Photon and Neutron Discrimination

Early calculations with 0.5-mm diameter fibers pointed to the economic feasibility of layering polyethylene between each of the ribbon layers while achieving an $\varepsilon \approx 50\%$ with a $\tau \approx 10\mu s$. As determined experimentally in Abel, *et al.* [4], the 1.0-mm boron-loaded plastic fibers were incapable of separating electron and neutron events on the basis of a threshold cut. This observation led to a study of energy deposition as a function of fiber diameter as shown in Figure 3. The capture of a neutron by ^{10}B and subsequent emission of an α-particle produces a 93-keV electron-equivalent light output [2]. The simulated data shown in Figure 3 is in terms of the number of photoelectrons produced by the photocathode of the photomultiplier tube (PMT). The figure points out that for a 1-MeV photon incident upon a bare fiber, the photon events in 1.0-mm and 0.50-mm fibers can not be separated from the neutron induced events. The 0.25-mm fiber shows promise. By reducing the diameter, however, the amount of energy the Compton scattered electrons can deposit decreases, thus allowing for the possible discrimination of

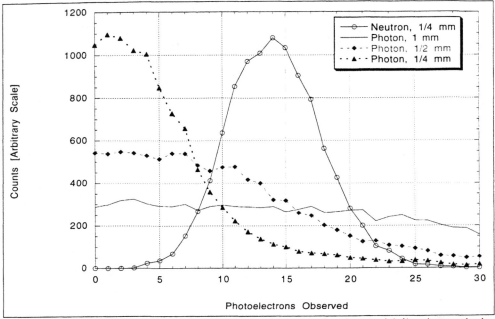

FIGURE 3. Simulated number of photoelectrons produced in the photomultiplier photocathode due to neutron and photon events produced in 1.0-mm, 0.50-mm, and 0.25-mm fibers. This model assumes that the light collected from neutron-induced events are independent of the fiber diameter as the range of the α-particle is a few μm.

photon and neutron events.

The model used to generate the number of photoelectrons assumes geometric (ray-like) light propagation. The events are randomly distributed along the 2.5-m length of the fiber. The attenuation length was taken to be 2.2-m, which has since been confirmed in a benchtop experiment for a 1.0-mm fiber to be $\approx$ 2.3-m. At this level of simulations, bending and coupling losses were assumed to be negligible compared to the light lost due to attenuation and reflection/transmission at the core/cladding interface. The interactions assumed single wavelength emission with a 20% quantum efficient tube. The model represents an idealistic result, not accounting for the PMT resolution and other factors which tend to smear the response and/or reduce the number of photoelectrons produced.

Efficiency and Die-Away Time

The efficiency and die-away time of a given detector geometry were simulated using MCNP. The neutrons were generated assuming a ^{252}Cf point source with a Watt's fission spectrum

TABLE 1. Summary of 0.25-mm fiber Simulations.

Case #	Inner Moderator Thickness [cm]	CH$_2$ Moderator Thickness [cm]	Efficiency ε [%]	Die-Away Time τ [μs]	Ribbon Stacks
1	1.0	0.2	47.6	18.3	2
2	1.0	0.1	37.3	13.5	2
3	1.0	0.1	54.9	12.4	4
4	0.5	0.1	55.3	11.9	4
5	0.5 0.041 Cd	0.1	51.3	10.1	4

$$f(E) = C exp(-E/a) sinh(bE)^{1/2}, \tag{1}$$

where $a = 1.025$ and $b = 2.926$. This differs slightly from the neutron spectra of plutonium, but the codes will initially be benchmarked with ^{252}Cf sources. The efficiency was calculated from the total number of neutrons captured by the ^{10}B. It was assumed that all of the captures resulted in a detectable signal. The fraction of capture events that are recorded by the PMT is a quantity that can be determined through benchtop experiments.

The die-away time can be calculated by taking the time history from the simulations. A single exponential was assumed for the function, and the die-away parameter was calculated from the fit. Table 1 lists some of the results from different configurations for 0.25-mm fibers. The detector optimization involved varying the thickness of polyethylene sheet between the ribbon layers as well as the thickness of the polyethylene inner shell, adding the layers of Pb and Cd in the well for shielding, adding graphite top and bottom plugs and introducing neutron reflecting material on the exterior. The graphite end plugs and external nickel reflector increased the overall efficiency without a large increase in neutron die-away time. The effect on die-away time and efficiency as a function of moderator thickness is shown in Figure 2.

TESTING AND BENCHMARK PLANS

Prior to investing the time and resources in building a prototype detector system, a number of tests have been considered to prove the feasibility of a boron-loaded scintillating fiber NCC system. Currently a set of 1.0-mm, 0.50-mm, and 0.25-mm singly clad boron-loaded fibers are being tested. The first goal of the project is to determine the idealized characteristics of neutron detection with independent fibers and to compare the characteristics for different fiber sizes and configurations. A validation of the calculations, including a determination of what, if any, bias exists, prompted the benchmark experiments which are being performed. The set of test fibers will allow for a comparison similar to that shown in Figure 3.

A set of 0.25-mm multiclad fibers will be compared to the singly clad fibers used to provide information on the benefits of the additional coatings. A 3.0-m ribbon of 0.25-mm fibers (200 fibers wide) will be used in the construction of a miniature well detector which will provide benchmark data and allow for experimenting with different layers of polyethylene and metal foils.

Preliminary Data

Initially the electronic equipment were setup with a 2.54-cm diam. x 1.27-cm thick bismuth germanate (BGO) crystal mounted on a Hamamatsu 5-cm R329-02 PMT. The preamplifier pulse passed through an ORTEC 855 dual spectroscopy amplifier to a Canberra 1510 ADC, which was fed into a Canberra S100 card and read into MCA software. The BGO provided easily resolved photoelectric peaks to test the electronics.

The BGO cylinder was replaced by a 2.54-cm diam. x 2.54-cm thick BC454 plastic cylinder to determine whether neutron events could be observed. Figure 4 shows the clear neutron peak from a ^{252}Cf source along with the scattering continuum underneath.

A cylinder was made from a hollow polyethylene annulus with ≈ 100 of the 1.0-mm fibers, 2.54-cm long, epoxied in the center and polished on both ends. The fiber cylinder was considered a first step in determining whether neutron events would stand out without the effects of light propagating through a wrapped 2.5 - 3.0-m long fiber. Going from a 2.54-cm aperture in the case of the BC454 crystal to a 1.0-mm aperture in the fiber should reduce the light collected by a factor $\approx 4 - 10$ (numerical aperture). The data indicate that the tubes currently used may not be optimal for this purpose, and that a lower noise level and better energy resolution is required. Several tubes are being considered, including the hybrid tubes.

REFERENCES

1. D. Reilly et al.,*Passive Nondestructive Assay of Nuclear Materials,* Nuclear Regulatory Commission Report NUREG/CR-5550 (March 1991), LA-UR-90-732.
2. M. Miller, *Neutron Detection and Multiplicity Counting Using a Boron-Loaded Plastic Scintillator/Bismuth Germanate Phoswhich Detector Array,*Ph.D. Thesis, University of New Mexico (1997), LA-13315-T.
3. Briesmeister J.F., Ed., *MCNP4B Monte Carlo N-Particle Transport Code System,* Los Alamos National Laboratory report LA-12625-M (November 1993).
4. K.H. Abel et al., "Performance and Applications of Scintillating-Glass-Fiber Neutron Sensors," *Proceedings of the SCIFI93 Conference Workshop on Scintillating Fiber Detectors,* October 24-28, 1993, Notre Dame, Indiana.

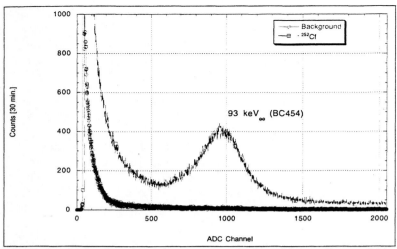

FIGURE 4. Background and moderated neutron spectra from 2.54-cm diam. x 2.54 cm BC454 cylinder. The neutron capture peak is clearly seen.

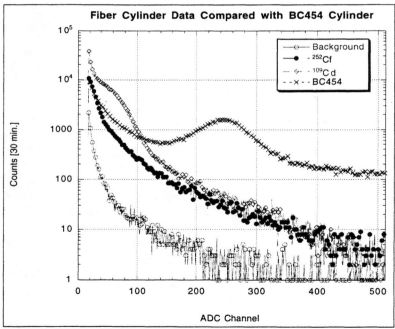

FIGURE 5. Preliminary spectra measured for BC454 cylinder and boron-loaded fiber cylinder. Although the ^{252}Cf spectra measured by the fibers does not exhibit an distinct peak, it does show a response to neutrons.

554

Development of FOND:
A Scintillating Fiber-Optic Neutron Detector

Daniel Holslin, Janis Baltgalvis, Aaron Polichar, David Shreve,
and Scott Smith

*Science Applications International Corporation
16701 West Bernardo Drive, San Diego, California 92127*

ABSTRACT. Science Applications International Corporation in San Diego has been developing and testing a plastic scintillating fiber-optic neutron detector (FOND) for various applications where detection of fast neutrons is required. The detector-converter component of the device is a fiber bundle constructed of plastic scintillating fibers each measuring 5 to 10 cm long and 100-500 μm in diameter. Bundles ranged in size from 2.5 cm by 2.5 cm in cross section up to 10 cm by 10 cm. The bundle is coupled to a set of electro-optic intensifiers whose output is recorded by a CCD camera (standard and high speed) directly coupled to the intensifiers. The FOND has been advanced for several applications, including the identification and location of sources emitting fast neutrons (such as special nuclear material) and for measuring the solar neutron energy spectrum (in collaboration with the University of New Hampshire).

INTRODUCTION AND BACKGROUND

During the last several years, we have constructed and tested several evolutions of a plastic fiber-optic scintillating fiber detector (FOND) in a number of studies supported by the U.S. Army Space and Strategic Defense Command, DARPA, and, most recently, NASA. The sensor is used to count fast neutrons emitted by a source and must be rugged, efficient, able to discriminate against background (such as gamma rays and cosmic-ray muons), and be gamma-ray insensitive. In order to detect fast neutrons within an intense gamma-ray background, the detector elements must be small compared to the range of the recoil electrons produced in the gamma-ray interactions. It must be large enough, however, to stop the recoil charged particles resulting from the neutron interactions. An ideal sensor for this application is the scintillating fiber detector whose neutron sensor is formed from long, thin plastic fibers.

A scaled version of the detector was first constructed and tested at SAIC to verify its feasibility for Army requirements. In two subsequent variations certain components were improved and additional ones were added to enhance the performance of the detector or to demonstrate upward scalability in the design. A smaller, compact version of FOND was built and tested for DARPA for locating weapons of mass destruction (mainly Pu based). In a recent collaboration with the University of New Hampshire, we built a prototype detector (SONTRAC) for measuring the energy spectrum of solar neutrons (1).

CP450, *SciFi97: Workshop on Scintillating Fiber Detectors*
edited by A. D. Bross, R. C. Ruchti, and M. R. Wayne
© 1998 The American Institute of Physics 1-56396-792-8/98/$15.00

Advances in scintillating fiber technology, which has been used typically to detect charged particles in astrophysics and high-energy experiments (2), and in CCD camera technology permit the development of the FOND for these applications. In these studies, square scintillating fibers, each measuring between 100-500 μm on a side and 5 to 10 cm long, were grouped together to form a square bundle. Each fiber is constructed of a polystyrene (CH) core doped with scintillating dyes and surrounded by an acrylic cladding with slightly lower refractive index to trap the scintillation light. Several such bundles with different dyes and with or without an EMA coating were characterized in each of the studies described here. The fiber bundle in a flyable SONTRAC detector is made of alternating orthogonal rows of 250 μm diameter fibers, thus allowing for 3-dimensional track analysis. The bundle is typically supported in a thin-walled, light-tight enclosure made of aluminum and coupled to a set of electro-optic intensifiers. The intensifiers amplify the light signals and reduce the image size so they can be coupled to a CCD camera. Output images from the CCD camera are acquired by a PC-based image processor for storage and analysis.

In this paper, we summarize the simulations performed to support the scintillating fiber concept, describe the various detector configurations, and present the results of experimental tests made with 14 MeV neutrons and an intense beam of high-energy gamma-rays, and with neutrons from a Cf-252 isotopic source. Please refer to ref. (1) for a description of the SONTRAC detector assembly and experiments to measure its response for detecting solar neutrons.

SIMULATIONS OF THE DETECTOR RESPONSE

The high-energy FOND concept (for 5-100 MeV neutron detection) was first demonstrated with Monte Carlo simulations to determine the neutron detection efficiency, directionality, and gamma-ray rejection capability (3). The simulations were used to optimize the detector's performance for 14 MeV neutrons since neutrons of this energy are readily available from d-t generators and this energy represented that of the target neutrons. To detect neutrons from the source, the fibers could be positioned so that the neutrons enter normal to the fiber axis as suggested in ref. (4). However, for the Army requirements, the bundle was positioned so that neutrons enter along the fiber axis. Events are simply counted if the brightest fiber within an event track exceeds a given threshold. This simple, yet effective, technique suggested by the authors in ref. (5) reduces the computational time and, therefore, the complexity of the detector system so that weight, size, and power constraints are diminished. The angular efficiency (or directionality) is extremely useful to discriminate against background neutrons entering non-parallel to the fiber axis.

The simulations were used to help determine the appropriate cross-sectional area size of the fibers. The length was fixed at 10 cm, corresponding to the attenuation length of 14 MeV neutrons in the fiber material. The simulations indicated that 500 μm fibers are optimal because the probability that the recoil proton remains in only one fiber is greater than for smaller fibers. Fibers much larger than 500 μm would not be useful since the recoil proton would tend to stay in one fiber independent of the incident neutron direction, thus reducing the directional sensitivity of the FOND.

The detection efficiencies for 14 MeV neutrons for thresholds near 3 and 8 MeV were calculated to be 20% and 5%, respectively. The angular efficiency response (directionality), shown in Figure 1 at a threshold of 8 MeV, indicates that neutrons entering at angles near 90 degrees are almost completely discriminated against. Results are shown for the large bundle with 300 µm diameter fibers.

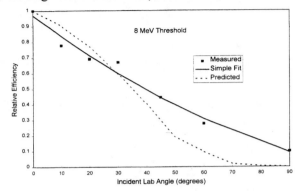

FIGURE 1. The neutron detection efficiency normalized to the 0 degree efficiency (directionality) for an 8 MeV threshold. The solid curve is a simple polynomial fit (quadratic) to the data.

For the low-energy (fission neutron) design of the FOND, simulations were made using a modified version of MCNP, a Monte Carlo neutron-photon transport code developed by Los Alamos National Laboratories. MCNP was modified to track the recoil proton as it passed through the polystyrene and cladding of the fibers.[1] The energy loss in the active regions of the fibers were determined and converted to light output using the data in ref. (6). Simulations of recoil proton tracks were made for various incident neutron energies and directions. The detection efficiency and directionality were predicted for fiber bundles with diameters of 50, 100, and 150 µm.

DETECTOR ASSEMBLY

High-Energy Design

For the Army-supported study, three versions of the FOND were assembled for testing. The first was a feasibility detector used to verify the physics of the performance capability of the fiber bundle. The feasibility detector was designed to be modular so that various combinations of intensifiers could be tested. The second version of the FOND was fitted with a high-speed CCD camera, and changes in the fiber bundles were made to improve their performance. The final version was a large area adaptation for

---

[1] MCNP was modified by Tom Jordan of Experimental and Mathematical Physics Consultants in Gaithersburg, Maryland.

addressing scalability issues.

The FOND is comprised of three major components: the fiber bundle, the intensifiers, and the CCD camera. The scintillating fibers measured 300- or 500-μm square and 10 cm long. The smaller bundles used for the initial feasibility studies were 2.5 cm by 2.5 cm in cross section (2500 fibers total). The larger bundles were assembled from 300 μm fibers and measured 10 cm by 10 cm in cross section (110,000 fibers total). The fiber bundle was attached using optical coupling grease to a set of image intensifiers.

For the smaller feasibility detectors, the intensifier system consisted of a GEN I (40 mm/13 mm), a proximity focus tube, and a GEN I (18 mm/7 mm) tube to provide adequate gain to amplify the signal well above the CCD noise. The large area version, shown in Figure 2, employed a 10-cm diameter GEN I tube at the bundle interface in place of the GEN I (40 mm/13 mm). Electro-optic GEN I type intensifier tubes were used because they are relatively insensitive to gammas. Phosphors with short decay times (P46 or P47) were chosen to minimize any potential complications arising from signal pile-up during gamma-ray sensitivity experiments.

FIGURE 2. Large-area FOND with a 10 cm cube fiber bundle, shown exposed to the left in the photo.

Two types of CCD cameras were used in this study: an interline Pulnix model TM-745E CCD with an electronic shutter and a high-speed (850 Hz) frame transfer model made by Dalsa. The shuttering capability of the Pulnix and the high-speed camera were used to test the gamma-ray response at high fluxes. The shuttering in the Pulnix controls the amount of time that the CCD is sensitive to the incident light. This gating can be adjusted in several increments, down to 1/10,000 sec per frame. Data from the camera is read out at standard video rates (30 Hz) so that as the integration time is shortened, the deadtime increases. In a deployed system, the camera would be read out at nearly the integration rate thus keeping the deadtime at a minimum (less than 10%); however, for the tests here, a higher deadtime had no affect on the outcome of the results. To demonstrate the feasibility of a high-speed camera, the Dalsa CA-D1 camera was used. This camera produces images consisting of 128 by 128 pixels per frame, at rates up to 850 Hz with dead times of less than 10%. Images from the cameras are acquired using a PC-based image processor.

Low-Energy Design

The FOND design which is sensitive to fission neutrons employed a 2.5 cm by 2.5 cm fiber bundle with 100 μm diameter fibers 5 cm long attached to a GEN II with dual microchannel plates (MCP) for high photon gains (greater than 10^4). The output of the GEN II was directly coupled to a Pulnix TM-745E CCD camera. A photo of the benchmodel is shown in Figure 3. As with the high-energy version of FOND, the CCD images are captured and analyzed by a PC-based, image-processing card.

This type of sensor has the advantage of detecting the fission neutrons directly without the need for heavy moderator material. It is compact, durable, and can be made into a self-contained package.

FIGURE 3. Benchmodel FOND tested for detecting sources emitting fission neutrons. The fiber bundle enclosure is shown to the left followed by the Gen II image intensifier and the CCD camera.

MEASUREMENTS AND RESULTS

High-Energy Design

For the Army's application, several experiments were performed to determine the performance capability of the FOND. The detection efficiency and directionality for 14 MeV neutrons and the gamma-ray response to high fluxes from a 10 Ci Co-60 and a 50 Ci Cs-137 source were measured. Data images were collected and analyzed using a PC-based Matrox IM-640 image processor. The results compared well with the Monte Carlo predictions, as discussed below.

In the evaluation of the neutron data, a blob analysis algorithm was used to locate individual event tracks and to sum the light within the brightest fiber of the event. A histogram of the "brightest fiber" light was generated and the light output converted to recoil proton energy using measured values from the literature (6). The energy scale was normalized to the maximum light output, which corresponded to a 14 MeV recoil proton stopping within one fiber.

At an 8 MeV threshold, the measured efficiency is about 30% less than predicted. This discrepancy is partly due to the lower effective area of the fiber because of the inactive regions taken by the acrylic cladding, which contributes about 20% of the total volume. A second consideration is a result of the recoil protons losing energy when they pass through the cladding and into another fiber. This second effect shifts the protons to lower light outputs, thus lowering the efficiency for higher energies and increasing it for the lower energies. Both effects can be reduced by using fibers with thinner cladding. The cladding and EMA coating were not included in the fiber composition for the simulations. Subsequently, fiber bundles with thinner cladding were obtained where active area of the fibers was increased by 20%. Tests showed the neutron efficiency to be 15-20% higher compared to the first set of bundles with thicker cladding and in better agreement with simulations, which neglected the cladding.

The measured and predicted directionality results are shown in Figure 1. In general, the measured results are higher than the predictions, but the shape of the curves are very similar. This discrepancy may be a consequence of the effects related to the cladding as described above. Further tests to determine the number of false events from pile-up or double scattering are under consideration. Regardless, there is more than a factor of 10 decrease in the efficiency at 90 degrees where the discrimination against background neutrons is most important for this application.

When the bundle is illuminated with an intense flux of gamma radiation, the light produced by the gamma-ray interactions act as an offset with a noise component. In the gamma-ray response evaluation, the important quantity for determining the false event rate is the rms fluctuation in the gamma-ray light produced in each fiber. When the flux is high, the light in a fiber can easily exceed that of the recoil protons produced by the incident neutrons of interest. However, the offset caused by the gamma rays can be taken into account when setting the neutron threshold as long as the number of false events caused by the gamma-ray light exceeding the threshold is tolerable. The false event rate was determined directly using gamma ray only and neutron plus gamma ray measurements. The gamma-ray only data were analyzed first to ascertain the number of false events as a function of energy threshold. Then the neutron plus gamma-ray data were analyzed to determine the number of false and true events as a function of energy threshold. As an example of the results, the number of false events for a large detector with an area of 1 m^2 is less than 10 for an 8 MeV threshold (about 3.2% neutron efficiency) and 1 msec readout time at a flux of about 10^7 gammas/cm^2-sec (about 20 rad/hour) from a Co-60 source. For a given gamma-ray flux, bundles made of 300 μm diameter fibers provided the lower false count rate when compared to those comprised of 500 μm fibers.

Low-Energy Design

The response of the low-energy FOND to fission neutrons was determined using a 1.5 μgm Cf-252 isotopic source. The source was placed at various locations around the detector and images collected and analyzed using the same system described above for the high-energy FOND. Background (such as cosmic-ray muons and electron emission from the photocathode) was determined by capturing images with no source present. Recoil proton tracks were counted using the blob analysis technique and normalized to the incident neutron flux to determine the detection efficiency.

At a threshold just above the noise level (equivalent to the light emitted by a 0.5 MeV proton), the detection efficiency was about 6%. Using this value and the same size detector, it is possible to detect a 4-kg, Pu-based warhead at a maximum distance of 22 meters in 10 minutes (95% confidence level). As a comparison, the counting efficiency-to-detector mass ratio for the FOND is over a factor of 10 greater than that for a Bonner sphere detector. Though the measurements showed directional sensitivity, its full capability was not tested because of programmatic constraints.

CONCLUSIONS

The FOND takes advantage of plastic scintillating fiber technology, electro-optics, and CCD camers to produce an efficient, fast neutron detector with imaging capability, directionality, and gamma-ray insensitivity. All of the detector components, except for the fiber bundles, were purchased off-the-shelf and the data acquisition and analysis performed with a PC-based image processing board. Improved bundles were fabricated and tested with thinner cladding and better EMA to improve the neutron detection efficiency. Several designs of the FOND were assembled and tested for various applications, including the detection and location of sources of fission neutrons emitted by Pu-based weapons of mass destruction. Experiments demonstrated the viability of using the FOND for high-energy neutron detection (including spectroscopy as with the SONTRAC detector) and as a compact unit for fission neutron identification.

ACKNOWLEDGEMENTS

We would like to thank Prof. Robert Binns of Washington University and Mike Kusner of Bicron for supplying the fiber bundles and for many discussions related to understanding the response and improving the quality of the bundles. We thank Larry Acton and Melisa Galba of the Electronic Vision Systems Division at SAIC for their expertise and help with the design and construction of the detector housing and operation of the electro-optic components. This work was sponsored by the U.S. Army Strategic Defense Command and DARPA.

REFERENCES

1. J. Ryan, paper presented at this conference.
2. "Proceedings of the Workshop on Scintillating Fiber Detector Development for the SSC" held at Fermi National Accelerator Laboratory, November 14-16, 1988 Volumes I and II.
3. T.W. Armstrong and B.L. Colborn, "Mass Sensor Simulations in Support of Neutral Particle Beam Technology", Science Applications International Corporation Report SAIC-TN-9201, Contract No. DASG60-89-C-0020 for U.S. Army Strategic Defense Command, March 1992.
4. Y. Yariv, R.C. Byrd, A. Gavron, and W.C. Sailor, "Simulations of Neutron Response and Background Rejection for a Scintillating-Fiber Detector", Nucl. Instr. Meth. A292 (1990) 351.
5. T.G. Miller and W. Gebhart, "A Neutron Sensor with Enhanced Directionality Capability -- The 3-D Sensor", General Research Corp. Report CR-459-1372, Rev. 2, July 1988.
6. V.V. Verbinski, et al., "Calibration of an Organic Scintillator for Neutron Spectrometry" Nucl. Instr. Meth. 65 (1968) 8-25.

Photon Position and Energy Reconstruction in a Cherenkov Hodoscope

Yuriy A.Bashmakov and Margaret S.Korbut

P.N.Lebedev Physical Institute Russian Academy of Sciences,
Leninsky Prospect 53, 117924 Moscow, Russia

Abstract. The iteration procedure has been developed for high energy electrons and photons coordinates reconstruction and energy determination in a cherenkov hodoscope. The procedure is based on the average radial response electromagnetic shower distribution function in terms of Cherenkov light entering PM obtained from an experimental data. Optimal particle impact point and energy are found automaticly. For a given photon impact point the energy deposition in each cell are determined by integration over cell cross-section. New impact point position is determined by the spatial distribution of differences between observed and calculated on this iteration step deposit energies taken for each hodoscope cell. The number of iteration depends on accuracy required and impact electron parameters. Experimental data for the photon hodoscope of the luminosity monitor of the H1 Detector at the electron-proton collider HERA was used for algorithm checking . High efficiency of the algorithm was demonstrated.

I INTRODUCTION

Total absorption hodoscope cherenkov spectrometers are widely used in high energy physics for photons and electrons coordinates and energy measurements. Plans for its usage on future e^+e^- linear colliders and on pp colliders on superhigh energy are now developed .

High energy particle, photon or electron, generates in the crystal an electromagnetic cascade shower. Shower charged particles emit cherenkov radiation with intensity proportional to the total length of the trajectories of all charged particles created in the crystal. For high energy resolution achievement crystal length is usually choose from requirement for absorption to be no less than 98% energy of electromagnetic shower. Cherenkov radiation is directed on detector input. As a detector compact photomultiplyers with small diameter photocathode are mostly used. Number of photoelectrons at PM entry is determined by the crystal transparency, efficiency of the cherenkov radiation collection and the spectral sensitivity of PM.

CP450, *SciFi97: Workshop on Scintillating Fiber Detectors*
edited by A. D. Bross, R. C. Ruchti, and M. R. Wayne
© 1998 The American Institute of Physics 1-56396-792-8/98/$15.00

The detector response can be used for energy determination and coordinates re-construction of the incident high energy photon [1]. For this object the tabular method of reconstruction is widely used, when data obtained from a mathematical simulation, for example, by means of GEANT code [2], are entering in a table. For a cherenkov hodoscope the simulation is essentially complicated by difficulty of reasonable accounting of light collection process [3].

In the present paper the special algorithm is suggested. The procedure is based on the average spatial response electromagnetic cascade shower distribution function extracted from the results of measurements for isolated hodoscope cells response induced by detecting particle. This distribution is used as standard one.

II BASIC CONCEPTS

The averaged transverse response function of the hodoscope on electromagnetic cascade shower can be presented in the form of the sum of two Gaussians [4]

$$F(x, y) = a_1 \cdot \exp(-(x^2 + y^2)/b_1^2) + a_2 \cdot \exp(-(x^2 + y^2)/b_2^2). \qquad (1)$$

Where the normalization is used $\int\limits_{-\infty}^{\infty} \int\limits_{-\infty}^{\infty} F(x,y)dxdy = 1$. The first term at $b_1 < b_2$ describes the shower narrow core, and the second one describes the shower periph-ery. This distribution takes into account both the electromagnetic shower develop-ment in the crystal and the light collection efficiency. The fraction of the energy released in the surface element $dxdy$ with x, y coordinates with respect to the shower axis is equal to $F(x,y)dxdy$. The hodoscope spatial resolution is determinated by the relationship between the characteristic electromagnetic shower width and the hodoscope cells transverse size. The hodoscopes composed of identical cells with cross-section in the form of square has now a most extension. Later on the side of this square will be designated as h.

At the direct problem solution at first on the given particle input point coordinates x, y and energy E_γ through the integration of the response function $F(x,y)$ over hodoscope cells cross-section set of values $e^{x,y} = \{e_1^{x,y}, e_2^{x,y}, ..., e_n^{x,y}, ...\}$ is found, where $e_i^{x,y}$ is normalized energy deposition in the cell with number i. The bound-aries of integrals according to the shower axis are depend on the impact photon parameters. Then by multiplication of $e_i^{x,y}$ on E_γ one find real energy deposition

$$E_i^{x,y} = E_\gamma e_i^{x,y} \qquad (2)$$

in every cell. Note that for energy deposit calculations for hodoscopes fabricated from the same material, but having cells with distinguished size, it is sufficient simply to change the limits of the integral over cells on new one.

Search of particle parameters is performed by means of analyzable distribution comparison with standard one. Extreme value of a criterion selected for comparison allows to make decision how distribution under study is close to the standard one. Experimental event of photon hitting in the hodoscope is specified by the collection of values $\mathbf{E} = \{E_1, E_2, ..., E_n, ...\}$, that gives in the GeV energy deposit in i crystal. Total energy deposition

$$E^{obs} = \sum_{i=1}^{n} E_i, \tag{3}$$

where summation on i is performed either on all hodoscope cells or on cluster of cells selected according to specific criterion.

Search of the particle input point and its energy is carried out by means of iteration procedure. As initial value coordinates of shower center of gravity are used $x_{cg} = \sum_{i=1}^{n} x_i E_i / E^{obs}$, where x_i - coordinates of i- hodoscope cell center. At the k iteration step by using the intermediate particle coordinates x_k, y_k by integration of $F(x, y)$ on cells cross-sections normalized energy depositions $\mathbf{e}^k = \{e_1^k, e_2^k, ..., e_n^k\}$ in hodoscope cells are calculated. Then summary normalized energy deposition is found $e^k = \sum_{i=1}^{n} e_i^k$. The particle energy on this step is determined as relation of total experimental deposit energy to adjusted energy calculated on this step

$$E^k = \frac{E^{obs}}{e^k}. \tag{4}$$

Therefore if in consequence of the edge effect there is a shower leakage $e^k < 1$, then reconstructed energy is in excess of the total energy deposition in hodoscope $E^k > E^{obs}$. Using calculated normalize energy deposition and energy one find energy deposition in each cell.

$$E_i^k = E^k e_i^k. \tag{5}$$

Observed and obtained on this step calculated distributions are found to be slightly displaced one relative to the other and are distinguished in the amplitude. The magnitude of this displacement can be characterized by means of the suitable dipole momentum. Transverse displacement on k iteration step is found automatically. For every cell difference between experimental and obtained by this method energies is found $DE_i^k = E_i^{obs} - E_i^k$. During translation on hodoscope area from cell to cell DE_i^k can alternate its sign. After that the dipole momenta of energy deposition distribution are calculated. For every cell for every transverse coordinate displacement of the center of this cell relative to obtained on the previous step photon input point in hodoscope and on energy deposition deviation is calculated for given cell. Then summation over all cells is performed

$$Dx^k = \sum_{i=1}^{n} (x_i - x_\gamma^k) DE_i^k. \tag{6}$$

As a result one find the desired displacement, which can be called dipole,

$$\Delta x_\gamma^k = \frac{D x^k}{E^{obs}}. \tag{7}$$

For the new input point coordinates determination obtained dipole displacements is adding to the coordinates of the previous input point

$$x_\gamma^{k+1} = x_\gamma^k + \Delta x_\gamma^k. \tag{8}$$

Absolute value of the transverse displacement of the input point is found. Besides for each cell second order deviation of deposit energy from calculated one is formed and these values are summing over all cells $\Delta E^2 = \sum_{i=1}^n \left(E_i^{obs} - E_i^{rec} \right)^2$. The calculation process is repeated until either the displacement become lower then the given value ε, or the iteration number N achieves the preassigned limiting number N_{max}. Displacement of the reconstructed coordinate in the detector transverse plane during iteration procedure is an oscillation in nature with die out amplitude. As a result of computations the code gives coordinates of the input particle point x_γ, y_γ, its energy E_γ and second order deviation of deposit energy ΔE^2, that can indicate the quality of the reconstruction.

III   EXPERIMENTAL DATA HANDLING

Numerical simulation was performed for the photon detector of the luminosity monitor of the H1 detector of the electron-proton storage ring HERA [5], [6]. The photon detector together with the electron tagger are used for the luminosity measurement of the HERA collider using the bremsstrahlung of electrons with the energy $E_e = 27.5 GeV$ on counter protons with the energy $E_p = 820 GeV$ [7]. By means of two total absorption cherenkov hodoscopes simultaneous detection of bremsstrahlung photons and scattered electrons is brought off. Detected photon energy ranges in value from 6 to 20 GeV .

The photon detector (PD) consists of 25 separated modules with KRS-15 crystal as the radiator and FEU-147 photomultiplier as the viewer assembled in the 5×5 matrix. Radiation length of KRS-15 crystal $X_0 = 0.92cm$, Moliere radius $R_m = 2.10cm$. Longitudinal radiator size ($l = 200mm, l > 20X_0$) provides nearly total absorption of electromagnetic shower in the detector. The transverse radiator sizes are $20 \times 20mm^2$ [6], [8].

Parameters of the response function (see (1)) were determined from experimental data given in [8], where the dependence of relation of deposit energy in neighbouring modules on electron input point distance relative to the modules boundary was investigated. These experimental data are described in the best way by means of the following set of parameters: $a_1/a_2 = 1.095/0.06, b_1 = 4.4mm, b_2 = 16.5mm$. Fig. 1 shows energy deposition in separated photon detector cells obtained from numerical simulation for the following coordinates of photon input point in detector: $x_\gamma = 2.5mm, y_\gamma = 27.5mm$. The origin of frame of reference coincides with the photon detector center. To visualize deposit in every cell is presented in the form of the parallelepiped. Parallelepiped centers coincide with cells centers. Its ground area as well as height are proportional to the energy deposition in the given cell.

For the developed algorithm checking the data sample recorded by the luminosity

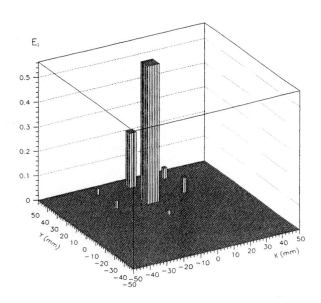

FIGURE 1. The energy deposition in the photon detector cells obtained as a result of simulation

monitor of the H1 detector at the collider HERA at the end of 1995 year with positron beam was used. Preliminary events with summary positron and photon energy $E_e + E_\gamma$ deviating from mean value no more then $\pm 3 GeV$ were selected. During reconstruction the iteration procedure was continued until displacement of reconstruction photon input point position with respect to obtained on previous step became less then $\varepsilon = 0.2$ mm or iteration number N was not exceeding $N_{max} = 20$. The energy and spatial distributions for photons entering PD were constructed. As an illustration Fig. 2 shows the ΔE^2 distribution for 1292 events taken by the photon detector. Mean value of ΔE^2 is equal to $1.68 GeV^2$, and root-mean-square deviation is $2.17 GeV^2$. These values demonstrate good quality of the reconstruction algorithm developed.

Note that developed approach can be used for energy deposit simulation and reconstruction of electron and photons parameters in wide class of cherenkov and scintillation hodoscopes, as well as for reconstruction of hadron parameters by using hadronic shower in the heavy crystals.

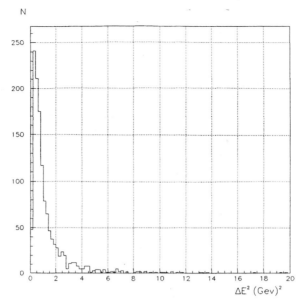

FIGURE 2. ΔE^2 distribution for the events recorded by the phonon detector

ACKNOWLEDGMENTS

Authors are thankful to H1-collaboration members: A.S. Belousov, A.M. Fomenko, A.I. Lebedev, E.I. Malinovsky and Ya.A. Vazdik and also to S.P. Kharlamov for their interest to this work and numerous useful discussions.

REFERENCES

1. Akopdjanov G.A. et al., *Nucl. Instrum. Methods* **140**, 441-445 (1977).
2. Detector description and simulation tool, GEANT 3.15, CERN Geneva, Switzerland, 1991.
3. Kalinin B.N. et al., Preprint 1/94 NIYaF TPI, 1994.
4. Review of particle properties, Particle Data Group, Hikasa K. *et al.*, *Phys. Rev.* D **45**, 1 (1992).
5. Wiik B.H., "HERA Operations and Physics," in Proceedings of the 1993 IEEE Particle Accelerator Conference. V.1, pp. 1-8.
6. H1-collaboration, I. Abt et al., *Nucl. Instrum. Methods* A **386**, 310-347 (1997).
7. Andreev V.F. et al., Preprint FIAN N 102, Moscow (1989).
8. Andreev V.F. et al., Preprint FIAN N 32, Moscow (1992).

SESSION 16: ASTROPHYSICS - II

Chair: W. R. Binns
Scientific Secretary: T. Lin

GLAST: Using Scintillation Fibers for Both the Tracker and the Calorimeter

Gerald J. Fishman

Space Sciences Laboratory
NASA/Marshall Space Flight Center
Huntsville, AL 35812 USA

-for the SIFTER Collaboration

University of Alabama in Huntsville
NASA/Marshall Space Flight Center
Washington University

Abstract.   The Gamma-Ray Large Area Space Telescope (GLAST) is planned to be the next major NASA mission in gamma-ray astronomy. It will operate at energies above 20 MeV to study the most energetic objects in the Universe. While the baseline tracker detector for GLAST during the study phase is based on silicon strip detectors, we believe that scintillating fibers have considerable advantages for this purpose. Among the scientific advantages are: larger effective area and better angular resolution at low energies. Practical advantages include: lower cost, the use of a common technology for both the tracker and the calorimeter, lower power consumption, and a simplified thermal design. Several alternative readout methods for the fibers are under study. A set of recent references is provided to indicate the current status of scintillation fiber technology, applications of scintillating fiber systems and readout methods.

INTRODUCTION AND HISTORY

Gamma-ray astronomy has emerged as an important new branch of astronomy, which has enabled detailed studies of the most energetic objects and phenomena in the Universe (1). Recent gamma-ray astronomy missions such as the Compton Gamma Ray Observatory and the SIGMA/GRANAT experiment have produced a wealth of new observational data in the field. Within a few years, the INTEGRAL mission (2), will be providing high resolution observations in the nuclear gamma-ray energy regime. At higher energies (>20 MeV), there are no experiments currently being developed to replace the highly successful Energetic Gamma-ray Experiment Telescope (EGRET) experiment on the Compton Observatory (3). That experiment is now completing its observational lifetime, constrained by a limited supply of gas for its spark chamber tracker.

CP450, *SciFi97: Workshop on Scintillating Fiber Detectors*
edited by A. D. Bross, R. C. Ruchti, and M. R. Wayne
1998 The American Institute of Physics 1-56396-792-8/98/$15.00

It was recommended by a NASA Headquarters panel (4) that the next new mission in this field should be a high-energy gamma-ray telescope which operates, like EGRET, on the pair tracking/calorimeter principle. This recommendation was made in recognition of the great observational advances made by EGRET in recent years, for example in the discovery and study of blazars and new gamma-ray pulsars, the discovery of high energy (>10 GeV) photons from gamma-ray bursts and the first comprehensive high-energy map of the entire sky.

A Mission Concept Study for gamma-ray astronomy was initiated by NASA in 1993, led by a collaboration including SLAC, Stanford, NRL, and several other institutions (5,6). That study effort was termed GLAST (Gamma-Ray Large Area Space Telescope). A telescope was baselined which included a silicon strip tracker and a CsI calorimeter. In 1996, a mission Facility Working Group was formed by NASA Headquarters, which included the original GLAST team and new representatives from other institutions. The name "GLAST" was kept to describe the large, facility-class mission. Other alternative concepts for the tracker and the calorimeter are being studied, as well as continuing work on the baseline design. These alternative detectors include a gas microstrip and scintillating fiber tracker detector and a scintillating fiber-based calorimeter. A mission study team at the NASA/Goddard Space Flight Center was formed in 1997 to further study the spacecraft and mission. More information on the GLAST mission study may be found at http://www-glast.stanford.edu.

The SIFTER (ScIntillation Fiber Telescope for Energetic Radiation) collaboration was formed in 1996 when a proposal was approved to study and demonstrate the potential for a gamma-ray astronomy experiment based on scintillating fibers in a combined tracker-calorimeter configuration. The GLAST mission is an obvious candidate for this application. These studies are being conducted primarily through simulations and accelerator tests. The SIFTER configuration and performance characteristics derived from simulations are presented by Pendleton, et al. (7). They are also available at the SIFTER Web site http://www.batse.msfc.nasa.gov/~pendleto/sifter/. The initial accelerator tests of a SIFTER fiber system for gamma rays are scheduled for Spring 1998 at the Thomas Jefferson Laboratory.

SCINTILLATING FIBERS VS. SILICON STRIP TRACKER

There are various trades that can be made in the GLAST tracker-calorimeter configuration between sensitive area, angular resolution, energy resolution, and field-of view. In astronomy in general, the most significant breakthroughs have come with increases in sensitivity rather than by improvements in angular resolution or energy resolution. In the case of EGRET, for example, the discovery of numerous blazars, the diffuse Galactic mapping, and high energy emission from gamma-ray bursts were made possible through increased sensitivity. Field-of view is likewise an important consideration for the detection of transient phenomena, since these scale directly as the solid angle observed.

572

Limitations in the total power, size and cost of the GLAST mission will undoubtedly limit the total channel count and number of layers in the tracker system. The baseline GLAST silicon strip tracker consists of a number of individual "towers". For a given geometric area, A, the number of readout channels in an x-y plane for a silicon strip array consisting of N^2 towers is $N^2 \times 2S$, where S is the number of channels along one tower side. By contrast, a scintillating fiber tracker plane of the same area would only have $N \times 2S$ readout channels, since the fibers can be N times as long as the silicon strips. For the present GLAST design, $N^2 = 25$, resulting in five times the number of readout channels per plane, for a given A and S (same sensitive area and tracking resolution). In addition, the use of towers introduces amplifier power and passive materials into the internal detector volume for mechanical support, wiring and heat conduction. These passive materials further limit the performance of the silicon tracker. Limitations on the total number of converter layers in the tracker inevitably results in a trade-off between tracker efficiency and electron scattering in the converter plates. Electron scattering is especially detrimental at the lower energies (<150 MeV), where there are the most photons from most sources. This scattering results in decreased angular resolution of the tracker system. The localization of most sources also benefits from improved photon statistics that could be afforded by a higher efficiency scintillating fiber system.

While there is little doubt that a silicon strip tracker for GLAST is viable, it appears that a scintillating fiber tracker for GLAST could have a larger sensitive area at considerably lower cost. For the baseline silicon strip tracker, the effective tracker area is only about one-half of the geometric area, whereas the effective area of a scintillating fiber tracker is nearly equal to the geometrical area. Concerning energy resolution, spectral measurements of most sources observed by EGRET have been constrained more by counting statistics than by intrinsic detector energy resolution. This is true for all but the brightest sources. It is expected that GLAST will be similarly constrained for the vast majority of sources that it detects.

The GLAST mission is also considering a scintillating fiber based calorimeter. Such calorimeters have seen widespread application in high-energy physics in recent years (8-12). In the SIFTER version of GLAST, the tracker and calorimeter are integrated into a single, all-fiber system, with some variations of the converter thickness and possibly the fiber diameter through the depth of the system. This would permit a single detector technology for GLAST, further simplifying the mission and reducing the cost.

SCINTILLATING FIBER TRACKERS IN HIGH ENERGY PHYSICS AND ASTROPHYSICS

In recent years, enormous improvements have been made in scintillating fiber systems and they have become the tracker of choice for high-energy physics experiments requiring large areas, large volume, and moderate resolution (13-22). These systems have additional advantages of radiation immunity, long-term stability and low cost. They have electronic readout systems that are fast and can be remotely located via optical couplers, and long, clear (non-scintillating) fibers. The attenuation length has been continually improving, thanks to improved optical fiber uniformity, clarity, and cladding techniques. Measured attenuation lengths of several meters are typical. Accompanying the intrinsic improvements in scintillation fiber efficiency and light, technical improvements in scintillation light collection and transport have also been made in recent years. These include improvements in multi-fiber ribbon fabrication, improved long-term fiber stability, and extra-mural attenuation. The readout devices have similarly improved greatly in recent years. A comprehensive set of references is provided in this paper to illustrate these recent advances in scintillating fiber technology, readout systems, and their use in large accelerator tracker systems (8, 12-34). As a result of these advances, large systems for high-energy physics of over a million fibers have been produced (33).

For high-energy astrophysics, although sci-fi systems were described almost ten years ago (Pendleton, priv. comm.; 36), the fiber efficiency only recently improved sufficiently to make these systems viable. They have been described for use in a variety of astrophysical observations, including both cosmic-ray research and high-energy gamma-ray astronomy (7,35-47). Several balloon-borne systems have been successfully flown to measure cosmic-ray electrons (37,39,43) and for hi-z cosmic-ray tracking (40,44,46). The first scintillation fiber tracker system was launched last year as part of the CRIS experiment on the ACE spacecraft and is showing outstanding performance (47). In contrast, we are unaware of any large silicon strip tracking systems that have been flown as part of a balloon-borne or space-borne system.

POTENTIAL READOUT SYSTEMS FOR A GLAST SCINTILLATION FIBER TRACKER

There are three general types of readout systems for the SIFTER-GLAST system that are under consideration: 1) Optical image intensifier/CCD systems, 2) Electron-bombarded silicon pixel or CCD arrays, and 3) Avalanche photodiodes. All three types of systems have been used successfully with scintillating fibers in ground-based tests, but only the optical systems have thus far been used in flight experiments or in large-scale accelerator experiments. Their design is relatively straight-forward, using readily available components. The relatively slow output of multiple phosphors and readout time of CCDs limit their livetime, although intensifier gating and special CCD read and fast-clear circuits can greatly improve their readout time (cf. 41).

In recent years, direct electron bombardment of a silicon pixel array (or CCD) device internal to an image intensifier have shown great promise as a scintillation fiber readout device. The direct production of electron-holes in the silicon by high-energy electrons accelerated from a photocathode results in superb resolution of individual photoelectrons (13,16,18,21,25,31,32,34; these proceedings). At least three companies now offer these devices as commercial products and others have been assembled as custom units from both standard and specially-designed components. In addition to their better photoelectron resolution, they are generally faster and less bulky than the optical systems. Spatial resolution is governed primarily by the silicon pixel size and the de-magnification ratio, rather than by the resolution of the optical components.

Avalanche photodiodes (APDs) have been used for many years with glass fiber optic systems for communications, LIDAR, and bio-luminescence research (cf. 48, and references therein) and their performance has been steadily improving (49). Arrays of APDs have also been successfully produced (50,51). Several groups have tested APDs with scintillating fibers (52-55). They are compact, have reasonable gain and high quantum efficiency, and they are very fast. However, their cost and stability with respect to temperature and bias voltage have prevented their use on a large scale. If signal proportionality is not required, operation in the Geiger mode permits operation at very high gains, while largely overcoming most of the stability problems (55,56). When used with active quenching circuits, Geiger-mode APDs are extremely fast (~1ns). However, high dark counts and nuclear radiation sensitivity in the Geiger mode limit their use and, in some applications, cooling of the devices may be required. The SIFTER program is studying APDs for space-borne applications.

CONCLUSIONS

We believe that a scintillating fiber tracker has many advantages over a silicon-strip tracker in large-area applications such as GLAST. These include performance, efficiency, simplicity of design, and cost. Others have come to similar conclusions (21, 25,38). Furthermore, the SIFTER concept combines a scintillating fiber tracker with a scintillating fiber calorimeter into one homogeneous unit with a single detector technology. The SIFTER collaboration will continue performance simulations, accelerator tests and advanced technology development for future gamma-ray astronomy missions.

REFERENCES

1. Dermer, C.D., Strickman, M.S., and Kurfess, J.D., Eds., Proc. The Fourth Compton Symposium, Williamsburg, VA, AIP Conf. Proc. #410 (1997).
2. Gehrels, N. and Winkler, C. "International Gamma-Ray Astrophysics Laboratory (INTEGRAL): A Future ESA Mission for Gamma-ray Astronomy", SPIE 2806, pp.210-216, (1996).
3. Thompson, D., et. al, "Supplement to the Second EGRET Catalog of High-Energy Gamma-Ray Sources" Ap. J. Supp. V. 107, pp. 227, (1996).
4. NASA Gamma Ray Astronomy Working Group, "Recommended Priorities for NASA's Gamma Ray Astronomy Program", 1996-2010, NASA/GSFC, NP-1997(03)-008 (1997).
5. Michelson, P.F., "GLAST: A Detector for High Energy Gamma Rays", SPIE Proc. 2806, pp.31-40 (1996).
6. Wood, K.S., "A Broadband High Energy Gamma-Ray Telescope Using Silicon Strip Detectors", in Imaging in High Energy Astronomy, L. Bassani and G. DiCocco, eds., Kluwer:Amsterdam, pp.287-91 (1995).

7. Pendleton, G.N. et al., "Development of a Gamma-Ray ScIntillation Fiber Telescope for Energetic Radiation (SIFTER) with Simultaneous Tracking and Calorimetry", SPIE 2806, pp.164-74, (1996).

SCINTILLATING FIBER CALORIMETER SYSTEMS:

8. Ambrosio, M., et al., "A New Design Scintillating Fiber Calorimeter to Seach for Neutrino Oscillations in Massive Underground Detectors", , Nucl. Instr. & Methods A, v. 363, pp.604-10 (1995).
9. St'avina, P., et al., "Simulation Studies of the Electromagnetic Energy Resolution of a Scintillating Fiber Calorimeter", Nucl. Instr. & Methods A, v. 364, pp.124-32 (1995).
10. Beck, M., et al., "Online Detection of Neutrons with a Lead/Scintillating Fiber Calorimeter and a Scintillating Tile Hodoscope", Nucl. Instr. & Methods A, v. 381, pp.330-37 (1996).
11. Ishii, K., et al., "Performance Studies of a Lead-Scintillation Fiber Calorimeter Prototype for the Linear Collider Detector" , Nucl. Instr. & Methods A, v. 385, pp.215-24 (1997).
12. Appuhn, R.-D., et al., " The H1 Lead/Scintillating Fiber Calorimeter", Nucl. Instr. & Methods A, v. 386, pp.397-408 (1997).

SCINTILLATING FIBER RESPONSE:

13. SPIE vol. 2007, "Scintillating Fiber Technology and Applications", E.J. Fenyves, Ed. (1993).
14. Cline, D., "New Developments in the Application of Scintillating Fiber Detectors" , SPIE Proc.2281, pp.89-94 (1994).
15. Margulies, S., and Chung, M., "Fabrication and Testing of Clear Lightguide Fiber Bundles for the D0 Prototype Fiber Tracker Cosmic-Ray Test", SPIE Proc.2281, pp.24-34 (1994).
16. SPIE vol. 2281, "Scintillating Fiber Technology and Applications II", E.J. Fenyves, Ed. (1994).
17. Baumbaugh, A.E., "Scintillating Fiber Detector Performance, Detector Geometries, Trigger and Electronics Issues for Scintillating Fiber Trackers" Nucl. Instr. & Methods A, v. 360, 1-6 (1995).
18. SPIE vol. 2551, "Photoelectronic Detectors, Cameras, and Systems", C.B. Johnson and E.J. Fenyves, Eds. (1995).
19. Leutz, H., " Scintillating Fibers", Nucl. Instr. & Methods A, v. 364, 422-48 (1995).
20. Mussa, R., et al., "Development of a Cylindrical Scintillating Fiber Tracker for Experiment E835 at FNAL", Nucl. Instr. & Methods A, v. 360, pp.13-16 (1995).
21. D'Ambrosio, C., Gys, T. Leutz, H. and Puertolas, D., "Particle Tracking with Scintillating Fibers", IEEE Trans. Nucl. Sci., v.43, 2115-27 (1996).
22. Antonelli, A., et al., "Measurements of Light Yield, Attenuation Length, and Time Response of Long Samples of Blue Scintillating Fibers", Nucl. Instr. & Methods A, v. 370, pp. 367-71 (1996).

SCINTILLATING FIBER TRACKER AND READOUT SYSTEMS:

23. Ruchti,R.C. and Wayne, M.R., "Studies of Scintillating Fiber Readout with visible light photon counters for the D0 Upgrade", SPIE Proc. 2281, pp.2-9 (1994).
24. Aoki, S., et al. "Scintillating Fiber Trackers with Optoelectronic Readout for the CHORUS Neutrino Experiment", Nucl. Instr. & Methods A, v. 344, pp. 143-48 (1994).
25. D'Ambrosio, C., et al. "First Beam Exposures of a Scintillating Fiber Tracker Read Out by an ISPA Tube", Nucl. Instr. & Methods A, v. 359, pp.618-21 (1995).
26. Bruckner, W., et al., "A Scintillating Fiber Hodoscope with Avalanche Photodiode Readout", Nucl. Instr. & Methods A, v. 313, pp.429-36 (1992).
27. Lanceri, L. and Vauagnin, G., "A Scintillating Fiber Hodoscope with Multi-anode Photomultiplier Readout" , Nucl. Instr. & Methods A, v. 357, pp.87-94 (1995).
28. Buontempo, S., et al. "Development of Large-volume, High-resolution Tracking Detectors Based on Capillaries Filled with Liquid Scintillator", Nucl. Instr. & Methods A, v. 360, pp. 7-12 (1995).
29. Agoritsas, V., et al., "Scintillating Fiber Hodoscopes Using Position-sensitive Photomultipliers", Nucl. Instr. & Methods A, v. 372, pp.63-69 (1996).
30. SPIE vol. 2806, "Gamma-ray and Cosmic-Ray Detectors, Techniques and Missions", B.D. Ramsey and T.A. Parnell, Eds. (1996).
31. Cushman, P., et al., "Multi-pixel Hybrid Photodiode Tubes for the CMS Hadron Calorimeter", Nucl. Instr. & Methods A, v. 387, pp.107-12 (1997).

32. DeSalvo, R., "Why People Like the Hybrid PhotoDiode", Nucl. Instr. & Methods A, v. 387, pp. 92-96 (1997).

33. Annis, P., et al., "The CHORUS Scintillating Fiber Tracker and Opto-electronic Readout System", Preprint CERN-PPE/97-100 (1997).

34. Datema, C.P., Meng, L-J., Ramsden, D., "The Detection of Minimum Ionizing Particles with Multi-Pixel Hybrid Photodiodes", SPIE Proc., Albuquerque, in press (1998).

SCINTILLATING FIBERS FOR ASTROPHYSICS APPLICATIONS:

35. Doke, T., et al. "A Scintillating Fiber Sampling Calorimater (SSCT-SF) for the Observation of Cosmic Gamma Rays in the GeV Region" Proc. Internat. Conf. On High Energy Gamma-ray Astronomy, Ann Arbor, MI, Oct. 1990 (1991).

36. Atac, M., Cline, D., Chrisman, D., Kolonko, J.,and Park, J., "High Resolution Gamma-Ray Telescope Using Scintillating Fibers and Position Sensitive Photomultiplier Tubes" Nucl. Phys. B, (Proc. Suppl.) v.10B, pp. 139-42 (1989).

37. Torii, S., et al., "Scintillating Fiber Calorimeter as a Detector for the Observation of GeV Cosmic Gamma Rays", SPIE Proc. 1734, pp.220-30 (1992).

38. Sako, T., et al., "R&D for High Energy Gamma-Ray Telescope Using Scintillation Fiber with Image Intensifies", Nucl. Instr. & Methods A, v. 378, pp.185-95 (1996).

39. Torii, S., et al., "Balloon-borne Electron Telescope with Scintillating Fibers", SPIE 2806, pp. 145-54 (1996).

40. Christl, M., et al., "The Scintillating Optical Fiber Calorimeter (SOFCAL) Instrument", SPIE 2806, pp. 155-63 (1996).

41. Hink, P., et al., "The ACE-CRIS Scintillating Optical Fiber (SOFT) Detector: A Calibration at the NSCL", SPIE 2806, pp. 155-63 (1996).

42. Ryan, J.M., et al., "A Prototype for SONTRAC, a Scintillating Plastic Fiber Detector for Solar Neutron Spectroscopy", SPIE Proc. 3114, pp.514-25 (1997).

43. Torii, S. -these Proceedings (1998).

44. Sposato, S. -these Proceedings (1998).

45. Ryan, J. -these Proceedings (1998).

46. Cristl, M. -these Proceedings (1998).

47. Binns, R. -these Proceedings (1998).

AVALANCHE PHOTODIODES:

48. Cova, S., et al. "Avalanche Photodiodes and Quenching Circuits for Single-photon Detection", Applied Optics, v. 35, pp. 1956-76 (1996).

49. Kirn, T., et al., "Properties of the Most Recent Hamamatsu Avalanche Photodiode", Nucl. Instr. & Methods A, v. 387, pp. 199-201 (1997).

50. Gramsch, E., et al., "Fast, High Density Avalanche Photodiode Array", IEEE Trans. Nucl. Sci., v.41, pp. 762-66 (1994).

51. Bruckner, W., et al., "A Scintillating Fiber Hodoscope with Avalanche Photodiode Readout", Nucl. Instr. & Methods A, v. 313, pp. 429-36 (1992).

52. Okumura, S., Okusawa, T. and Yoshida, T. "Readout of Scintillating Fibers by Avalanche Photodiodes Operated in the Normal Avalanche Mode", Nucl. Instr. & Methods A, v. 388, pp.235-40 (1997).

53. Vasile, S., Gordon, J.S., Farrell, R., Squillante, M., and Entine, G., "Fast Avalanche Photodiode Detectors for Scintillating Fibers", in Proc. SCIFI 93: Workshop on Scintillating Fiber Detectors" Notre Dame, Oct. 1993, World Scientific Press (1994).

54. Farrell, R., Olschner, F., Shah, K., Squillante, M., "Advances in Semiconductor Photodetectors for Scintillators", Nucl. Instr. & Methods A, v. 387, pp. 194-98 (1997).

55. Farrell, et al. -these Proceedings (1998).

56. Vasile, S., et al. -these Proceedings (1998).

A Pb-SciFi Imaging Calorimeter for High Energy Cosmic Electrons

S.Torii*, N.Tateyama*, T.Tamura*, K.Yoshida*, T.Yamagami[†],
H.Murakami[‡], T.Kobayashi[¶], T.Yuda[§] and J.Nishimura[#]

*Kanagawa University, Yokohama, JAPAN
[†] Institute of Space and Astronautical Science, Sagamihara, JAPAN
[‡] Rikkyo University, Tokyo, JAPAN
[¶] Aoyama-gakuin University, Tokyo, JAPAN
[§] Institute for Cosmic Ray Research, University of Tokyo, Tanashi, JAPAN
[#] Yamagata Institute of Technology, Yamagata, JAPAN

Abstract. The BETS (balloon-borne electron telescope with scintillating fiber) detector has been developed for high-altitude balloon flights to observe high-energy cosmic-electrons. The detector consists of an imaging calorimeter and a trigger system for particle identification and energy measurement. The calorimeter is composed of scintillating fibers and leads of a total thickness of ~ 8 r.l. Two sets of an image-intensifier and CCD camera system are adopted for read-out of 10,080 scintillating fibers. The accelerator tests were carried out to study performance of the detector by the CERN-SPS electron and proton beams. It is demonstrated in the flight data that a reliable identification of the electron component has been successfully achieved up to 100 GeV, and the energy spectrum has been measured.

INTRODUCTION

It is well known that the cosmic-ray electron observation is very difficult especially in the high energy region since the background protons become relatively dominant with the increase of energy. The detector must, therefore, meet the requirements of a large exposure area and a high capability of electon/proton separation. For the purpose, a detector applying the mass difference between electron and proton has widely been used in lower energies. The TOF technique is useful in the GeV region, and Čerenkov counter [1] in the 10 GeV. The transition radiation detector [2] is very effective for the electron observation up to a few 100 GeV. They are, however, invalid in rejection of the background protons over several 100 GeV. Emulsion chamber (EC) technique has uniquely measured the electron flux up to the TeV region [3] by the excellent capability of track recognition. EC has still difficulties in detection of the electrons below 100 GeV, in long-term exposure, and in time resolution.

CP450, SciFi97: Workshop on Scintillating Fiber Detectors
edited by A. D. Bross, R. C. Ruchti, and M. R. Wayne
© 1998 The American Institute of Physics 1-56396-792-8/98/$15.00

The Balloon-borne Electron Telescope with Scintillating fibers (BETS) has been developed for observing high-energy cosmic-electrons by imaging shower profiles [4]. The accelerator tests have been carried out to study the performance by using electron beams in 1996 and proton beams in 1997 by CERN-SPS. The high-altitude balloon observations demonstrated that the detector has a reliable performance in observing the electron component in the energy region of 10 GeV to 100 GeV.

THE DETECTOR SYSTEM

The BETS instrument consists of a shower detector incorporates an imaging calorimeter and a trigger system, a data recording system and a telemetry system. The geometrical factor (~ 500 cm^2 sr) is relatively large with a weight of ~ 250 kg. In order to keep the temperature and to avoid the discharge by high voltages , the whole instrument is contained in a pressurized vessel with a diameter of 1 m and a height of 1.7 m. The vessel made of aluminum has a thickness of 4 mm (i.e. 0.045 r.l.) at the top part to avoid the effects of secondaries produced in the covering material.

Imaging Calorimeter

Figure 1 shows a schematic cross-section of the BETS detector. SciFi belts , with a width of 28 cm each, are placed alternately with the lead plates of a thickness of 5 mm. The total thickness of lead is ~ 8 r.l. and the number of SciFi layer is nine. In each layer, two belts are set in right angles with each other to observe the projected profile in the x and y direction. The effective area is 28 cm $\times$ 28 cm. Three plastic scintillators with a thickness of 10 mm for each, are placed to make a trigger for an electron-induced shower by three-fold coincidence of these signals.

A SciFi , Kuraray SCSF77, has one millimeter diameter and involves a poly-styrene core ($n=1.59$) surrounded by a poly-methylmetacrylate (PMMA) clad ($n=1.49$). These 280 SciFi's form a belt in one millimeter pitch. In the converter part, a layer is composed of four of the belt in order to ensure the detection of minimum ionizing particles at the pre-shower stage, and these 6 layers (24 belts in total) are used. A layer in the colorimeter part is of one , and the total number is 12. Each SciFi splices a clear fiber one by one at the edge of detector.

A clear-fiber belt is equally divided to four parts (70 mm width for each), and these four are piled up to make a belt of four layers, as presented in Figure 2. The number of layer of SciFi in one direction is 18, and 72 layers of the clear-fiber belts are coupled together with input window of an image intensifier (I.I) system with a diameter of 100 mm. One end of each belt makes contact with the input window of I.I; the other at right angles with another I.I. The total number of SciFi's is 10,080 (5,040 for each direction).

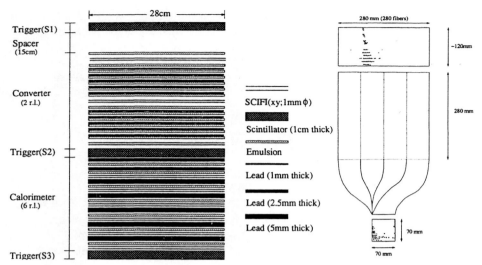

FIGURE 1. Schematic side view of the balloon-borne electron telescope (BETS). The effective area is 28 cm×28 cm and the thickness is about 10 cm except the space of 15 cm. Emulsions are used for confirming capability of SciFi .

FIGURE 2. Configuration of SciFi belt coupled with I.I window. The belt is divided by four and these are piled up to be attached to the window.

Image Intensifier and CCD system

The I.I unit is Hamamatsu II-100U, and it consists of an image reducer and connected two I.I's which are successively joined with a CCD camera. The image reducer (V4440) is a first generation inverter-type with a 4 inch-diameter fiber-optics window. The input image is reduced by a quarter in length on the phosphor output-window. Each of connected I.I (V1366P) is a 1 inch inverter-type incorporating a micro-channel plate and the P-20 phosphor screen. The output of the last I.I is connected with a taper fiber to a CCD camera (Sony XC77RR-CE).

The camera has 756(H) × 581(V) picture elements with a cell size of 11(H) × 11(V) μm. The SciFi area (70 mm × 70 mm) corresponds approximately to 512 × 512 picture elements on the CCD camera when the image in detector is reduced by one-twelfth in the CCD image. A video module reads 512 × 256 picture elements by the 2:1 interlace CCD scanning method and the signals which exceed a noise level are digitized with an 8-bit flash ADC. By averaging the intensities in horizontally-adjacent picture elements, these of 256 × 256 picture elements are obtained.

Trigger signal makes a gate pulse with a width of 7 μsec to activate the last I.I and open the shutter of the CCD camera with a speed of 1/1360 sec. These "gating" procedures can effectively remove noises by the instruments and by the accidental events.

Trigger System

Event trigger is performed by three-fold coincidence of signals by the scintillators ($S_1 \sim S_3$) presented in Fig. 1. Each scintillator (Bicron BC-408, 10 mm thick) is viewed through a light guide by a Hamamatsu photomultiplier tube (H1949 for S_1, H1161 for S_2 and S_3). The anode signals are employed to make the trigger signals after shaping the pulse forms. The dynode signals are used to measure the pulse heights. The trigger system was optimized by simulation to enhance as large as possible the ratio of electrons in the triggered events. For the observation of electrons over 10 GeV and with a zenith angle lower 30 degrees, it is found that the highest proton-rejection power of ~ 150 at an electron detection efficiency of 85% is expected if we impose the numbers of particles observed in S_1, S_2 and S_3 be 0.7 ~ 5, $10 \sim 100$, and ≥ 40 (in unit of single minimum ionizing particle of vertical incident), respectively.

Simulation for Electron Selection

Figure 3 shows the results of simulations for the distributions of ratios of energy deposition in the SciFi's within the distance of 5 mm from the shower axis (RE). The ratio of proton events in the region of $RE \geq 0.7$ is only 5.6%, while 85% of the electron events are in the region. As the simulation events are "triggered" by the same condition with the observation, the selection by the criterion that RE is greater than 0.7 gives additionally the rejection power of ~ 20.

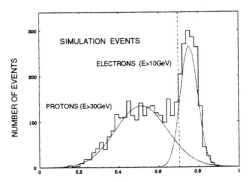

RATIO OF ENERGY DEPOSITION WITHIN 5mm FROM SHOWER AXIS

FIGURE 3. Simulation distributions of the ratios of energy deposition in the SciFi's within 5 mm from shower axis for electrons over 10 GeV and for protons over 30 GeV.

The total rejection-power combining the on-board trigger and the off-line analysis is, therefore, predicted to be about 3,000 in the energy region over 10 GeV. It is, moreover, expected from a neural network analysis of the shower profiles that the rejection power by the imaging might be increased by five times.

CALIBRATION BY ACCELERATOR BEAM

Performance of the detector has been studied by using electron beams with CERN-SPS. The energy range was from 5 GeV to 100 GeV, and it covered the whole range of observation. Figure 4 presents raw images of the electron showers with energy of 10 GeV, 30 GeV and 100 GeV. The RE distribution was made by same analysis with the balloon experiment mentioned in next section. The distribution presented in Fig. 5 is consistent with the observed one (Fig. 6) in the region of $RE \geq 0.7$ in which electrons exceedingly dominate. The distributions for each energy are plotted with weights of the event numbers assuming a cosmic-electon energy spectrum. The total is mostly contributed from the lower energy electrons.

The pulse height of the bottom scintillator is used for estimation of the energy, since it is nearly proportional to the energy of shower. Relation of the pulse height and the energy is obtained for the different incident angles and for the incident positions. The energy resolution which weakly depends on the energy ranges from 15 % to 18 % .

The beam test by protons with energy of 60 GeV to 250 GeV has also been carried out to establish the rejection power expected by simulation. The data analysis is now going on, and the results will be reported in elsewhere. It will be tried to get a rejection power better than 10^4 which is necessary for observing the TeV electrons.

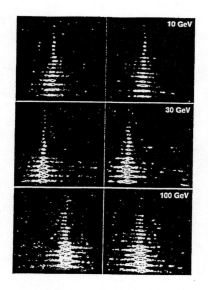

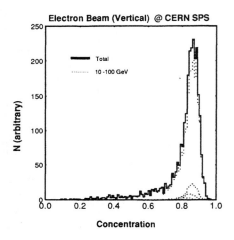

FIGURE 4. Examples of raw CCD images of electron showers obtained by CERN-SPS. Two images for each energy are presented in the x and y direction.

FIGURE 5. RE distributions for electrons by CERN-SPS. The distributions for each energy are plotted with weights of the event number assuming a cosmic-electron energy spectrum for comparison with the balloon data.

PERFORMANCE IN BALLOON OBSERVATION

The instrument of a total weight of 320 kg and a power consumption of 130 W (in max) was launched from the Sanriku Balloon Center in Japan, and it was flown for 7 hours at an average altitude of 35 km. The event trigger was carried out either by three-hold coincidence of S_1, S_2 and S_3 (SH mode) or by two-hold of S_1 and S_2 (SI mode). The SH mode for shower events was used mostly in the electron observation, and the SI mode was for heavy particles. The observation of heavy tracks was done for the position calibration of SciFi's by the correspondence of the track positions in emulsion plates.

The trigger rate for showers over a threshold energy of 10 GeV was $\sim$ 1.6 Hz at the level flight, and the dead time was negligible. Almost all triggered events ($\sim$ 28,000) were stored in the EXB 8mm tape on board.

Image data analysis

The configuration of SciFi belts deforms the shower image in the detector when it is observed on the CCD image as presented in Fig. 2. In order to recover the shower image from the CCD image, it is necessary to define the positions of each SciFi on the CCD image. The positions of all SciFi's are measured by the detection of cosmic-muon tracks at the ground. If muon passes over one SciFi within the belt, the SciFi emits photons which make a bright spot on the CCD image. The spot consists of a cluster of pixels with a signal over the threshold. The cluster consists of nearly 3×3 picture elements which corresponds to the cross section of a SciFi . Center of the cross section was obtained by the two-dimensional Gaussian fit for the area corresponds to one SciFi . By applying this analysis for all SciFi's , these centers were defined within an accuracy of one pixel. The absolute positions of fiber were calibrated every 100 events by illuminating a part of fibers by LED during observation.

The signal intensity from one SciFi is obtained by summing up the intensities of pixels in the corresponding cluster. The showers which show typical development of electro-magnetic cascade, as presented in Fig. 4, usually fulfill the criterion for electrons that $RE \geq 0.7$. The others which have sometimes secondary hadron-tracks in the converter have a wider spread, and the values of RE are mostly lower than 0.7.

Electron selection

The three-dimensional shower axis of incident particle is obtained from the projected directions observed with SciFi's in right angles with each other. Reduction of the events were carried out by the following sequence:

1. The shower passes over all layers of detector.

2. The zenith angle is less than 30 degree.

3. The charge of incident particle is lower than two.

Figure 6 shows the RE distributions of the events surviving the reduction. The distribution in the level flight is very similar to that by the simulation presented in Fig. 3. The peak region of distribution in $RE \geq 0.7$ might, therefore, correspond to electrons. The ratio of the number of events with $RE \geq 0.7$ to that of the others is consistent with the simulation. The proton acceptance is, therefore, predicted to be 8×10^{-3} at the electron detection efficiency of ~ 85 %.

Additional rejection against the background protons by selecting the events with $RE \geq 0.7$ is ~ 0.05, and the combined proton acceptance is obtained to be about 4×10^{-4}. It means that the remaining background in the electron sample is a few % of the number of electrons. The flux is derived with a better accuracy since the contamination of protons can statistically corrected from the distribution of REs.

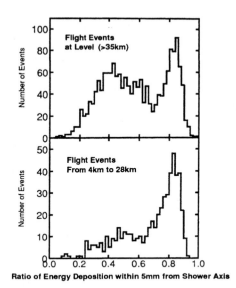

FIGURE 6. RE distributions at the level flight (upper) and during the ascending period (lower) observed by '97 flight.

SUMMARY AND DISCUSSION

A new telescope for the high-energy electron observation has been successfully developed. Analysis of the balloon experiment in 1997 has shown that it has the performance which has been expected by simulations; the geometrical factor is as large as 500 cm^2sr in spite of the light weight, and the rejection power against the background protons is enough to measure the electron flux in energies of 10 GeV to 100 GeV. The results of electron flux and the astrophysical significance are reported in Ref. [5].

Future long-term observation should follow to get final conclusion on the effects of near-by sources including the anisotropy by a supernova remnant, Vela [6,7]. We are planning the development of a new detector with $S\Omega$ of 0.5 m^2sr, which will be used for an observation in space. The Japanese Experiment Module (JEM) exposure facility in the International Space Station might be available for the observation. As presented in Table 1, by the observation, more than 500 electrons will be observed in one year in the energy region over 1 TeV, and we could clearly identify the acceleration cites of the electrons.

Table 1: Expected number of electrons in the one-year observation with a detector of $S\Omega = 0.5\ m^2 sr$

Energy	> 10 GeV	> 100 GeV	> 1,000 GeV	> 3,000 GeV
Electron Number	2.0×10^7	1.0×10^5	5.2×10^2	4.0×10^1
Electron/Proton	8.5×10^{-3}	2.4×10^{-3}	6.8×10^{-4}	3.6×10^{-4}

ACKNOWLEDGMENTS

We are indebted to the crews of the Sanriku Balloon Center, ISAS for successful balloon flight. We express sincere thanks to the technical staffs in Kanagawa University for their helps in manufacturing of the detector. This work is partially supported by Grant-in-Aid for Scientific Research(A) and Grants-in-Aid for International Scientific Research (Field Research) from the Ministry of Education, Science, Sports and Culture in Japan.

REFERENCES

1. Golden R.L., *Ap. J.* **287**, 622 (1980)
2. Tang K.K., *Ap. J.* **278**, 881 (1984)
3. Nishimura J.*et al.*, *Ap. J.* **238**, 394 (1980)
4. Torii S. *et al.*, *Proc. of the 24 th International Cosmic Ray Conference* **3**, 575 (1995)
   Torii S. *et al.*, Proc. of UCLA Int. Conf. on Imaging Detector in High Energy and Astrophysics (Singapore: World Scientific) , 35 (1995)
   Torii S. *et al.*, Proc. of SPIE, Gamma-Ray and Cosmic-Ray Detectors, Techniques, and Missions **2806**, 145 (1996).
5. Taira T. *et al.*, *Proc. of the 23 rd International Cosmic Ray Conference* **3**, 128 (1993)
   Nishimura J. *et al.*, *Proc. of the 24 th International Cosmic Ray Conference* **3**, 29 (1995)
   Torii S. *et al.*, *Proc. of the 25 th International Cosmic Ray Conference* **3**, 117 (1997)
6. Nishimura J., Proc. of Towards a Major Atmospheric Cherenkov Detector III (Tokyo: Universal Academy Press) , 1 (1994)
7. Ptuskin V.S. and Ormes J.F, *Proc. of the 24 th International Cosmic Ray Conference* **3**, 56 (1995)

A Hard X-Ray
Solar Flare Polarimeter Design
Based on Scintillating Fibers

J.M. Ryan, M.L. McConnell, D.J. Forrest,
J. Macri, M. McClish and W.T. Vestrand

Space Science Center, University of New Hampshire, Durham, NH 03824

Abstract. We have developed a design for a Compton scatter polarimeter to measure the polarization of hard X-rays (50–300 keV) from solar flares. The modular design is based on an annular array of scintillating fibers coupled to a 5-inch position-sensitive PMT. Incident photons scatter from the fiber array into a small array of NaI detectors located at the center of the annulus. The location of the interactions in both the fiber array and in the NaI array can be used to measure the linear polarization of the incident flux. This compact design may be well-suited to a variety of astrophysical applications. An extensive series of Monte Carlo simulations has been performed to characterize this design.

INTRODUCTION

The measurement of hard X-ray polarization in solar flares would provide insights into the geometry of the electron acceleration process. In particular, such polarization measurements would indicate the extent to which the electrons are beamed. Here we report on the development of a hard X-ray polarimeter for solar flares that is based on the use of scintillating fibers. Due to its relatively large FoV, this design may also be useful in studies of γ-ray bursts.

The basic physical process used to measure linear polarization of hard X-rays (100–300 keV) is Compton scattering. The measurement is based on the fact that the incident photons tend to be scattered at right angles to the incident electric field vector. A Compton scatter polarimeter consists of two detectors that are used to determine the energies of both the scattered photon and the scattered electron. One detector (the *scattering detector*) provides the medium for the Compton interaction to take place. This detector must be designed to maximize the probability of a single Compton interaction with a subsequent escape of the scattered photon. The primary purpose of the

CP450, *SciFi97: Workshop on Scintillating Fiber Detectors*
edited by A. D. Bross, R. C. Ruchti, and M. R. Wayne
© 1998 The American Institute of Physics 1-56396-792-8/98/$15.00

second detector (the *calorimeter*) is to absorb the full energy of the scattered photon. To be recorded as a polarimeter event, an incident photon Compton scatters from one (and only one) of the scattering detectors into the central calorimeter. The incident photon energy can be determined from the sum of the energy losses in both detectors and the scattering angle can be determined by the azimuthal angle of the associated scattering detector. When the polarimeter is arranged so that the incident flux is parallel to the symmetry axis, unpolarized radiation will produce an axially symmetric coincidence rate. If the incident radiation is linearly polarized, then the coincidence rate will show an azimuthal asymmetry whose phase depends on the position angle of the incident radiation's electric vector and whose magnitude depends on the degree of polarization.

LABORATORY PROTOTYPE

In an earlier paper, we discussed a polarimeter design consisting of a ring of twelve individual scattering detectors (composed of low-Z plastic scintillator) surrounding a single NaI calorimeter [1]. The characteristics of this design were investigated using a series of Monte Carlo simulations (based on a modified version of GEANT). We have recently prototyped this design in the laboratory to validate our Monte Carlo code. For prototype testing, we set up a semicircular array around a central NaI detector, eliminating the redundancy and simplifying the hardware and associated electronics. Seven plastic scintillators (each 5.5 cm $\times$ 5.5 cm $\times$ 7.0 cm in size) were positioned at a radius of 15 cm from a 7.6 cm diameter $\times$ 7.6 cm high cylindrical NaI(Tl) detector.

Polarized photons were generated by Compton scattering photons from a radioactive source [2]. The exact level of polarization is dependent on both the initial photon energy and the photon scatter angle. The use of plastic scintillators as a scattering block permits the electronic tagging of the scattered (polarized) photons. This is used to provide a coincidence signal to the polarimeter. For our laboratory measurements we used a ^{137}Cs source to generate a beam of polarized 288 keV photons.

The laboratory data (Figure 1) led to a measured polarization value of 64.0%($\pm$3.0%), in good agreement with the estimated value of 50-60% based on analytical estimates [3]. This result demonstrates: a) the ability of a simple Compton scatter polarimeter to measure hard X-ray polarization; b) the ability of our Monte Carlo code to predict the polarimeter response; and c) the ability to generate a source of polarized photons using a simple scattering technique. In another laboratory measurement (Figure 2), the plane of polarization of the incident beam was rotated $\sim$ 45° with respect to that used in the first set of data. The measured shift of 50.4° in the polarization vector is consistent with the uncertainties in our experimental setup.

587

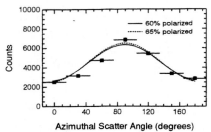

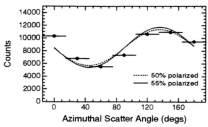

FIGURE 1. The prototype response to an on-axis polarized beam. The smooth curves represent simulation results.

FIGURE 2. The prototype response with the polarization vector rotated 45° with respect to that in Figure 1.

A NEW DESIGN CONCEPT

There are at least two possible means of improving the polarimeter performance: 1) by more precisely measuring the scattering geometry of each event; and 2) by rejecting those events that undergo multiple Compton scattering within the scattering elements. (Our simulations indicate that roughly 30-40% of the events recorded in the prototype polarimeter as valid events involved multiple scattering within a single scatter element.) Improvements in either area will lead directly to a more clearly defined modulation and, therefore, a better polarization sensitivity.

We have developed a new design that places an entire device on the front end of a single 5-inch diameter position-sensitive PMT (PSPMT) [4]. A bundle of scintillation fibers (each with a cross section of 4 mm × 4 mm) provides the improved spatial resolution in the scattering elements. The bundle is in the form of an annulus with an outside diameter of 10 cm and an inside diameter of 4 cm. A 2 × 2 array of 1 cm inorganic scintillators is positioned within the annulus, each scintillator being coupled to its own independent PMT for light collection and signal timing. Figure 3 shows a schematic view of such an assembly.

Monte Carlo simulations have been used to determine the characteristics of this design. Figure 4 shows the modulation curves that result from completely polarized incident radiation at two different energies (100 keV and 300 keV). Figures 5 and 6 show the modulation factor and the effective area, respectively, as a function of energy. The low energy response is very sensitive to the energy threshold in the fiber array. Figure 7 shows the off-axis response of the design, which suggests a useful FoV of at least one steradian.

We are currently involved in a series of laboratory tests designed to evaluate the characteristics of a PSPMT / fiber bundle detector system. These tests will help determine the achievable energy resolution and, more importantly, the achievable energy threshold of the fiber array. The precise energy threshold level of the fiber bundle will have a major impact on the final characteristics of

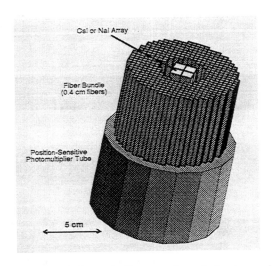

FIGURE 3. Schematic diagram of a polarimeter module.

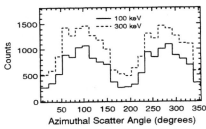

FIGURE 4. Simulated modulation curves (counts versus azimuthal scatter angle) at energies of 100 and 300 keV.

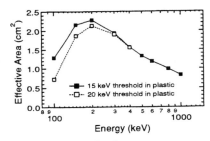

FIGURE 5. Effective area versus energy.

the polarimeter module. In particular, the fiber bundle threshold will dictate the low energy threshold of the polarimeter module and also the characteristics of the polarimeter module (in terms of effective area and modulation factor) at low energies. Our goal is to achieve an energy threshold in the range of 10–30 keV for the fiber bundle.

SUMMARY

We anticipate that this design would be used in the context of a (not necessarily contiguous) array of polarimeter modules. In the case of solar flares, we calculate that an array of 4 modules would be capable of measuring polarization levels down to a few percent in X-class flares. A larger array of 16

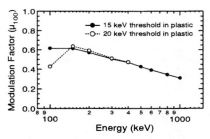

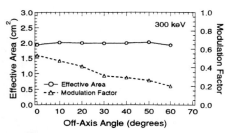

FIGURE 6. Modulation factor versus energy.

FIGURE 7. The modulation factor and effective area as a function of incidence angle for a photon energy of 300 keV.

modules would be capable of measuring solar flare polarization levels below 1% for the largest events and would also be capable of measuring polarization levels down to about 15% in some of the largest γ-ray bursts [12]

The use of polarimetry in X-ray and γ-ray astronomy has so far been largely limited to energies below 100 keV [5–7,2,8], with an emphasis on the study of non-transient sources. Several higher energy experiments offer polarimetry as a secondary capability [9,10]. Although designs similar to that proposed here have been discussed in the literature [6,11], we are unaware of any other *active* effort to specifically measure polarization in solar flares and γ-ray bursts at energies above 100 keV.

ACKOWLEDGEMENT

This work has been supported by NASA grant NAGW-5704.

REFERENCES

1. M. McConnell, D. Forrest, K. Levenson, and W.T. Vestrand, "The design of a gamma-ray burst polarimeter", in AIP Conf. Proc. 280, *Compton Gamma-Ray Observatory*, M. Friedlander, N. Gehrels and D.J. Macomb, Eds. New York: AIP, 1993, pp. 1142-1146.
2. H. Sakurai, M. Noma, and H. Niizeki, "A hard x-ray polarimeter utilizing Compton scattering", in *SPIE Conf. Proc.*, vol. 1343, pp.512-518, 1990.
3. W.H. McMaster, "Matrix representation of polarization", *Reviews of Mod. Phys.*, vol. 33, no. 1, pp. 8-28, January 1961.
4. M.L. McConnell, et al., "Development of a hard X-ray polarimeter for solar flares and gamma-ray bursts", submitted to *IEEE Trans. Nucl. Sci.*, 1998.
5. R. Novick, "Stellar and solar X-ray polarimetry", *Space Science Reviews*, vol. 18, pp. 389-408, 1975.

6. G. Chanan, A.G. Emslie, and R. Novick, "Prospects for solar flare X-ray polarimetry", *Solar Physics*, vol. 118, pp. 309-319, 1988.

7. P. Kaaret, et al., "The Stellar X-ray Polarimeter - a focal plane polarimeter for the Spectrum X-Gamma mission", *Optical Engineering*, vol. 29, pp. 773-780, July 1990.

8. E. Costa, M.N. Cinti, M. Feroci, G. Matt, and M. Rapisarda, "Scattering polarimetry for X-ray astronomy by means of scintillating fibers", *SPIE Conf. Proc*, vol. 2010, pp. 45-56, 1993.

9. E. Aprile, A. Bolotnikov, D. Chen, R. Mukherjee and F. Xu, "The polarization sensitivity of the liquid zenon imaging telescope", *ApJ Supp*, vol. 92, pp. 689-692, June 1994.

10. T.J. O'Neill, et al., "Tracking, imaging and polarimeter properties of the TIGRE instrument", *Astron. Astrophys. Suppl. Ser.*, vol., 120, pp. C661-C664.

11. T.L. Cline, et al., "A Gamma-Ray Burst Polarimeter Study", in *Proceedings of the 25th Internat. Cosmic Ray Conf.*, vol. 5, pp. 25-28, 1997.

12. M.L. McConnell, et al., "Development of a hard X-ray polarimeter for gamma-ray bursts", to be published in AIP Conf. Proc., *4th Gamma Ray Bursts Symposium*, 1998.

SONTRAC—A Low Background, Large Area Solar Neutron Spectrometer

J.M. Ryan[1], D. Holslin[2], J.R. Macri[1], M.L. McConnell[1], C.B. Wunderer[1]

[1]Space Science Center, University of New Hampshire, Durham, NH 03824
[2]Science Applications International Corporation, San Diego, CA 92121

Abstract. SONTRAC is a scintillating fiber neutron detector designed to measure solar flare neutrons from a balloon or spacecraft platform. The instrument is comprised of alternating orthogonal planes of scintillator fibers viewed by photomultiplier tubes and image intensifier/CCD camera optics. It operates by tracking the paths of recoil protons from the double scatter of 20 to 200 MeV neutrons off hydrogen in the plastic scintillator, thereby providing the necessary information to determine the incident neutron direction and energy. SONTRAC is also capable of detecting and measuring high-energy gamma rays >20 MeV as a "solid-state spark chamber." The self-triggering and track imaging features of a prototype for tracking in two dimensions have been demonstrated in calibrations with cosmic-ray muons, 14 to ~ 65 MeV neutrons and ~ 20 MeV protons.

INTRODUCTION AND MOTIVATION

Neutrons above the nuclear binding energy are ubiquitous in cosmic ray interactions, whether those reactions occur on the surface of the Sun or in the earth's atmosphere. These secondary neutrons are difficult to measure because they lack charge, and thus, do not interact readily with detector material.

When high-energy charged particle reactions occur on the surface of the Sun, neutrons carry away information about the spectrum of ions that produced them. They can be used as diagnostic measures of the accelerated ion spectrum in solar flares (1,2). As illustrated in Fig. 1, gamma rays are produced efficiently by protons below 20 MeV and above 300 MeV. The intervening region of the proton spectrum is most effective in producing neutrons of comparable energy. To determine the proton spectrum in this region secondary neutrons must be measured. This measurement must take place from a space platform or a balloon. In either environment the background is high making the measurement difficult.

To perform these measurements neutron telescopes based on double scatters have been used (3-6). Because of its directional properties a neutron telescope one can reject

CP450, *SciFi97: Workshop on Scintillating Fiber Detectors*
edited by A. D. Bross, R. C. Ruchti, and M. R. Wayne
© 1998 The American Institute of Physics 1-56396-792-8/98/$15.00

much of the background, thereby enhancing the signal-to-noise ratio, making these instruments effective in high background environments. Such instruments have applications in other disciplines including atmospheric physics, radiation therapy and nuclear materials monitoring.

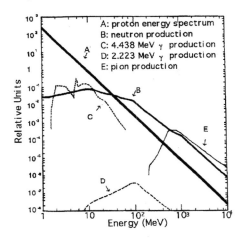

FIGURE 1. Spectrum-weighted emissions of solar flare gamma rays and neutrons.

TRACKING DETECTOR CONCEPT

We have been studying a tracking detector design for measuring neutrons in the difficult range of 20 to 250 MeV. The design is based around a closely packed bundle of square cross section plastic-scintillator fibers. The fibers are arranged in stacked planes with the fibers in each plane orthogonal to those in adjacent planes. This configuration allows one to record stereoscopic images of ionization tracks.

Neutrons undergo elastic scattering off hydrogen within the organic plastic-scintillator fibers, scattering at right angles with respect to the scattered proton at non-relativistic energies. The Bragg peak (greater ionization at the end of the track) is used to determine proton track direction. A second proton scatter of the neutron provides the spatial information that is necessary and sufficient to determine the incident neutron energy and direction. An image and spectrum of the neutron source can be constructed from these data. The angular and energy resolution of the instrument depend upon our ability to track the recoil protons and measure the scintillation light.

A functional diagram of an experiment utilizing the SONTRAC concept is shown in Fig. 2. The detector's spectroscopic, track detection and imaging components cover the entire light-emitting area of the fiber bundle and are duplicated in the orthogonal dimension (not shown). Scintillation light signal is collected and processed at both ends of the fiber bundle. At one end a signal above threshold from a photomultiplier tube (PMT) gates the trigger logic circuitry. At the other end, fiber-optic tapers and a pair of image intensifiers demagnify, capture and hold the scintillation-light image of the track(s) for the CCD camera. The first image intensifier is always ON, holding the image for ~ 1 ms. The second image intensifier is normally in the gated-OFF condition. However, with the particle coincidence requirements satisfied, the track image and PMT pulse height data are acquired, combined with any auxiliary data and recorded.

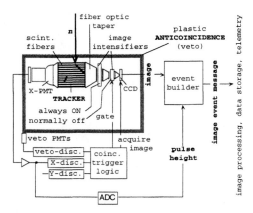

FIGURE 2. Complete instrument functional and logic diagram.

A 4π plastic scintillator shield surrounds the tracker to reject incoming cosmic rays. Other charged particles produced in the detectors cannot escape the detector volumewithout being registered. They represent lost energy and can be rejected.

The initial design of the SONTRAC instrument suffered at the time from a lack of affordable technology and existed in simulations only (7-9). Due to advancements in technology, we have been able to construct a simple prototype of the instrument to verify the concept and identify any fundamental technical limitations.

Fig. 3 is a schematic of the SONTRAC prototype operating at UNH. The prototype is a 10 cm long bundle of 250 μm-square (230 μm active), multiclad, organic, scintillating, plastic fibers (Bicron BCF-99-55) within a 12.7 mm square envelope. It is, therefore, limited to tracking in two dimensions. The fiber thickness was chosen such that a 10 MeV recoil proton traverses several fibers. The fiber pitch is 300 μm (including cladding and EMA) and the calculated range of a 10 MeV proton (50% of the proposed neutron threshold energy) is 1.25 mm (~ 4 fibers). The PMT is a bialkali photocathode device from Thorn EMI. Two 18 mm diameter, single-MCP, generation-2 image intensifiers from DEP are employed. The S20 photocathode for the first image intensifier was selected for its response to scintillation light. The P43 phosphor holds the image for ~ 1 ms. The second intensifier's photocathode (S25) and phosphor (P43) provide good spectral matches to the output of the first intensifier and the input to the CCD sensor, respectively. The CCD camera (Pulnix TM-9701) is an inexpensive, progressive scanning camera with digital readout and control, asynchronous external trigger and full frame shutter capability. The Matrox frame grabber and image processor operate within a Pentium PC.

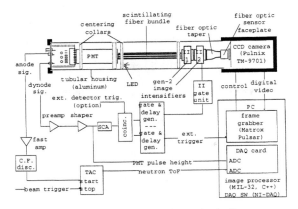

FIGURE 3. Prototype functional and logic diagram.

PROTOTYPE PERFORMANCE AND FLIGHT DESIGN

Tracks of recoil protons near our trigger threshold energy (14 MeV) have been published elsewhere (10). These data show that protons as low as ~ 10 MeV can be tracked over 4 or 5 fibers with a clear Bragg peak. The scintillation light also conforms to the expected distribution (within errors) of the ionization rate.

The SONTRAC prototype was also exposed to 20 MeV and 27 MeV neutrons and to ~ 20 MeV protons at the Crocker Nuclear Laboratory at the University of California at Davis. Additional measurements were made with ~ 65 MeV neutrons, although it was not possible to get a clean beam in the time available. Fig. 4 shows two CCD images of neutron interactions in the prototype fiber bundle. On the left is a double scatter event displaying two recoil proton tracks from a neutron (~ 65 MeV) incident from the top of the figure. On the right is a single proton recoil track from a 28 MeV neutron incident from the top left of the figure. The individual fibers are seen in the ionization tracks. Note that the Bragg peak is evident establishing the proton recoil direction. Gaps in the track images are due the ionizing particles traversing the passive cladding and EMA materials.

FIGURE 4. Double scatter of a ~65 MeV neutron (left) and a 28 MeV proton (right).

Fig. 5 is a cosmic-ray muon image recorded with the SONTRAC prototype, demonstrating the detector's ability to track minimum-ionizing particles. (The fiber mask is shown for reference.) This is necessary to track conversion electrons in high energy gamma detectors. We calculate that ~ 4 photoelectrons/fiber are generated for amplification within the image intensifier.

FIGURE 5. A cosmic-ray muon track.

A sketch of a larger volume detector for particle tracking in three dimensions is shown in Fig. 6. It employs orthogonal layers of scintillating fibers in a $10 \times 10 \times 10$ cm bundle with one PMT (left) and one image intensifier/CCD chain (right) in each dimension. The fiber block would be surrounded by charged particle detectors to (1) reject cosmic ray protons and electrons and (2) to detect the escape of secondary charged particles from internal reactions. The flight instrument would occupy a space less than $50 \times 50 \times 20$ cm^3 with a total mass less than 26 kg. We estimate that the proposed instrument will have an effective area for detecting and measuring 15 to 60 MeV neutrons of ~ 2 cm^2, approximately the same as that of the imaging Compton telescope, COMPTEL, a 1460 kg instrument (6).

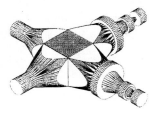

FIGURE 6. Sketch of a $10 \times 10 \times 10$ cm 3-d imager.

The typical energy resolution is on the order of 10% or better for the majority of neutron events. The angular resolution is largely determined by the pitch of the fibers, i.e., the uncertainty in the end points of the tracks. For a 45° scatter and 300 μm fiber pitch it ranges from 23° at 20 MeV to 5° at 50 MeV to 0.7° at 200 MeV. This angular resolution is the basis for our high signal-to-noise ratio and thus our good sensitivity.

CONCLUSIONS

We have demonstrated the important features of the SONTRAC instrument with a laboratory prototype. This will help to determine the engineering parameters important for the SONTRAC application (e.g., scintillating fiber pitch, light yields, gains, photocathode and phosphor selection, gating delays and intervals). Although the prototype is limited to tracking in two dimensions, we obtained self-triggered images of tracks of ~ 20 MeV protons, recoil protons from 14 to ~65 MeV neutrons and minimum ionizing tracks of cosmic ray muons. The track images and the associated pulse height information provide good resolution measurements of both the direction and energy of the incident radiation. An extension to 3-dimensional tracking promises to provide unprecedented measurement capabilities for studies in solar physics and other fields.

ACKNOWLEDGMENTS

We wish to thank Prof. Patrick Papin and David Smith for their assistance with the 14 MeV neutron calibration. We also wish to thank Drs. Carlos Castaneda and Juan Romero with the high energy neutron and proton calibrations and instrument simulations. We also thank Mr. Tim O'Gorman for simulations and Drs. Glenn Frye and Tom Jenkins for their support. This work is supported under NASA's Space Physics Supporting Research and Technology program.

REFERENCES

1. H. Hudson and J. Ryan, *Annual Review Astronomy and Astrophysics* **33**, 239-82 (1995).
2. W. T. Vestrand and J. A. Miller, "Particle Acceleration During Solar Flares," in *The Many Faces of the Sun: The Scientific Results of the Solar Maximum Mission*, edited by B. Haisch et al. (Springer Verlag, New York, 1998), in press.
3. A. M. Preszler, G. M. Simnett and R. S. White, *J. Geophys. Res.* **79**, 17-22 (1974).
4. G. Kanbach, C. Reppin and V. Schönfelder, *J. Geophys. Res.* **79**, 5159-265 (1974).
5. J. A. Lockwood, C. Chen, L. A. Friling and R. N. St. Onge, *J. Geophys. Res.* **81**, 6211-6216 (1976).
6. J. M. Ryan et al., "COMPTEL as a Solar Gamma Ray and Neutron Detector," in *Data Analysis in Astronomy*, Plenum Press: New York, (1992), pp. 261-270.
7. G. M. Frye, T. L. Jenkins and A. Owens, "SONTRAC: A Solar Neutron Track Chamber Detector," in *19th ICRC*, **5**, (1985), pp. 498-501.
8. G. M. Frye, C. J. Hall, T. L. Jenkins and G. N. Pendleton, "Predicted Performance of Solar Neutron Track Chamber Detector (SONTRAC)," in *20th ICRC*, **4**, (1987), pp. 392-394.
9. G. N. Pendleton, "Predicted performance of a prototype solar neutron detector," Ph.D. Thesis, Case Western Reserve Univ. (1988).
10. J. M. Ryan et al., "A Prototype for SONTRAC, a Scintillating Plastic Fiber Detector for Solar Neutron Spectroscopy," in *SPIE*, **3114**, (1997), pp. 514-525.

The SOFCAL Experiment

C.M.Benson[a], F.A.Berry[a], M.J.Christl[a], W.F.Fountain[a], J.C.Gregory[b],
J.S.Johnson[a*], T.A.Parnell[a], Y.Takahashi[b], J.W.Watts[a]

[a] Marshall Space Flight Center/NASA, Huntsville AL. 35812
[b] University of Alabama in Huntsville, Dept of Physics, 35899
*University Space Research Association

Abstract. SOFCAL is a hybrid instrument with both active and passive chambers to measure the proton and helium cosmic ray spectra from 0.2–10 TeV. An emulsion/x-ray film chamber is situated between a Cerenkov counter and an imaging calorimeter. Scintillating fibers (Sci-Fi) measure the electromagnetic(EM) cascades that develop in the calorimeter and identify the trajectory. The emulsion/x-ray film data provide an in-flight calibration of the Sci-Fi calorimeter. The data reduction techniques will be discussed and interim results of the analysis from a balloon flight will be presented.

INTRODUCTION

The Scintillating Optical Fiber Calorimeter (SOFCAL) is a balloon-borne instrument for making *direct* measurements of the galactic cosmic-ray(GCR) spectra. The instrument consists of four separate chambers(Fig.1) used in a complementary scheme to measure the energy and charge(Z) of GCRs. SOFCAL incorporates a *thin* ionization calorimeter employing small scintillating optical fibers (Sci-Fi), that measure the energy of Electro-Magnetic(EM) cascades from interactions occurring in the instrument. A test flight of the instrument was completed in Sept.1995 and the primary flight was completed in May 1997.

Measuring the GCR spectra is difficult because of the wide range of energies involved and the low flux at higher energies. Large area fully passive emulsion chambers ($>1.0m^2$) have been exposed in a series of balloon flights resulting in the highest-energy ($>10^{14}$eV) *direct* measurements of GCR spectra with a cumulative exposure 644 m^2-hrs (Cherry 97). GCRs exhibit power law spectra ($\propto E^{-2.7}$) with the flux decreasing sharply with energy and requiring large exposure factors to accumulate significant high energy data. The known GCR spectrum covers 12 decades of energy (10^8-10^{20}eV) and has a chemical composition somewhat similar to universal abundances, but with a few important difference. The most distinct feature of the all-particle spectrum is a bend at ~10^{15}eV which has been measured by *indirect* techniques (air-showers).

Figure 1 The SOFCAL Instrument comprises 4 units: Cerenkov, Target/EmCal, Sci-Fi,EmCal.

CP450, *SciFi97: Workshop on Scintillating Fiber Detectors*
edited by A. D. Bross, R. C. Ruchti, and M. R. Wayne
© 1998 The American Institute of Physics 1-56396-792-8/98/$15.00

The most promising source models are centered on supernova remnants (SNR). These models are attractive because the GCR composition has similar features to slow plus rapid nucleosynthesis; only ~10% of the SNR energy is needed to maintain the GCR density; and acceleration is known to occur at shock waves(Ellison 91). A recent multi-component source model (Biermann 93) uses Type I and II SNR to account for the GCRs. In this model SN-I are predominately responsible for producing the proton spectrum by accelerating the interstellar medium (ISM) while SN-II preferentially produce the heavier nuclei spectra because the local ISM has been enriched by the massive progenitor star before exploding. The spectra from these two different sites is likely to show some differences also.

SOFCAL will measure the absolute intensity of the proton and helium spectra from 0.2 to >10 TeV, and a limited number of heavy nuclei. Passive emulsion chambers lose effectiveness below 10 TeV because of the threshold limit for detecting EM cascades with x-ray films. The data from SOFCAL is well suited for comparison with the data from the emulsion chambers because the calorimeter configurations are similar. SOFCAL will check the GCR intensity measured with passive chambers near their threshold energy. Also, extending the measurement lower by a factor of 50 in energy with SOFCAL will enable a more thorough examination of the proton and helium spectra. The present data in this range is sparse and combined from several different experiments and techniques. The P/He ratio at high energies is different from lower energy data (Asakimori 93) and needs additional measurements to clarify the behavior. The new data provided by SOFCAL will further define this energy dependence. The hybrid design of this instrument will also aid in evaluating the application of Sci-Fi calorimeters to *direct* GCR measurements. Several events(~200) from the recent flight were detected with both the Sci-Fi calorimeter and emulsion calorimeter(EmCal) A subset of these will be used for calibration of the Sci-Fi calorimeter.

The following sections contain a summary of the instrument configuration, energy measurement technique, details of the May '97 flight and details of the analysis completed so far. This includes measurements in EmCal and systematic corrections related to the imaging system. These corrections are emphasized here because they must be completed before any detailed analysis of the cascades in the Sci-Fi calorimeter can proceed to estimate their energy. A discussion follows that summarizes the plan for the continued analysis of the current data and future use of this method to study GCR spectra.

SOFCAL INSTRUMENT

The SOFCAL instrument consists of four chambers assembled in a vertical stack to measure the energy and atomic number(Z) of incident GCR(Fig1). Over 140 layers of material are used throughout the 4 chambers, which perform different functions. The entire instrument is mounted in a thin-wall pressure vessel that serves as the mechanical structure and flight gondola. A summary description of the instrument is given here and more details can be found in (Christl 96).

599

At the top of the instrument is a Cerenkov counter to measure the charge(Z) of the incident GCR. A Teflon slab 50×60 cm^2, 1.27cm thick serves as the radiator and is inside a light box coated with reflective paint (BaSO$_4$). Six 5" photo-multiplier tubes (PMT) are used to detected the Cerenkov signal. Teflon has a refractive index of 1.36 with a threshold velocity of $\beta_t \cong 0.74$, and it has a saturated Cerenkov response above $\beta_s \cong 0.97$. The saturated signal is proportional to $Z^2 \times$pathlength and was calibrated with ground level muons which produced an average 26 photo-electrons per cm.. These PMT signals are also used in the event trigger generator.

Next is the upper emulsion chamber which has a 40×50 cm^2 cross section. It contains thin nuclear-emulsion plates, interspersed with Lucite plates and lead sheets of varying thickness. The emulsion plates were fabricated with 50μm of Fuji nuclear track gel on each side of a clear acrylic base 0.5mm thick. This chamber is nominally divided into three sections: Primary, Target and EmCal. The "Primary" section, designed for particle identification is 0.4cm thick and contains 4 thick emulsion plates and two sheets of nuclear plastic track detector(CR39). The emulsion plates have alternating high and low sensitivity gel to assist in discriminating between proton and helium primaries. The "Target" section, designed to enhance the probability of interactions occurring, is 3.6 cm. thick and consists of emulsion plates interspersed with thin 0.5mm lead sheets and 2.0mm Lucite plates. The Target section has 0.11 proton mean free paths(MFP) and 1.1 radiation lengths(rl). Below the Target is a *thin* emulsion calorimeter(EmCal). This section has 15 lead plates (1 & 2 mm thick) separated by emulsion and x-ray films. It is designed to develop and measure the EM cascades and has a vertical thickness of 4.7 rl and 0.13 proton MFP. Both emulsions and x-ray films record the cascades from high energy GCRs that interact in the chamber. The x-ray films serve as a passive event trigger by registering small visible dark spots in several layers of the calorimeter. The x-ray film sets a detection limit of ~10 TeV primary energy for protons.

Immediately below EmCal is the Sci-Fi calorimeter. This chamber has a cross section of 50×50 cm^2 and is 5.8 cm thick. It is a monolithic unit comprised of 10 lead plates 4mm thick, each separated by two orthogonal layers of Sci-Fi. It contains 0.41 proton MFP and is 7.1 rl thick. This arrangement further develops and samples the EM cascades from interactions in the EmCal. The 20,000 Sci-Fi are square 0.5mm BCF-12(Bicron) with a polystyrene core and have a peak emission at 435nm. They are coated with an extramural absorber to reduce *cross talk* between fibers. The attenuation length is >1m and the geometrical efficiency for total internal reflection is 4.4%. The fibers for each axis are grouped together and folded(coded) into two square arrays and optically coupled to Image-Intensified CCDs(II-CCD) made by Hamamatsu Photonics. The Sci-Fi perform a similar function as the x-ray films in EmCal. The light generated in the fibers form a two-dimensional intensity profile of the cascade which is recorded by the II-CCD. Two additional fiber layers are included in the Sci-Fi calorimeter and coupled to PMTs for triggering purposes. One layer at the top and the other near mid-depth (3.8rl). The trigger planes and Cerenkov counter define an event trigger and the geometry factor for the instrument. The close proximity to EmCal is necessary for this hybrid design to accomplish the calibration of the Sci-Fi calorimeter.

The last chamber in the instrument is another EmCal. Its configuration is similar to the upper one but has only 5 repetitions of lead and emulsion plates. This chamber contributes data on the highest energy cascades that traverse the entire thickness of SOFCAL and those that originate deep in the instrument.

Energy Measurement

The measurement technique used by SOFCAL is based on *thin* ionization calorimetry which has been successfully used by balloon-borne emulsion experiments (Burnett 86). In contrast to *total absorption* calorimetry used for low energy applications or where neither weight nor volume restrictions apply, *thin* calorimetry measures characteristics of the early EM cascades from the *first* interaction to estimate the energy content. This process has been studied analytically(Greisen 41,Burnett 86,Fuki 87) and with numerous Monte Carlo simulations to demonstrate the relationship between the behavior of the cascade and its energy. In SOFCAL, emulsions, x-ray films and the Sci-Fi's in the ionization calorimeters measure the longitudinal and lateral behavior of the EM cascade and are used to analyze the cascade energy.

The response of the Sci-Fi calorimeter to cascades has been simulated using the electron-gamma ray shower(EGS) program. An event generator using the multi-chain model of nuclear interaction, (Fuki 87) interacts primaries in the detector and follows all produced particles. EGS uses gamma-rays from the π^{o} decay and a detailed model of the instrument to simulate the resulting EM cascade. The cut-off energies used in EGS are 1.0 MeV for electrons and positrons and 0.1 MeV for gamma rays. A transition curve of the average signal from many similar cascades is shown in Fig2. A complete set of these curves will be used to estimate energy from the Sci-Fi data. The electron transition curves(Fig3) have also been simulated for these same events, and are used with the emulsion and x-ray film data to determine cascade energies. The trajectory provided by the Sci-Fi is used to locate the cascade in EmCal even if it is below the trigger threshold of the x-ray film. This allows the same cascade to be measured in both chambers. Using the emulsion technique of electron counting within 100μm of the cascade axis a 1 TeV proton reaches a maximum development at ~5rl(Fig3) which is typically in EmCal. The same cascade

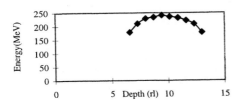

Figure 2. Simulated transition curve in Sci-Fi calorimeter for proton primary with E~ 1 TeV

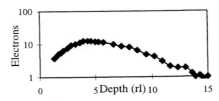

Figure 3 Transition curve for protons with E~1 TeV. Electrons counts within 100μm.

601

reaches a maximum deeper in the instrument when using the full energy deposition recorded by the Sci-Fi (Fig2). About 25 events will be used for this initial cross calibration.

The energy spectrum measured with a *thin* calorimeter is not the true GCR spectrum but it is related. The thinnest calorimeters measure only the energy of neutral pions (EM cascade) produced by the *first* inelastic interaction of the GCR in the detector. One technique to recover the primary spectrum is by normalizing the energy scale to equivalent primary energy(Burnett 86). This conversion effectively shifts the measured spectrum up in energy by an amount determined through detailed simulations of the π^0 production in the detector.

Flight Details

The flight occurred on May 19 and lasted for 22 hours. It was launched from Ft. Sumner N.M. and was recovered in southwest Arizona. The total weight of the apparatus was 2600 lbs. and the average float altitude was 125kft (~3 mbar). After an initial checkout period of the instrument, data was collected for 18 hours with the trigger threshold set to ~500 GeV equivalent energy. The threshold was lowered by a factor of 2 for the last 3 hours of the flight to collect the more abundant lower energy particles. A total of 8000 triggers were detected. The trigger rate for the first period was 5 per minute and for the second was ~22 per minute. The trigger system has a maximum rate of 1000 triggers per second, but the image acquisition and data storage has a maximum recording rate of ~20 events per minute. Because of this, a substantial fraction of events were missed in the low energy mode, but the total number of triggers was monitored independently, so the events recorded will reflect the true GCR intensity. A sample of the image data for a proton with energy ~5 TeV is displayed in Fig. 4. The 10 distinct layers of data are seen for each view of the cascade and the core of the cascade which shows the trajectory is also visible. Some broadening of the signal by the imaging system was observed which degrades the position resolution slightly but will not appreciably effect the analysis. The detector's performance was monitored by flashing an LED occasionally to trigger the detector and record the signals for reference. Also alpha particles excited a few fibers of each axis to monitor the gain during the flight.

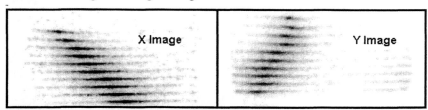

Figure 4 Raw image of a cascade in the Sci-Fi calorimeter. The primary is a proton with E~5 TeV.

DATA ANALYSIS, PASSIVE CHAMBER

The measurements of the x-ray films has been completed and those for the emulsion plates will proceed in parallel with the analysis of the image data. Correctly identifying proton and helium primaries is critical to study the P/He ratio and the emulsions will be used to confirm the primary charge for the highest energy events. For these events especially, the emulsions will play an important role because *backscattered* electrons from the cascade are likely to contaminate the Cerenkov signal and cause protons to be misidentified as helium.

All events above the x-ray film threshold have been located in EmCal and 3-dimensional maps created to determine their trajectory. Events are selected manually by locating 3 or more co-linear spots on consecutive x-ray films. The upper chamber has 161 events and the lower chamber has 114 events total. The optical density of each event has been measured and those with data

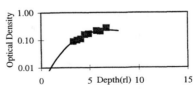

Figure 5 Optical density for an EmCal event . It is inclined with a slope of 0.48 and has a $D_{max} = 0.23$

on 6 or more layers have been fitted to determine the maximum density (D_{max}). A sample of the density data and fitted curve is shown in Fig5. The fitted curve is based on an analytical approximation for the number of electrons in cascades initiated by photons (Greisen 41). The D_{max} value is closely related to the energy of the cascade but must be calibrated using electron counts in the emulsion plates. Fig6 shows the differential density spectrum for all events. At low densities (<0.1 OD) the effect of the x-ray film threshold can be seen by the sharp decrease in apparent flux. For the highest energy cascades there is a systematic under-estimation of the D_{max} (energy) because EmCal is too thin to contain the cascade maximum. The Sci-Fi calorimeter will fill-in both high and low energy regions because it has a threshold ~50 times lower in energy and the added depth (7.1rl) will augment the EmCal data for high energy cascades, providing a better energy estimate.

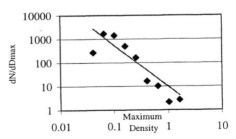

Figure 6 Spectrum of events from EmCal chamber analysis. The line shows a spectrum with index 2.75.

DATA ANALYSIS, SCI-FI CALORIMETER

The image system includes a de-magnifying tapered bundle(×4.2) coupled to the front-end of the dual-image-intensifier(II) and followed by a tandem set of Nikkor reducing(×3.8) lenses focused onto the CCD Chip. Each stage of the dual-II has a bi-alkali cathode, single MCP, and a P20 phosphor screen with fiber optic faceplates. The gain of each stage is 3×10^3 and the unit is capable of detecting minimum ionizing particles traversing 1mm Sci-Fi. The

CCD is an interlace device based on the Sony Corp XC-77CE chip with the output digitized by a DT-55 frame grabber from Data Translation Inc. The total de-magnification is ×16 which produced a Sci-Fi footprint of 4 to 6 pixels at the CCD. The systematic corrections for the Sci-Fi calorimeter fall into two categories: imaging and instrumental. The imaging corrections must be completed first and are reported here(temperature effects, timing biases, distortion). Instrument biases(trajectory, misalignment) will be covered later in the data analysis.

The temperature variation during data acquisition was 17°C. The dark signal for the imaging system includes; photo-cathode thermal emissions, signals generated in the CCD at both the photo-sensitive sites and in the readout registry, and temperature dependence of the gain and off-set of the ADC used by the frame grabber. The total dark signal was determined by using the field-of-view region not occupied by any Sci-Fi. There was no observable temperature dependence in the gain of the II-CCD. The system gain was monitored by using an [241]Am foil to excite a small number of fibers throughout the exposure.

Another image correction was required because the II-CCD uses a standard interlaced video format and not the more sophisticated slow-scan modes. When an event trigger was issued to the frame grabbers, they digitized the two fields (even+odd or odd+even) of the video frame exposed when the trigger occurred. The exposure time of the two fields is equal but they are offset by ½ the frame period(16.6 msec) to allow for the readout time. This timing difference together with the P20 decay creates an inequity between the two fields with the freshest field having a brighter image. The field fraction $\left(\dfrac{total\_FIELD\_signal}{total\_FRAME\_signal}\right)$ distribution for 1500 events is shown in Fig7. Three different factors contribute to this distribution. The field exposed to the phosphor screen longer will have a greater intensity and produces the bi-modal feature(~2% split) of the distribution. The second factor depends on the location of the

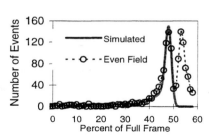

Figure 7. Comparison of data and simulation for image correction factors: Peak split 2%, Peak width 3.5%, Decay constant 2.0 msec.

event and is responsible for the width of each bi-modal peak (3.5% FWHM). There are ten layers of fiber data per image and an average of 2.5 field lines per layer giving a total of 25 field lines with the majority of the signal. Events going through the detector will sometimes be viewed by slightly more odd or even lines depending on the event trajectory. Consequently, one expects a spread in the peaks to be on the order of 1/25(4%) which agrees with the data. The last contributing factor is the decay time of the P20 screen itself and produces the long tail of the distribution. To minimize the effect of all these factors, events with a field fraction less than 40% will be analyzed using only the dominate field together with the data from the other axis(the two cameras were operated independently).

The last imaging correction is for distortion and intensity variation induced by the optical train. The principal source for both of these is the tapered fiber bundle at the front-end of the Il-CCD. The taper bundle has glass fibers that are pulled from 100μm dia. to 25μm and introduces a *pin-cushion* effect. The Sci-Fi cannot be resolved individually for a precise mapping so a 1.0 mm optical scale was used to register this effect which allowed spatial mapping to associate CCD pixels with the Sci-Fi arrays. This approximate technique has been sufficient to determine the cascade trajectories and is limited by image blurring. Intensity variations at the output of the image system were determined using a flat-field light source at the input. The contribution by the Il-CCD itself is minimal since the fiber bundle occupies only a small central area (1.5 cm^2) of the input window where gain and flatness are fairly uniform.

A preliminary inspection of the Cerenkov data reveals proton and helium charge peaks. Once the cascade trajectories have been determined by the Sci-Fi, maps developed using muons will be used to improve charge measurements. The charge resolution of the Cerenkov counter will be evaluated (including *backscatter* effects) using the primaries identified by the emulsion technique.

DISCUSSION

The present plan to complete the analysis and address the scientific objectives is the first priority because this instrument has unique capabilities to re-visit the GCR proton and helium spectra over an extensive energy range, possibly containing relevant signatures for recent GCR source models. The advances made in Sci-Fi technology enabled us to use the *thinnest* calorimetry method which was previously only practical with emulsion chambers. However, the present hybrid instrument does rely on the emulsion chambers for the energy calibration. A follow-on balloon instrument utilizing only a electronic instruments is being considered to make additional GCR spectra measurements. To utilize this technique at higher energies the Il-CCD system and the fiber layout must be redesigned improving the image resolution for finer measurements of the cascade core development. The ultimate application of this technique would be a large area detector on a space based platform, and very long exposure. A program has been initiated to define a detector to study GCR from the International Space Station(ACCESS). At least one of the detector designs being considered is based on Sci-Fi and thin calorimetry.

REFERENCES

(Asakimori 93) Asakimori K. et al., Proc. 23rd ICRC(Calgary), **2**,21 & 25 (1993)
(Biermann 93) Biermann P.L. et al., Astron. & Astrophys. , **271**, 649 (1993)
(Burnett 86) Burnett T.H. et al., NIM., **A251**, 583(1986)
(Cherry 97) Cherry M. L., et al, Proc. of 25th ICRC(Durban), **4**, 1 (1997)
(Christl 96) Christl M.J. et al., Proceedings of SPIE, **V2806**, 155 (1996)
(Ellison 91) Ellison D.C., and Jones F.C., Space Science Reviews, **58**, 259 (1991)
(Fuki 87) Fuki M., The Bull. Of the Okayama Univ. of Sci., No. **22**, 167 (1987)
(Greisen 41) Greisen K. and Rossi B., Rev Mod. Phys., **13**, 240 (1941)
(ACCESS-ICA) Parnell T.A., these proceedings, this volume, (1997)

THE SEARCH FOR DARK MATTER
PARTICLES IN THE UNIVERSE

David B. Cline

Department of Physics and Astronomy, Box 951547
University of California Los Angeles
Los Angeles, CA 90095-1547, USA

Abstract. We briefly review the latest evidence for non-baryonic dark matter in the Universe, including cosmological constants. We also review the most recent estimates for SUSY dark-matter particle interaction with matter. We then provide a critique of the current search for the dark matter with various detectors and describe one very promising method using powerful background discrimination in liquid xenon.

INTRODUCTION

The latest evidence for dark matter in the Universe has been reviewed recently at two University of California Los Angeles (UCLA) symposiums (1). Remarkably, even in the 1920s some evidence had been found and of course in the 1930s, F. Zwicky provided perhaps the first definitive evidence for dark or non-luminous matter in galaxies (2).

While no one knows the exact cause of dark matter, there is a reasonable likelihood that new elementary particles play some role in this phenomenon. Of all of the current ideas in this regard, many feel supersymmetry (SUSY) is the most "natural." Our viewpoint is to take the SUSY model seriously and to see what level of detection and discrimination is required to observe such particles. While even the SUSY model is not fully predictive, it would appear to be better than other even more ad hoc models. The current search for dark matter particles includes many detector concepts such as those associated with liquid noble gas detectors (3), as well as theoretical predictions of dark matter rates (4–6). Other detector concepts are discussed in Refs. 7–9 and in the text.

CP450, *SciFi97: Workshop on Scintillating Fiber Detectors*
edited by A. D. Bross, R. C. Ruchti, and M. R. Wayne
© 1998 The American Institute of Physics 1-56396-792-8/98/$15.00

EVIDENCE FOR NON-BARYONIC DARK MATTER

The evidence for non-baryonic dark matter is getting even stronger than it has been. Perhaps the best evidence comes from three different measurements of the gravitational potential in galactic clusters using
 1. Dynamic methods (galaxy motion),
 2. Hot X-rays (hot gas in the gravitational potential well),
 3. Weak gravitational lensing (sensitive to all gravitating matter).
These three methods now give convincing evidence for dominant non-baryonic dark matter in these clusters.

There is also strong evidence for dark matter in dwarf galaxies, which appear to totally dominate the total matter on all scales. Finally, it is now agreed that $\Omega > 0.4$, which is definitely larger than the limits for $\Omega_B < 0.1$ from recent nucleosynthesis studies (Fig. 1).

Recent results on the cosmological parameters of the Universe are given in Table 1. While this is a poor way to measure Ω_0, it shows that there is no longer any "age" crisis and that $\Omega_0 = 1$ is still possible.

METHODS AND RATES TO DETECT SUSY WIMPs

There are many estimates for the cross section of SUSY WIMPs with various targets. We believe this illustrates the difficulty, as well as the promise, for the search for SUSY WIMPs. In this report, we follow the recent work of Nath and Arnowitt (4) and the references cited therein. Without getting into the details of the assumptions in this calculation, we note that the range of rates goes from a few events/kg·d to 10^5 events/kg·d (see Figs. 2 and 3). Although the results are for Ge and Pb, we expect similar results for liquid Xe. These results, if taken at face value, suggest that the detection of SUSY WIMPs could be very difficult, requiring large detectors of certainly 100 kg and possibly tons of detector. In this case, the rejection of background is even more important. Table 2 gives the schematic for WIMP detectors. Figures 4–6 show some detection methods that will be discussed below.

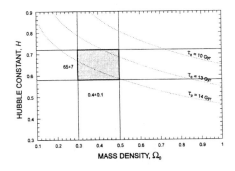

Figure 1. Recent estimates of the mass density as a function of Hubble's constant, and τ_0 is the age of the Universe.

Table 1. Recent Results on Cosmological Parameters

For Einstein–deSitter Universe, Ω_m, $\Omega_\lambda = 0$, we have the relationship, $H_0\tau_0 = 2/3$.

Examples:	H_0 (km/s/mpc)	τ_0 (GYN)
	70	10.5
	65	11.5
	60	12.5
	50	15.0

Current age of the Universe estimates: 12^{+3}_{-1} Gyr.

For SNIa determination of H_0 (ITP at UCSB, Aug. 1997), $H_0 = 65 \pm 7$ km/s/mpc and $H_0\tau_0 = 2/3$.

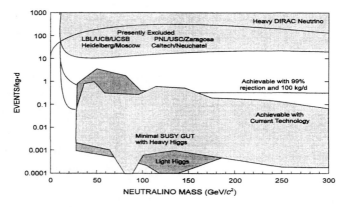

Figure 2. Regions of the event rate as a function of mass that were excluded in 1995 and the expected ranges for SUSY WIMPs.

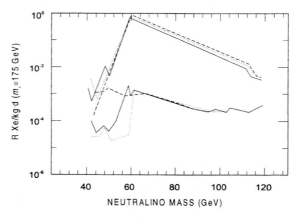

Figure 3. Very recent estimates of the allowed regions of event rates (from Ref. 4).

608

TABLE 2. WIMP Direct Detection Schematic

	T	T*

1. W + [atom/nucleus] → W + [atom]* ~ elastic scattering with excitation of atom ~10 nm

2. Form factors for recoil [atom/nucleus] $\Delta E \leq 30$ keV measure $n + A \rightarrow n + A^*$

3. Kinematics

4. Rate $\leq (1 - 10^{-5})$ kg^{-1} d^{-1}

5. Background ~ (10 - 100) kg^{-1} d^{-1}

6. Discrimination against background

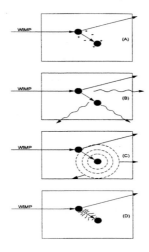

Figure 4. Schematic of WIMP detectors that use (A) semi-conductors, (B) scintillation light, (C) bolometers, and (D) etched mica.

In order to discover WIMPs interacting in a medium, the response of the medium must be extremely well understood. Discoveries are not made by removing backgrounds but by identifying a unique signature for the process. Table 3 lists many of the low temperature detectors in the world. This information builds on more than two decades of study of the excitations in liquid Xe. The UV scintillation light in xenon is produced by the formation of Excimer states, which are bound states of ion–atom systems (see Fig. 7). There are extensive studies in the use of this process for Excimer lasers, as well as for many other applications (10–12).

A successful test of the detection of a recoil Xe nucleus using neutron scattering has been recently carried out, and it shows clear evidence that SUSY WIMPs will give a strong, unique signal on a discriminating liquid-Xe detector (Figs. 8–10). A 2-kg detector will be installed at the Mt. Blanc Underground Laboratory (UL) to perform a first search for SUSY WIMPs using this tchnique.

In Figs. 5–7, we show the schematic of several other methods of searching for WIMPs and some current results.

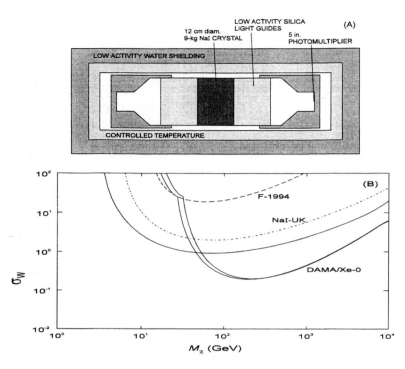

Figure 5. (A) The UK dark-matter-group NaI WIMP detector (from Ref. 7); (B) some recent limits on the WIMP cross section from NaI and Xe detectors (from Ref. 8).

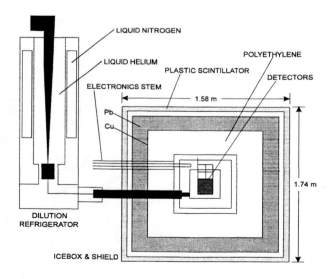

Figure 6. The CDMS detector at SLAC that uses the bolometer method (from Ref. 9).

610

Table 3. Various Low-Temperature Detectors Around the World

Name	Detector	Mass	Sensor	Discrimination	cts keV^{-1} kg^{-1} d^{-1}
EDELWEISS	Al$_2$O$_3$	24 g	NTD Ge	No	25
Milano	TeO$_2$	340 g	NTD Ge	No	8
CDMS	Ge	1 kg	NTD Ge	Yes	0.2
	Si & Ge	1 kg	W/Al QET	Yes	0.07
	^{73}Ge & ^{76}Ge	1 kg	NTD Ge	Yes	
CRESST	Al$_2$O$_3$	1 kg	W SPT	No	
Berne	Sn	13 g	Grains	No	
Tokyo	LiF	1 kg	NTD Ge	No	
HERON	^{4}He	10,000 kg	SPT	Yes	

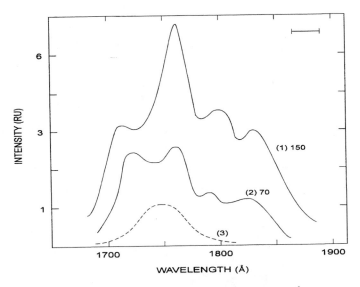

Figure 7. Emission spectrum of liquid Xe: (1 and 2) pump densities in Å/cm^2 and (3) emission spectrum at low excitation density. The resolution of the monochromator is shown in the upper right-hand corner. (From Ref. 10.)

CURRENT STATUS OF THE SEARCH FOR SUSY WIMPs

In the early days of the direct search for WIMPs, the motivation was the detection of a massive dirac neutrino, which is now excluded by the data (see Fig. 2). In this search, the expected rate was so large (~ 1000 events/kg·day) that the detector background was not extremely important. In the search for SUSY WIMPs, as was discussed earlier (Fig. 3), we expect much lower rates and either very massive detectors are needed (for the background limited search, which goes like $\sqrt{N}$) or powerful discrimination. The best current limits come from the study of the pulse-shape distribution in NaI or liquid Xe. One NaI detector (the UK group) is shown in Fig. 5, where the current best limits (from the Rome group) are also shown.

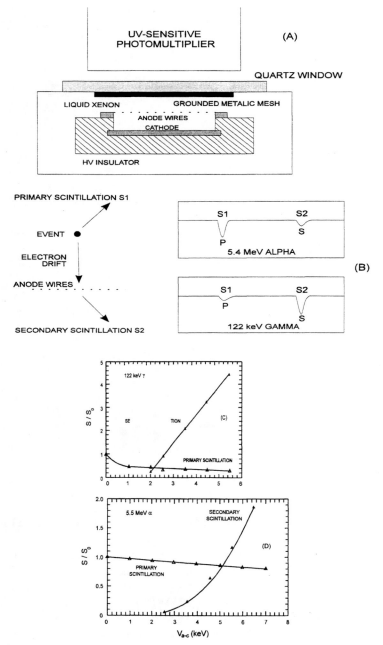

Figure 8. (A) Geometry
of liquid-xenon test chamber, (B) observed primary and secondary scintillation signals showing S1/S2 >>
1 for α events and << 1 for γ events, (C) variations of the secondary scintillation intensity as a function of
$V_{a\text{-}c}$ for photons, and (D) for α particles. (From Refs. 11–13.)

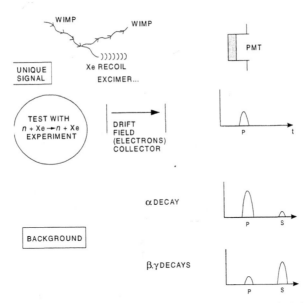

FIGURE 9. Schematic of the background rejection method in liquid Xe (rejection of 10^{-3}).

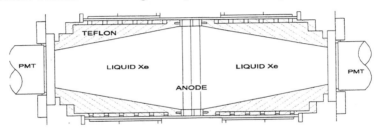

FIGURE 10. A 2-kg detector that has been constructed for tests at Mt. Blanc and a possible WIMP search by the UCLA–Torino ICARUS team.

Discriminating Liquid-Xenon Detector[1]

To allow lower limits to be reached, it is essential to develop methods of differentiating the desired nuclear recoil events from γ- and β-decay backgrounds. At the same time, it is desirable to develop techniques capable of being substantially scaled up in target mass. The need for targets in the 100–1000-kg region would arise, particularly in searches for the 5% "annual modulation" of any true dark-matter signal (due to the Earth's motion combined with the solar motion through the Galaxy). Large-mass targets would also be needed for heavier WIMP masses (> 50 GeV), because of the correspondingly smaller flux of such particles.

---

[1] This section is adapted from a recent report (12) by the ZEPLIN group.

Liquid Xe satisfies all of the above requirements for a dark matter detector because:

1. It is available in sufficiently large quantities with high purity.

2. It scintillates via two mechanisms, which are stimulated to different extents by nuclear-recoil and background electron-recoil events.

3. Its natural form consists of isotopes with and without nuclear spin, so it is suitable as a detector for both spin-independent and -dependent interactions.

The larger nuclear mass of Xe also makes it a better match to heavier WIMPs but, at the same time, the larger nuclear radius introduces a significant form-factor correction unless the energy threshold is low (1–10 keV). Efficient light collection is, therefore, of prime importance in a liquid-Xe detector. Figure 11(A) shows a schematic of the proposed 30-kg ZEPLIN-II detector.

There are two distinct approaches to discriminating nuclear-recoil events in liquid xenon:

1. Analyzing the total scintillation pulse shape or, at low energy, the individual photon arrival times, which will differ significantly for nuclear- and electron-recoil events;

2. Applying an electric field to prevent recombination and measuring (i) the primary scintillation and (ii) the ionization component by drifting and producing "secondary scintillation" (see Figs. 7 and 8). The lowest possible backgrounds [Fig. 11(B)] may be achieved at the Boulby Mine site.

In Fig. 12 we show some estimates of how well the search for SUSY WIMPs can go using the Xe detectors discussed here.

SUMMARY

The search for SUSY WIMPs must now be one of the most important activities in particle physics. Table 4 summarizes the key issues. Recent estimates of the rate put it into the $10^{-1} - 10^{-5}$ event/kg·d level, which is an extremely difficult task. We believe the search is well worth the effort needed.

ACKNOWLEDGMENTS

I wish to thank Hanguo Wang for his excellent work on the ICARUS–WIMP detector and the other members of the ICARUS and ICARUS–WIMP groups (P. Picchi), as well as members of the proposed ZEPLIN group (P. Smith in particular); my thanks also to P. Nath and R. Arnowitt for discussions on the theory of SUSY–WIMP detection and for Fig. 3.

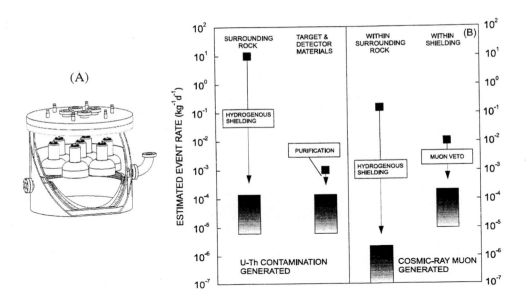

Figure 11. (A) Conceptual design for UCLA–Torino 30-kg ZEPLIN-II detector; (B) estimated background at the Boulby dark-matter laboratory in the UK (14).

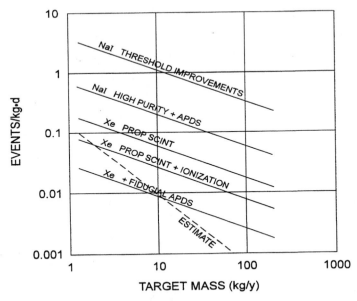

Figure 12. Possible sensitivity that can be reached with the Xe detetor discussed here (adapted from Ref. 14). The dashed lines show the sensitivity if the signal exceeds the background, as we expect for a discriminating liquid Xe detector.

Table 4. Status of the Search for Dark Matter Particles

- Strong evidence for non-baryonic dark matter in galaxies and superclusters.
- Recent cosmological parameters $H_0 = 65 \pm 7$ (SNIa) and $\tau_0 = {}^{+3}_{-1}$ Gyr are still consistent with $\Omega_0 = 1$.
- Supersymmetric particles are a "natural" candidate for dark matter, but detection cross sections are small.
- New WIMP detectors with powerful discrimination have been developed and are starting to operate.
- It is possible that LSP-neutralinos will be discovered even before the LHC operates.

REFERENCES

1. See the proceedings of "Sources of Dark Matter in the Universe," Santa Monica, 1996, ed. Cline, D., *Nucl. Phys. B* (PS) **51B** (1996); and Aller, L. H., and Trimble, V., in *Sources of Dark Matter in the Universe*, ed. Cline D., Singapore: World Scientific, 1994, pgs. 3 and 9.
2. Zwicky, F., *Ap. J.* **86**, 217 (1937).
3. CERN-UCLA-INFN Group, ICARUS Proposal (1993) unpublished.
4. Nath, P., and Arnowitt, R., *Phys. Rev. Lett.* **74**, 4592 (1995).
5. Jungman, G., *et al.*, *Phys. Reports* **267**, 195 (1996).
6. Bottino, A., *et al.*, *Astropart. Phys.* **2**, 77 (1994).
7. Smith, P. F., *et al.*, *Phys. Lett. B* **379**, 299 (1996).
8. Bernabei, R., *et al.*, *Phys. Lett. B* **389**, 757 (1996).
9. Cabrera, B., *Nucl. Phys. B* (PS) **51B**, 294-303 (1996).
10. Basou, N., *et al.*, *JETP Lett.* **12**, 329-331 (1970).
11. Benetti, P., *et al.*, *Nucl. Instrum. Methods A* **327**, 203-206 (1993).
12. Park, J., in *Sources of Dark Matter in the Universe*, ed. Cline, D., Singapore: World Scientific, 1994, p. 288.
13. Wang, H., CERN-UCLA, private communication (1996).
14. Smith, P. F., *et al.*, *Nucl. Phys. B* (PS) **51B**, 284-293 (1996).

LIST OF PARTICIPANTS

Nural Akchurin	University of Iowa
Katsushi Arisaka	University of California, Los Angeles
Muzaffer Atac	Fermilab/UCLA
Jurgen Baehr	DESY - JFH Zeuthen
Austin Ball	University of Maryland
Virgil Barnes	Purdue University
Yuriy Bashmakov	P. N. Lebedev Physical Institute
Barry Baumbaugh	University of Notre Dame
W. R. Binns	Washington University, St. Louis
Jim Bishop	University of Notre Dame
Nripendra Biswas	University of Notre Dame
Leon Bosch	Delft Electron Products B. V.
Alan Bross	Fermi National Accelerator Laboratory
Kees Brouwer	Delft Electron Products B. V.
Howard Budd	University of Rochester
Steve Carabello	Purdue University
Neal Cason	University of Notre Dame
Mark Christl	NASA/Marshall Space Flight Center
Manho Chung	University of Illinois at Chicago
David Cline	University of California, Los Angeles
Hans Cohn	University of Tennessee
Linda Coney	University of Notre Dame
William Cooper	Fermi National Accelerator Laboratory
Priscilla Cushman	University of Minnesota
Henri Dautet	E G & G Optoelectronics
Mario David	LIP, Lisbon
Paul Davison	Electron Tubes, Inc.
Pawel de Barbaro	University of Rochester

Ken Del Signore	University of Michigan
Dmitri Denisov	Fermi National Accelerator Laboratory
Casey Durandet	University of Virginia
John Elias	Fermi National Accelerator Laboratory
John Ellis	Collimated Holes, Inc.
Valeri Evdokimov	IHEP
Walter Fabian	Karl Franzens Universitaet
Richard Farrell	Radiation Monitoring Devices, Inc.
Gerald Fishman	NASA/Marshall Space Flight Center
Jim Freeman	Fermi National Accelerator Laboratory
Yauo Fukui	KEK/Fermilab
Tony Gerig	Taylor University
George Ginther	University of Rochester
Alexander Gorin	IHEP
Stefan Gruenendahl	Fermi National Accelerator Laboratory
Gaston Gutierrez	Fermi National Accelerator Laboratory
Vasken Hagopian	Florida State University
Kazu Hanagaki	Osaka University
William Harder	Bicron, Inc.
Kerstin Hoepfner	DESY, Hamburg
Henry Hogue	Boeing North America, Inc.
Dan Holslin	Science Applications International Corporatior
Ryan Hooper	University of Notre Dame
Yasuyuki Horiuchi	Hamamatsu Photonics K. K.
Todd Hossbach	University of Evansville
Dave Hurd	LeCroy Corporation
Joey Huston	Michigan State University
Tony Hyder	University of Notre Dame
Mikhail Ignatenko	Fermilab/JINR, Dubna
Marco Incagli	INFN, Pisa

Emil Ivanov	University of Notre Dame
C. B. Johnson	Photek, Inc.
Joel Kauffman	Philadelphia College of Pharmacy and Science
V. Paul Kenney	University of Notre Dame
Joel Kindem	University of Minnesota
Igor Kreslo	JINR, Dubna
Steve Kuhlmann	Argonne National Laboratory
Shuichi Kunori	University of Maryland
Karol Lang	University of Texas
Tony Lin	University of Notre Dame
Don Lincoln	University of Michigan
A. Maio	LIP, Lisbon
Jeffrey Marchant	University of Notre Dame
Matt Matsumoto	Hamamatsu Corporation
Douglas Mayo	Los Alamos National Laboratory
Sean McGinnis	Bicron, Inc.
Kerry Mellott	Fermi National Accelerator Laboratory
Douglas Michael	California Institute of Technology
Wayne Moser	Bicron, Inc.
William Moses	Lawrence Berkeley National Laboratory
Paul Mulligan	P. M. Manufacturing Services, Inc.
Pat Mulligan	P. M. Manufacturing Services, Inc.
Roberto Mussa	INFN, Ferrara
Rolf Nahnhauer	DESY - JFH Zeuthen
Gary Nelson	Polymicro Technologies, Inc.
Yasar Onel	University of Iowa
Nathan Pace	Fermi National Accelerator Laboratory
Adam Para	Fermi National Accelerator Laboratory
Aldo Penzo	INFN, Trieste
Michael Petroff	Photonics Consulting

Yuriy Pischalnikov	University of California, Los Angeles
Anna Pla-Dalmau	Fermi National Accelerator Laboratory
Evgeny Popkov	University of Notre Dame
Chris Quigg	Fermi National Accelerator Laboratory
Erik Ramberg	Fermi National Accelerator Laboratory
Ron Ray	Fermi National Accelerator Laboratory
Kirsten Reynolds	Utah State University
Ron Richards	Michigan State University
Dirk Rondeshagen	University of Munster
Anatoly Ronzhin	Fermi National Accelerator Laboratory
Randy Ruchti	University of Notre Dame
Roger Rusack	University of Minnesota
Douglas Ruuska	Northeastern University
James Ryan	University of New Hampshire
Rob Schomaker	Delft Electron Products B. V.
Yiping Shao	UCLA School of Medicine
William Shephard	University of Notre Dame
Stephanie Sposato	Washington University, St. Louis
Giulio Stancari	University of Ferrara/INFN
Sean Stoll	Brookhaven National Laboratory
Atsumu Suzuki	Kobe University
Richard Talaga	Argonne National Laboratory
Michelle Thompson	University of California, Irvine
Shoji Torii	Kanagawa University
Juan Valls	Rutgers University/Fermilab
Maria Varanda	LIP/University of Lisbon
Stefan Vasile	Radiation Monitoring Devices, Inc.
Jadwiga Warchol	University of Notre Dame
Mitch Wayne	University of Notre Dame
David Winn	Fairfield University

Stefan Wirth	University of Erlangen - Nurnberg
Thomas Wolff	University of Munster
Kin Yip	Fermi National Accelerator Laboratory
Hiroyuki Yonekura	Kuraray America, Inc.
Takuo Yoshida	Osaka University
Yuji Yoshizawa	Hamamatsu Photonics K. K.
J. C. Yun	Fermi National Accelerator Laboratory
Valery Zavarzin	Boston University
Hai Zheng	University of Notre Dame
Ren-yuan Zhu	California Institute of Technology
Ralf Ziegler	Bonn University

AUTHOR INDEX